Lecture Notes in Computer Science

982

Edited by G. Goos, J. Hartmanis and J. van Leeuwen

Advisory Board: W. Brauer   D. Gries   J. Stoer

Springer
Berlin
Heidelberg
New York
Barcelona
Budapest
Hong Kong
London
Milan
Paris
Santa Clara
Singapore
Tokyo

Manuel Hermenegildo
S. Doaitse Swierstra (Eds.)

Programming Languages: Implementations, Logics and Programs

7th International Symposium, PLILP '95
Utrecht, The Netherlands, September 20-22, 1995
Proceedings

 Springer

Series Editors

Gerhard Goos, Karlsruhe University, Germany

Juris Hartmanis, Cornell University, NY, USA

Jan van Leeuwen, Utrecht University, The Netherlands

Volume Editors

Manuel Hermenegildo
Facultad de Informática, Universidad Politécnica de Madrid
E-28660 Boadilla del Monte, Madrid, Spain

S. Doaitse Swierstra
Department of Computer Science, Utrecht University
P.O. Box 80.089, 3508 TB Utrecht, The Netherlands

Cataloging-in-Publication data applied for

Die Deutsche Bibliothek - CIP-Einheitsaufnahme

Programming languages: implementations, logics and programs
: 7th international symposium ; proceedings / PLILP '95,
Utrecht, The Netherlands, September 20 - 22, 1995. S. Doaitse
Swierstra ; Manuel Hermenegildo (ed.). - Berlin ; Heidelberg ;
New York : Springer, 1995
  (Lecture notes in computer science ; Vol. 982)
  ISBN 3-540-60359-X
NE: Swierstra, S. Doaitse [Hrsg.]; PLILP <7, 1995, Utrecht>; GT

CR Subject Classification (1991): D.1.1, D.1.6, D.3.1, D.3.4,F.3.3, F.4.1-3,
I.1.3, I.2.1

ISBN 3-540-60359-X Springer-Verlag Berlin Heidelberg New York

© Springer-Verlag Berlin Heidelberg 1995
Printed in Germany

Typesetting: Camera-ready by author
SPIN 10485668   06/3142 - 5 4 3 2 1 0   Printed on acid-free paper

Preface

This volume contains the papers accepted for presentation at the *Seventh International Symposium on Programming Languages: Implementations, Logics and Programs*, PLILP'95, held near Utrecht (The Netherlands), September 20–22, 1995. The symposium was preceded by six previous meetings in: Orléans, France (1988), Linköping, Sweden (1990), Passau, Germany (1991), Leuven, Belgium (1992), Tallinn, Estonia (1993) and Madrid, Spain (1994). All proceedings were published by Springer-Verlag as Lecture Notes in Computer Science, volumes 348, 456, 528, 631, 714 and 844 respectively.

The PLILP symposium aims at stimulating research on declarative programming languages, and seeks to disseminate insights in the relation between the logics of those languages, implementation techniques and the use of these languages in constructing real programs. Keeping the PLILP acronym it was decided for this edition to change the wording behind it in order to reflect the broadening of the scope of the symposium. It was felt that treating any one of these subjects in isolation often does too little justice to the other two aspects, and especially papers dealing with more than one of these subjects were invited.

On this occasion, PLILP was held concurrently with the *Fifth International Workshop on Logic Program Synthesis and Transformation*, LoPSTr'95, and the ESPRIT CompuLog Network Area Workshops for the areas of Program Development and Parallelism and Implementation Technologies.

In response to the call for papers, 84 papers were submitted, all electronically. The assignment of papers to program committee members was done based on preferences expressed by committee members after having seen a list of anonymous abstracts. Furthermore program committee members discussed, shortly before the program committee meeting took place, their comments on those papers on which there was a significant disagreement about the quality of the paper. The Program Committee met in Utrecht on May 13, 1995, and selected 23 papers, as well as eight posters and demos. In addition to the selected papers, the scientific program included also the three invited lectures, in common with LoPSTr, by Kim Marriott, Oege de Moor, and Mark Jones.

On behalf of the Program Committee, the Program Chairmen would like to thank all those who submitted papers, posters, and system demonstrations and all the referees for their careful work in the reviewing and selection process.

The support of the several funding agencies, listed in the volume, which have provided part of the funds for the organization of the conference is also gratefully acknowledged. Finally, we would like to express our gratitude to the members of the local Organizing Committees of PLILP'95 and LoPSTr'95 for the invaluable help and enthusiasm provided throughout the preparation and organization of the event and in the composition of the proceedings.

Utrecht
Madrid
July 1995

Doaitse Swierstra
Manuel Hermenegildo
Program Committee Chairmen

Conference Chairmen

Manuel Hermenegildo   Universidad Politécnica de Madrid (UPM), Spain
Doaitse Swierstra   Universiteit Utrecht, The Netherlands

Program Committee

Mats Carlsson	Swedish Institute of Computer Science, Sweden
Michael Codish	Ben-Gurion University, Israel
Patrick Cousot	École Normale Supérieure Paris, France
Bart Demoen	KU Leuven, Belgium
Pierre Deransart	INRIA Rocquencourt, France
Moreno Falaschi	Università di Padova, Italy
Michael Hanus	RWTH Aachen, Germany
Neil Jones	University of Copenhagen, Denmark
Herbert Kuchen	RWTH Aachen, Germany
Alexander Letichevsky	Ukrainian Academy of Sciences Kiev, Ukraine
Rita Loogen	Philipps-Universität Marburg, Germany
Jan Małuszyńsky	Linköping University, Sweden
Erik Meijer	Universiteit Utrecht, The Netherlands
Juan J. Moreno Navarro	Universidad Politécnica de Madrid (UPM), Spain
Hiroshi Nakashima	Kyoto University, Japan
Rinus Plasmeijer	KU Nijmegen, The Netherlands
Laurence Puel	CNRS Université Paris Sud, France
Mario Rodríguez Artalejo	Universidad Complutense de Madrid, Spain
Francesca Rossi	Università di Pisa, Italy
Peter Stuckey	University of Melbourne, Australia
Will Winsborough	Pennsylvania State University, U.S.A.

Organizing Committee

Jeroen Fokker
Erik Meijer
Margje Punt
Doaitse Swierstra

Sponsors (of PLILP and LoPSTr)

European Comission - Esprit Basic Research (CompuLog-Net)
Royal Dutch Academy of Sciences (KNAW)
The Netherlands Computer Science Research Foundation (SION)
The Association of Logic Programming
Italian National Research Centre (IASI-CNR)
Universiteit Utrecht

List of Referees

Many other referees helped the Program Committee in evaluating papers. Their assistance is gratefully acknowledged.

Mina Abdiche, Peter Achten, Yohji Akama, María Alpuente, Nils Andersen, Antoy, Roberto Bagnara, Anindya Banerjee, Jonas Barklund, Alexander Bockmayr, Frank de Boer, Roland Bol, George Botorog, Boudet, Johan Boye, Silvia Breitinger, Maurice Bruynooghe, Francisco Bueno, Daniel Cabeza, Björn Carlson, Manuel Carro, Manuel Chakravarty, Baudouin Le Charlier, Takashi Chikayama, Charles Consel, Evelyne Contejean, Roberto Di Cosmo, Radhia Cousot, Mads Dam, Philip Dart, Saumya Debray, Lars Degerstedt, Włodzimierz Drabent, Dirk Dussart, Marko van Eekelen, Michael Elhadad, Emmanuel Engel, Sandro Etalle, Heinz Faßbender, José-Alb. Fernández, Maribel Fernandez, Gérard Ferrand, Maria Ferreira, Jeroen Fokker, Martin Fränzle, Laurent Fribourg, Julio García-Martín, María García de la Banda, A. Gavilanes, Simon Gay, Roberto Giacobazzi, Robert Giegerich, Ana Gil-Luezas, Katia Gladitz, Ursula Goltz, J.C. González-Moreno, Jean Goubault, John van Groningen, Gudjon Gudjonsson, Mírian Halfeld Ferrari, Kiyoharu Hamaguchi, Michael Hansen, Werner Hans, James Harland, John Hatcliff, Morten Heine Sorensen, Angel Herranz-Nieva, Guido Hogen, Teresa Hortala, Graham Hutton, Nick Ioffe, Gerda Janssens, Bharat Jayaraman, Thomas Jensen, Martin Jourdan, David Kemp, Marco Kesseler, Jan Komorowski, Jeroen Krabbendam, Shigeru Kusakabe, Pedro López-García, Vitaly Lagoon, Michael Leushel, Francisco Lopez-Frag., Andy Mück, P. Malacaria, Julio Mariño, Kim Marriott, Bern Martens, Narciso Marti-Oliet, Satoshi Matsuoka, Chiara Meo, Spiro Michaylov, Odile Millet-Botta, Markus Mohnen, Bruno Monsuez, Johan Montelius, Remco Moolenaar, Anne Mulkers, Eric Nöcker, Robert Nieuwenhuis, Ulf Nilsson, Shin-ya Nishizaki, Thomas Noll, Mikael Pettersson, Ernesto Pimentel, Alberto Policriti, Andrés del Pozo-Prieto, Christian Prehofer, Germán Puebla, Jean-Hugues Réty, Femke van Raamsdonk, María José Ramírez, Jakob Rehof, Jean-Hugues Rety, Kristoffer Rose, Albert Rubio, Salvatore Ruggieri, Harald Søndergaard, Y. Sagiv, Dan Sahlin, Dave Sands, Sjaak Smetsers, Terrance Swift, Peter Thiemann, Eyal Tuvi, Kazunori Ueda, Franck Vedrine, Arnaud Venet, Germán Vidal, F. Voisin, Christel Vrain, Uwe Waldmann, Mitchell Wand, Harvey Warwick, Stephan Winkler, Roland Yap, Enea Zaffanella, Frank Zartmann.

Table of Contents

Abstract Interpretation

Transformation

Partial Evaluation

Graphical User Interfaces

Logic Programming Theory

Posters and Demonstrations

A Generic Program for
Sequential Decision Processes

Oege de Moor

Programming Research Group, Oxford University Computing Laboratory,
Wolfson Building, Parks Road, Oxford OX1 3QD, UK

1 Motivation

A generic program in a class of problems is a program which can be used to solve all problems in that class by suitably instantiating its parameters. The idea of a generic program is the idea that of Abu Hofstroh: to program for the algebraic class problem [1]: then we have a single program, parameterised on a number of operators, that solves instances concerning properties. Furthermore, the applicability of the generic program is always guaranteed as the condition that the parameters form a check structure.

Examples of such generic algorithms have recently been, and one might be forgiven to believe that the majority of programs cannot be expressed in a generic manner. I tentatively am hopeful that even for the industrious, the ...

... quite little effort has been attempted to classes concerning algorithms ... and it is that transparency statements, as rather elsewhere materialises. In the awareness of new algorithms search, the organisation of existing knowledge ... a certain unpolished gloss makes important blocks are indispensable. For only given the design of architecture independent parallel algorithms, famous algorithms have been discovered, and form the core of the subject [12]. Secondly, to some arguments also those one requires a higher-level notation in which both programs and programs can be expressed. In other programs. Such higher-order notations encourage the programmer to explore generality where possible, and indeed many functional programmers now do so as a matter of course. Most traditional programming notations do not offer similar support for writing generic programs.

This paper is an attempt to persuade you of my viewpoint by presenting a novel generic program for a certain class of optimisation problems, named sequential decision processes. This class was originally identified by Richard Bellman in his pioneering work on dynamic programming [4]. It is a perfect example of a class of problems which are very much alike but which has until now escaped solution by a single program.

Those readers who have followed some of the work that Richard Bird and I have been doing over the last few years [5, 7] will recognise many individual examples, all of these have now been unified. The point of this observation is that everywhere you are on the lookout for generic programs it can take rather long time to discover them. This presentation below will follow that earlier work.

A Generic Program for Sequential Decision Processes

Oege de Moor

Programming Research Group, Oxford University Computing Laboratory
Wolfson Building, Parks Road, Oxford OX1 3QD, UK

1   Motivation

A *generic program* for a class of problems is a program which can be used to solve
each problem in that class by suitably instantiating its parameters. The most
celebrated example of a generic program is the algorithm of Aho, Hopcroft and
Ullman for the *algebraic path problem* [1]. Here we have a single program, parameterised by a number of operators, that solves numerous seemingly disparate
problems. Furthermore, the applicability of the generic program is elegantly
stated as the condition that the parameters form a *closed semi-ring*.

Examples of such generic algorithms are admittedly scarce, and one might
be led to believe that the majority of programs cannot be expressed in such
a generic manner. I personally do not share that view, for the following two
reasons:

Firstly, little effort has gone into attempts to classify existing algorithms,
reflecting the fact that the computing community as a whole places more value on
the invention of new algorithms than on the organisation of existing knowledge.
In certain specialised areas where organisational tools are indispensable (for
example in the design of architecture-independent parallel algorithms), numerous
generic algorithms have been discovered, and form the core of the subject [13].

Secondly, to express generic algorithms one requires a higher-order notation
in which both programs and types can be parameters to other programs. Such
higher-order notations encourage the programmer to exploit genericity where
possible, and indeed many functional programmers now do so as a matter of
course. More traditional programming notations do not offer similar support for
writing generic programs.

This paper is an attempt to persuade you of my viewpoint by presenting a
novel generic program for a certain class of optimisation problems, named *sequential decision processes*. This class was originally identified by Richard Bellman
in his pioneering work on dynamic programming [4]. It is a perfect example of
a class of problems which are very much alike, but which has until now escaped
solution by a single program.

Those readers who have followed some of the work that Richard Bird and
I have been doing over the last five years [6, 7] will recognise many individual
examples: all of these have now been unified. The point of this observation is
that even when you are on the lookout for generic programs, it can take a rather
long time to discover them. The presentation below will follow that earlier work,

by referring to the calculus of relations and the relational theory of data types. I shall however attempt to be light on the formalism, as I do not regard it as essential to the main thesis of this paper. Undoubtedly there are other (perhaps more convenient) notations in which the same ideas could be developed. This paper does assume some degree of familiarity with a lazy functional programming language such as *Haskell, Hope, Miranda*[1] or *Gofer*. In fact, the LaTeX file used to produce this paper is itself an executable Gofer program.

2 Sequential Decision Processes

Many optimisation problems can be specified in the form

```
spec r p gen = listmin r . filter p . gen
```

Here the *generator* gen generates a list of candidate solutions, `filter p` selects those solutions which are feasible (those that satisfy the predicate p), and finally the *selector* `listmin r` picks a minimum solution according to the order relation r.

   Sequential decision processes are distinguished from other optimisation problems in the way the generator is formulated. The *sequential* nature of the generator is captured by expressing it as an instance of the function fold[2]. Informally, fold is defined by the equation

```
fold add e  [a1,a2,...,an] = a1 'add' (a2 'add' ... (an 'add' e))
```

In words one might say that `fold add e x` traverses the list x from right to left, starting with the seed value e, and summing with the operator add at each step. The recursive definition of fold is

```
>fold add e [] = e
>fold add e (a:x) = a 'add' (fold add e x)
```

In a sequential decision process, we have

```
gen = fold f e
```

---

[1] Miranda is a trademark of Research Software Ltd.
[2] The function fold is normally called foldr. Because I shall deviate from the common definition of foldr1, I have decided to use fold and fold1 to avoid confusion.

and at each step, the operator f offers a nondeterministic choice between a number of alternatives. It is this choice that accounts for the term *decision process*.

The alternatives are presented as a list of operators, and the specification of a sequential decision process therefore reads

```
>sdp_spec r p fs c = listmin r . filter p . fold (choice fs) [c]
>choice fs a xs     = [ f a x | f <- fs, x <- xs ]
```

That is, at each step we have the choice of applying any of the operators from fs to any of the results so far produced in xs. Cognoscenti will recognise a *relational fold* here [2].

The above can be varied slightly according to the kind of lists one works with. For instance, for non-empty lists the counterpart of fold is

```
>fold1 f g [a]   = g a
>fold1 f g (a:x) = f a (fold1 f g x)
```

That is, the function g is applied to the last element of the list, and then we sum from right to left using the function f. Modifying the notion of sequential decision process accordingly, one obtains

```
>sdp1_spec r p fs l = listmin r . filter p . fold1 (choice fs) l'
>                     where l' a = [l a]
```

Note that this differs from the definition of sdp_spec in the base case only.

Below we shall also have a use for lists with at least two elements. Again fold can be modified for this particular data type:

```
>fold2 f g [a,b] = g a b
>fold2 f g (a:x) = f a (fold2 f g x)
```

In words, g is applied to the last two elements of the list, and then we sum (as before) from right to left, using the function f. The corresponding notion of sequential decision process is given by

```
>sdp2_spec r p fs l = listmin r . filter p . fold2 (choice fs) l'
>                     where l' a b = [l a b]
```

These variations already hint at the possibility that there exists a notion of decision process for other data types besides lists. We shall consider an example of this phenomenon at the end of this paper.

3 Preorders, Minning, Squeezing, Merging

We have not yet discussed the function `listmin` and the order relation `r` on which it operates. As might be expected, it is the notion of order and various ways of selecting minimum elements that are at the heart of our generic program for sequential decision processes. For this reason, we shall now look at a number of combinators for manipulating orders (to be precise, *preorders*) and for rejecting non-minimum elements. These combinators are the building blocks for the generic program presented in the next section.

Preorders. A *preorder* is a relation of type

```
>type Rel a = a -> a -> Bool
```

A preorder `r` is furthermore required to be *reflexive* (`r a a` for all a) as well as *transitive* (`r a b && r b c` implies `r a c`). A trivial example of a preorder is the relation

```
>top a b = True
```

New preorders can be created from functions whose target type carries an implicit order or an equality test:

```
>leq :: Ord b => (a -> b) -> Rel a
>leq f a b = f a <= f b

>geq :: Ord b => (a -> b) -> Rel a
>geq f a b = f a >= f b

>eq :: Eq b => (a -> b) -> Rel a
>eq f a b = f a == f b
```

For instance, `leq length` is the length preorder on lists. We shall be combining preorders built from the above primitives using two combinators, named meet (also called intersection) and sequential composition (also called lexicographical composition) respectively:

```
>meet :: Rel a -> Rel a -> Rel a
>meet r s a b = r a b && s a b

>seq :: Rel a -> Rel a -> Rel a
>seq r s a b = r a b && (not (r b a) || s a b)
```

It is easy to check that the meet of two preorders is again a preorder, and sequential composition also preserves preorders. To illustrate sequential composition, consider `r = seq (leq length) (geq sum)`. We have `r a b` precisely when

```
(length a < length b) || (length a == length b && sum x >= sum y)
```

The degenerate preorder `top` is the identity element of both `meet` and `seq`, in the sense that

```
    top 'meet' r = r = r 'meet' top
```
and
```
    top 'seq' r = r = r 'seq' top
```

Both `meet` and `seq` are associative.

listmin. Let us now take a closer look at the selection of a minimum element by the function `listmin r`. It is defined by

```
>listmin :: Rel a -> [a] -> a
>listmin r = foldl m id
>           where m a b = a, if r a b
>                       = b, otherwise
```

Note that `listmin r` can only be applied to non-empty arguments. Also, the definition implicitly assumes that r is *connected*: `r a b` or `r b a` for all a and b. It is important to realise that `listmin r x` returns the first minimum element (reading from left to right) in x. In particular, `listmin top = head`.

squeeze. It is sometimes desirable to reject non-optimal elements while comparing with preorders that are not connected. This can be achieved, at least to some extent, by the function `squeeze r`. It removes an element from a list if one of its neighbours is lesser in the preorder r:

```
>squeeze :: Rel a -> [a] -> [a]
>squeeze r []      = []
>squeeze r [a]     = [a]
>squeeze r (a:b:x) = squeeze r (a:x),   if r a b
>                  = squeeze r (b:x),   if r b a
>                  = a:squeeze r (b:x), otherwise
```

When r is connected, and x is a non-empty list, we have

```
     squeeze r x = [listmin r x]
```

In this sense squeeze is a proper generalisation of listmin.

mmerge. Another operator for manipulating lists and preorders is mmerge[3]. It takes a preorder r, two lists x and y that are ordered with respect to r, and it merges the two lists to produce another ordered list whose elements are precisely those of x and y. The definition of mmerge is

```
>mmerge :: Rel a -> [a] -> [a] -> [a]
>mmerge r [] y        = y
>mmerge r x []        = x
>mmerge r (a:x) (b:y) = a:mmerge r x (b:y), if r a b
>                     = b:mmerge r (a:x) y, otherwise
```

Note that mmerge has been defined so that mmerge top x y = x++y.

purge. The combination of squeeze and merge is called purge. It takes two ordered lists, merges them, and then squeezes the result:

```
>purge :: Rel a -> Rel a -> [a] -> [a] -> [a]
>purge r s x y = squeeze r (mmerge s x y)
```

Note that if r is connected, we have purge r top x y = [listmin r (x++y)]. Curiously, the discovery of the generic algorithm in the next section was held up for over a year because an earlier version of this work used a slightly less general definition of purge. As every functional programmer knows, the design of one's primitive combinators can make all the difference...

4   A Generic Program

Consider the sequential decision process

```
sdp_spec r p fs c = listmin r . filter p . fold g [c]
                    where g a xs = [ f a x | f <- fs, x <- xs ]
```

---

[3] The double m in mmerge is just there to avoid a name clash with the existing **merge** function in Gofer, which differs from our definition in that it does not take the preorder as an explicit argument.

Bellman's original insight was that, provided the principle of optimality is satisfied, this sequential decision process can be solved by dynamic programming [4]. Below we formalise the principle of optimality in three parts, and give a single program that captures the dynamic programming solution. This program is the novel contribution of the present paper. To apply the program, one has to invent a predicate q and preorders s and t that satisfy the following three requirements:

1. Let q be a predicate such that for each function f in fs

    p (f a x) = q (f a x) && p x

2. Furthermore, let s be a preorder such that for each f in fs

    r x y   &&   s x y   &&   q (f a y)
    implies
    r (f a x) (f a y)   &&   s (f a x) (f a y)   &&   q (f a x)

3. Finally, let t be a connected preorder such that for each f in fs

    t x y   implies   t (f a x) (f a y)

If furthermore we have that p c holds, the function sdp r s t q fs c x defined below computes a solution to sdp_spec r p fs c x, for all lists x:

```
>sdp r s t q fs c = listmin r . fold (step (meet s r) t q fs) [c]
>step v t q fs a xs = foldl (purge v t) id
>                          [filter q [f a x | x <- xs] | f <- fs ]
```

It is worthwhile to consider the conditions of this result in some detail. The first condition, in conjunction with the requirement that p c holds, says that filter p can be *promoted* into the generator part of the sequential decision process. Readers who are familiar with Bird's theory of lists [5] may recognise that when f is the cons operator (:), this is another way of expressing that p is *suffix-closed*. Similar conditions can be found in other work involving *filter promotion*.

The second property is a so-called *dominance* criterion. Informally, it says that when x is better than y (that is (r x y) && (s x y)), then the extension of x is better than the extension of y; furthermore, the extension of x is a feasible solution if the extension of y is. The exploitation of dominance criteria is a well-known strategy for speeding up dynamic programming algorithms [25].

The third property states that each f in fs is monotonic on the connected preorder t. This condition gives little guidance for choosing a preorder t, but in most cases t is a simple combination of r and s.

It would be nice if these three properties amounted to a well-understood mathematical structure, in analogy with closed semi-rings in the algebraic path problem. Unfortunately, I have been unable to identify such a connection with well-established mathematics, perhaps because the conditions are phrased as implications rather than as algebraic properties.

The program sdp generates a list of candidates which is ordered according to the preorder t. Because each f is monotonic on t, we can maintain this order at each stage by merging ordered lists. The result is then squeezed using the dominance criterion meet r s: this is where the improvement over the original formulation of sdp_spec occurs. If the squeezing is successful, only a handful of intermediate solutions will be kept at each stage. Finally, an optimum solution is selected using listmin r.

Some readers may wonder why the conclusion in the above result is not phrased as

```
sdp_spec r p fs c x  =  sdp r s t q fs c x
```

instead using the awkward phrase '... sdp computes a solution to sdp_spec ...'. The reason is that the above equation is false. The new program sdp, when applied to x, will yield a result that is equivalent to that of sdp_spec, in the sense that it is a minimum element of filter p (fold (choice fs) [c] x), but it need not be the same minimum element as the one returned by listmin r. It is here that we pay the price for discussing these ideas within a purely functional framework. A truly satisfactory solution requires a programming notation based on relations (or predicates) rather than functions. One such notation is expounded in [8], and that book contains detailed proofs of the claims in this paper.

Again we remark that with minor changes, the above goes through for non-empty lists. Consider the specification

```
sdp1_spec r p fs l = listmin r . filter p . foldl (choice fs) l'
                         where l' a = [l a]
```

The only change to the above result concerns the requirement that p c holds. For non-empty lists, the corresponding condition is

Suppose that for each a, *we have* p (l a).

Conditions (1-3) (which are concerned with the newly invented functions q, s and t) are the same as for possibly empty lists. The generic program for non-empty lists is

```
>sdp1 r s t q fs l = listmin r . fold1 (step (meet s r) t q fs) l'
>                    where l' a = [l a]
```

Analogously, we have a generic program for lists that have at least two elements, namely

```
>sdp2 r s t q fs l = listmin r . fold2 (step (meet s r) t q fs) l'
>                    where l' a b = [l a b]
```

and here the requirement that p c holds is replaced by

Suppose that for each a *and* b, *we have* p (l a b).

Conditions (1-3) are the same for all three kinds of list.

5 Examples

Below we apply the generic algorithm to three different programming exercises. In each exercise, most of the effort goes into phrasing the problem as a sequential decision process. Once the problem is in the right form, inventing the additional predicate q and the preorders s and t is easy. Three exercises is all that space allows in this paper; [8] presents seven more applications, and that number is still growing.

5.1 Knapsack

In the 0/1 knapsack problem, we are given a sequence x of items and a non-negative capacity c. Each item is a pair of non-negative numbers, called the value and the weight of that item. The objective is to compute a subsequence of x whose total weight does not exceed c, and whose total value is as large as possible. The total weight of a sequence is defined as the sum of the weights of the items it contains, and the total value of a sequence is the sum of the individual values.

Formal specification. An item is a pair of non-negative numbers

```
>type Item = (Int, Int)
```

whose first component signifies the value of the item, and the second component the weight:

```
>v,w :: Item -> Int
>v (a,b) = a
>w (a,b) = b
```

Our aim is to give an efficient implementation of

```
>kp_spec :: Int -> [Item] -> [Item]
>kp_spec c   =   listmin rval . filter (ok c) . subs
```

The function **subs** is the generator: it generates all subsequences of its argument, in some order. One can define **subs** as

```
>subs  =  fold (choice [(:),rhs]) [[]]
>          where rhs a x = x
```

This definition reflects the observation that, at each stage, we have the choice of either including a in a subsequence by applying (:), or excluding it by applying **rhs**. The predicate **ok c** checks whether the total weight of a subsequence does not exceed the capacity c. The function **weight** returns the total weight of a sequence, so the definition of **ok** is

```
>ok :: Int -> [Item] -> Bool
>ok c x  =  weight x <= c

>weight :: [Item] -> Int
>weight  =  sum . map w
```

Finally, we are interested in a subsequence whose value is as large as possible. Taking the maximum in the value preorder means taking the minimum in the *reverse* value preorder, which is given by

```
>rval :: Rel [Item]
>rval = geq value

>value :: [Item] -> Int
>value = sum . map v
```

Here **value** is the function that returns the total value of a sequence. Summarising, the knapsack problem can be formulated as a sequential decision process:

```
kp_spec c   =   sdp rval (ok c) [(:),rhs] []
```

In order to apply the generic algorithm, we need to determine a predicate q, and preorders s and t that satisfy the three applicability conditions.

Checking the conditions. First, note that ok c [] holds because c is non-negative. We shall take q = (ok c). This does satisfy the promotion condition on q

```
    ok c (a:x)   implies   ok c x
```

This implication is valid because weights are non-negative. For the dominance criterion, observe that

```
  value x >= value y   &&   weight x <= weight y
  && ok c (a:y)
implies
  value (a:x) >= value (a:y) && weight (a:x) <= weight (a:y)
  && ok c (a:x)
```

It follows that we can take s = leq weight. Finally, observe that both (:) and rhs are monotonic on rval, so we can take t = rval = geq value. We conclude that the generic algorithm is applicable.

Program. There is still one minor source of inefficiency, namely the repeated computation of the total value and weight of a subsequence. The solution is to store the total value and weight of a sequence together with the sequence itself, and maintain them incrementally. The resulting program is displayed below:

```
>kp c =
> the.sdp (geq value) (leq weight) (geq value) ok [cons,rhs] empty
> where value (x,v,w) = v
>       weight (x,v,w) = w
>       the (x,v,w) = x
>       ok x = weight x <= c
>       cons (a,b) (x,v,w) = ((a,b):x,a+v,b+w)
>       rhs (a,b) (x,v,w) = (x,v,w)
>       empty = ([],0,0)
```

This is the standard pseudo-polynomial time $\mathcal{O}(cn)$ algorithm for the knapsack problem. Unlike the program found in many textbooks, however, it does not depend on the values and weights being integers: it is equally applicable to floating point numbers. Starting from the above algorithm as a *specification*, my research student Michael Zhu Ning has developed a novel, highly optimised algorithm that outperforms all other methods in practice [22]. The main improvement in his work comes from a very simple change to the generic program: because we have chosen t = r = **geq value**, the maximum value solution appears at the head of the list produced by **fold** and so — for this particular application — **listmin r** in the generic program can be replaced by **head**. Lazy evaluation then leads to a substantial time saving.

5.2   Bitonic Tours

Our next programming exercise is taken from [10]:

> The *Euclidean travelling salesman* problem is the problem of determining a shortest closed tour that connects a given set of n points in the plane. The left-hand side of Figure 1 shows the solution to a 7-point problem. The general problem is NP-complete, and its solution is therefore believed to require more than polynomial time.
>
> J.L. Bentley has suggested that we simplify the problem by restricting our attention to *bitonic tours*, that is, tours that start at the leftmost point, go strictly left to right to the rightmost point, and then go strictly right to left back to the starting point. The right-hand side of Figure 1 shows the shortest bitonic tour of the same 7 points. In this case, a polynomial-time algorithm is possible.
>
> Describe an $O(n^2)$-time algorithm for determining an optimal bitonic tour. You may assume that no two points have the same x-coordinate. (*Hint:* Scan right to left, maintaining optimal possibilities for the two components of the tour.)

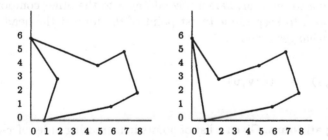

Fig. 1. An optimal tour and an optimal *bitonic* tour.

Representation of tours. We represent a bitonic tour by a pair of lists (u, v). The components u and v are the outward and return journeys of the tour. Note that the roles of u and v are symmetrical: if (u, v) is a bitonic tour, so is (v, u), and these tours are essentially the same. To prevent the existence of different representations of the same tour, we stipulate that the first element of u is the leftmost point of the tour. For instance, the pair (u, v) represents the optimal bitonic tour in Figure 1:

```
u = [(0,6),(1,0),(6,1)]
v = [(2,3),(5,4),(7,5),(8,2)]
```

Note that u and v are chosen to be disjoint non-empty lists.

Representation of input. It does not make sense to talk about a bitonic tour of one point because such a tour cannot be split into two disjoint unidirectional components, so we assume that there are at least two points. Furthermore, the input list is assumed to be presented in increasing order of x-coordinate. Because the input list is a list of length at least two, we shall be using the **fold2** operator, and the sequential decision process skeleton **sdp2_spec**.

Constructing tours. The operator **wrap** takes two input points, and turns them into the only possible bitonic tour, namely

```
>wrap a b    = ([a],[b])
```

The result of **wrap** is a correct representation of a bitonic tour provided that the x-coordinate of a is strictly smaller than the x-coordinate of b.

The binary operator **addl** takes an element and a tour, and adds the element to the left-hand component of the tour:

```
>addl a (u,v) = (a:u,v)
```

This yields a correct representation of a tour only if a has an x-coordinate strictly smaller than the elements of u and v. If this condition is satisfied, there is another way of adding a to the tour, namely by adding a to the other component of the tour. As we wish to keep the leftmost point of the tour at the head of the first list, this interchanges u and v:

```
>addr a (u,v) = (a:v,u)
```

Generating all tours. Given a list of points in increasing order of x-coordinate, all tours are returned by

```
>tours = fold2 (choice [addl,addr]) wrap'
>       where wrap' a b = [wrap a b]
```

That is, the sequence is scanned from right to left, each point being appended with either **addl** or **addr**.

Length of a tour. Let `path_len` be the function that returns the length of a path:

```
>path_len x = sum (zipWith dist x (tail x))
```

Here `dist a b` returns the distance between two points a and b. The Euclidean distance measure is coded as

```
>dist (a0,a1) (b0,b1) = sqrt ((a0 - b0)^2 + (a1 - b1)^2)
```

The total length of a tour can then be defined by

```
>len  (u,v)  =  path_len u + path_len v +
>                 dist (head u) (head v) + dist (last u) (last v)
```

So the formal specification of Bentley's exercise reads

```
>mt_spec  =  listmin (leq len) . tours
```

This specification is a sequential decision process:

```
mt2 =  sdp2_spec (leq len) (const True) [addl,addr] wrap
```

where `const True` is the predicate that is always true.

Checking the conditions. Again we need to find a predicate and two preorders to satisfy the applicability conditions of the generic algorithm. In this case, the choice of predicate is trivial, as in the original sequential decision process, p is the constant function returning True; we take q to be (`const True`) as well.

For the dominance criterion, observe that

```
  len (u,v) <= len (u',v')  &&   head u == head u'
  && head v == head v'
implies
  len (a:u,v) <= len (a:u',v') && head (a:u) == head (a:u')
  && head v == head v'
```

which gives the dominance criterion for addl; a similar property holds for addr. We can therefore take s = eq heads, where heads(u,v) = (head u,head v).

Finally, to satisfy the last condition we need a connected preorder t on which addl and addr are monotonic. There is no obvious choice for such a preorder dictated by the problem, so we might as well set t = top. Trivially, any total function is monotonic on the degenerate preorder top.

Program. In analogy with the knapsack example, it makes sense to maintain the length of a tour incrementally, so that it does not have to be recomputed at each comparison. A tour (u,v) will be represented by the three-tuple

```
(u, v, path_len u + path_len v + dist (last u) (last v))
```

The resulting program is shown below:

```
>mt =
> the.sdp2 (leq len) (eq heads) top (const True) [addl,addr] wrap
> where len (a:x,b:y,n) = dist a b + n
>       heads (a:x,b:y,n) = (a,b)
>       the (a:x,b:y,n) = (a:x,b:y)
>       wrap a b = ([a],[b],dist a b)
>       addl a (b:x,c:y,n) = (a:b:x,c:y,n+dist a b)
>       addr a (b:x,c:y,n) = (a:c:y,b:x,n+dist a c)
```

This program has the required $\mathcal{O}(n^2)$ time complexity.

5.3 Bus stops

Many ideas in this paper have been gleaned from the operations research literature, which has been much more concerned with exposing connections between algorithms than the computing literature. It seems fitting, therefore, to illustrate our generic algorithm with a typical operations research problem. The following exercise occurs in [11]:

A long one-way street consists of m blocks of equal length. A bus runs "uptown" from one end of the street to the other. A fixed number n of bus stops are to be located so as to minimise the total distance walked by the population. Assume that each person taking an uptown bus trip walks to the nearest bus stop, gets on the bus, rides, gets off at the stop nearest his destination, and walks the rest of the way. During the day, exactly B_j people from block j start uptown bus trips, and C_j complete uptown bus trips at block j. Write a program that finds an optimal location of bus stops.

Representing the input. We shall represent the input as a list

```
[B1 + C1, B2 + C2, ... , Bm + Cm]
```

Here each item Bj + Cj represents the number of people passing through block
j. If B and C are originally given as separate lists, one may convert them into
the above representation by evaluating

```
zipWith (+) [B1,B2,...,Bm] [C1,C2,...,Cm]
```

Representing bus stop arrangements. The original formulation of the problem is not quite clear about this, but we shall assume that bus stops are placed
in front of blocks, and not at block boundaries. Furthermore, if a bus stop is
placed in front of block j, people passing through j do not have to walk at all
in order to take the bus. An example arrangement of bus stops is shown below.
The first row are the numbers of blocks, and the second row shows how many
people pass through each block. In the last row, a tick ($\sqrt{}$) indicates that a bus
stop has been placed at a certain block, and the arrows show the direction by
which people walk to bus stops.

$$
\begin{array}{ccccccccccc}
1 & 2 & 3 & 4 & 5 & 6 & 7 & 8 & 9 & 10 & 11 \\
50 & 12 & 25 & 33 & 15 & 19 & 22 & 58 & 12 & 42 & 30 \\
\rightarrow & \rightarrow & \sqrt{} & \leftarrow & \leftarrow & \rightarrow & \sqrt{} & \leftarrow & \rightarrow & \sqrt{} & \leftarrow
\end{array}
$$

For instance, people passing through block 5 have to walk two blocks to the left
in order to take the bus at block 3. It follows that the total walking distance
required by the above arrangement is

$$2*50+1*12+0*25+1*33+2*15+1*19+0*22$$
$$+1*58+1*12+0*42+1*30$$

We shall represent an arrangement of bus stops for a sequence x of blocks as
a partition of x++[0], with a bus stop at the end of each partition component
except the last. The role of 0 at the end is to allow the placement of a bus stop at
the very last block in the street. The above example would thus be represented
as

```
[[50,12,25],[33,15,19,22],[58,12,42],[30,0]]
```

Having the decided on this representation, we have the following specification:

```
>buses_spec n x = (listmin(leq walk) . filter(atmost n) . parts)
>                 (x++[0])
```

where the function **parts** generates all partitions of its argument, the filter removes partitions that have more than n+1 components (n bus stops), and the selector picks a partition that realises the minimum walking distance. The definition of **atmost** is easy:

```
>atmost n xs  =  length xs <= n+1
```

The definitions of **parts** and **walk** are elaborated below.

Total walking distance. To give a formal definition of the total walking distance, we first introduce the *convoluted sum* cnv x of a list of numbers x. Informally, it is given by

$$\text{cnv } [a1, a2, \ldots , an] = 1*a1 + 2*a2 + \ldots + n*an$$

and that translates to the Gofer definition

```
>cnv x = sum (zipWith (*) [1..] x)
```

The total walking distance is then defined

```
>walk [x]       =  cnv x
>walk (x:xsy)   =  cnv (reverse (init x)) + sum (map wd xs) + cnv y
>                  where xs = init xsy
>                        y  = last xsy
```

The walking distance of a singleton partition is the convoluted sum of its single component: everybody has to walk to the very end of the street, in fact to a bus stop at the imaginary block that has 0 people passing through it. In the first partition component x of a non-singleton partition, people have to walk to the last block in that component. The function **init** x returns all of x except the last element. The contribution to the total walking distance is therefore cnv (**reverse** (**init** x)). The cost of the last component y is similarly computed, except that we do not have to strip off the last element: it is zero anyway. The walking distance contributed by a non-extremal partition component is defined by

```
>wd x = cnv y + cnv (reverse z)
>        where (y,z) = split (init x)
```

The function `split` splits its argument x into two approximately equal halves.
Using the standard function `splitAt`, which splits a list at a specified position,
it can be defined by

```
>split x = splitAt (length x / 2) x
```

Generating partitions. We know that the input list for `parts` is non-empty
(its last element is 0), so we aim for a definition of `parts` in terms of `fold1`. The
function `wwrap` takes an element and creates a singleton partition:

```
>wwrap a   =   [[a]]
```

Given an element a and a partition `xs`, we can either add a as a new partition
component, or glue a to the first component of `xs`:

```
>new a xs   =   [a]:xs
>glue a (x:xs)  =   (a:x):xs
```

This leads us to the following definition of *parts*:

```
>parts  =   fold1 (choice [new,glue]) wwrap'
>            where wwrap' a = [wwrap a]
```

In summary, we have specified the bus stop problem as a sequential decision
process

```
  buses_spec  n x  =  sdp (leq walk) (atmost n) [new,glue] (x++[0])
```

Checking the conditions. In analogy with the knapsack example, we take the
predicate q to be `atmost n` itself. This satisfies the requirements because we
have `atmost n (wwrap a)` for any a, and

```
    (atmost n ([a]:xs))   implies   (atmost n xs)
```
and
```
    (atmost n ((a:x):xs)) implies   (atmost n (x:xs))
```

For the dominance criterion, we take our cue from the bitonic tours example, and observe that the following property holds for **new**:

```
walk xs <= walk ys && length xs <= length ys
&& head xs == head ys && atmost n ([a]:ys)
implies
    walk([a]:xs) <= walk([a]:ys) && length([a]:xs) <= length([a]:ys)
&& head([a]:xs) == head([a]:ys) && atmost n ([a]:xs)
```

The analogous property for **glue** is also easily verified. It follows that we can take s = meet (leq length) (eq head). In fact, since we are comparing partitions of the same sequence, we can actually use the computationally more efficient definition s = meet (leq length) (eq (length . head)).

It remains to find a connected preorder t on which both **new** and **glue** are monotonic. For efficiency reasons it is best to take a preorder that is as restrictive as possible, and a heuristic that often succeeds is to take a relaxation of s. In this case, that heuristic suggests a sequential composition of leq length and leq (length.head):

```
t = seq (leq length) (leq (length . head))
```

Program. Each partition xs will be represented as a 5-tuple

```
(xs, walk(tail xs), cnv(reverse(init(head xs))),
  length xs, length (head xs))
```

This facilitates constant time comparison in all three preorders defined above. The resulting program, which is displayed below, has time complexity $\mathcal{O}(nm^2)$, where n is the number of bus stops and m the number of blocks in the street:

```
>buses n x =
>    the (sdp1 r s t atmost [new,glue] wwrap (x++[0]))
>    where r = leq walk
>          s = leq length 'meet' eq length_head
>          t = leq length 'seq' leq length_head
>          atmost xs = length xs <= n+1
>          the (xs,s,c,n,m) = xs
>          walk (xs,s,c,n,m) = s+c
>          length (xs,s,c,n,m) = n
>          length_head (xs,s,c,n,m) = m
>          wwrap a = ([[a]],0,0,1,1)
```

```
>           new a (xs,s,c,1,m) = ([a]:xs,0,cnv(head xs),2,1)
>           new a (xs,s,c,n,m) = ([a]:xs,0,wd(head xs)+c,n+1,1)
>           glue a (x:xs,s,c,n,m)= ((a:x):xs,m*a+s,c,n,m+1)
```

Note that we have not made essential use of the fact that the blocks are all of equal size; it would be quite easy to adjust the program to cope with different sizes.

6 Why polytypism matters

Some readers may feel that I have cheated in the above examples: after all, I am not using a single program but three different programs which operate on possibly empty lists, non-empty lists, and lists with at least two elements. Surely it would have been possible to code the problems in such a way that only one kind of lists and only one program sufficed?

While this is probably true of the specific examples shown above, I would rather say that I have presented a single program that is parameterised by the type on which it operates. Such a program is said to be *polytypic*. To reinforce the point, let me show you one more instance of the generic program for sequential decision processes, this time applied to trees rather than lists. The following problem is the topic of [18]:

Let G be an acyclic directed graph with weights and values assigned to its vertices. In the partially ordered knapsack problem we wish to find a maximum-valued subset of vertices whose total weight does not exceed a given knapsack capacity, and which contains every predecessor of a vertex if it contains the vertex itself. We consider the special case where G is an out-tree. Even though this special case is still NP-complete, we observe how dynamic programming techniques can be used to construct pseudopolynomial time optimisation algorithms.

The data structure described as an *out-tree* in the above quotation can be defined in a functional programming style as

```
>data Tree a = Node a [Tree a] | Null
```

where every node has at least one descendant. A *pruning* of a tree t is a copy of t where some of its subtrees have been replaced by Null. Given a tree t :: Tree (Int,Int) of (value,weight) pairs, we can define the total value of t as the sum of the values it contains, and similarly the total weight is defined as the sum of the individual weights. The *tree knapsack* problem is to compute a pruning whose total weight does not exceed a given capacity c, and whose value is as large as possible. This problem has sinister applications in cutting a hierarchical workforce; here it is merely a programming challenge.

Clearly we have a decision process here, albeit not a sequential one: at every node we have the choice between including that node or leaving it out. Indeed, one can define a notion of *fold* over trees, and replay the discussion that led us to a generic list-based algorithm, to get a generic tree-based algorithm. These algorithms will be 'essentially the same'; in fact, if we had a programming notation that allowed the parameterisation of programs by type structure, both programs would be instances of the same *polytypic* program. Theories and notations for reasoning about polytypic programs are under construction, but it is beyond the scope of this paper to go into details. The specific instance of that polytypic program, specialised for the above programming exercise, is shown below:

```
>foldTree f c Null = c
>foldTree f c (Node a x) = f a (map (foldTree f c) x)

>treekp c =
>   the . listmin (geq value) . foldTree f [nul]
>   where f a tss = foldl step (filter ok . map (add a)) tss++[nul]
>         step ts ss = foldl (purge dom (geq value)) id
>                      [filter ok [glue t s | t<- ts] | s <- ss]
>         the (x,v,w) = x
>         weight (x,v,w) = w
>         value (x,v,w) = v
>         nul = (Null,0,0)
>         ok x = weight x <= c
>         add (a,b) (t,v,w) = (Node (a,b) [t],v+a,w+b)
>         glue (t,v,w) (Node a ts,vs,ws) = (Node a(t:ts),v+vs,w+ws)
>         dom = geq value 'meet' leq weight
```

Hopefully this example gives some indication of the possibilities for writing generic, type independent programs. Recently a number of papers have suggested that Gofer's *constructor classes* are the appropriate medium for polytypic programming [20, 21]. The examples in those papers certainly show the great flexibility of constructor classes, and they present an interesting experiment in pushing the limits of genericity in programming. I would, however, like to add a word of caution. Constructor classes are essentially a means of introducing *ad hoc* polymorphism into the Hindley-Milner type system. The polytypic programs that I have in mind are — if we ignore exponential types — parametric in the type constructors. To find a type system that reflects such parametricity would, in my view, be a major step towards a truly generic style of programming. Jay's exploration of *shapely types* may be a step in the right direction [16].

7   Related work

The view of sequential decision processes put forward here is, by the standards of operations research, rather restrictive: many experts in that community would classify all applications of dynamic programming as sequential decision processes. It then becomes difficult to give a precise algorithmic definition of the concept. Several semi-algorithmic definitions have however been put forward, and especially the work of Helman and Rosenthal [14] has been a major influence on the results reported here. Our view of sequential decision processes is quite close to that in the seminal paper by Karp and Held [19].

In the algorithm design community, a slightly less general class of problems has received considerable attention under the name *least weight subsequence* problems [15]. In terms of the terminology introduced here, all of these problems are concerned with list partitions. Most recent work in this area has been concerned with improving the time complexity of naive dynamic programming solutions, by exploiting special properties of the cost function [12].

In programming methodology, our work is very much akin to that of Smith and Lowry [23, 24]. Smith's notion of *problem reduction generators* is quite similar to the generic algorithm presented here, but his results are more concerned with similarities in the derivation process, and not with a single generic program.

There has recently been a surge of interest in polytypic programming. Initially most of this work was concerned with generalising combinators such as *fold* and *zip* to arbitrary data types [3]. Starting with [9], more algorithmic problems have been considered in a polytypic setting. The most spectacular example so far is Jeuring's polytypic pattern matching algorithm [17].

Acknowledgements

Much of the work reported here was done in collaboration with Richard Bird, and I am grateful to him for his support, and his insistence that there is always a better, more efficient, and more elegant solution. Sharon Curtis suggested an improvement over my original solution to the Bitonic Tours problem. Roland Backhouse, Richard Bird, Jeroen Fokker and Doaitse Swierstra suggested substantial improvements over a first draft.

References

1. A. V. Aho, J. E. Hopcroft, and J. D. Ullman. *The Design and Analysis of Computer Algorithms*. Addison–Wesley, 1974.
2. R. C. Backhouse, P. De Bruin, G. Malcolm, T. S. Voermans, and J. C. S. P. Van der Woude. Relational catamorphisms. In B. Möller, editor, *Proceedings of the IFIP TC2/WG2.1 Working Conference on Constructing Programs*, pages 287–318. Elsevier Science Publishers B.V., 1991.
3. Roland Backhouse, Henk Doornbos, and Paul Hoogendijk. Commuting relators. Technical Report. Available by anonymous ftp from ftp.win.tue.nl, directory pub/math.prog.construction., 1992.
4. R. Bellman. *Dynamic Programming*. Princeton University Press, 1957.
5. R. S. Bird. An introduction to the theory of lists. In M. Broy, editor, *Logic of Programming and Calculi of Discrete Design*, volume 36 of *NATO ASI Series F*, pages 3–42. Springer–Verlag, 1987.

6. R. S. Bird and O. de Moor. List partitions. *Formal Aspects of Computing*, 5(1):61–78, 1993.

7. R. S. Bird and O. de Moor. Solving optimisation problems with catamorphisms. In *Mathematics of Program Construction*, volume 669 of *Springer Lecture Notes in Computer Science*, 1993.

8. R. S. Bird and O. de Moor. *Algebra of Programming*. Prentice-Hall, To appear, 1996.

9. R. S. Bird, P. Hoogendijk, and O. de Moor. Generic programming with relations and functors. Submitted to Journal of Functional Programming, 1993.

10. T. H. Cormen, C. E. Leiserson, and R. L. Rivest. *Introduction to Algorithms*. The MIT electrical engineering and computer science series. MIT Press, Cambridge, Mass. ; London / McGraw-Hill, New York, 1990.

11. E. V. Denardo. *Dynamic Programming — Models and Applications*. Prentice-Hall, 1982.

12. D. Eppstein, Z. Galil, R. Giancarlo, and G. F. Italiano. Sparse dynamic programming II: Convex and concave cost functions. *Journal of the ACM*, 39:546–567, 1992.

13. A. Gibbons and W. Rytter. *Efficient Parallel Algorithms*. Cambridge University Press, 1988.

14. P. Helman and A. Rosenthal. A comprehensive model of dynamic programming. *SIAM Journal on Algebraic and Discrete Methods*, 6(2):319–334, 1985.

15. D. S. Hirschberg and L. L. Larmore. The least weight subsequence problem. *SIAM Journal on Computing*, 16(4):628–638, 1987.

16. B. Jay. Matrices, monads and the Fast Fourier transform. Available by anonymous ftp from ftp.socs.uts.edu.au directory pub/jay, August 1993.

17. J. Jeuring. Polytypic pattern matching. In S. Peyton Jones, editor, *Proceedings of the 7th Conference on Functional Programming Languages and Computer Architecture, FPCA '91*, 1995.

18. D. S. Johnson and K. A. Niemi. On knapsacks, partitions, and a new dynamic programming technique for trees. *Mathematics of Operations Research*, 8(1):1–15, 1983.

19. R.M. Karp and M. Held. Finite-state processes and dynamic programming. *SIAM Journal on Applied Mathematics*, 15(3):693–718, 1967.

20. E. Meijer and G. Hutton. Bananas in space: extending fold and unfold to exponential types. In S. Peyton Jones, editor, *Proceedings of the 7th Conference on Functional Programming Languages and Computer Architecture, FPCA '91*, 1995.

21. E. Meijer and M. Jones. Gofer goes bananas. Unpublished manuscript, 1994.

22. M. Z. Ning. A lazy algorithm for the 0/1 knapsack problem. Technical Report (to appear), Oxford University Computing Laboratory, 1995.

23. D. R. Smith and M. R. Lowry. Algorithm theories and design tactics. *Science of Computer Programming*, 14:305–321, 1990.

24. D.R. Smith. Structure and design of problem reduction generators. In B. Möller, editor, *Proc. of the IFIP TC2 Working Conference on Constructing Programs from Specifications*. North-Holland, 1991.

25. F. F. Yao. Speed–up in dynamic programming. *SIAM Journal on Algebraic and Discrete Methods*, 3(4):532–540, 1982.

Parsing Visual Languages with Constraint Multiset Grammars

Kim Marriott

Department of Computer Science
Monash University
Australia
marriott@cs.monash.edu.au

Visual languages, that is two dimensional languages, such as electric circuit diagrams, musical notation, structural-chemical formulae, or mathematical equations have two main differences to one-dimensional text languages. The first is that the basic structure for organizing text, the sequence, is not relevant for two-dimensional input. Although, one might sequence input gestures according to the order in which they were entered, this seems too restrictive - there is no one natural order in which people draw a complex diagram. The second is that in text the only relationship between tokens is adjacency, whereas in a diagram the possible relationships are much richer and include geometric relationships such as containment and adjacency.

For these reasons we have introduced a new formalism, *constraint multiset grammars*, (CMGs) [7, 10] which gives a general, high-level framework for the definition of visual languages. Productions in a constraint multiset grammar have the form

$$P ::= P_1, ..., P_n \text{ where } C$$

indicating that the non-terminal token P can be rewritten to the multiset of tokens $P_1, ..., P_n$ whenever the attributes of all tokens satisfy the constraints C. The constraints enable information about spatial layout to be naturally encoded in the grammar. In particular, constraints may also refer to other sub-diagrams in the picture by requiring the existence, or non-existence, of tokens.

Constraint multiset grammars are a refinement of definite set grammars used in visual language generation [5, 6] which were, in turn, based on logic grammars used in natural language parsing [1]. Constraint multiset grammars are related to visual mappings [9], attributed multi-set grammars [4] and relation grammars [3]. We will sketch the theory of constraint multiset grammars, their relationship with text grammars and with constraint logic programming. Then we will look at an application of constraint multiset grammars to user interface construction.

The *membership problem* for constraint multiset grammars, that is, determining if a particular multiset of gestures is in the language of a grammar, is NP-hard even when the grammar has no attributes and no existentially quantified symbols. This is somewhat surprising, and indicates that parsing with multisets is intrinsically harder than parsing with formalisms based on sequence rewriting, such as text grammars, and formalisms based on set rewriting such as logic programming. However, despite this theoretical complexity, we give a practical algorithm for parsing diagrams which is based on the differential algorithm used in bottom-up evaluation of logic programs. To efficiently handle the geometric constraints, data structures and query optimization techniques developed for spatial databases are used [8]. In general, the parsing algorithm has exponential complexity because of possible ambiguities in the grammar, but for deterministic grammars it has polynomial complexity.

Constraint multiset grammars have obvious applications in human-computer interaction. In particular to user-interfaces for pen-based computers. We have developed two tools – a parser generator and a graphics editor – which allow the automatic generation of a sophisticated user interfaces based on an application specific visual language [2]. The parser generator takes a grammar specifying the language and produces a parser which is combined with a constraint-based graphics editor to give the user interface. The resultant user interface allows the user to construct diagrams in the visual language from text, lines, rectangles, or circles. These are incrementally parsed into sub-diagrams. During parsing, automatic error correction removes geometric errors in the original diagram, providing feedback about what has been recognized. The user may add or delete tokens and may manipulate components of the diagram. During manipulation, the user interface takes account of the semantics of the diagram, preserving the geometric relationships dictated by the visual language.

References

1. H. Abramson and V. Dahl. *Logic Grammars*. Springer-Verlag, New York, 1989.
2. S.S. Chok and K. Marriott. Automatic Construction of User Interfaces from Constraint Multiset Grammars. To appear *IEEE Symposium on Visual Languages*. Darmstadt, Germany. Sep. 1995.
3. F. Ferrucci, G. Tortora, M. Tucci, G. Vitiello. A Predictive Parser for Languages Specified by Relation Grammars. In *Proc. IEEE Symposium on Visual Languages*, pages 245-252. Oct 1994.
4. E. J. Golin. *A method for the Specification and Parsing of Visual Languages*. Ph.D Dissertation, Brown University, 1991.
5. A. R. Helm and K. Marriott. Declarative graphics. In *International Conference on Logic Programming*, volume 225, pages 513–527. Springer Verlag, Lecture Notes in Computer Science, 1986.
6. R. Helm and K. Marriott. A declarative specification and semantics for visual languages. *Journal of Visual Languages and Computing* 2, 1991.
7. R. Helm, K. Marriott and M. Odersky. Building visual language parsers. *Proc. ACM Conf. Human Factors in Computing*, pages 105-112, CHI'91.
8. R. Helm, K. Marriott and M. Odersky. Constraint-based query optimization for spatial databases. *Proc. Tenth ACM Symp. Principles of Database Systems*, pages 181–191. Denver, Colorado, 1991.
9. T. Kamada and S. Kawai. A General Framework for Visualizing Abstract Objects and Relations. *ACM Transactions on Graphics*, 10 (1):1–39, Jan. 1991.
10. K. Marriott. Constraint Multiset Grammars. In *Proc. IEEE Symposium on Visual Languages*, pages 118-125. Oct 1994.

Programming with Constructor Classes

Mark P. Jones

Department of Computer Science
University of Nottingham
University Park, Nottingham NG7 2RD, UK
mpj@cs.nott.ac.uk

Abstract

Polymorphic type systems offer both type security and flexibility, allowing the definition and use of values which behave uniformly across a range of types. Constructor classes are one attempt to increase the expressiveness of such systems without losing the benefits of effective type inference. Combining overloading in the style of Haskell type classes, and a simple form of higher-order polymorphism, constructor classes encourage the definition and use of general purpose operations that behave uniformly across a range of type constructors. We describe a number of examples, concentrating in particular on the application to programming with monads in a functional language, to illustrate the benefits of a language with support for constructor classes.

Towards a Taxonomy of Functional Language Implementations

Rémi Douence   *Pascal Fradet*

INRIA/IRISA

Campus de Beaulieu, 35042 Rennes Cedex, France

[douence,fradet]@irisa.fr

Abstract

We express implementations of functional languages as a succession of program transformations in a common framework. At each step, different transformations model fundamental choices or optimizations. A benefit of this approach is to structure and decompose the implementation process. The correctness proofs can be tackled independently for each step and amount to proving program transformations in the functional world. It also paves the way to formal comparisons by estimating the complexity of individual transformations or compositions of them. We focus on call-by-value implementations, describe and compare the diverse alternatives and classify well-known abstract machines. This work also aims to open the design space of functional language implementations and we suggest how distinct choices could be mixed to yield efficient hybrid abstract machines.

1 Introduction

One of the most studied issues concerning functional languages is their implementation. Since the seminal proposal of Landin, 30 years ago [18], a plethora of new abstract machines or compilation techniques have been proposed. The list of existing abstract machines includes (but is surely not limited to) the SECD [18], the FAM [6], the CAM [7], the CMCM [20], the TIM [10], the ZAM [19], the G-machine [15] and the Krivine-machine [8]. Other implementations are not described via an abstract machine but as a collection of transformations or compilation techniques such as CPS-based compilers [1][12][17]. Furthermore, numerous papers present optimizations often adapted to a specific abstract machine or a specific approach [3][4][16]. Looking at this myriad of distinct works, obvious questions spring to mind: what are the fundamental choices? What are the respective benefits of these alternatives? What are precisely the common points and differences between two compilers? Can a particular optimization, designed for machine A, be adapted to machine B? One finds comparatively very few papers devoted to these questions. There have been studies of the relationship between two individual machines [25][21] but, to the best of our knowledge, no global approach to describe, classify and compare implementations.

This paper presents an advance towards a general taxonomy of functional language implementations. Our approach is to express in a common framework the whole compilation process as a succession of program transformations. The framework considered is a hierarchy of intermediate languages all of which are subsets of the lambda-calculus. Our description of an implementation consists of a series of transformations $\Lambda \xrightarrow{\mathcal{T}_1} \Lambda_1 \xrightarrow{\mathcal{T}_2} \dots \xrightarrow{\mathcal{T}_n} \Lambda_n$ each one compiling a particular task by mapping an expression from one intermediate language into another. The last language Λ_n consists of functional expressions which can be seen as machine code (essentially, combinators with explicit sequencing and calls). For each step, different transformations are designed to represent fundamental choices or optimizations. A benefit of this approach is to structure and decompose the implementation process. Two seemingly disparate implementations can be found to share some compilation steps. This approach has also interesting payoffs as far as correctness proofs and comparisons are concerned. The correctness of each step can be tackled independently and amounts to proving a program transformation in the functional world. It also paves the way to formal comparisons by estimating the complexity of individual transformations or compositions of them.

The two steps which cause the greatest impact on the compiler structure are the implementation of the reduction strategy (searching for the next redex) and the environment management (compilation of β-reduction). Other steps include implementation of control transfers (calls & returns), representation of components like data stack or environments and various optimizations.

The task is clearly huge and our presentation is by no means complete (partly because of space concerns, partly because some points are still under study). First, we concentrate on pure λ-expressions and our source language Λ is $E ::= x \mid \lambda x.E \mid E_1 E_2$. Most fundamental choices can be described for this simple language. Second, we focus on the call-by-value reduction strategy and its standard implementations. In section 2 we describe the framework used to model the compilation process. In section 3 (resp. section 4) we present the alternatives and optimizations to compile call-by-value (resp. the environment management). Each section is concluded with a comparison of the main options. Section 5 is devoted to two other simple steps leading to machine code. In section 6, we describe how this work can be easily extended to deal with constants, primitive operators, fix-point and call-by-name strategies. We also mention what remains to be done to model call-by-need and graph reduction. Finally, we indicate how it would be possible to mix different choices within a single compiler (section 8) and conclude by a short review of related works.

2 General Framework

The transformation sequence presented is this paper involves four intermediate languages (very close to each other) and can be described as $\Lambda \rightarrow \Lambda_s \rightarrow \Lambda_e \rightarrow \Lambda_k$. The first one, Λ_s, bans unrestricted applications and makes the reduction strategy explicit using a sequencing combinator. The second one Λ_e excludes unrestricted uses of variables and encodes environment management. The last one Λ_k handles control transfers by using calls and returns. This last language can be seen as a machine code. We focus here on the first intermediate language; the others (and an overview of their use) are briefly described in 2.5.

2.1 The control language Λ_s

Λ_s is defined using the combinators o, push_s, and $\lambda_s x. E$ (this last construct can be seen as a shorthand for a combinator applied to $\lambda x. E$). This language is a subset of λ-expressions therefore substitution and the notion of free or bound variables are the same as in λ-calculus.

$$\Lambda_s \qquad E ::= x \mid \text{push}_s E \mid \lambda_s x. E \mid E_1 o E_2 \qquad x \in \textit{Vars}$$

The most notable syntactic feature of Λ_s is that it rules out unrestricted applications. Its main property is that the choice of the next redex is not relevant anymore (all redexes are needed). This is the key point to compile evaluation strategies which are made explicit using the primitive o. Intuitively, o is a sequencing operator and $E_1 o E_2$ can be read "evaluate E_1 then evaluate E_2", push_s E returns E as a result and $\lambda_s x. E$ binds the previous intermediate result to x before evaluating E.

These combinators can be given different definitions (possible definitions are given at the end of this section (**DEF1**) and in sub-section 5.2). We do not pick a specific one up at this point; we simply impose that their definitions satisfy the equivalent of β-and η-conversions

$$(\beta_s) \qquad (\text{push}_s F) o (\lambda_s x. E) = E[F/x]$$

$$(\eta_s) \qquad \lambda_s x.(\text{push}_s x o E) = E \quad \textit{if } x \textit{ does not occur free in } E$$

As the usual imperative sequencing operator ";", it is natural to enforce the associativity of combinator o. This property will prove especially useful to transform programs.

$$\text{(assoc)} \quad (E_1 \circ E_2) \circ E_3 = E_1 \circ (E_2 \circ E_3)$$

We often omit parentheses and write e.g. $\mathbf{push}_s E \circ \lambda_s x.F \circ G$ for $(\mathbf{push}_s E) \circ (\lambda_s x.(F \circ G))$.

2.2 A typed subset

We are not interested in all the expressions of Λ_s. Transformations of source programs will only produce expressions denoting results (i.e. which can be reduced to expressions of the form $\mathbf{push}_s$ F). In order to express laws more easily it is convenient to restrict Λ_s using a type system (Figure 1). This does not impose any restrictions on source programs. For example, we can allow reflexive types $(\alpha=\alpha\to\alpha)$ to type any source λ-expression. The restrictions enforced by the type system are on how results and functions are combined. For example, composition $E_1 \circ E_2$ is restricted so that E_1 must denote a result (i.e. has type Rσ, R being a type constructor) and E_2 must denote a function.

$$\frac{\Gamma \vdash E : \sigma}{\Gamma \vdash \mathbf{push}_s E : \text{R}\sigma} \qquad \frac{\Gamma \cup \{x:\sigma\} \vdash E : \tau}{\Gamma \vdash \lambda_s x. E : \sigma \to_s \tau} \qquad \frac{\Gamma \vdash E_1 : \text{R}\sigma \quad \Gamma \vdash E_2 : \sigma \to_s \tau}{\Gamma \vdash E_1 \circ E_2 : \tau}$$

Figure 1 Λ_s **typed subset**

2.3 Reduction

We consider only one reduction rule corresponding to the classical β-reduction:

$$\mathbf{push}_s F \circ \lambda_s x. E \Rightarrow E[F/x]$$

As with all standard implementations, we are only interested in modelling weak reductions. Subexpressions inside $\mathbf{push}_s$'s and λ_s's are not considered as redexes and from here on we write "redex" (resp. reduction, normal form) for weak redex (resp. weak reduction, weak normal form).

Any two redexes are clearly disjoint and the β_s-reduction is left-linear so the term rewriting system is orthogonal (hence confluent). Furthermore any redex is needed (a rewrite cannot suppress a redex) thus all reduction strategies are normalizing.

Property 1 *If a closed expression E:Rσ has a normal form, there exist V such that $E \overset{*}{\Rightarrow} \mathbf{push}_s V$*

Due to the lack of space we do not display proofs here and refer the interested reader to a companion paper [9].

The reduction should be done modulo associativity if we allow an unrestricted use of (assoc) which may produce ill-typed programs. The rule (β_s) along with (assoc) specifies a string reduction confluent modulo (assoc).

2.4 Laws

This framework enjoys a number of algebraic laws useful to transform the functional code or to prove the correctness or equivalence of program transformations. We list here only two of them.

(L1) *if x does not occur free in F* $\qquad\qquad\qquad (\lambda_s x.E) \circ F = \lambda_s x. (E \circ F)$

(L2) $\forall E_1$:Rσ, $\forall E_2$:Rτ, *if x does not occur free in E_2* $\qquad E_1 \circ (\lambda_s x. E_2 \circ E_3) = E_2 \circ E_1 \circ (\lambda_s x.E_3)$

These rules permit code to be moved inside or outside function bodies or to invert the evaluation of two results. Their correctness can be established very simply. For example (L1) is sound since x does not occur free in $(\lambda_s x.E)$ nor, by hypothesis, in F and

$$(\lambda_s x.E) \circ F = \lambda_s x. \ \mathbf{push}_s \ x \circ ((\lambda_s x.E) \circ F) \qquad (\eta_s)$$

$$= \lambda_s x. \ ((\mathbf{push}_s \ x \circ (\lambda_s x.E)) \circ F) \qquad (\text{assoc})$$

$$= \lambda_s x. \ (E[x/x] \circ F) \qquad (\beta_s)$$

$$= \lambda_s x. \ (E \circ F) \qquad (\text{subst})$$

In the rest of the paper, we introduce other laws to express optimizations of specific transformations.

2.5 Overview of the compilation phases

Before describing implementations formally, let us first give an idea of the different phases, choices and the hierarchy of intermediate Λ-languages.

The first phase is the compilation of control which is described by transformations $(\mathcal{V})$ from Λ to Λ_s. The pair $(\mathbf{push}_s, \lambda_s)$ specifies a component storing intermediate results (e.g. a data stack). The main choice is using the eval-apply model $(\mathcal{V}a)$ or the push-enter model $(\mathcal{V}m)$. For the $\mathcal{V}a$ family we describe other minor options such as avoiding the need for a stack $(\mathcal{V}a_s, \mathcal{V}a_f)$ or right-to-left $(\mathcal{V}a)$ vs. left-to-right evaluation $(\mathcal{V}a_L)$.

Transformations $(\mathcal{A})$ from Λ_s to Λ_e are used to compile β-reduction. The language Λ_e avoids unrestricted uses of variables and introduces the pair $(\mathbf{push}_e, \lambda_e)$. They behave exactly as $\mathbf{push}_s$ and λ_s and corresponding properties (β_e, η_e) hold. They just act on a (at least conceptually) different component (e.g. a stack of environments). The main choice is using list-like environments $(\mathcal{A}s)$ or vector-like environments $(\mathcal{A}c)$. For the latter choice, there are several transformations depending on the way environments are copied $(\mathcal{A}c1, \mathcal{A}c2, \mathcal{A}c3)$.

A last transformation $(\mathcal{S})$ from Λ_e to Λ_k is used to compile control transfers (this step can be avoided by using a transformation $(\mathcal{S}l)$ on Λ_s-expressions). The language Λ_k makes calls and returns explicit. It introduces the pair $(\mathbf{push}_k, \lambda_k)$ which specifies a component storing return addresses.

Control	Λ_s $(\mathbf{push}_s, \lambda_s)$	$\mathcal{V}a$ $\mathcal{V}a_L$ $\underline{\mathcal{V}a_s}$ $\underline{\mathcal{V}a_f}$ $\mathcal{V}m^{\#}$	$(+\ \underline{\mathcal{S}l}^*)$
Abstraction	Λ_e $(\mathbf{push}_e, \lambda_e)$	$\mathcal{A}s$ $\mathcal{A}c1$ $\mathcal{A}c2$ $\mathcal{A}c3^{\#}$	
Transfers	Λ_k $(\mathbf{push}_k, \lambda_k)$	$\mathcal{S}^*$	

Figure 2   **Summary of the Main Compilation Steps and Options**

Figure 2 gathers the different options described in the three following sections. Any two transformations of different phases can be combined except those with the same superscript (# or *). Stack-like components are avoided by underlined transformations.

Combinators, expressed in terms of $\mathbf{push}_x$ and λ_x, are described along with transformations. To simplify the presentation, we also use syntactic sugar such as tuples $(x_1,...,x_n)$ and pattern-matching $\lambda_x(x_1,...,x_n).E$.

This hierarchy of Λ-languages is a convenient abstraction to express the compilation process. But recall that they are made of combinators and therefore subsets of the λ-calculus. An important point is that we do not have to give a precise definition to pairs $(\mathbf{push}_x, \lambda_x)$. We just enforce that they respect properties (β_x) and (η_x). Definitions do not have to be chosen until the very last step. For example, definitions of $\circ$ and $\mathbf{push}_s$ would be of the form

$$o\ E_1\ E_2\ X_1\ \dots\ X_n \rightarrow \dots \qquad\qquad \mathbf{push_s}\ V\ X_1\ \dots\ X_n \rightarrow \dots$$

where $X_1,\dots,X_n$ are components on which the code acts (e.g. control or data stack, registers,...). In other words, $X_1,\dots,X_n$ along with the Λ_x-code can be seen as the state of an abstract machine. We do not want to commit ourselves to a precise definition of combinators, however we want to ensure that the reduction from left to right using the rules of combinators simulates the reduction $\blacktriangleright$. In order to enforce this property, it is possible to state a few conditions that the standard reduction of combinators must verify. We do not expound on this issue, but a possible definition for Λ_s is

(DEF1) $\quad \mathbf{push_s}\ E = \lambda c.\ c\ E \qquad \lambda_s x.\ E = \lambda c.\lambda x.\ E\ c \qquad E_1 \circ E_2 = \lambda c.\ E_1\ (E_2\ c) \qquad (c\ \text{fresh})$

and we can easily check that β-reduction simulates reduction $\blacktriangleright$. Alternative definitions are presented in section 5.2.

3 Compilation of Control

We do not consider left-to-right *vs.* right-to-left as a fundamental choice to implement call-by-value. A more radical dichotomy is *explicit applies* vs. *marks*. The first option is the standard technique (e.g. used in the SECD or CAM) while the second was hinted at in [10] and used in ZINC.

3.1 Compilation of control using apply

Applications $E_1\ E_2$ are compiled by evaluating the argument E_2, the function E_1 and finally applying the result of E_1 to the result of E_2. The compilation of right-to-left call-by-value is described in Figure 3. Normal forms denote results so λ-abstractions and variables (which, in strict languages, are always bound to a normal forms) are transformed into results (i.e. $\mathbf{push_s}\ E$).

$$\mathcal{V}a : \Lambda \rightarrow \Lambda_s$$

$$\mathcal{V}a\ [\![x]\!] = \mathbf{push_s}\ x$$

$$\mathcal{V}a\ [\![\lambda x.E]\!] = \mathbf{push_s}\ (\lambda_s x.\ \mathcal{V}a\ [\![E]\!])$$

$$\mathcal{V}a\ [\![E_1\ E_2]\!] = \mathcal{V}a\ [\![E_2]\!] \circ \mathcal{V}a\ [\![E_1]\!] \circ \mathbf{app} \quad \text{with } \mathbf{app} = \lambda_s x.\ x$$

Figure 3   **Compilation of Right-to-Left CBV with Explicit Applies($\mathcal{V}a$)**

The rules can be explained intuitively by reading "return the value" for $\mathbf{push_s}$, "evaluate" for $\mathcal{V}a$, "then" for $\circ$ and "apply" for $\mathbf{app}$. $\mathcal{V}a$ produces well-typed expressions. Its correctness is stated by Property 2 which establishes that the reduction of transformed expressions ($\overset{*}{\blacktriangleright}$) simulates the call-by-value reduction (CBV) of source λ-expressions.

Property 2 $\quad \forall E \in \Lambda,\ CBV(E) \equiv V \Leftrightarrow \mathcal{V}a\ [\![E]\!] \overset{*}{\blacktriangleright} \mathcal{V}a\ [\![V]\!]$

It is clearly useless to store a function to apply it immediately after. This optimization is expressed by the following law

(L3) $\qquad \mathbf{push_s}\ E \circ \mathbf{app} = E \qquad\qquad\qquad (\mathbf{push_s}\ E \circ \lambda_s x.\ x =_{\beta_s} x[E/x] = E)$

Example. Let $E \equiv (\lambda x.x)((\lambda y.y)(\lambda z.z))$ then after simplifications

$$\mathcal{V}a\,[\![E]\!] \equiv \mathbf{push_s}(\lambda_s z.\ \mathbf{push_s}\ z) \circ (\lambda_s y.\ \mathbf{push_s}\ y) \circ (\lambda_s x.\ \mathbf{push_s}\ x)$$

$$\blacktriangleright \mathbf{push_s}(\lambda_s z.\ \mathbf{push_s}\ z) \circ (\lambda_s x.\ \mathbf{push_s}\ x) \blacktriangleright \mathbf{push_s}(\lambda_s z.\ \mathbf{push_s}\ z) \equiv \mathcal{V}a\,[\![\lambda z.z]\!]$$

The choice of redex in Λ_s does not matter anymore. The illicit (in call-by-value) reduction $E \rightarrow (\lambda y.y)(\lambda z.z)$ cannot occur within $\mathcal{V}a\,[\![E]\!]$. $\qquad\qquad\qquad\qquad\qquad\qquad\qquad\qquad\square$

To illustrate possible optimizations, let us take the standard case of a function applied to all of its arguments $(\lambda x_1....\lambda x_n.E_0) E_1 ... E_n$, then

$$\mathcal{V\!a} \; [\![(\lambda x_1....\lambda x_n.E_0) E_1 ... E_n]\!]$$

$$= \mathcal{V\!a} \; [\![E_n]\!] \circ ... \circ \mathcal{V\!a} \; [\![E_1]\!] \circ \mathbf{push}_s \; (\lambda_s x_1...(\mathbf{push}_s \; (\lambda_s x_n. \, \mathcal{V\!a} \; [\![E_0]\!])...) \circ \mathbf{app} \circ ... \circ \mathbf{app}$$

$$= \mathcal{V\!a} \; [\![E_n]\!] \circ ... \circ \mathcal{V\!a} \; [\![E_1]\!] \circ (\lambda_s x_1.\lambda_s x_2....\lambda_s x_n. \, \mathcal{V\!a} \; [\![E_0]\!])$$

All the **app** combinators have been statically removed using associativity, (L1) and (L3). In doing so, we have avoided the construction of n intermediary closures corresponding to the n unary functions denoted by $\lambda x_1....\lambda x_n.E_0$. This optimization can be generalized to implement the *decurryfication* phase present in many implementations. In our framework, $\lambda_s x_1....\lambda_s x_n.E$ denotes a function always applied to at least n arguments (otherwise there would be **push**'s between the λ_s's). More sophisticated optimizations could be designed. For example, if a closure analysis ensures that a set of binary functions are bound to variables always applied to at least two arguments, more **app** and **push**$_s$ combinators can be eliminated. Such information requires a potentially costly analysis and still, many functions or application contexts might not satisfy the criteria. Usually, implementations assume that higher order variables are bound to unary functions. That is, functions passed in arguments are considered unary and compiled accordingly.

The transformation $\mathcal{V\!a}_L$ describing left-to-right call-by-value can be derived from $\mathcal{V\!a}$. It is expressed as before except the rule for composition which becomes

$$\mathcal{V\!a}_L \; [\![E_1 \; E_2]\!] = \mathcal{V\!a}_L \; [\![E_1]\!] \circ \mathcal{V\!a}_L \; [\![E_2]\!] \circ \mathbf{app}_L \qquad with \quad \mathbf{app}_L = \lambda_s x.\lambda_s y. \; \mathbf{push}_s \; x \circ y$$

Property 2 still holds for $\mathcal{V\!a}_L$. Decurryfication can also be expressed although it involves slightly more complicated shifts. The equivalent of the rule (L3) is

(L4) $E : \mathsf{R}\sigma$ $\mathbf{push}_s \; F \circ E \circ \mathbf{app}_L = E \circ F$

Transformations $\mathcal{V\!a}$ and $\mathcal{V\!a}_L$ may produce expressions such as $\mathbf{push}_s \; E_1 \circ \mathbf{push}_s \; E_2 \circ ... \circ \mathbf{push}_s \; E_n \circ ....$ The reduction of such expressions requires a structure (such as a stack) able to store an arbitrary number of intermediate results. Some implementations make the choice of not using a data stack, hence they disallow several pushes in a row. In this case the rule for compositions of $\mathcal{V\!a}$ must be changed into

$$\mathcal{V\!a}_s \; [\![E_1 \; E_2]\!] = \mathcal{V\!a}_s \; [\![E_2]\!] \circ (\lambda_s m. \; \mathcal{V\!a}_s [\![E_1]\!] \circ \lambda_s n. \; \mathbf{push}_s \; m \circ n)$$

This new rule is easily derived from the original. Similarly the rule for compositions of $\mathcal{V\!a}_L$ can be changed into

$$\mathcal{V\!a}_f [\![E_1 \; E_2]\!] = \mathcal{V\!a}_f \; [\![E_1]\!] \circ (\lambda_s m. \; \mathcal{V\!a}_f [\![E_2]\!] \circ m)$$

For these expressions, the component on which **push**$_s$ and λ_s act may be a single register. Another possible motivation for these transformations is that the produced expressions now possess a unique redex throughout the reduction. The reduction sequence must be sequential and is unique.

3.2 Compilation of control using marks

Instead of evaluating the function and its argument and then applying the results, another solution is to evaluate the argument and to apply the unevaluated function right away. Actually, this implementation is very natural in call-by-name when a function is evaluated only when applied to an argument. With call-by-value, a function can also be evaluated as an argument and in this case it cannot be immediately applied but must be returned as a result. In order to detect when its evaluation is over, there has to be a way to distinguish if its argument is present or absent: this is the role of marks. After a function is evaluated, a test is performed: if there is a mark, the function is returned as a result (and a closure is built), otherwise the argument is present and the function is applied. This technique avoids building some closures but at the price of dynamic tests.

The mark ε is supposed to be a value which can be distinguished from others. Functions are transformed into **grab** E with the intended reduction rules

$$\textbf{push}_s \, \varepsilon \circ \textbf{grab} \, E \Rightarrow \textbf{push}_s \, E$$

and $\quad \textbf{push}_s \, V \circ \textbf{grab} \, E \Rightarrow \textbf{push}_s \, V \circ E \quad (V \not\equiv \varepsilon)$

Combinator **grab** and the mark ε can be defined in Λ_s. In practice, it should be implemented using a conditional which tests the presence of a mark. The transformation of right-to-left call-by-value is described in Figure 4.

$$\mathcal{V}m : \Lambda \rightarrow \Lambda_s$$

$$\mathcal{V}m[\![x]\!] = \textbf{grab} \, x$$

$$\mathcal{V}m[\![\lambda x.E]\!] = \textbf{grab} \, (\lambda_s x. \; \mathcal{V}m[\![E]\!])$$

$$\mathcal{V}m[\![E_1 \, E_2]\!] = \textbf{push}_s \, \varepsilon \circ \mathcal{V}m[\![E_2]\!] \circ \mathcal{V}m[\![E_1]\!]$$

Figure 4   **Compilation of Right-to-Left Call-by-Value with Marks($\mathcal{V}m$)**

The correctness of $\mathcal{V}m$ is stated by Property 3 which establishes that the reduction of transformed expressions simulates the call-by-value reduction of source λ-expressions.

Property 3 $\forall E \in \Lambda, \, CBV(E) \equiv V \Leftrightarrow \mathcal{V}m[\![E]\!] \overset{*}{\Rightarrow} \mathcal{V}m[\![V]\!]$

There are two new laws corresponding to the reduction rules of **grab**:

(L5) $\qquad\qquad \textbf{push}_s \, \varepsilon \circ \textbf{grab} \, E = \textbf{push}_s \, E$

(L6) $\quad E : \text{R}\sigma \qquad E \circ \textbf{grab} \, F = E \circ F$

Example. Let $E \equiv (\lambda x.x)((\lambda y.y)(\lambda z.z))$ then after simplifications

$\mathcal{V}m[\![E]\!] \equiv \textbf{push}_s \varepsilon \circ \textbf{push}_s(\lambda_s z.\textbf{grab} \, z) \circ (\lambda_s y.\textbf{grab} \, y) \circ (\lambda_s x.\textbf{grab} \, x)$

$\qquad \Rightarrow \textbf{push}_s \, \varepsilon \circ \textbf{grab} \, (\lambda_s z. \; \textbf{grab} \, z) \circ (\lambda_s x. \; \textbf{grab} \, x)$

$\qquad \Rightarrow \textbf{push}_s \, (\lambda_s z. \; \textbf{grab} \, z) \circ (\lambda_s x. \; \textbf{grab} \, x) \Rightarrow \textbf{grab} \, (\lambda_s z. \; \textbf{grab} \, z) \equiv \mathcal{V}m[\![\lambda z.z]\!] \qquad \square$

As before, when a function is known to be applied to n arguments, the code can be optimized to save n dynamic tests. Actually, it appears that $\mathcal{V}m$ is subject to the same kind of optimizations as $\mathcal{V}a$. Decurryfication and related optimizations can be expressed based on rule (L6).

It would not make much sense to consider a left-to-right strategy here. The whole point of this approach is to prevent building some closures by testing if the argument is present. Therefore the argument must be evaluated before the function.

3.3 Comparison

We compare the efficiency of codes produced by both transformations. We saw before that both transformations are subject to identical optimizations and we examined unoptimized codes only. A code produced by $\mathcal{V}m$ builds less closures than the corresponding $\mathcal{V}a$-code. A mark can be represented by one bit so $\mathcal{V}m$ is likely to be on average less greedy on space resources. Concerning time efficiency, the size of compiled expressions gives a first approximation of the overhead entailed by the encoding of the reduction strategy. It is easy to show that code expansion is linear with respect to the size of the source expression. More precisely

If $Size[\![E]\!] = n$ then $Size(\mathcal{V}[\![E]\!]) < 3n$ (for $\mathcal{V} = \mathcal{V}a$ or $\mathcal{V}m$)

This upper bound can be reached by taking for example $E \equiv \lambda x.x \ldots x$ (n occurrences of x). A more thorough investigation is possible by associating costs with the different combinators encoding the control: *push* for the cost of "pushing" a variable or a mark, *clos* for the cost of building a closure (i.e. **push**$_s$ E), *app* and *grab* for the cost of the corresponding combinators. If we take n_λ for the number of λ-abstractions and n_v for the number of occurrences of variables in the source expression, we have

$$Cost(\mathcal{V}a[\![E]\!]) = n_\lambda \, clos + n_v \, push + (n_v\text{-}1) \, app \quad \text{and} \quad Cost(\mathcal{V}m[\![E]\!]) = (n_\lambda + n_v) \, grab + (n_v\text{-}1) \, push$$

The benefit of $\mathcal{V}m$ over $\mathcal{V}a$ is to sometimes replace a closure construction and an **app** by a test and an **app**. So if *clos* is comparable to a test (for example, when returning a closure amounts to build a pair as in section 4.1) $\mathcal{V}m$ will produce more expensive code than $\mathcal{V}a$.

If closure building is not a constant time operation (as in section 4.2) $\mathcal{V}m$ can be arbitrarily better than $\mathcal{V}a$. Actually, it can change the program complexity in pathological cases. In practice, however, the situation is not so clear. When no mark is present a **grab** is implemented by a test followed by an **app**. If a mark is present the test is followed by a **push**$_s$ (for variables) or a closure building (for λ-abstractions). So we have

$$Cost(\mathcal{V}m[\![E]\!]) = (n_\lambda + n_v) \, test + \bar{p} \, (n_\lambda + n_v) \, app + p \, n_\lambda \, clos + p \, n_v \, push + (n_v\text{-}1) \, push$$

with p (resp. $\bar{p}$) representing the likelihood ($p + \bar{p} = 1$) of the presence (resp. absence) of a mark which depends on the program. The best situation for $\mathcal{V}m$ is when no closure has to be built, that is $p = 0$ & $\bar{p} = 1$. If we take some reasonable hypothesis such as *test=app* and $n_\lambda < n_v < 2n_\lambda$ we find that the cost of closure construction must be 3 to 4 times more costly than *app* or *test* to make $\mathcal{V}m$ advantageous. With less favorable odds such as $p = \bar{p} = 1/2$, *clos* must be worth up to 6 *app*.

We are lead to conclude that $\mathcal{V}m$ should be considered only with a copy scheme for closures. Even so, tests may be too costly in practice compared to the construction of small closures. The best way would probably be to perform an analysis to detect cases when $\mathcal{V}m$ is profitable. Such information could be taken into account to get the best of each approach. We present in section 8.1 how $\mathcal{V}a$ and $\mathcal{V}m$ could be mixed.

4 Compilation of the β-Reduction

This transformation step implements the substitution. Variables are replaced by combinators acting on environments. The value of a variable is fetched from the environment when needed. Because of the lexical scope, paths to values in the environment are static. Compared to Λ_s, Λ_e adds the pair (**push**$_e$, λ_e) which is used to define combinators.

4.1 Shared environments

The denotational-like transformation $\mathcal{A}s$ is widely used among the functional abstract machines [7][18][19]. The structure of the environment is a tree of closures. A closure is added to the environment in constant time. On the other hand, a chain of links has to be followed when accessing a value. The access time complexity is $O(n)$ where n is the number of λ_s's from the occurrence to its binding λ_s (i.e. its de Bruijn number). The transformation (Figure 5) is done relatively to a compile-time environment ρ made of pairs. The integer i in x_i denotes the rank of the variable in the environment.

$$As: \Lambda_s \to env \to \Lambda_e$$

$$As[\![E_1 \circ E_2]\!] \, \rho = \mathbf{dupl}_e \circ As[\![E_1]\!] \, \rho \circ \mathbf{swap}_{se} \circ As[\![E_2]\!] \, \rho$$

$$As[\![\mathbf{push}_s \, E]\!] \, \rho = \mathbf{push}_s \, (As[\![E]\!] \, \rho) \circ \mathbf{mkclos}$$

$$As[\![\lambda_s x.E]\!] \, \rho = \mathbf{bind} \circ As[\![E]\!] \, (\rho,x)$$

$$As[\![x_i]\!] \, (...((\rho,x_i),x_{i-1})...,x_0) = \mathbf{fst}^i \circ \mathbf{snd} \circ \mathbf{appclos}$$

Figure 5   Abstraction with Shared Environments (As)

As needs seven new combinators to express saving and restoring environments (**dupl$_e$**, **swap$_{se}$**), closure building and opening (**mkclos**, **appclos**), access to values (**fst**, **snd**), adding a binding (**bind**). They are defined in Λ_e by

$$\mathbf{dupl}_e = \lambda_e e. \, \mathbf{push}_e \, e \circ \mathbf{push}_e \, e \qquad\qquad \mathbf{swap}_{se} = \lambda_s x. \, \lambda_e e. \, \mathbf{push}_s \, x \circ \mathbf{push}_e \, e$$

$$\mathbf{mkclos} = \lambda_s x. \, \lambda_e e. \, \mathbf{push}_s \, (x,e) \qquad\qquad \mathbf{appclos} = \lambda_s (x,e). \, \mathbf{push}_e \, e \circ x$$

$$\mathbf{fst} = \lambda_e (e,x). \, \mathbf{push}_e \, e \qquad\qquad\qquad \mathbf{snd} = \lambda_e (e,x). \, \mathbf{push}_s \, x$$

$$\mathbf{bind} = \lambda_e e. \, \lambda_s x. \, \mathbf{push}_e \, (e,x)$$

As correctness is stated by Property 4.

Property 4 $\forall E$: $R\sigma$ *closed*, $E \overset{*}{\twoheadrightarrow} V \Rightarrow As[\![E]\!] \, () =_\beta As[\![V]\!] \, ()$

Transformation As can be optimized by adding the rules

$$As[\![\mathbf{app}]\!] \, \rho = As[\![\lambda_s x.x]\!] \, \rho = \mathbf{bind} \circ \mathbf{snd} \circ \mathbf{appclos} = \mathbf{appclos'}$$

$$\textit{with } \mathbf{appclos'} = \lambda_e z.\lambda_s(x,e). \, \mathbf{push}_e \, e \circ x$$

$$As[\![\lambda_s x.E]\!] \, \rho = \mathbf{pop}_{se} \circ As[\![E]\!] \, \rho \qquad\qquad \textit{with } \mathbf{pop}_{se} = \lambda_e e. \, \lambda_s x. \, \mathbf{push}_e \, e \textit{ and } x \textit{ is not free in } E$$

$$As[\![\mathbf{push}_s \, x_i]\!] \, (...((\rho,x_i),x_{i-1})...,x_0) = \mathbf{fst}^i \circ \mathbf{snd}$$

Variables are bound to closures stored in the environment. With the original rules, $As[\![\mathbf{push}_s x_i]\!]$ would build yet another closure. This useless "boxing" is avoided by the above rule.

Example. $As[\![\lambda_s x_1.\lambda_s x_0. \, \mathbf{push}_s \, E \circ x_1]\!] \, \rho = \mathbf{bind} \circ \mathbf{bind} \circ \mathbf{dupl}_e \circ \mathbf{push}_s \, (As[\![E]\!] \, ((\rho,x_1),x_0))$

$$\circ \, \mathbf{mkclos} \circ \mathbf{swap}_{se} \circ \mathbf{fst} \circ \mathbf{snd} \circ \mathbf{appclos}$$

Two bindings are added (**bind** $\circ$ **bind**) to the current environment and the x_1 access is now coded by **fst** $\circ$ **snd**.   □

In our framework, $\lambda_s x_1....\lambda_s x_n.E$ denotes a function always applied to at least n arguments. So the corresponding links in the environment can be collapsed without any loss of sharing [8]. The list-like environment can become a vector locally and variable accesses have to be modified consequently.

Also, the combinator **mkclos** can be avoided by an abstraction which unfolds the pair (code,env) in the environment itself, as in TIM [10].

4.2 Copied environments

Another choice is to provide a constant access time [1][12]. In this case, the structure of the environment must be a vector of closures. Code which copies the environment (a O(*length* ρ) operation) has to be inserted in As in order to avoid links.

The macro-combinator **Copy** ρ produces code that copies an environment according to ρ's structure.

Copy $(...((),x_n),...,x_0) = \lambda_e e.$ **push**$_e$ () o (**push**$_e$ e o **get**$_n$ o **bind**) o ... o (**push**$_e$ e o **get**$_0$ o **bind**)

Combinators **get**$_i$ are a shorthand for **fst**^{i} o **snd**. However, if environments are represented by vectors, **get**$_i$ can be considered as a constant time operation and **bind** can be seen as adding a binding in a vector.

There are several abstractions according to the time of the copies. We present only the rules differing from $\mathcal{A}s$ scheme. A first solution (Figure 6) is to copy the environment just before adding a new binding (as in [10]). From the first step we know that n-ary functions ($\lambda_s x_1....\lambda_s x_n.E$) are never partially applied and cannot be shared: they need only one copy of the environment. The overhead is placed on function entry and closure building remains a constant time operation. This transformation produces environments which can be shared by several closures but only as a whole. So, there must be an indirection when accessing the environment.

$\mathcal{A}c1 [\![\lambda_s x_i...x_0.E]\!] \rho =$ **Copy** $\bar{\rho}$ o **bind**$^{i+1}$ o $\mathcal{A}c1 [\![E]\!] (...(\bar{\rho},x_i)...,x_0)$

$\mathcal{A}c1 [\![x_i]\!] (...(\rho,x_i),...,x_0) =$ **get**$_i$ o **app**

Figure 6   **Copy at Function Entry ($\mathcal{A}c1$ Abstraction)**

The environment $\bar{\rho}$ represents ρ restricted to variables occurring free in the subexpression E.

Example. $\mathcal{A}c1 [\![\lambda_s x_1.\lambda_s x_0.$ **push**$_s$ E_1 o $x_1]\!]$ ρ

= **Copy** $\bar{\rho}$ o **bind**^{2} o **dupl**$_e$ o **push**$_s$ $(\mathcal{A}c1 [\![E]\!] ((\bar{\rho},x_1),x_0)))$ o **mkclos** o **swap**$_{se}$ o **get**$_1$ o **appclos**

The code builds a vector environment made of a specialized copy of the previous environment and two new bindings (**bind**^{2}) ; the x_1 access is now coded by **get**$_1$. $\qquad\square$

A second solution (Figure 7) is to copy the environment when building and opening closures (as in [12]). The copy at opening time is necessary in order to be able to add new bindings in contiguous memory (the environment has to remain a vector). This transformation produces environments which cannot be shared but may be accessed directly (they can be packed with a code pointer to form a closure).

$\mathcal{A}c2 [\![$**push**$_s$ $E]\!] \rho =$ **Copy** $\bar{\rho}$ o **push**$_s$(**Copy** $\bar{\rho}$ o $\mathcal{A}c2 [\![E]\!] \bar{\rho}$) o **mkclos**

$\mathcal{A}c2 [\![x_i]\!] (...((\rho,x_i),x_{i-1})...,x_0) =$ **get**$_i$ o **appclos**

Figure 7   **Copy at Closure Building and Opening ($\mathcal{A}c2$ Abstraction)**

A third solution is to copy the environment only when building closures (as in [6]). In order to be able to add new bindings after closure opening, a local environment ρ_L is needed. When a closure is built, the concatenation of the two environments (ρ_G++ρ_L) is copied. The code for variables now has to specify which environment is accessed. Although the transformation scheme remains similar, every rule must be redefined to take into account the two environments. We list here only two of them.

$\mathcal{A}c3 [\![$**push**$_s$ $E]\!] (\rho_G,\rho_L) =$ **Copy** $(\bar{\rho}_G$++$\bar{\rho}_L)$ o **push**$_s$ $(\mathcal{A}c3 [\![E]\!] (\bar{\rho}_G$++$\bar{\rho}_L,()))$ o **mkclos**

$\mathcal{A}c3 [\![\lambda_s x.E]\!] (\rho_G,\rho_L) =$ **bind3** o $\mathcal{A}c3 [\![E]\!] (\rho_G,(\rho_L,x))$ with **bind3** $= \lambda_e(e_g,e_l).\lambda_s x.$ **push**$_e$ $(e_g,(e_l,x))$

Figure 8   **Abstraction with Local Environments ($\mathcal{A}c3$ Abstraction)**

Local environments are not compatible with $\mathcal{V}m$: $\mathcal{A}c3 [\![$**grab** $E]\!]$ would generate two different versions of $\mathcal{A}c3 [\![E]\!]$ since E may appear in a closure or may be applied. This code duplication is obviously not realistic.

4.3 Refinements

The sequencing can be exploited by the abstraction process. Instead of saving and restoring the environment (as in $As[\![E_1 \circ E_2]\!]$), we can pass it to E_1 which may add new bindings (**bind**) but has to remove them (using **fst**) before passing the environment to E_2. For example the rules for sequences and λ_s-abstractions might be

$$Aseq[\![E_1 \circ E_2]\!] \ \rho = Aseq[\![E_1]\!] \ \rho \circ Aseq[\![E_2]\!] \ \rho$$

and $\qquad Aseq[\![\lambda_s x.E]\!] \ \rho = \textbf{bind} \circ Aseq[\![E]\!] \ (\rho,x) \circ \textbf{fst}$

Many other refinements are possible. For example, environments can be unfolded so that the environment stack becomes a closure stack [12]. This avoids an indirection and provides a direct access to values.

4.4 Comparison

The size of the abstracted expressions gives a first approximation of the overhead entailed by the encoding of the β-reduction. It is easy to show that code expansion is quadratic with respect to the size of the source expression. More precisely

$$\text{if } Size[\![E]\!] = n \text{ then } Size(As(Va[\![E]\!])) \le n_\lambda n_v - n_v + 6n + 6$$

with n_λ the number of λ-abstractions and n_v the number of variable occurrences ($n = n_\lambda + n_v$) of the source expression. This expression reaches a maximum with $n_v = (n-1)/2$. This upper bound can be approached with, for example, $\lambda x_1 \dots \lambda x_{n\lambda}. \ x_1 \dots x_{n\lambda}$. The product $n_\lambda n_v$ indicates that the efficiency of As depends equally on the number of accesses (n_v) and their length (n_λ). For $Ac1$ we have

$$\text{if } Size[\![E]\!] = n \text{ then } Size(Ac1(Va[\![E]\!])) \le 6n_\lambda^2 - 6n_\lambda + 7n + 6$$

which makes clear that the efficiency of $Ac1$ is not dependent of accesses. The abstractions have the same complexity order, nevertheless one may be more adapted than the other to individual source expressions. These complexities highlight the main difference between shared environments that favors building, and copied environments that favors access. Let us point out that these bounds are related to the quadratic growth implied by Turner's abstraction algorithm [29]. Balancing expressions reduces this upper bound to $O(n \log n)$ [16]. It is very likely that this technique could also be applied to λ-expressions to get a $O(n \log n)$ complexity for environment management.

The abstractions can be compared according to their memory usage too. $Ac2$ copies the environment for every closure, where $Ac1$ may share a bigger copy. So, the code generated by $Ac2$ consumes more memory and implies frequent garbage collections whereas the code generated by $Ac1$ may create space leaks and needs special tricks to plug them (see [25] section 4.2.6).

5 Compilation To Machine Code

In this section, we make explicit control transfers and propose combinator definitions. After these steps the functional expressions can be seen as realistic machine code.

5.1 Control transfers

A conventional machine executes linear code where each instruction is basic. We have to make explicit calls and returns. In our framework reducing expressions of the form **appclos** $\circ$ E involves evaluating a closure and returning to E. There are two solutions to save the return address.

We model the first one with a transformation $\mathcal{S}$ on Λ_e-expressions. It shifts the code following the function call using $\mathbf{push_k}$, and returns to it with $\mathbf{rts_s}$ ($= \lambda_s x.\lambda_k f.\ \mathbf{push_s}\ x \circ f$) when the function ends (as in [12][18][19]). Intuitively these combinators can be seen as implementing a control stack. Compared to Λ_e, Λ_k-expressions do not have $\mathbf{appclos} \circ E$ code sequences.

$$S: \Lambda_e \to \Lambda_k$$

$$S[\![\mathbf{dupl_e} \circ E_1 \circ \mathbf{swap_{se}} \circ E_2]\!] = \mathbf{dupl_e} \circ \mathbf{push_k}\ (\mathbf{swap_{se}} \circ S[\![E_2]\!]) \circ \mathbf{swap_{ke}} \circ S[\![E_1]\!]$$

$$S[\![\mathbf{push_s}\ E \circ \mathbf{mkclos}]\!] = \mathbf{push_s}\ S[\![E]\!] \circ \mathbf{mkclos} \circ \mathbf{rts_s}$$

$$S[\![\mathbf{bind} \circ E]\!] = \mathbf{bind} \circ S[\![E]\!]$$

$$S[\![E \circ \mathbf{appclos}]\!] = \mathbf{push_k}\ (\mathbf{appclos}) \circ \mathbf{swap_{ke}} \circ S[\![E]\!]$$

$$S[\![\mathbf{fst}^i \circ \mathbf{snd}]\!] = \mathbf{fst}^i \circ \mathbf{snd} \circ \mathbf{rts_s}$$

Figure 9   Compilation of Control Transfers ($\mathcal{S}$)

The combinator $\mathbf{swap_{ke}} = \lambda_k x.\ \lambda_e e.\ \mathbf{push_k}\ x \circ \mathbf{push_e}\ e$ is necessary in order to mix the new component k with the other ones. The resulting code can be simplified to avoid useless sequence breaks. To get a real machine code a further step would be to introduce labels to name sequences of code (such as E in $\mathbf{push_x}\ E$).

A second solution uses a transformation $\mathcal{S}l$ between the control and the abstraction phases. It transforms the expression into continuation passing style. The continuation encodes return addresses and will be then abstracted in the environment as any variable. This solution, known as stackless, is chosen in Appel's ML compiler [1]. It prevents the use of a control stack but relies heavily on the garbage collector. Appel claims that it is simple, not inefficient and well suited to implement *callcc*.

$$\mathcal{S}l: \Lambda_s \to \Lambda_s$$

$$\mathcal{S}l[\![E_1 \circ E_2]\!] = \lambda_s k.\ \mathbf{push_s}\ (\mathbf{push_s}\ k \circ \mathcal{S}l[\![E_2]\!]) \circ \mathcal{S}l[\![E_1]\!]$$

$$\mathcal{S}l[\![\mathbf{push_s}\ E]\!] = \lambda_s k.\ \mathbf{push_s}\ (\mathcal{S}l[\![E]\!]) \circ k$$

$$\mathcal{S}l[\![\lambda_s x.E]\!] = \lambda_s k.\lambda_s x.\ \mathbf{push_s}\ k \circ \mathcal{S}l[\![E]\!]$$

$$\mathcal{S}l[\![x]\!] = x$$

$$\mathcal{S}l[\![\mathbf{app}]\!] = \mathcal{S}l[\![\lambda_s x.\ x]\!] = \lambda_s k.\lambda_s x.\ \mathbf{push_s}\ k \circ x = \mathbf{app_k}$$

Figure 10   Compilation of the Control as a Standard Argument ($\mathcal{S}l$)

The following optimization removes unnecessary manipulations of the continuation k :

$$\mathbf{push_s}\ E_1 \circ (\lambda_s k.\ \mathbf{push_s}\ E_2 \circ k) = \mathbf{push_s}\ E_2 \circ E_1$$

5.2 Separate vs. merged components

The pairs of combinators $(\lambda_s, \mathbf{push_s})$, $(\lambda_e, \mathbf{push_e})$, and $(\lambda_k, \mathbf{push_k})$ do not have definitions yet. Each pair can be seen as encoding a component of an underlying abstract machine and their definitions specify the state transitions. We can now choose to keep the components separate or merge (some of) them. Both options share the same definition $\circ = \lambda xyz.\ x\ (y\ z)$.

Keeping the components separate brings new properties, allowing code motion and simplifications. The sequencing of two combinators on different components is commutative and administrative combinators such as $\mathbf{swap_{se}}$ are useless. Possible definitions (c, s, e being fresh variables) follow

$$\lambda_s x.X = \lambda c.\lambda(s,x).\ X\ c\ s \qquad\qquad \mathbf{push}_s\ N = \lambda c.\lambda s.c\ (s,N)$$

$$\lambda_e x.X = \lambda c.\lambda s.\lambda(e,x).\ X\ c\ s\ e \qquad\qquad \mathbf{push}_e\ N = \lambda c.\lambda s.\lambda e.c\ s\ (e,N)$$

and similarly for $(\lambda_k, \mathbf{push}_k)$. The reduction of our expressions can be seen as state transitions of an abstract machine, e.g. :

$$\mathbf{push}_s\ N\ C\ S\ E\ K \to C\ (S,N)\ E\ K \qquad\qquad \mathbf{push}_e\ N\ C\ S\ E\ K \to C\ S\ (E,N)\ K$$

A second option is to merge all components. Here, administrative combinators remain necessary.

$$\lambda_a x.X = \lambda c.\lambda(z,x).X\ c\ z \qquad\qquad \mathbf{push}_a\ N = \lambda c.\lambda z.\ c\ (z,N) \qquad with\ (\mathrm{a} \equiv \mathrm{s}, \mathrm{e}\ or\ \mathrm{k})$$

6 Extensions

We describe here several extensions needed in order to handle realistic languages and to describe a wider class of implementations.

6.1 Constants, primitive operators & data structures

We have only considered pure λ-expressions because most fundamental choices can be described for this simple language. Realistic implementations also deal with constants, primitive operators and data structures. Concerning basic constants, a question is whether base-typed results are of the form $\mathbf{push}_s\ n$ or another component is introduced (e.g. $\mathbf{push}_b$, λ_b). Both options can be chosen. The latter has the advantage of marking a difference between pointers and values which can be exploited by the garbage collector. But in this case, type information must also be available to transform variables and λ-abstractions correctly. The conditional, the fix-point operator, and primitive operators acting on basic values are introduced in our language in a straightforward way. As far as data structures are concerned we can again choose to treat them as closures or separately. A more interesting choice is whether we represent them using tags or higher-order functions [10].

$$\mathcal{V}[\![rec\,f\,(\lambda x.E)]\!] = \mathbf{push}_s\ (rec_s f\,(\lambda_s x.\ \mathcal{V}[\![E]\!]\,))$$

$$\mathcal{V}[\![\mathbf{if}\ E_1\ \mathbf{then}\ E_2\ \mathbf{else}\ E_3]\!] = \mathcal{V}[\![E_1]\!] \circ \mathbf{cond}_s\ (\mathcal{V}[\![E_2]\!],\ \mathcal{V}[\![E_3]\!]\,)$$

$$\mathcal{V}[\![E_1 + E_2]\!] = \mathcal{V}[\![E_2]\!] \circ \mathcal{V}[\![E_1]\!] \circ \mathbf{plus}_s \qquad\qquad \mathcal{V}[\![n]\!] = \mathbf{push}_s\ n$$

$$\mathcal{V}[\![cons\ E_1\ E_2]\!] = \mathcal{V}[\![E_2]\!] \circ \mathcal{V}[\![E_1]\!] \circ \mathbf{cons}_s \qquad\qquad \mathcal{V}[\![head]\!] = \mathbf{head}_s$$

Figure 11   An extension with constants, primitive operators and lists

A possible extension using the component defined by $(\mathbf{push}_s, \lambda_s)$ to store constants and tagged cells of lists is described in Figure 11 with

$$\mathbf{cons}_s = \lambda_s h.\lambda_s t.\ \mathbf{push}_s(tag,h,t) \qquad\qquad \mathbf{head}_s = \lambda_s(tag,h,t).\ \mathbf{push}_s\ h$$

$$\mathbf{push}_s\ n_2 \circ \mathbf{push}_s\ n_1 \circ \mathbf{plus}_s \Rightarrow \mathbf{push}_s\ n_1 + n_2$$

6.2 Call-by-name & mixed evaluation strategies

Many of the choices discussed before remain valid for call-by-name implementations. Only the compilation of the computation rule has to be described. Figure 12 presents two possible transformations. The first one considers λ-abstractions as values and evaluates the function before applying it to the unevaluated argument. The second one (used by the TIM and Krivine machine) directly applies the function to the argument. In this scheme functions are not considered as results.

$$\mathcal{N}a : \Lambda \rightarrow \Lambda_s \qquad\qquad\qquad\qquad \mathcal{N}m : \Lambda \rightarrow \Lambda_s$$

$$\mathcal{N}a \; [\![x]\!] = x \qquad\qquad\qquad\qquad\qquad \mathcal{N}m \; [\![x]\!] = x$$

$$\mathcal{N}a \; [\![\lambda x.E]\!] = \textbf{push}_s \, (\lambda_s x. \; \mathcal{N}a[\![E]\!]) \qquad\qquad \mathcal{N}m \; [\![\lambda x.E]\!] = \lambda_s x. \; \mathcal{N}m \; [\![E]\!]$$

$$\mathcal{N}a \; [\![E_1 \, E_2]\!] = \textbf{push}_s \, (\mathcal{N}a \; [\![E_2]\!]) \circ \mathcal{N}a \; [\![E_1]\!] \circ \textbf{app} \qquad \mathcal{N}m \; [\![E_1 \, E_2]\!] = \textbf{push}_s (\mathcal{N}m \; [\![E_2]\!]) \circ \mathcal{N}m \; [\![E_1]\!]$$

Figure 12 Two Transformations for Call-by-Name ($\mathcal{N}a$ & $\mathcal{N}m$)

The transformation $\mathcal{N}m$ is simpler and avoids some overhead of $\mathcal{N}a$. On the other hand, making $\mathcal{N}m$ lazy is problematic: it needs marks to be able to update closures [10][8][27]. This is exactly the same problem as with $\mathcal{V}m$; without marks we cannot know if a function represents a result or has to be applied. In the first case, we have to return it (cbv, $\mathcal{V}m$) or update a closure (cbn, $\mathcal{N}m$).

Strictness analysis can be taken into account in order to produce mixed evaluation strategies. In fact, the most interesting optimization brought by strictness information is not the change of the evaluation order but avoiding thunks using unboxing [5]. If we assume that a strictness analysis has annotated the code by $E_1 \, E_2$ if E_1 denotes a strict function and $\underline{x}$ if the variable is defined by a strict λ-abstraction then $\mathcal{N}a$ can be extended as follows

$$\mathcal{N}a \; [\![\underline{x}]\!] = \textbf{push}_s \, x \qquad\qquad\qquad \mathcal{N}a \; [\![E_1 \, E_2]\!] = \mathcal{N}a \; [\![E_2]\!] \circ \mathcal{N}a \; [\![E_1]\!] \circ \textbf{app}$$

Underlined variables are known to be already evaluated; they are represented as unboxed values. For example, without any strictness information, the expression $(\lambda x. \, x+1) \, 2$ is compiled into $\textbf{push}_s \, (\textbf{push}_s \, 2) \circ (\lambda_s x. \; x \circ \textbf{push}_s \, 1 \circ \textbf{plus}_s)$. The code $\textbf{push}_s \, 2$ will be represented as a closure and evaluated by the call x; it is the boxed representation of 2. With strictness annotations we have $\textbf{push}_s \, 2 \circ (\lambda_s x. \; \textbf{push}_s \, x \circ \textbf{push}_s \, 1 \circ \textbf{plus}_s)$ and the evaluation is the same as with call-by-value (no closure is built). Actually, more general forms of unboxing and optimizations (as in [26]) could be expressed as well.

6.3 Call-by-need and graph reduction

Call by need brings yet other options. The update mechanism can be implemented by self-updatable closures (as in [24]), by modifying the continuation (as in [12]). Updating is also central in implementations based on graph reduction. Expressing redex sharing and updating is notoriously difficult. In our framework, a straightforward idea is to add a store component along with new combinators. Each expression takes and returns the store; the sequencing ensures that the store is single-threaded. We suspect that adding store and updates in our framework will complicate correctness proofs. On the other hand, this can be done at a very late stage (e.g. after the compilation of call-by-name and β-reduction). All the transformations, correctness proofs, optimizations previously described would remain valid. The complications involved by updating would be confined in a single step. We are currently working on this issue.

7 Classical Functional Implementations

Descriptions of functional compilers often hide their fundamental structure behind implementation tricks and optimizations. Figure 13 states the main design choices which represent the skeleton of several classical implementations.

There are cosmetic differences between our description and the real implementation. Also, some extensions and optimizations are not described here. Let us state precisely the differences for the categorical abstract machine. Let $\mathcal{CAM} = \mathcal{A}s \bullet \mathcal{V}a_L$ as stated in Figure 13, by simplifying this composition of transformations we get

$$CAM \; [\![x_i]\!] \; \rho = \mathbf{fst}^i \circ \mathbf{snd}$$

$$CAM \; [\![\lambda x.E]\!] \; \rho = \mathbf{push}_s \; (\mathbf{bind} \circ (CAM[\![E]\!] \; (\rho,x))) \circ \mathbf{mkclos}$$

$$CAM \; [\![E_1 \; E_2]\!] \; \rho = \mathbf{dupl}_e \circ (CAM[\![E_1]\!] \; \rho) \circ \mathbf{swap}_{se} \circ (CAM[\![E_2]\!] \; \rho) \circ \mathbf{appclos}$$

The **fst, snd, dupl**$_e$ and **swap**$_{se}$ combinators match with CAM's **Fst, Snd, Push** and **Swap**. The sequence **push**$_s$ (E) ∘ **mkclos** is equivalent to CAM's **Cur**(E). The only difference comes from the place of **bind** (at the beginning of each closure in our case). Shifting this combinator to the place where the closures are evaluated (i.e. merging it with **appclos**), we get $\lambda_s(x,e)$. **push**$_e$ e ∘ **bind** ∘ x, which is exactly CAM's sequence **Cons;App**.

Compiler	Transformations				Components
SECD	Va	Id	As	S	s (e ≡ k)
CAM	Va$_L$	Id	As	Id	s ≡ e
ZAM	Vm	Id	As	S	s (e ≡ k)
SML-NJ	Va$_f$	Sl	(Ac3+As)	Id	s e (registers)
TABAC (cbv)	Va	Id	Ac2	S	(s ≡ e) k
TABAC (cbn)	Na	Id	Ac2	S	(s ≡ e) k
TIM (cbn)	Nm	Id	Ac1	Id	s e

Figure 13   Several Classical Compilation Schemes

Let us quickly review the other differences between Figure 13 and real implementations. The SECD machine [18] saves environments a bit later than in our scheme. Furthermore, the control stack and the environment stack are gathered in a component called dump. The data stack is also (uselessly) saved in the dump. Actually, our replica is closer to the idealized version derived in [13]. The ZAM [19] uses a slightly different compilation of control than Vm and has an accumulator and registers. The SML-NJ compiler [1] uses only the heap which is represented in our framework by a unique environment e. It also includes registers and many optimizations not described here. The TABAC compiler is a by-product of our work in [12] and has greatly inspired this study. It implements strict or non-strict languages by program transformations. Compared to the description above the environments are unfolded in the environment/data stack. The call-by-name TIM [10] unfolds closures in the environment as mentioned in 4.1. The transformation $Ac1$ has the same effect as the preliminary lambda-lifting phase of TIM.

8 Towards Hybrid Implementations

The study of the different options proved that there is no universal best choice. It is natural to strive to get the best of each world. Our framework makes intricate hybridizations and related correctness proofs possible. We first describe how Va and Vm could be mixed and then how to mix shared and copied environments. In both cases, mixing is a compile time choice and we suppose that a static analysis has produced an annotated code indicating the chosen mode for each subexpression.

8.1 Mixing different control schemes

The annotations are of the form of types $T ::= a \mid m \mid T_1 \overset{a/m}{\to} T_2$ with a (resp. m) for apply (resp. marks) mode. Intuitively a function E: $\alpha \overset{\delta}{\to} \beta$ takes an argument which is to be evaluated in the α-mode whereas the body is evaluated in the δ-mode. This style of annotation imposes that each variable is evaluated in a fixed mode.

$$\mathit{MixV}[\![x^\alpha]\!] = \mathbf{X}_\alpha\, x$$

$$\mathit{MixV}[\![\lambda x.E^{\alpha \overset{\Delta}{\to} \beta}]\!] = \mathbf{X}_\delta\, (\lambda_s x.\ \mathit{MixV}[\![E]\!])$$

$$\mathit{MixV}[\![E_1{}^{\alpha \overset{\Delta}{\to} \beta} E_2{}^\alpha]\!] = \mathbf{Y}_\alpha \circ \mathit{MixV}[\![E_2]\!] \circ \mathit{MixV}[\![E_1]\!] \circ \mathbf{Z}_\delta$$

with	$\mathbf{X}_a = \mathbf{push}_s$	$\mathbf{Y}_a = \mathbf{Id}$	$\mathbf{Z}_a = \mathbf{app}$
	$\mathbf{X}_m = \mathbf{grab}$	$\mathbf{Y}_m = \mathbf{push}_s\ \varepsilon$	$\mathbf{Z}_m = \mathbf{Id}$

Figure 14   Hybrid Compilation of Right to Left Call-by-Value

We suppose, as in 3.2, that it is possible to distinguish the special closure ε from the others. The values produced by each mode are of the same form and no coercion is necessary. MixV (Figure 14) just adds $\mathbf{push}_s$ ε before the evaluation of an argument in mode m and $\mathbf{app}$ after the evaluation of a function in mode a. Results are returned using $\mathbf{push}_s$ or $\mathbf{grab}$ according to their associated mode.

8.2 Mixing different abstraction schemes

One solution uses coercion functions which fit the environment into the chosen structure (vector or linked list). The compilation can then switch from one world to another. In particular, switching from As to Acl creates a kind of strict display (by comparison to the lazy display of [22]).

$$\mathit{As}[\![E]\!]\ \rho = \mathbf{List2Vect}\ \rho \circ \mathit{Acl}[\![E]\!]\ \rho$$

Another solution uses environments mixing lists and vectors (as in [28]).

$$\mathit{MixA}[\![\lambda_s x.E^{\theta,\oplus}]\!]\ \rho = \mathbf{Mix}\ \rho\ \theta \circ \mathbf{bind}_\oplus \circ \mathit{MixA}[\![E]\!]\ (\theta \oplus x)$$

$$\mathit{MixA}[\![x_i]\!]\ (\ldots(\rho,\rho_i),\ldots,\rho_0) = \mathbf{fst}^i \circ \mathbf{snd} \circ \mathit{MixA}[\![x_i]\!]\ \rho_i \qquad\qquad \text{with } x_i \text{ in } \rho_i$$

$$\mathit{MixA}[\![x_i]\!]\ [\rho:\rho_i:\ldots:\rho_0] = \mathbf{get}_i \circ \mathit{MixA}[\![x_i]\!]\ \rho_i \qquad\qquad \text{with } x_i \text{ in } \rho_i$$

$$\mathit{MixA}[\![x_i]\!]\ (\ldots(\rho,x_i),\ldots,x_0) = \mathbf{fst}^i \circ \mathbf{snd} \circ \mathbf{app}$$

$$\mathit{MixA}[\![x_i]\!]\ [\rho:x_i:\ldots:x_0] = \mathbf{get}_i \circ \mathbf{app}$$

Figure 15   Hybrid Abstraction

Each λ-abstraction is annotated by a new mixed environment structure θ and $\oplus$ which indicates how to bind the current value (as a vector ":" or as a link ","). Mixed structures are built by **bind:**, **bind,** and the macro-combinator **Mix** which copies and restructures the environment ρ according to the annotation θ (Figure 15). Paths to values are now expressed by sequences of $\mathbf{fst}^i$ $\circ$ $\mathbf{snd}$ and $\mathbf{get}_j$. The abstraction algorithm distinguishes vectors from lists in the compile time environment using constructors ":" and ",".

9 Conclusion

In this paper, we have presented a framework to describe, prove and compare functional implementation techniques and optimizations (see Figure 2 in 2.5 for a summary). Our first intermediate language Λ_s bears strong similarities with CPS-expressions. Indeed, if we take combinator definitions (**DEF1**) (section 2.5) we naturally get Fischer's CPS transformation [11] from $\mathit{Va_f}$ (section 3.1). On the other hand, our combinators are not fully defined (they just have to respect a few properties) and we avoid issues such as administrative reductions. We see Λ_s as a powerful and more abstract framework than CPS to express different reduction strategies. As pointed out by Hatcliff & Danvy [14], Moggi's computational metalanguage [23] is also a more abstract alternative language to CPS. Arising from different roots, Λ_s is surprisingly close to Moggi's. In particular, we may interpret the monadic constructs [E] as **push** E and (**let** $x \Leftarrow E_1$ **in** E_2) as E_1 o $\lambda_s x.E_2$ and get back the monadic laws (let.β), (let.η) and (ass) [23]. On the other hand, we disallow unrestricted applications and Λ_s-expressions are more general than merely combinations of [ ] and **let**'s.

Related work also includes the derivation of abstract machines from denotational [30] or operational semantics [13] [27]. Their goal is to provide a methodology to formally derive implementations for a (potentially large) class of programming languages. A few works explore the relationship between two abstract machines such as TIM and the G-Machine [4][25] and CMCM and TIM [21]. The goal is to show the equivalence between seemingly very different implementations. Also, let us mention Asperti [2] who provides a categorical understanding of the Krivine machine and an extended CAM.

Our approach focuses on the description and comparison of fundamental options. The use of program transformations appeared to be suited to model precisely and completely the compilation process. Many standard optimizations (decurryfication, unboxing, hoisting, peephole optimizations) can be expressed as program transformations as well. This unified framework simplifies correctness proofs and makes it possible to reason about the efficiency of the produced code as well as about the complexity of transformations themselves. Our mid-term goal is to provide a general taxonomy of known implementations of functional languages. The last tricky task standing in the way is the expression of destructive updates. This is crucial in order to completely describe call-by-need and graph reduction machines. We hinted in section 6.3 how it could be done and we are currently investigating this issue. Still, as suggested in section 6, many options and optimizations (more than we were able to describe in this paper) are naturally expressed in our framework. Nothing should prevent us from completing our study of call-by-value and call-by-name implementations.

Acknowledgments. Thanks to Charles Consel, Luke Hornof and Daniel Le Métayer for commenting an earlier version of this paper.

References

[1]   A. W. Appel. *Compiling with Continuations.* Cambridge University Press. 1992.

[2]   A. Asperti. A categorical understanding of environment machines. *Journal of Functional Programming*, 2(1), pp.23-59,1992.

[3]   G. Argo. Improving the three instruction machine. In *Proc. of FPCA'89*, pp. 100-115, 1989.

[4]   G. Burn, S.L. Peyton Jones and J.D. Robson. The spineless G-machine. In *Proc. of LFP'88*, pp. 244-258, 1988.

[5]   G. Burn and D. Le Métayer. Proving the correctness of compiler optimisations based on a global analysis. *Journal of Functional Programming*, 1995. (to appear).

[6]   L. Cardelli. Compiling a functional language. In *Proc. of LFP'84*, pp. 208-217, 1984.

[7]   G. Cousineau, P.-L. Curien and M. Mauny, The categorical abstract machine. *Science of Computer Programming*, 8(2), pp. 173-202, 1987.

[8]   P. Crégut. *Machines à environnement pour la réduction symbolique et l'évaluation partielle.* Thèse de l'université de Paris VII, 1991.

[9]   R. Douence and P. Fradet. A taxonomy of functional language implementations. Part I: Call-by-Value, *INRIA Research Report*, 1995. (to appear)

[10] J. Fairbairn and S. Wray. Tim: a simple, lazy abstract machine to execute supercombinators. In *Proc of FPCA'87*, LNCS 274, pp. 34-45, 1987.

[11] M. J. Fischer. Lambda-calculus schemata. In *Proc. of the ACM Conf. on Proving Properties about Programs*, Sigplan Notices, Vol. 7(1), pp. 104-109,1972.

[12] P. Fradet and D. Le Métayer. Compilation of functional languages by program transformation. *ACM Trans. on Prog. Lang. and Sys.*, 13(1), pp. 21-51, 1991.

[13] J. Hannan. From operational semantics to abstract machines. *Math. Struct. in Comp. Sci.*, 2(4), pp. 415-459, 1992.

[14] J. Hatcliff and O. Danvy. A generic account of continuation-passing styles. In *Proc. of POPL'94*, pp. 458-471, 1994.

[15] T. Johnsson. *Compiling Lazy Functional Languages*. PhD Thesis, Chalmers University, 1987.

[16] M. S. Joy, V. J. Rayward-Smith and F. W. Burton. Efficient combinator code. *Computer Languages*, 10(3), 1985.

[17] D. Kranz, R. Kesley, J. Rees, P. Hudak, J.Philbin, and N. Adams. ORBIT: An optimizing compiler for Scheme. *SIGPLAN Notices, 21(7)*, pp.219-233, 1986.

[18] P. J. Landin. The mechanical evaluation of expressions. *The Computer Journal*, 6(4), pp.308-320, 1964.

[19] X. Leroy. The Zinc experiment: an economical implementation of the ML language. *INRIA Technical Report 117*, 1990.

[20] R. D. Lins. Categorical multi-combinators. In *Proc. of FPCA'87*, LNCS 274, pp. 60-79, 1987.

[21] R. Lins, S. Thompson and S.L. Peyton Jones. On the equivalence between CMC and TIM. *Journal of Functional Programming*, 4(1), pp. 47-63, 1992.

[22] E. Meijer and R. Paterson. Down with lambda lifting. copies available at: erik@cs.kun.nl, 1991.

[23] E. Moggi. Notions of computation and monads. *Information and Computation*, 93:55-92, 1991.

[24] S.L. Peyton Jones. Implementing lazy functional languages on stock hardware: the spineless tagless G-machine. *Journal of Functional Programming*, 2(2):127-202, 1992.

[25] S. L. Peyton Jones and D. Lester. *Implementing functional languages, a tutorial*. Prentice Hall, 1992.

[26] S. L. Peyton Jones and J. Launchbury. Unboxed values as first class citizens in a non-strict functional language. In *Proc. of FPCA'91*, LNCS 523, pp.636-666, 1991.

[27] P. Sestoft. Deriving a lazy abstract machine. *Technical Report 1994-146, Technical University of Denmark*, 1994.

[28] Z. Shao and A. Appel. Space-efficient closure representations. In *Proc. of LFP'94*, pp. 150-161,1994.

[29] D.A. Turner. A new implementation technique for applicative languages. *Soft. Pract. and Exper.*, 9, pp. 31-49, 1979.

[30] M. Wand. Deriving target code as a representation of continuation semantics. *ACM Trans. on Prog. Lang. and Sys.*, 4(3), pp. 496-517, 1982.

A λ-calculus à la de Bruijn with explicit substitutions

Fairouz Kamareddine and Alejandro Ríos

Department of Computing Science, 17 Lilybank Gardens, University of Glasgow, Glasgow G12 8QQ, Scotland, fax: +44 41 330 4913, *email*: fairouz@dcs.gla.ac.uk and rios@dcs.gla.ac.uk

Abstract. The aim of this paper is to present the λs-calculus which is a very simple λ-calculus with explicit substitutions and to prove its confluence on closed terms and the preservation of strong normalisation of λ-terms. We shall prove strong normalisation of the corresponding calculus of substitution by translating it into the λσ-calculus [ACCL91], and therefore the relation between both calculi will be made explicit. The confluence of the λs-calculus is obtained by the "interpretation method" ([Har89], [CHL92]). The proof of the preservation of normalisation follows the lines of an analogous result for the λυ-calculus (cf. [BBLRD95]). The relation between λs and λυ is also studied.

1 Introduction

Most literature on the λ-calculus considers substitution as an implicit operation. It means that the computations to perform substitution are usually described with operators which do not belong to the language of the λ-calculus. There has however been an interest in formalising substitution explicitly; various calculi including new operators to denote substitution and new rules to handle these operators have been proposed. Amongst these calculi we mention $C\lambda\xi\phi$ (cf. [dB78b]); the calculi of categorical combinators (cf. [Cur86]); $\lambda\sigma$, $\lambda\sigma_{\Uparrow}$, $\lambda\sigma_{SP}$ (cf. [ACCL91], [CHL92], [Río93]) referred to as the λσ-family; $\lambda\upsilon$ (cf. [BBLRD95]), a descendant of the λσ-family and $\varphi\sigma BLT$ (cf. [KN93]). The basic features of these systems of substitution depart quite extensively from the classical λ-calculus while in this paper we propose a system which remains as close as possible to it.

Furthermore, for the above systems either strong normalisation (SN) has not been studied (as for $C\lambda\xi\phi$ and $\varphi\sigma BLT$) or negative results (cf. [Mel95]) have been established concerning the preservation of SN (for the λσ-family). In particular, these negative results imply that the simplest typed versions of these calculi are not SN. One positive and recent result concerning the preservation of SN is that for $\lambda\upsilon$ (cf. [BBLRD95]) for which, as far as we know, there is still work in progress.

As stated in [ACCL91], the λσ-calculi and the calculi of combinators give full formal accounts of the process of computation and they make it easy to derive machines for the λ-calculus and to show the correctness of these machines. Hence, the λσ-calculus is proposed as a step in closing the gap between the classical λ-calculus and concrete implementations. We believe that the λs-calculus presented in this paper offers another possibility for closing this gap and, being closer to the λ-calculus, it preserves strong normalisation. Furthermore, we think that in the presence of the negative results of [Mel95] calculi like λs are worth studying.

The main interest in introducing the λs-calculus is to provide a calculus of explicit substitutions which would have both the property of preserving strong normalisation and a confluent extension on open terms. As far as we know no such calculus has yet been proposed. There are calculi of explicit substitutions which are confluent on open terms: the λσ$_⇑$- calculus (cf. [HL89] and [CHL92]), but, as mentioned above, the non-preservation of strong normalisation for λσ$_⇑$ has recently been proved. There are also calculi which satisfy the preservation property: the λυ-calculus (cf. [BBLRD95]), but this calculus is not confluent on open terms. Moreover, in order to get a confluent extension, the introduction of a composition operator for substitutions seems unavoidable, but precisely this operator is the cause of the non-preservation of strong normalisation as shown in [Mel95]. We believe that the λs-calculus, while preserving strong normalisation, could admit a confluent extension on open terms thanks to the fact that composition of substitutions (in the sense of the λσ-calculi) could be handleld indirectly and in a very subtle way via a new family of rules mimicking the substitution lemma for the classical λ-calculus (see lemma 4 below).

Mention to a very close calculus to the λs-calculus can be already found in [Cur86], exercise 1.2.7.2, where reference to previous unpublished notes of Y. Lafont is given. The $\varphi\sigma BLT$-calculus is also of this kind but the essential difference is that the redex is preserved when the β-rule is applied. The calculus we are going to study, we call it λs, is obtained in a very natural way from the classical λ-calculus in de Bruijn notation: we just orientate the equalities defining the meta-operators of substitution and include them as new operators of the language.

We prove in this paper the confluence (CR) of the λs-calculus on closed terms (these terms contain all terms of the classical λ-calculus) and the preservation of strong normalisation (terms which are strongly normalising in the λ-calculus are also strongly normalising in λs). We also compare the λs-calculus to λσ and λυ via translation functions.

2 Preliminaries

We begin by giving a quick presentation of the λ-calculus à la de Bruijn and the λσ-calculus.

2.1 The classical λ-calculus in de Bruijn notation

We shall assume the reader familiar with de Bruijn indices (see [dB72] and [dB78a]) which can be explained via the following two examples: $\lambda x \lambda y.xy$ is written using de Bruijn indices as $\lambda\lambda(21)$ and $\lambda x \lambda y.(x(\lambda z.zx))y$ is written as $\lambda\lambda(2(\lambda(13))1)$.

Remark here that variables are removed and are replaced by natural numbers. These numbers are informative as to the λ which binds the occurrence of the variable. Hence in the second example, the same x was translated into 2 and 3 according to the different positions, whereas z and y become the same de Bruijn index, 1.

The interest in introducing de Bruijn indices is that they avoid clashes of variable names and therefore neither α-conversion nor Barendregt's convention are needed. Here is the λ-calculus à la de Bruijn.

Definition 1 *We define* Λ, *the set of terms with de Bruijn indices, as follows:*

$$\Lambda ::= \mathbf{N} \mid (\Lambda\Lambda) \mid (\lambda\Lambda)$$

We use $a, b, \ldots$ *to range over* Λ *and* $m, n, \ldots$ *to range over* $\mathbf{N}$ *(positive natural numbers). Furthermore, we assume the usual conventions about parentheses and avoid them when no confusion occurs. Throughout the whole article,* $a = b$ *is used to mean that* a *and* b *are syntactically identical.*

When rewriting a term a with variable names into its de Bruijn version, we consider a to be a subterm of $\lambda x_1 \ldots x_k.a$ where $x_1 \ldots x_k$ are all the free variables of a. For instance: $\lambda x.xyz$ becomes $\lambda 123$ (or $\lambda 132$) and $(\lambda x.xy)y$ becomes $(\lambda 12)1$. In order for this to work independently of the order in which the free variables appear, we assume that the set of variable names is ordered and call this ordered set *the free variable list*. For example, if the list was $\cdots, z, y, x$ then the term to be translated should be prefixed with $\cdots, \lambda z, \lambda y, \lambda x$ before its translation. Thus, $\lambda x.yz$ translates as $\lambda 34$ whereas $\lambda x.zy$ translates as $\lambda 43$. Now check that $(\lambda x\lambda y.zxy)(\lambda x.yx)$ translates as $(\lambda\lambda 521)(\lambda 31)$ and that $\lambda u.z(\lambda x.yx)u$ translates to $\lambda 4(\lambda 41)1$.

In order to define β-reduction à la de Bruijn, we must define the substitution of a variable by a term b in a term a. Therefore, we must identify amongst the numbers of a term a those that correspond to the variable that is being substituted for and we need to update the term to be substituted in order to preserve the correct bindings of its variables.

For example, translating $(\lambda x\lambda y.zxy)(\lambda x.yx) \rightarrow_\beta \lambda u.z(\lambda x.yx)u$ to de Bruijn notation we get $(\lambda\lambda 521)(\lambda 31) \rightarrow_\beta \lambda 4(\lambda 41)1$. But if we simply replace 2 in $\lambda 521$ by $\lambda 31$ we get $\lambda 5(\lambda 31)1$, which is not correct. We needed to decrease 5 as one λ disappeared and to increment the free variables of $\lambda 31$ as they occur within the scope of one more λ.

For incrementing the free variables we need a family of updating functions:

Definition 2 *The* updating functions $U_k^i : \Lambda \to \Lambda$ *for* $k \geq 0$ *and* $i \geq 1$ *are defined inductively as follows:*

$$U_k^i(ab) = U_k^i(a)\,U_k^i(b) \qquad\qquad U_k^i(\mathbf{n}) = \begin{cases} \mathbf{n+i-1} & \text{if } n > k \\ \mathbf{n} & \text{if } n \leq k. \end{cases}$$
$$U_k^i(\lambda a) = \lambda(U_{k+1}^i(a))$$

The intuition behind U_k^i is the following: k tests for free variables and $i - 1$ is the value by which a variable, if free, must be incremented.

Now we define the family of meta-substitution functions:

Definition 3 *The* meta-substitutions at level i, *for* $i \geq 1$, *of a term* $b \in \Lambda$ *in a term* $a \in \Lambda$, *denoted* $a\{\!\{i \leftarrow b\}\!\}$, *is defined inductively on* a *as follows:*

$$(a_1 a_2)\{\!\{i \leftarrow b\}\!\} = (a_1\{\!\{i \leftarrow b\}\!\})(a_2\{\!\{i \leftarrow b\}\!\}) \qquad\qquad \mathbf{n}\{\!\{i \leftarrow b\}\!\} = \begin{cases} \mathbf{n-1} & \text{if } n > i \\ U_0^i(b) & \text{if } n = i \\ \mathbf{n} & \text{if } n < i. \end{cases}$$
$$(\lambda a)\{\!\{i \leftarrow b\}\!\} = \lambda(a\{\!\{i + 1 \leftarrow b\}\!\})$$

Ultimately, the intention is to define $(\lambda a)b \rightarrow_\beta a\{\!\{1 \leftarrow b\}\!\}$ (see definition 4 below). The first two equalities propagate the substitution through applications and abstractions and the last one carries out the substitution of the intended variable (when $n = i$) by the updated term. If the variable is not the intended one it must be decreased by 1 if it is free (case $n > i$) beacuse one λ has disappeared, whereas if it is bound (case $n < i$) it must remain unaltered.

It is easy to check that $(\lambda 521)\{\!\{1 \leftarrow (\lambda 31)\}\!\} = \lambda 4(\lambda 41)1$. This will mean $(\lambda\lambda 521)(\lambda 31) \to_\beta \lambda 4(\lambda 41)1$.

The following lemmas establish the properties of the meta-substitutions and updating functions. The Meta-substitution and Distribution lemmas are crucial to prove the confluence of λs. The proofs of lemmas 1 - 6 are obtained by induction on a. Furthermore, the proof of lemma 3 requires lemma 2 with $p = 0$; the proof of lemma 4 uses lemmas 1 and 3 both with $k = 0$; finally, lemma 5 with $p = 0$ is needed to prove lemma 6.

Lemma 1 *For $k < n \leq k + i$ we have:* $U_k^i(a) = U_k^{i+1}(a)\{\!\{n \leftarrow b\}\!\}$.

Lemma 2 *For $p \leq k < j + p$ we have:* $U_k^i(U_p^j(a)) = U_p^{j+i-1}(a)$.

Lemma 3 *For $i \leq n - k$ we have:* $U_k^i(a)\{\!\{n \leftarrow b\}\!\} = U_k^i(a\{\!\{n - i + 1 \leftarrow b\}\!\})$.

Lemma 4 (Meta-substitution lemma) *For $1 \leq i \leq n$ we have:*
$$a\{\!\{i \leftarrow b\}\!\}\{\!\{n \leftarrow c\}\!\} = a\{\!\{n + 1 \leftarrow c\}\!\}\{\!\{i \leftarrow b\{\!\{n - i + 1 \leftarrow c\}\!\}\}\!\}$$

Lemma 5 *For $m \leq k + 1$ we have:* $U_{k+p}^i(U_p^m(a)) = U_p^m(U_{k+p+1-m}^i(a))$.

Lemma 6 (Distribution lemma) *For $n \leq k + 1$ we have:*
$$U_k^i(a\{\!\{n \leftarrow b\}\!\}) = U_{k+1}^i(a)\{\!\{n \leftarrow U_{k-n+1}^i(b)\}\!\} .$$

Definition 4 *β-reduction is the least compatible relation on Λ generated by:*

    (β-rule) $(\lambda a)\,b \to_\beta a\{\!\{1 \leftarrow b\}\!\}$

The λ-calculus à la de Bruijn, abbreviated λ-calculus is the reduction system whose only rewriting rule is β.

Theorem 1 *The λ-calculus à la de Bruijn is confluent.*

Proof: The λ-calculus with de Bruijn indices and the classical λ-calculus with variable names are isomorphic (cf. [Mau85]). The confluence of the latter (cf. [Bar84] thm. 3.2.8) is hence transportable to the λ-calculus à la de Bruijn.

A proof which does not use the mentioned isomorphism is given in [Río93] (corol. 3.6) as a corollary of a more general result concerning the $\lambda\sigma$-calculus. $\square$

Finally, the following lemma ensures the good passage of the β-rule through the meta-substitutions and the U_k^i. It is crucial for the proof of the confluence of λs.

Lemma 7 *Let $a, b, c, d \in \Lambda$.*

1. *If $c \to_\beta d$ then $U_k^i(c) \to_\beta U_k^i(d)$.*
2. *If $c \to_\beta d$ then $a\{\!\{i \leftarrow c\}\!\} \twoheadrightarrow_\beta a\{\!\{i \leftarrow d\}\!\}$.*
3. *If $a \to_\beta b$ then $a\{\!\{i \leftarrow c\}\!\} \to_\beta b\{\!\{i \leftarrow c\}\!\}$.*

Proof:

1. Induction on c. We just check the interesting case which arises when $c = c_1 c_2$ and the reduction takes place at the root, i.e. $c_1 = (\lambda a)$, $c_2 = b$ and $d = a\{\!\{1 \leftarrow b\}\!\}$:

$$U_k^i((\lambda a)b) = (\lambda(U_{k+1}^i(a)))U_k^i(b) \to_\beta U_{k+1}^i(a)\{\!\{1 \leftarrow U_k^i(b)\}\!\} \overset{L\,6}{=} U_k^i(a\{\!\{1 \leftarrow b\}\!\})$$

2. Induction on a using 1 above.
3. Induction on a. The interesting case is again $a = (\lambda d)e$ and $b = d\{\!\{1 \leftarrow e\}\!\}$:

$$((\lambda d)e)\{\!\{i \leftarrow c\}\!\} = (\lambda(d\{\!\{i+1 \leftarrow c\}\!\}))(e\{\!\{i \leftarrow c\}\!\}) \to_\beta$$

$$(d\{\!\{i+1 \leftarrow c\}\!\})\{\!\{1 \leftarrow e\{\!\{i \leftarrow c\}\!\}\}\!\} \overset{L\,4}{=} (d\{\!\{1 \leftarrow e\}\!\})\{\!\{i \leftarrow c\}\!\} \qquad \square$$

2.2 The $\lambda\sigma$-calculus

The $\lambda\sigma$-calculus ([ACCL91]) is a formalism which enables explicit substitution. Its syntax is two-sorted: the sort **term** of *terms* and the sort **substitution** of *explicit substitutions*. These can be interpreted as a sequence of terms and the result of executing a substitution in a term can be interpreted as the term obtained by replacing the occurrences of the n-th index of de Bruijn in the term by the n-th term of the sequence. This intuitive interpretation is developped and illustrated with many examples in [ACCL91].

Here are the syntax and the rules of the calculus:

Definition 5 *The syntax of the $\lambda\sigma$-calculus is given by:*

$$\text{Terms} \qquad \Lambda\sigma^t ::= 1 \mid \Lambda\sigma^t \Lambda\sigma^t \mid \lambda\Lambda\sigma^t \mid \Lambda\sigma^t[\Lambda\sigma^s]$$
$$\text{Substitutions } \Lambda\sigma^s ::= id \mid \uparrow \mid \Lambda\sigma^t \cdot \Lambda\sigma^s \mid \Lambda\sigma^s \circ \Lambda\sigma^s$$

The set, denoted $\lambda\sigma$, of rules of the $\lambda\sigma$-calculus is the following:

(Beta)	$(\lambda a)\, b \longrightarrow a\,[b \cdot id]$
(VarId)	$1\,[id] \longrightarrow 1$
(VarCons)	$1\,[a \cdot s] \longrightarrow a$
(App)	$(a\,b)[s] \longrightarrow (a\,[s])\,(b\,[s])$
(Abs)	$(\lambda a)[s] \longrightarrow \lambda(a\,[1 \cdot (s \circ \uparrow)])$
(Clos)	$(a\,[s])[t] \longrightarrow a\,[s \circ t]$
(IdL)	$id \circ s \longrightarrow s$
(ShiftId)	$\uparrow \circ\, id \longrightarrow\, \uparrow$
(ShiftCons)	$\uparrow \circ (a \cdot s) \longrightarrow s$
(Map)	$(a \cdot s) \circ t \longrightarrow a\,[t] \cdot (s \circ t)$
(Ass)	$(s_1 \circ s_2) \circ s_3 \longrightarrow s_1 \circ (s_2 \circ s_3)$

The set of rules of the σ-calculus is $\lambda\sigma - \{(Beta)\}$. We use $a, b, c, \ldots$ to range over $\Lambda\sigma^t$ and $s, t, \ldots$ to range over $\Lambda\sigma^s$.

Notation 1 *For a given set of rules $\mathcal{R}$ we take $\rightarrow_{\mathcal{R}}$ to be the reduction relation of the $\mathcal{R}$-calculus (i.e. the least compatible relation containing the rules of $\mathcal{R}$).*

We take $\twoheadrightarrow_{\mathcal{R}}$ to be the derivation relation of the $\mathcal{R}$-calculus (i.e. the least reflexive and transitive relation containing $\rightarrow_{\mathcal{R}}$) and we denote by $\rightarrow^+_{\mathcal{R}}$ the transitive closure of $\rightarrow_{\mathcal{R}}$ (i.e. the least transitive relation containing $\rightarrow_{\mathcal{R}}$).

For any two relations $\rightarrow$ and $\rightarrow'$, by $a \rightarrow \cdot \rightarrow' b$ we mean $(\exists c)(a \rightarrow c \rightarrow' b)$.

Finally, we write $a \twoheadrightarrow^n_{\mathcal{R}} b$ to mean that the derivation from a to b consists of n steps of $\mathcal{R}$-reduction.

When it will be clear from the context, we may omit the subscript $\mathcal{R}$.

We recall that a is a $\mathcal{R}$-normal form if there exists no b such that $a \rightarrow_{\mathcal{R}} b$. We say that c is a $\mathcal{R}$-normal form of d if c is a $\mathcal{R}$-normal form and $d \twoheadrightarrow_{\mathcal{R}} c$.

Notation 2 *For every substitution s we define the iteration of the composition of s inductively as $s^1 = s$ and $s^{n+1} = s \circ s^n$. We use the convention $s^0 = id$.*

Note that the only de Bruijn index used is 1, but we can code n by the term $1[\uparrow^{n-1}]$. By so doing, we have $\Lambda \subset \Lambda\sigma^t$.

β-reduction of the λ-calculus is interpreted in the $\lambda\sigma$-calculus in two steps. The first, obtained by the application of *(Beta)*, consists in generating the substitution. The second step executes the propagation of this substitution, using the set of the σ-rules, until the concerned variables are reached. The reader is invited to check that $(\lambda\lambda 521)(\lambda 31) \twoheadrightarrow_{\lambda\sigma} \lambda 4(\lambda 41)1$.

We summarize now the properties of the σ- and $\lambda\sigma$-calculi:

Theorem 2 *The σ-calculus is strongly normalising (SN) and confluent (CR).*

Proof: We know three proofs of strong normalisation: [HL86], [CHR92] and [Zan94]. Local confluence (WCR) is ensured by analysis of critical pairs (cf. [Río93], annex B), and the Knuth-Bendix theorem ([KB70] or [Hue80]). Now Newman's lemma, which states that SN+WCR yields CR ([Bar84], prop. 3.1.25), guarantees confluence. $\square$

Theorem 3 *The $\lambda\sigma$-calculus is confluent.*

Proof: See [ACCL91], theorem 3.2. This proof is based on the confluence of σ, that of the λ-calculus and the technique of interpretation. $\square$

3 The λs-calculus and its confluence

The idea is to handle explicitly the meta-operators defined in definitions 2 and 3. Therefore, the syntax of the λs-calculus is obtained by adding to the syntax of the λ-calculus à la de Bruijn two families of operators :

- $\{\sigma^i\}_{i\geq 1}$ This family is meant to denote the explicit substitution operators. Each σ^i is an infix operator of arity 2 and $a\,\sigma^i b$ has as intuitive meaning the term a where all free occurrences of the variable corresponding to the de Bruijn number i are to be substituted by the term b.
- $\{\varphi^i_k\}_{k\geq 0\ i\geq 1}$ This family is meant to denote the updating functions necessary when working with de Bruijn numbers to fix the variables of the term to be substituted.

Definition 6 *The set of* terms *of the* λs-*calculus, noted* Λs *is given as follows:*

$$\Lambda s ::= \mathbf{N} \mid \Lambda s \Lambda s \mid \lambda \Lambda s \mid \Lambda s \, \sigma^i \Lambda s \mid \varphi^i_k \Lambda s \quad \text{where} \quad i \geq 1, \; k \geq 0.$$

We take a, b, c *to range over* Λs. *A term of the form* $a \, \sigma^i b$ *is called a* closure. *Furthermore, a term containing neither* σ*'s nor* φ*'s is called a* pure term.

The λs-calculus should carry out, besides β-reduction, the computations of updating and substitution explicitly. For that reason we include, besides the rule mimicking the β-rule (σ-*generation*), a set of rules which are the equations in definitions 2 and 3 oriented from left to right.

Definition 7 *The* λs-*calculus is given by the following rewriting rules:*

σ-*generation*	$(\lambda a)\, b \longrightarrow a\,\sigma^1 b$
σ-λ-*transition*	$(\lambda a)\, \sigma^i b \longrightarrow \lambda(a\,\sigma^{i+1} b)$
σ-*app-transition*	$(a_1 \, a_2)\, \sigma^i b \longrightarrow (a_1 \, \sigma^i b)\,(a_2 \, \sigma^i b)$
σ-*destruction*	$\mathbf{n}\,\sigma^i b \longrightarrow \begin{cases} \mathbf{n-1} & \text{if } n > i \\ \varphi^i_0 b & \text{if } n = i \\ \mathbf{n} & \text{if } n < i \end{cases}$
φ-λ-*transition*	$\varphi^i_k(\lambda a) \longrightarrow \lambda(\varphi^i_{k+1} a)$
φ-*app-transition*	$\varphi^i_k(a_1 \, a_2) \longrightarrow (\varphi^i_k \, a_1)\,(\varphi^i_k \, a_2)$
φ-*destruction*	$\varphi^i_k \, \mathbf{n} \longrightarrow \begin{cases} \mathbf{n+i-1} & \text{if } n > k \\ \mathbf{n} & \text{if } n \leq k \end{cases}$

We use λs *to denote this set of rules. The calculus of substitutions associated with the* λs-*calculus is the rewriting system whose rules are* $\lambda s - \{\sigma\text{-}generation\}$ *and we call it* s-*calculus.*

In order to give the translation into the $\lambda\sigma$-calculus we give the following two definitions.

Definition 8 *For* $k \geq 0$ *and* $i \geq 1$ *we define* $s_{k\,i} = 1 \cdot 2 \cdot \ldots \cdot \text{k} \cdot \uparrow^{k+i-1}$ *(we use the convention* $s_{0\,i} = \uparrow^{i-1}$ *and hence* $s_{0\,1} = id$ *).*

Definition 9 *Let* $b \in \Lambda\sigma^t$, *we define a family of substitutions* $(b_k)_{k \geq 1}$ *as follows:*

$$b_1 = b[id] \cdot id \qquad b_2 = 1 \cdot b[\uparrow] \cdot \uparrow \quad \ldots \quad b_{i+1} = 1 \cdot 2 \cdot \ldots \cdot i \cdot b[\uparrow^i] \cdot \uparrow^i \quad \ldots$$

Using the rules *(Map)*, *(Clos)*, *(Ass)* and *(IdL)* it is easy to verify that:

Remark 1 $1 \cdot (b_i \circ \uparrow) \twoheadrightarrow_\sigma b_{i+1}$ *and* $1 \cdot (s_{k\,i} \circ \uparrow) \twoheadrightarrow_\sigma s_{k+1\,i}$.

Definition 10 *The translation function* $T : \Lambda s \to \Lambda\sigma^t$ *is defined by:*

$$T(\mathbf{n}) = \mathbf{n} \quad T(a\,b) = T(a)T(b) \quad T(a\,\sigma^i b) = T(a)[T(b)_i]$$
$$T(\lambda a) = \lambda(T(a)) \qquad T(\varphi^i_k a) = T(a)[s_{k\,i}]$$

Theorem 4 *If $a \to_s b$ then $T(a) \overset{+}{\to}_\sigma T(b)$.*

Proof: Induction on a. We just check, as an example, the case $a = n\,\sigma^i c$ when the reduction takes place at the root:

$$T(n\,\sigma^i c) = n[T(c)_i] \overset{+}{\to}_\sigma \begin{cases} n - 1 = T(n-1) & \text{if } n > i \\ T(c)[\uparrow^{i-1}] = T(\varphi_0^i c) & \text{if } n = i \\ n = T(n) & \text{if } n < i \end{cases} \qquad \square$$

Corollary 1 *The reduction $\to_s$ is strongly normalising.*

Proof: Use Theorem 2. $\qquad \square$

Remark 2 *The reduction $\to_s$ is locally confluent.*

Proof: There are no critical pairs and the theorem of Knuth-Bendix applies trivially.
$\qquad \square$

Corollary 2 *The reduction $\to_s$ is confluent.*

Proof: Newman's Lemma (see proof of thm. 2) yields CR. $\qquad \square$

These corollaries guarantee the existence and unicity of s-normal forms (s-nf), which we shall use to interpret the λs-calculus in the λ-calculus. We shall denote the s-nf of a term a by $s(a)$. The following lemma characterizes s-normal forms.

Lemma 8 *The set of s-normal forms is exactly Λ.*

Proof: Check first by induction on a that $a\,\sigma^i b$ and $\varphi_k^i a$ are not normal forms. Then check by induction on a that if a is an s-nf then $a \in \Lambda$. Conclude by observing that every term in Λ is in s-nf. $\qquad \square$

As there are no s-rules whose left-hand side is an application or an abstraction, the following properties of s-normal forms (which will be used throughout without explicit mention) are immediate.

Lemma 9 *For all $a, b \in \Lambda s$: $s(a\,b) = s(a)s(b)$ and $s(\lambda a) = \lambda(s(a))$.*

We establish now the relation between the operators σ^i and φ_k^i and the meta-operators of the classical de Bruijn setting: $\_\{\!\{i \leftarrow \_\}\!\}$ and U_k^i.

Lemma 10 *For all $a, b \in \Lambda s$ we have:*

$$s(\varphi_k^i a) = U_k^i(s(a)) \quad \text{and} \quad s(a\,\sigma^i b) = s(a)\{\!\{i \leftarrow s(b)\}\!\}.$$

Proof: Prove the first equality for terms in s-nf, i.e. use an inductive argument on $c \in \Lambda$ to show $s(\varphi_k^i c) = U_k^i(s(c))$. Let now $a \in \Lambda s$, $s(\varphi_k^i a) = s(\varphi_k^i s(a)) = U_k^i(s(s(a))) = U_k^i(s(a))$.
Prove the second claim similarly using the first claim. $\qquad \square$

We give now the key result that allows us to use the Interpretation Method in order to get confluence: the good passage of the σ-generation rule to the s-normal forms.

Proposition 1 *Let* $a, b \in \Lambda s$, *if* $a \to_{\sigma-gen} b$ *then* $s(a) \twoheadrightarrow_\beta s(b)$.

Proof: Induction on a. We just study the interesting cases.

$a = c\,d$: If the reduction takes place within c or d just use the inductive hypothesis (IH). The interesting case is when $c = \lambda e$ and hence $b = e\,\sigma^1 d$:

$$s((\lambda e)d) = (\lambda s(e))(s(d)) \to_\beta s(e)\{\!\{1 \leftarrow s(d)\}\!\} \overset{L\,10}{=} s(e\,\sigma^1 d)$$

$a = c\,\sigma^i d$: If the reduction takes place within c, i.e. $c \to_{\sigma-gen} e$ and $b = e\,\sigma^i d$, then

$$s(c\,\sigma^i d) \overset{L\,10}{=} s(c)\{\!\{i \leftarrow s(d)\}\!\} \overset{IH\,\&\,L\,7.3}{\twoheadrightarrow_\beta} s(e)\{\!\{i \leftarrow s(d)\}\!\} \overset{L\,10}{=} s(e\,\sigma^i d)$$

If the reduction takes place within d, lemma 7.2 applies.

$a = \varphi_k^i c$: The reduction must take place within c. Use lemma 10 and lemma 7.1. $\square$

Now, the following corollaries are immediate.

Corollary 3 *Let* $a, b \in \Lambda s$, *if* $a \twoheadrightarrow_{\lambda s} b$ *then* $s(a) \twoheadrightarrow_\beta s(b)$.

Corollary 4 (Soundness) *Let* $a, b \in \Lambda$, *if* $a \twoheadrightarrow_{\lambda s} b$ *then* $a \twoheadrightarrow_\beta b$.

Finally, before proving confluence, we verify that the λs-calculus is powerful enough to simulate β-reduction.

Lemma 11 (Simulation of β-reduction) *Let* $a, b \in \Lambda$, *if* $a \to_\beta b$ *then* $a \twoheadrightarrow_{\lambda s} b$.

Proof: Induction on a. As usual the interesting case is when $a = (\lambda c)d$ and $b = c\{\!\{1 \leftarrow d\}\!\}$:

$$(\lambda c)d \to_{\sigma-gen} c\,\sigma^1 d \twoheadrightarrow_s s(c\,\sigma^1 d) \overset{L\,10}{=} s(c)\{\!\{1 \leftarrow s(d)\}\!\} \overset{c,d\in\Lambda}{=} c\{\!\{1 \leftarrow d\}\!\} \qquad \square$$

Theorem 5 *The λs-calculus is confluent.*

Proof: We interpret the λs-calculus into the λ-calculus via s-normalisation. We have:

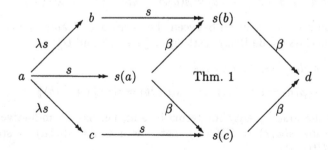

The existence of the arrows $s(a) \twoheadrightarrow_\beta s(b)$ and $s(a) \twoheadrightarrow_\beta s(c)$ is guaranteed by corollary 3. We can close the diagram thanks to the confluence of λ-calculus and finally lemma 11 ensures $s(b) \twoheadrightarrow_{\lambda s} d$ and $s(b) \twoheadrightarrow_{\lambda s} d$ proving thus CR for the λs-calculus. $\square$

4 The λs-calculus preserves strong normalisation

In this section we shall prove that every term a which is strongly normalising (all its derivations are finite) in the λ-calculus (denoted $a \in \beta$-SN) is also strongly normalising in the λs-calculus (denoted $a \in \lambda s$-SN). In particular, pure simply typed terms will be strongly normalising in λs.

This result is not valid for the $\lambda\sigma$-calculus, neither for its confluent version $\lambda\sigma_{\Uparrow}$, neither for the calculus of categorical combinators, as was recently proved by the counterexamples of Melliès (see [Mel95]). But there is work in progress to prove it for the λv-calculus (cf. [BBLRD95]).

The natural translation of λs into λv which we shall give in this section is good enough to ensure the preservation of strong normalisation for λs as soon as the result will be obtained for λv. However, the general idea in [BBLRD95] can be adapted for the preservation of strong normalisation of λs.

We begin by presenting the λv-calculus and the translation.

Definition 11 *The terms of the λv-calculus are given by the following syntax:*

$$\textbf{Terms} \qquad \Lambda v^t ::= \mathbf{N} \mid \Lambda v^t \Lambda v^t \mid \lambda \Lambda v^t \mid \Lambda v^t [\Lambda v^s]$$
$$\textbf{Substitutions } \Lambda v^s ::= \uparrow \mid \Uparrow (\Lambda v^s) \mid \Lambda v^t /$$

The set, denoted λv, of rules of the λv-calculus is the following:

(Beta)	$(\lambda a)\, b \longrightarrow a\,[b/]$
(App)	$(a\,b)[s] \longrightarrow (a\,[s])\,(b\,[s])$
(Abs)	$(\lambda a)[s] \longrightarrow \lambda(a\,[\Uparrow (s)])$
(FVar)	$1\,[a/] \longrightarrow a$
(RVar)	$\mathbf{n}+1\,[a/] \longrightarrow \mathbf{n}$
(FVarLift)	$1\,[\Uparrow (s)] \longrightarrow 1$
(RVarLift)	$\mathbf{n}+1\,[\Uparrow (s)] \longrightarrow \mathbf{n}[s][\uparrow]$
(VarShift)	$\mathbf{n}[\uparrow] \longrightarrow \mathbf{n}+1$

We use $a, b, c, \ldots$ to range over Λv^t and $s, t, \ldots$ to range over Λv^s.

This choice of operators and rules is based on the idea of expressing the *(Beta)*-rule as economically as possible. In the $\lambda\sigma$-calculus it reads $(\lambda a)\, b \to a[b \cdot id]$ and requires the introduction of the operators $\cdot$ and id. Just one unary operator can do the job, this operator is denoted by $/$ in the λv-syntax. Hence $a/$ plays the role of the $\lambda\sigma$-term $a \cdot id$. Now the *(Abs)*-rule, which in $\lambda\sigma$ reads $(\lambda a)[s] \to \lambda(a\,[1 \cdot (s \circ \uparrow)])$, must be modified to avoid the use of $\cdot$ which is no longer available. Hence the introduction of $\Uparrow$ and the intuitive interpretation of $\Uparrow (s)$ as the $\lambda\sigma$-term $1 \cdot (s \circ \uparrow)$.

Notation 3 *For $a \in \Lambda v^t$ and $s \in \Lambda v^s$ we denote:*
- $\Uparrow^i (s) = \Uparrow (\Uparrow (\ldots \Uparrow (s)\ldots))$ *(i times). By $\Uparrow^0 (s)$ we mean s.*
- $a[s]^i = a[s][s]\ldots[s]$ *(i times). By $a[s]^0$ we mean a.*

Definition 12 *The "natural" translation $S : \Lambda s \to \Lambda v^t$ is given by:*

$$S(\mathbf{n}) = \mathbf{n} \quad S(a\,b) = S(a)S(b) \quad S(a\,\sigma^i b) = S(a)[\Uparrow^{i-1}(S(b)/)]$$
$$S(\lambda a) = \lambda(S(a)) \quad S(\varphi_k^i a) = S(a)[\Uparrow^k(\uparrow)]^{i-1}$$

It is easy to check by induction on a that $a \to_{\sigma-gen} b$ implies $S(a) \overset{+}{\to}_{\lambda v} S(b)$ and that $a \to_s b$ implies $S(a) \twoheadrightarrow_{\lambda v} S(b)$. Therefore, preservation of SN for λv yields preservation of SN for λs.

Notation 4 *We write $a \underset{p}{\to} b$ in order to denote that p is the occurrence of the redex which is contracted. Therefore $a \underset{\epsilon}{\to} b$ means that the reduction takes place at the root. If no specification is made the reduction must be understood as a λs-reduction.*

Furthermore, we denote by $\prec$ the prefix order between occurrences of a term. Therefore if p, q are occurrences of the term a such that $p \prec q$, and we write a_p (resp. a_q) for the subterm of a at occurrence p (resp. q), then a_q is a subterm of a_p.

For example, if $a = 2\sigma^3((\lambda 1)4)$, we have $a_1 = 2$, $a_2 = (\lambda 1)4$, $a_{21} = \lambda 1$, $a_{211} = 1$, $a_{22} = 4$. Since, for instance, $2 \prec 21$, a_{21} is a subterm of a_2.

The aim of the three following lemmas is to assert that all the σ's in the last term of a derivation beginning with a λ-term must have been created at some previous step by a σ-generation and to trace the history of these closures. The first of them explains this situation for a one-step derivation where the redex is at the root:

Lemma 12 *If $a \underset{\epsilon}{\to} C[d\sigma^i e]$ then one of the following must hold:*

1. *$a = (\lambda d)e$, $C = \Box$ (a hole) and $i = 1$.*
2. *$a = C'[d'\sigma^j e]$ for some context C', some term d' and some natural j.*

Proof: We must check for every rule $a \to b$ in λs that if $d\sigma^i e$ occurs in b then $a = (\lambda d)e$ or $d'\sigma^j e$ occurs in a. We just check the interesting rules:

(σ-gen) : If $b = d\sigma^i e$ then $i = 1$ and $a = (\lambda d)e$. Otherwise $b = b_1\sigma^1 b_2$ and $d\sigma^i e$ occurs either in b_1 or in b_2, both cases are immediate since now $a = (\lambda b_1)b_2$.
(σ-λ-trans) : If $b = \lambda(d\sigma^i e)$ then $i > 1$ and $a = (\lambda d)\sigma^{i-1}e$; take $d' = \lambda d$ and $j = i - 1$. If the occurrence of $d\sigma^i e$ is in a deeper position (i.e. if $d\sigma^i e$ is a proper subterm of b), proceed as in the previous case.
(σ-app-trans) : If, for instance, $b = (c\,\sigma^i e)(d\sigma^i e)$ then $a = (c\,d)\,\sigma^i e$; take $d' = c\,d$. For deeper positions the result is straightforward. $\quad\Box$

The second lemma generalizes the previous one.

Lemma 13 *If $a \to C[d\sigma^i e]$ then one of the following must hold:*

1. *$a = C[(\lambda d)e]$ and $i = 1$.*
2. *$a = C'[d'\sigma^j e']$ where $e' = e$ or $e' \to e$.*

Proof: Induction on a, using lemma 12 for the reductions at the root. $\quad\Box$

Finally, the third lemma gives the result for arbitrary derivations.

Lemma 14 *Let* $a_1 \to \ldots \to a_n \to a_{n+1} = C[d\sigma^i e]$ *then* $a_1 = C'[d'\sigma^j e']$ *or there exists* $k \le n$ *such that* $a_k = C'[(\lambda d')e']$ *and* $a_{k+1} = C'[d'\sigma^1 e']$. *In both cases* $e' \twoheadrightarrow e$.

Proof: Induction on n and use the previous lemma. $\qquad\qquad\qquad\qquad\square$

We shall define now the notions of internal and external reductions. The intuitive meaning of an internal reduction is a reduction that takes place somewhere at the right of a σ^i operator. An external reduction is a reduction that is not internal.

We give a definition by induction. Another possibility is to define first the notion of internal and external position (occurrence) as is done in [BBLRD95].

Definition 13 *The reduction* $\xrightarrow{\text{int}}_{\lambda s}$ *is defined by the following rules:*

$$\frac{a \longrightarrow_{\lambda s} b}{c\sigma^i a \xrightarrow{\text{int}}_{\lambda s} c\sigma^i b} \qquad \frac{a \xrightarrow{\text{int}}_{\lambda s} b}{a c \xrightarrow{\text{int}}_{\lambda s} b c} \qquad \frac{a \xrightarrow{\text{int}}_{\lambda s} b}{c a \xrightarrow{\text{int}}_{\lambda s} c b}$$

$$\frac{a \xrightarrow{\text{int}}_{\lambda s} b}{\lambda a \xrightarrow{\text{int}}_{\lambda s} \lambda b} \qquad \frac{a \xrightarrow{\text{int}}_{\lambda s} b}{a \sigma^i c \xrightarrow{\text{int}}_{\lambda s} b \sigma^i c} \qquad \frac{a \xrightarrow{\text{int}}_{\lambda s} b}{\varphi_k^i a \xrightarrow{\text{int}}_{\lambda s} \varphi_k^i b}$$

Therefore, $\xrightarrow{\text{int}}_{\lambda s}$ *is the least compatible relation closed under* $\dfrac{a \longrightarrow_{\lambda s} b}{c\sigma^i a \xrightarrow{\text{int}}_{\lambda s} c\sigma^i b}$.

Definition 14 *The reduction* $\xrightarrow{\text{ext}}_{s}$ *is defined by induction. The axioms are the rules of the s-calculus and the inference rules are the following:*

$$\frac{a \xrightarrow{\text{ext}}_{s} b}{a c \xrightarrow{\text{ext}}_{s} b c} \qquad \frac{a \xrightarrow{\text{ext}}_{s} b}{c a \xrightarrow{\text{ext}}_{s} c b} \qquad \frac{a \xrightarrow{\text{ext}}_{s} b}{\lambda a \xrightarrow{\text{ext}}_{s} \lambda b} \qquad \frac{a \xrightarrow{\text{ext}}_{s} b}{a \sigma^i c \xrightarrow{\text{ext}}_{s} b \sigma^i c} \qquad \frac{a \xrightarrow{\text{ext}}_{s} b}{\varphi_k^i a \xrightarrow{\text{ext}}_{s} \varphi_k^i b}$$

Analogously, an external σ-generation is defined by the axiom $(\lambda a)b \xrightarrow{\text{ext}}_{\sigma-gen} a\sigma^1 b$ *and the five inference rules stated above where* $\xrightarrow{\text{ext}}_{s}$ *is replaced by* $\xrightarrow{\text{ext}}_{\sigma-gen}$.

Note that the inference rules $\dfrac{a \xrightarrow{\text{ext}}_{s} b}{c\sigma^i a \xrightarrow{\text{ext}}_{s} c\sigma^i b}$ and $\dfrac{a \xrightarrow{\text{ext}}_{\sigma-gen} b}{c\sigma^i a \xrightarrow{\text{ext}}_{\sigma-gen} c\sigma^i b}$ are excluded from the definitions of external s-reduction and external σ-generation, respectively. Thus, as we expected, external reductions will not occur at the right of a σ^i operator. This will permit us to write $\xrightarrow{+}_{\beta}$ instead of $\twoheadrightarrow_{\beta}$ in proposition 2.

Remark 3 *By inspecting the inference rules one checks immediately that:*

1. *If* $a \xrightarrow{\text{int}}_{\lambda s} \lambda b$ *then* $a = \lambda c$ *and* $c \xrightarrow{\text{int}}_{\lambda s} b$.
2. *If* $a \xrightarrow{\text{int}}_{\lambda s} b c$ *then* $a = d e$ *and* $((d \xrightarrow{\text{int}}_{\lambda s} b \text{ and } e = c) \text{ or } (e \xrightarrow{\text{int}}_{\lambda s} c \text{ and } d = b))$.
3. $a \xrightarrow{\text{int}}_{\lambda s} \mathsf{n}$ *is impossible.*

The following lemma is a slight but essential variation of proposition 1. A step of *external* σ-generation is studied now and the lemma ensures that we have *at least one step* of β-reduction between the corresponding s-normal forms.

Proposition 2 *Let $a, b \in \Lambda s$. If $a \xrightarrow{\text{ext}}_{\sigma-gen} b$ then $s(a) \xrightarrow{+}_{\beta} s(b)$.*

Proof: Induction on a. The lines of this proof follow the proof of proposition 1. Now, the point is that in the case $a = c\sigma^i d$, the reduction cannot take place within d because it is external, and this is the only case that forced us to consider the reflexive-transitive closure because of lemma 7.2. □

The following lemma plays a fundamental rôle in lemma 16 and hence in the Preservation theorem.

Lemma 15 (Commutation lemma) *Let $a, b \in \Lambda s$ such that $s(a) \in \beta\text{-}SN$ and $s(a) = s(b)$. If $a \xrightarrow{\text{int}}_{\lambda s} . \xrightarrow{\text{ext}}_{s} b$ then $a \xrightarrow{\text{ext}}_{s}^{+} . \xrightarrow{\text{int}}_{\lambda s} b$.*

Proof: By a careful induction on a while analysing the positions of the redexes. The detailed proof is given in the appendix. □

Lemma 16 *Let a be a strongly normalising term of the λ-calculus. For every infinite λs-derivation $a \to_{\lambda s} b_1 \to_{\lambda s} \cdots \to_{\lambda s} b_n \to_{\lambda s} \cdots$, there exists N such that for $i \geq N$ all the reductions $b_i \to_{\lambda s} b_{i+1}$ are internal.*

Proof: An infinite λs-derivation must contain infinite σ-generations, since the s-calculus is SN. The first rule must also be a σ-generation beacuse a is a pure term. We can thus write the derivation as follows:

$$a = a_1 \to_{\sigma-gen} a_1' \twoheadrightarrow_s a_2 \to_{\sigma-gen} a_2' \twoheadrightarrow_s \cdots \twoheadrightarrow_s a_n \to_{\sigma-gen} a_n' \twoheadrightarrow_s \cdots$$

By proposition 2, there must be only a finite number of external σ-generations (otherwise we can construct an infinite β-derivation contradicting the hypothesis $a \in \beta\text{-SN}$). Therefore there exists P such that for $i \geq P$ we have $a_i \xrightarrow{\text{int}}_{\sigma-gen} a_i'$. Furthermore, by proposition 1, $s(a_i) \twoheadrightarrow_{\beta} s(a_i')$ for all i, and therefore

$$a = s(a) = s(a_1) \twoheadrightarrow_{\beta} s(a_1') = s(a_2) \twoheadrightarrow_{\beta} s(a_2') = \cdots = s(a_n) \twoheadrightarrow_{\beta} s(a_n') = \cdots$$

Since a is β-SN, we conclude that $s(a_i)$ is β-SN for all i and that therefore there exists $M \geq P$ such that for $i \geq M$ we have $s(a_i) = s(a_i')$. We claim that there exists $N \geq M$ such that for $i \geq N$ all the s-rewrites are also internal. Otherwise, there would be an infinity of external s-rewrites and at least one copy of each of these external rewrites can be brought, by the Commutation Lemma, in front of a_M, and so generate an infinite s-derivation beginning at a_M, which is a contradiction. This intuitive idea can be formally stated as:

Fact: *If there exists an infinite derivation $a_M \xrightarrow{\text{ext}}_{s}^{n} b \twoheadrightarrow^m c \xrightarrow{\text{ext}}_{s} d \longrightarrow \cdots$ where all the rewrites in $b \longrightarrow\!\!\!\twoheadrightarrow c$ are either $\xrightarrow{\text{ext}}_{s}$ or $\xrightarrow{\text{int}}_{\lambda s}$, then there exists an infinite derivation $a_M \xrightarrow{\text{ext}}_{s}^{n+1} b' \longrightarrow\!\!\!\twoheadrightarrow d \longrightarrow \cdots$.*

Proof of Fact: The fact is easily proved by induction on m, using the Commutation lemma. We remark that M has been so chosen in order to satisfy the hypothesis of this lemma. □

In order to prove the Preservation Theorem we need two definitions.

Definition 15 *An infinite λs-derivation $a_1 \to \cdots \to a_n \to \cdots$ is minimal if for every step of reduction $a_i \underset{p}{\to}_{\lambda s} a_{i+1}$, every other derivation beginning with $a_i \underset{q}{\to}_{\lambda s} a'_{i+1}$ where $p \prec q$, is finite.*

The intuitive idea of a minimal derivation is that if one rewrites at least one of its steps within a subterm of the actual redex, then an infinite derivation is impossible.

Definition 16 Skeletons *are defined by the following syntax:*

$$\text{Skeletons } K ::= \mathbb{N} \mid K\,K \mid \lambda K \mid K\,\sigma^i[.] \mid \varphi_k^i K$$

The skeleton of a term a is defined by induction as follows:

$$Sk(n) = n \quad Sk(a\,b) = Sk(a)Sk(b) \quad Sk(a\,\sigma^i b) = Sk(a)\,\sigma^i[.]$$
$$Sk(\lambda a) = \lambda Sk(a) \qquad Sk(\varphi_k^i a) = \varphi_k^i Sk(a)$$

Remark 4 (Properties of the skeleton)

1. *Each occurrence of $[.]$ in the skeleton of a corresponds to an external closure of the term a (by external closure, we mean a closure that is not at the right of any other closure), and this correspondence is a bijection.*
2. *Internal closures (those which are at the right of another closure) vanish in the skeleton.*
3. *If $a \xrightarrow{\text{int}}_{\lambda s} b$ then $Sk(a) = Sk(b)$.*

Theorem 6 (Preservation of strong normalisation) *If a pure term is strongly normalising in the λ-calculus, then it is strongly normalising in the λs-calculus.*

Proof: Suppose a is a strongly normalising term in the λ-calculus, but not λs-SN. Let us consider a minimal infinite λs-derivation $\mathcal{D} : a \to a_1 \to \cdots \to a_n \to \cdots$. By lemma 16, there exists N, such that for $i \geq N$, $a_i \to a_{i+1}$ is internal. Therefore, by the previous remark, $Sk(a_i) = Sk(a_{i+1})$ for $i \geq N$. As there are only a finite number of closures in $Sk(a_N)$ and as the reductions within these closures are independent, an infinite subderivation of $\mathcal{D}$ must take place within the same and unique closure in $Sk(a_N)$ and , evidently, this subderivation is also minimal. Let us call it $\mathcal{D}'$ and let C be the context such that $a_N = C[c\,\sigma^i d]$ and $c\,\sigma^i d$ is the closure where $\mathcal{D}'$ takes place. Therefore we have:

$$\mathcal{D}' : a_N = C[c\,\sigma^i d] \xrightarrow{\text{int}}_{\lambda s} C[c\,\sigma^i d_1] \xrightarrow{\text{int}}_{\lambda s} \cdots \xrightarrow{\text{int}}_{\lambda s} C[c\,\sigma^i d_n] \xrightarrow{\text{int}}_{\lambda s} \cdots$$

Since a is a pure term, lemma 14 ensures the existence of $I \leq N$ such that

$$a_I = C'[(\lambda c')d'] \to a_{I+1} = C'[c'\sigma^1 d'] \text{ and } d' \twoheadrightarrow d.$$

But let us consider the following derivation:

$$\mathcal{D}'' : a \twoheadrightarrow a_I = C'[(\lambda c')d'] \twoheadrightarrow C'[(\lambda c')d] \to C'[(\lambda c')d_1] \to \cdots \to C'[(\lambda c')d_n] \to \cdots$$

In this infinite derivation the redex in a_I is within d' which is a proper subterm of $(\lambda c')d'$, whereas in $\mathcal{D}$ the redex in a_I is $(\lambda c')d'$ and this contradicts the minimality of $\mathcal{D}$. $\qquad \square$

5 Conclusion

There are two unsolved problems concerning λs which we want to discuss briefly.

The first of them is the confluence of λs on open terms. We remind that by open terms we mean terms which admit variables of sort **term**, namely open terms are given by the following syntax:

$$\Lambda s_{op} ::= \mathbf{V} \mid \mathbb{N} \mid \Lambda s_{op} \Lambda s_{op} \mid \lambda \Lambda s_{op} \mid \Lambda s_{op} \, \sigma^i \Lambda s_{op} \mid \varphi_k^i \Lambda s_{op} \qquad where \quad i \geq 1, \ k \geq 0$$

and where $\mathbf{V}$ stands for a set of variables, over which X, Y, ... range.

Working with open terms one loses confluence as shown by:

$$((\lambda X)Y)\sigma^1 1 \to (X\sigma^1 Y)\sigma^1 1 \qquad ((\lambda X)Y)\sigma^1 1 \to ((\lambda X)\sigma^1 1)(Y\sigma^1 1)$$

Moreover, local confluence is lost. Since $((\lambda X)\sigma^1 1)(Y\sigma^1 1) \twoheadrightarrow (X\sigma^2 1)\sigma^1(Y\sigma^1 1)$, the solution to the problem seems easy: add to λs rules obtained by orienting the equalities given by the lemmas 1 - 6. For instance, the rule corresponding to the Meta-substitution lemma (lemma 4) would solve the critical pair in our counterexample.

We believe that by adding these rules to λs we could get local confluence. But the problem of confluence of the extended system seems a difficult one. As a matter of fact it was one of the open questions in Lafont's notes.

We note here that the same problem for $\lambda \sigma$ leads to the introduction of $\lambda \sigma_{SP}$, the latter being locally confluent on open terms but confluent only on semi-closed terms (terms with variables of sort **term** but without variables of sort **substitution**)(cf. [Río93]).

The second problem we wish to present is the strong normalisation of the typed version of λs. We believe, as suggested by P.-L. Curien, that a translation, say $\mathcal{T}$, similar to the one presented in [CR91] would allow us to use the preservation of SN obtained in this paper to get the desired result. But this translation should have the additional property $\mathcal{T}(a) \twoheadrightarrow_{\lambda s} a$, and this is not the case, but possibly could be if the system considered is the extension of λs suggested by the first problem. This would permit to get SN for this extended calculus, and get as a corollary SN for the simply typed λs.

Acknowledgements. We thank Pierre-Louis Curien for his useful discussions and comments. We are also grateful for the interactions we had with Pierre Lescanne, Roel Bloo and Antonio Bucciarelli. Last but not least, we would like to thank Jeroen Krabbendam and the anonymous referees for their useful remarks.

This work was carried out under EPSRC grant GR/K25014.

References

[ACCL91] M. Abadi, L. Cardelli, P.-L. Curien, and J.-J. Lévy. Explicit Substitutions. *Journal of Functional Programming*, 1(4):375–416, 1991.

[Bar84] H. Barendregt. *The Lambda Calculus : Its Syntax and Semantics (revised edition)*. North Holland, 1984.

[BBLRD95] Z. Benaissa, D. Briaud, P. Lescanne, and J. Rouyer-Degli. λv, a calculus of explicit substitutions which preserves strong normalisation. *Personal communication*, 1995.

[CHL92] P.-L. Curien, T. Hardin, and J.-J. Lévy. Confluence properties of weak and strong calculi of explicit substitutions. Technical Report RR 1617, INRIA, Rocquencourt, 1992. To appear in the JACM.

[CHR92]   P.-L. Curien, T. Hardin, and A. Ríos. Strong Normalization of Substitutions in Proceedings of MFCS'92. In I.M. Havel and V. Koubek, editors, *Lecture Notes in Computer Science 629*, pages 209–217, Prague, 1992. Springer-Verlag.

[CR91]   P.-L. Curien and A. Ríos. Un résultat de Complétude pour les substitutions explicites. *Comptes Rendus de l'Académie des Sciences*, 312, I:471–476, 1991.

[Cur86]   P.-L. Curien. *Categorical Combinators, Sequential Algorithms and Functional Programming*. Pitman, 1986. Revised edition : Birkhäuser (1993).

[dB72]   N. de Bruijn. Lambda-Calculus notation with nameless dummies, a tool for automatic formula manipulation, with application to the Church-Rosser Theorem. *Indag. Mat.*, 34(5):381–392, 1972.

[dB78a]   N. de Bruijn. Lambda-Calculus notation with namefree formulas involving symbols that represent reference transforming mappings. *Indag. Mat.*, 40:348–356, 1978.

[dB78b]   N. G. de Bruijn. A namefree lambda calculus with facilities for internal definition of expressions and segments. Technical Report TH-Report 78-WSK-03, Department of Mathematics, Eindhoven University of Technology, 1978.

[Har89]   T. Hardin. Confluence Results for the Pure Strong Categorical Logic CCL : λ-calculi as Subsystems of CCL. *Theoretical Computer Science*, 65(2):291–342, 1989.

[HL86]   T. Hardin and A. Laville. Proof of Termination of the Rewriting System SUBST on CCL. *Theoretical Computer Science*, 46:305–312, 1986.

[HL89]   T. Hardin and J.-J. Lévy. A Confluent Calculus of Substitutions. *France-Japan Artificial Intelligence and Computer Science Symposium*, December 1989.

[Hue80]   G. Huet. Confluent Reductions: Abstract Properties and Applications to Term Rewriting Systems. *Journal of the Association for Computing Machinery*, 27:797–821, October 1980.

[KB70]   D. Knuth and P. Bendix. Simple Word Problems in Universal Algebras. In J. Leech, editor, *Computational Problems in Abstract Algebra*, pages 263–297. Pergamon Press, 1970.

[KN93]   F. Kamareddine and R. P. Nederpelt. On stepwise explicit substitution. *International Journal of Foundations of Computer Science*, 4(3):197–240, 1993.

[Mau85]   M. Mauny. *Compilation des langages fonctionnels dans les combinateurs catégoriques. Application au langage ML*. PhD thesis, Université Paris VII, Paris, France, 1985.

[Mel95]   P.-A. Melliès. Typed λ-calculi with explicit substitutions may not terminate in Proceedings of TLCA'95. *Lecture Notes in Computer Science*, 902, 1995.

[Río93]   A. Ríos. *Contribution à l'étude des λ-calculs avec substitutions explicites*. PhD thesis, Université de Paris 7, 1993.

[Zan94]   H. Zantema. Termination of term rewriting: interpretation and type elimination. *J. Symbolic Computation*, 17(1):23–50, 1994.

Appendix

Proof of lemma 15 : By induction on a. The basic case which is $a = n$ is trivial.

$a = a_1 a_2$: Since we are dealing with an internal reduction there are only two possibilities: $a_1 \xrightarrow{\text{int}}_{\lambda_s} a_1'$ or $a_2 \xrightarrow{\text{int}}_{\lambda_s} a_2'$. Let us study, for instance, the first one. Since the external reduction cannot take place at the root (because no s-rule contracts an application), there are only two cases left:

- $a = a_1 a_2 \xrightarrow{\text{int}}_{\lambda_s} a_1' a_2 \xrightarrow{\text{ext}}_s a_1'' a_2$ and $a_1 \xrightarrow{\text{int}}_{\lambda_s} a_1' \xrightarrow{\text{ext}}_s a_1''$. Let us verify that the hypotheses are valid for a_1 and a_1'' in order to use the IH.

1. If $s(a_1 a_2) = s(a_1)s(a_2)$ is β-SN, so is $s(a_1)$.
2. If $s(a_1 a_2) = s(a_1'' a_2)$ then $s(a_1)s(a_2) = s(a_1'')s(a_2)$ and so $s(a_1) = s(a_1'')$.

Therefore, by IH $a_1 \xrightarrow{\text{ext}}{}_s^+ \cdot \xrightarrow{\text{int}}{}_{\lambda s} a_1''$, and then $a_1 a_2 \xrightarrow{\text{ext}}{}_s^+ \cdot \xrightarrow{\text{int}}{}_{\lambda s} a_1'' a_2$.

$-\ a = a_1 a_2 \xrightarrow{\text{int}}{}_{\lambda s} a_1' a_2 \xrightarrow{\text{ext}}{}_s a_1' a_2'$ with $a_1 \xrightarrow{\text{int}}{}_{\lambda s} a_1'$ and $a_2 \xrightarrow{\text{ext}}{}_s a_2'$. We can simply commute the reductions: $a = a_1 a_2 \xrightarrow{\text{ext}}{}_s a_1 a_2' \xrightarrow{\text{int}}{}_{\lambda s} a_1' a_2'$.

$a = \lambda a_1$: The reduction must take place within a_1. There is no difficulty in checking the hypotheses and then using the IH.

$a = a_1\, \sigma^i a_2$: Again, as we are analysing an internal reduction, two cases arise:

$a_1 \xrightarrow{\text{int}}{}_{\lambda s} a_1'$: The external reduction can only take place within a_1 or at the root:

$-\ a = a_1\, \sigma^i a_2 \xrightarrow{\text{int}}{}_{\lambda s} a_1'\, \sigma^i a_2 \xrightarrow{\text{ext}}{}_s a_1''\, \sigma^i a_2$ and $a_1 \xrightarrow{\text{int}}{}_{\lambda s} a_1' \xrightarrow{\text{ext}}{}_s a_1''$. Let us verify that the hypotheses are valid for a_1 and a_1'' in order to use the IH.

1. We know that $s(a_1\, \sigma^i a_2)$ is β-SN. If we suppose that $s(a_1)$ is not β-SN, there exists an infinite derivation $s(a_1) \to_\beta c_1 \to_\beta \ldots \to_\beta c_n \to_\beta \ldots$ in Λ, i.e. with $c_k = s(c_k)$ for all k. Now, by lemma 7.3 we obtain an infinite derivation:

$$s(a_1\, \sigma^i a_2) = s(a_1)\{\!\{ i \leftarrow s(a_2) \}\!\} \to_\beta s(c_1)\{\!\{ i \leftarrow s(a_2) \}\!\} \to_\beta \ldots$$

which is a contradiction. Therefore $s(a_1)$ is β-SN.

2. We know that $s(a_1\, \sigma^i a_2) = s(a_1''\, \sigma^i a_2)$ and, by corollary 3, $s(a_1) \twoheadrightarrow_\beta s(a_1'')$. If $s(a_1) \xrightarrow{+}{}_\beta s(a_1'')$, then by lemma 7.3 we have

$$s(a_1)\{\!\{ i \leftarrow s(a_2) \}\!\} \xrightarrow{+}{}_\beta s(a_1'')\{\!\{ i \leftarrow s(a_2) \}\!\}.$$

But this is a contradiction since

$$s(a_1)\{\!\{ i \leftarrow s(a_2) \}\!\} = s(a_1\, \sigma^i a_2) = s(a_1''\, \sigma^i a_2) = s(a_1'')\{\!\{ i \leftarrow s(a_2) \}\!\}$$

and we are assuming that $s(a_1\, \sigma^i a_2)$ is β-SN. Therefore, $s(a_1) = s(a_1'')$. We can now apply the IH to $a_1 \xrightarrow{\text{int}}{}_{\lambda s} a_1' \xrightarrow{\text{ext}}{}_s a_1''$ to obtain $a_1 \xrightarrow{\text{ext}}{}_s^+ \cdot \xrightarrow{\text{int}}{}_{\lambda s} a_1''$, and hence $a_1\, \sigma^i a_2 \xrightarrow{\text{ext}}{}_s^+ \cdot \xrightarrow{\text{int}}{}_{\lambda s} a_1''\, \sigma^i a_2$.

$-\ a = a_1\, \sigma^i a_2 \xrightarrow{\text{int}}{}_{\lambda s} a_1'\, \sigma^i a_2 \xrightarrow{\text{ext}}{}_s b$, $a_1 \xrightarrow{\text{int}}{}_{\lambda s} a_1'$, and the external reduction takes place at the root. We study the three possible rules:

- $(\sigma\text{-}\lambda\text{-}trans)$: We have $a_1' = \lambda c'$ and $b = \lambda(c'\sigma^{i+1} a_2)$. Remark 3.1 ensures that $a_1 = \lambda c$ and $c \xrightarrow{\text{int}}{}_{\lambda s} c'$. We can then commute:

$$a = a_1\, \sigma^i a_2 = (\lambda c)\, \sigma^i a_2 \xrightarrow{\text{ext}}{}_s \lambda(c\sigma^{i+1} a_2) \xrightarrow{\text{int}}{}_{\lambda s} \lambda(c'\sigma^{i+1} a_2) = b$$

- $(\sigma\text{-}app\text{-}trans)$: We have $a_1' = c'd'$ and $b = (c'\, \sigma^i a_2)(d'\, \sigma^i a_2)$. Remark 3.2 ensures that $a_1 = cd$ and, either $c \xrightarrow{\text{int}}{}_{\lambda s} c'$ and $d = d'$, or $d \xrightarrow{\text{int}}{}_{\lambda s} d'$ and $c = c'$. In both cases we can commute as in the previous case.

- $(\sigma\text{-}dest)$: We have $a_1' = \mathbf{n}$ and this is impossible by remark 3.3.

$a_2 \to_{\lambda s} a_2'$: As in the previous case, the external reduction can take place within a_1 or at the root:

$-\ a = a_1\, \sigma^i a_2 \xrightarrow{\text{int}}{}_{\lambda s} a_1\, \sigma^i a_2' \xrightarrow{\text{ext}}{}_s a_1'\, \sigma^i a_2'$, $a_2 \to_{\lambda s} a_2'$ and $a_1 \xrightarrow{\text{ext}}{}_s a_1'$. We can commute to obtain: $a = a_1\, \sigma^i a_2 \xrightarrow{\text{ext}}{}_s a_1'\, \sigma^i a_2 \xrightarrow{\text{int}}{}_{\lambda s} a_1'\, \sigma^i a_2'$.

$-\ a = a_1\, \sigma^i a_2 \xrightarrow{\text{int}}{}_{\lambda s} a_1\, \sigma^i a_2' \xrightarrow{\text{ext}}{}_s b$, $a_2 \to_{\lambda s} a_2'$ and the external reduction takes place at the root. We study the three possible rules:

- $(\sigma\text{-}\lambda\text{-}trans)$: We have $a_1 = \lambda c$ and $b = \lambda(c\sigma^{i+1} a_2')$. We can commute:

$$a = a_1\, \sigma^i a_2 = (\lambda c)\, \sigma^i a_2 \xrightarrow{\text{ext}}{}_s \lambda(c\sigma^{i+1} a_2) \xrightarrow{\text{int}}{}_{\lambda s} \lambda(c\sigma^{i+1} a_2') = b$$

- *(σ-app-trans)* : We have $a_1 = c\,d$ and $b = (c\,\sigma^i a_2')(d\,\sigma^i a_2')$. We can commute generating two internal steps:

$$a = a_1\,\sigma^i a_2 = (c\,d)\,\sigma^i a_2 \xrightarrow{\text{ext}}_s (c\,\sigma^i a_2)(d\,\sigma^i a_2) \xrightarrow{\text{int}}_{\lambda s}$$

$$(c\,\sigma^i a_2')(d\,\sigma^i a_2) \xrightarrow{\text{int}}_{\lambda s} (c\,\sigma^i a_2')(d\,\sigma^i a_2') = b$$

- *(σ-dest)* : We have $a_1 = \text{n}$.

  If $n > i$ then $b = \text{n} - 1$. But $\text{n}\,\sigma^i a_2 \xrightarrow{\text{ext}}_s \text{n} - 1$.

  If $n < i$ then $b = \text{n}$. But $\text{n}\,\sigma^i a_2 \xrightarrow{\text{ext}}_s \text{n}$.

  If $n = i$ then $b = \varphi_0^i a_2'$. We must now consider wether $a_2 \to_{\lambda s} a_2'$ is external or internal. If it is internal we can commute to obtain:

$$a = a_1\,\sigma^i a_2 = \text{n}\,\sigma^i a_2 \xrightarrow{\text{ext}}_s \varphi_0^i a_2 \xrightarrow{\text{int}}_{\lambda s} \varphi_0^i a_2' = b\,.$$

But if it is external we must check that it is in fact an s-reduction to conclude:

$$a = a_1\,\sigma^i a_2 = \text{n}\,\sigma^i a_2 \xrightarrow{\text{ext}}_s \varphi_0^i a_2 \xrightarrow{\text{ext}}_s \varphi_0^i a_2' = b$$

giving us $a \xrightarrow{\text{ext}}{}_s^+ b \xrightarrow{\text{int}}_{\lambda s} b$. So we must prove that $a_2 \xrightarrow{\text{ext}}_{\sigma-gen} a_2'$ is impossible. It is for the treatment of this case that our additional hypotheses are necessary.

Suppose that $a_2 \xrightarrow{\text{ext}}_{\sigma-gen} a_2'$, then $\varphi_0^i a_2 \xrightarrow{\text{ext}}_{\sigma-gen} \varphi_0^i a_2'$. By proposition 2, $s(\varphi_0^i a_2) \xrightarrow{+}_\beta s(\varphi_0^i a_2')$. But by hypothesis, $s(\text{n}\,\sigma^i a_2) = s(a) = s(b) = s(\varphi_0^i a_2')$ and, since $n = i$, $s(\text{n}\,\sigma^i a_2) = s(\varphi_0^i a_2)$. This contradicts the hypothesis $s(a)$ is β-SN.

Therefore $a_2 \xrightarrow{\text{ext}}_s a_2'$ and we are done.

$a = \varphi_k^i a_1$: Two possibilities according to the position of the external reduction.

- $\varphi_k^i a_1 \xrightarrow{\text{int}}_{\lambda s} \varphi_k^i a_1' \xrightarrow{\text{ext}}_s \varphi_k^i a_1''$ and $a_1 \xrightarrow{\text{int}}_{\lambda s} a_1' \xrightarrow{\text{ext}}_s a_1''$. Let us check the hypotheses:
  1. We know that $s(\varphi_k^i a_1) = U_k^i(s(a_1))$ is β-SN. To prove that $s(a_1)$ is β-SN, suppose it is not, and use lemma 7.1 to find an infinite derivation for $s(\varphi_k^i a_1)$.
  2. We know that $s(\varphi_k^i a_1) = s(\varphi_k^i a_1'')$, hence $U_k^i(s(a_1)) = U_k^i(s(a_1''))$. We also know, by corollary 3, that $s(a_1) \twoheadrightarrow_\beta s(a_1'')$. If $s(a_1) \xrightarrow{+}_\beta s(a_1'')$, then by lemma 7.1 we have $U_k^i(s(a_1)) \xrightarrow{+}_\beta U_k^i(s(a''))$. But this is a contradiction since $U_k^i(s(a_1)) = s(\varphi_k^i a_1) = s(\varphi_k^i a_1'') = U_k^i(s(a''))$ and we are assuming that $s(\varphi_k^i a_1)$ is β-SN. Therefore, $s(a_1) = s(a_1'')$.

  We can then apply IH and conclude easily.

- $\varphi_k^i a_1 \xrightarrow{\text{int}}_{\lambda s} \varphi_k^i a_1' \xrightarrow{\text{ext}}_s b$, $a_1 \xrightarrow{\text{int}}_{\lambda s} a_1'$ and the external reduction takes place at the root. Three rules are possible:

  - *(φ-λ-trans)* : We have $a_1' = \lambda c'$ and $b = \lambda(\varphi_{k+1}^i c')$. Remark 3.1 ensures that $a_1 = \lambda c$ and $c \xrightarrow{\text{int}}_{\lambda s} c'$. We can then commute:

$$a = \varphi_k^i a_1 = \varphi_k^i(\lambda c) \xrightarrow{\text{ext}}_s \lambda(\varphi_{k+1}^i c) \xrightarrow{\text{int}}_{\lambda s} \lambda(\varphi_{k+1}^i c') = b\,.$$

  - *(φ-app-trans)* : We have $a_1' = c'd'$ and $b = (\varphi_k^i c')(\varphi_k^i d')$. Remark 3.2 ensures that $a_1 = c\,d$ and, either $c \xrightarrow{\text{int}}_{\lambda s} c'$ and $d = d'$, or $d \xrightarrow{\text{int}}_{\lambda s} d'$ and $c = c'$. In both cases we can commute as in the previous case.

  - *(φ-dest)* : We have $a_1' = \text{n}$ and this is impossible by remark 3.3. □

A Verified Implementation of Narrowing

Heinz Faßbender

Universität Ulm, Abt. Theoretische Informatik, D-89069 Ulm, Germany,
e-mail: fassbend@informatik.uni-ulm.de

Abstract. Although there exist a lot of deterministic implementations
of functional logic programming languages, up to now, none of them has
been verified. For abolishing this grievance, we present a simple implementation of the leftmost outermost narrowing strategy for a restricted
functional logic programming language and prove its correctness.

1 Introduction

In recent years many implementations of functional logic programming languages
have been developed (cf. [12] for a survey). Usually, these implementations realize
a deterministic version of a nondeterministic narrowing relation by performing a
depth-first left-to-right traversal over the introduced narrowing tree. The concept
of *narrowing* which also serves as a mechanism for solving E-unification problems
[9, 13], was firstly used to formalize the operational semantics of a functional
logic programming language in [22].

Although there exist a lot of deterministic implementations, none of them
has been verified. Up to now, there exist only the following two approaches
which are related to this sphere. In [20] a proof sketch for the correctness of a
nondeterministic implementation of a functional logic language on an abstract
machine, called *CAMEL* [19] is presented. Furthermore, a first step for proving
the correctness of the implementation in [15] has been formalized in [3] by describing the deterministic narrowing semantics by *evolving algebras* [10].

In this paper we present a deterministic graph-based implementation of a
particular narrowing relation and prove the correctness of the implementation.
Since such a kind of investigation is very costly for a complete language, we
only consider a restricted one, the programs of which are particular term rewriting systems which are called *modular tree transducers (for short: mt's)* [6].
Nevertheless, every primitive recursive tree function can be described by an mt
[6] and therefore most of the functions occurring in functional logic programs
are implemented.

An mt is a *constructor-based term rewriting system*, the rules of which have
the following structure. For every function symbol f of rank $n + 1$ and every
constructor σ of rank k there exists exactly one rule with the left hand side
$f(\sigma(x_1, \ldots, x_k), y_1, \ldots, y_n)$; the right hand side is a term over constructors, variables from the left hand side, and recursive function calls which have to fulfill some
conditions that imply the termination of the reduction relation. Furthermore, the
function symbols are partitioned into modules. For example, the set R_1 of rules

of the mt $M_1 = (F_1, \Delta_1, R_1)$, where $F_1 = \{mult^{(2)}, add^{(2)}\}$, $\Delta_1 = \{\gamma^{(1)}, \alpha^{(0)}\}$, and M_1 has two modules, is shown in Figure 1. By M_1 the multiplication and addition for natural numbers are defined in the usual primitive recursive way, where the constructors α and γ represent zero and the successor, respectively.

Module 1:	$mult(\alpha, y_1)$	$\rightarrow$ α	(1)
	$mult(\gamma(x_1), y_1)$	$\rightarrow$ $add(mult(x_1, y_1), y_1)$	(2)
Module 2:	$add(\alpha, y_1)$	$\rightarrow$ y_1	(3)
	$add(\gamma(x_1), y_1)$	$\rightarrow$ $\gamma(add(x_1, y_1))$	(4)

Fig. 1. Rules of the modular tree transducer M_1.

The restriction to mt's has the following advantages:

Since mt's are canonical, totally-defined, not strictly sub-unifiable term rewriting systems (cf. [5, 8]), every narrowing strategy is complete [5]. Hence, we can fix a narrowing strategy and there only remains the nondeterminism of the rule selection. This leads to a simple formalization of the deterministic narrowing semantics which is usually not defined, although it is implemented, and which has to be defined in this paper, because it serves as a base for the correctness proof of the implementation. Especially, the deterministic narrowing semantics formalizes a depth-first left-to-right traversal through narrowing trees which are introduced by the leftmost outermost (for short: lo-) narrowing relation.

A very simple implementation of the deterministic narrowing semantics can be formalized because of the structure of the rewrite rules' left hand sides. Since every function symbol f is strict exactly in its first argument, only this argument has to be evaluated into head normal form. After that, either one or every rule for f can be applied. Hence, a very simple mechanism has to be implemented and therefore we are not confronted with problems which occur in implementations of, e.g. lazy narrowing [11].

Nevertheless, if we restrict to mt's, then lo-narrowing is exactly the same as *lazy narrowing* [22], *needed narrowing* [1], and *outer narrowing* [24]. Thus, the presented formalisms can serve for formalizing deterministic versions of these narrowing semantics as well as for correctness proofs of their implementations in the context of a complete functional logic language.

This paper is organized in six sections, where the second section contains preliminaries. In Section 3 we recall the concept of mt. Then we introduce the deterministic lo-narrowing semantics for mt's in Section 4. In Section 5 we formalize and verify its implementation on the graph narrowing machine. Finally, Section 6 contains some concluding remarks and it indicates further research topics. Since the complete formalization of this framework is still very expensive, we leave out the formal definitions of some auxiliary functions.

2 Preliminaries

We recall and collect some notations, basic definitions, and terminology which will be used in the rest of the paper.

For $j \in \mathbb{N}$, $[j]$ denotes the set $\{1, \ldots, j\}$; thus $[0] = \emptyset$. For a set A, $\mathcal{P}(A)$ is the *set of finite subsets* of A and A^* denotes the set of words over A. The empty word is denoted by Λ. If A is finite, then $card(A)$ denotes the cardinality of A.

A finite indexed set Ω, where every element σ in Ω is indexed by its unique rank $rank(\sigma) \in \mathbb{N}$, is called a *ranked alphabet*. The subset $\Omega^{(m)}$ of Ω consists of all symbols of rank m ($m \geq 0$). We write an element $\sigma \in \Omega^{(k)}$ as $\sigma^{(k)}$.

In the rest of the paper we let $\mathcal{V}$ denote a fixed enumerable set of *variables* which is divided into three disjoint sets $X = \{x_1, x_2, \ldots\}$, $Y = \{y_1, y_2, \ldots\}$, and $LV = \{z_1, z_2, \ldots\}$ of *recursion variables*, *parameter variables*, and *logic variables*, respectively.

Let Ω be a ranked alphabet and let S be an arbitrary set. Then the set of *terms over Ω indexed by S*, denoted by $T\langle\Omega\rangle(S)$, is defined inductively as follows: (i) $S \subseteq T\langle\Omega\rangle(S)$ and (ii) for every $f \in \Omega^{(k)}$ with $k \geq 0$ and $t_1, \ldots, t_k \in T\langle\Omega\rangle(S)$: $f(t_1, \ldots, t_k) \in T\langle\Omega\rangle(S)$.

For a term $t \in T\langle\Omega\rangle(\mathcal{V})$, the set of *occurrences of t*, denoted by $O(t)$, is a subset of $\mathbb{N}^*$ and it is defined inductively on the structure of t as follows: (i) If $t \in \mathcal{V} \cup \Omega^{(0)}$, then $O(t) = \{\Lambda\}$ and (ii) if $t = f(t_1, \ldots, t_n)$, where $f \in \Omega^{(n)}$, $n > 0$, and for every $i \in [n] : t_i \in T\langle\Omega\rangle(\mathcal{V})$, then $O(t) = \{\Lambda\} \cup \bigcup_{i \in [n]} \{iu \mid u \in O(t_i)\}$.

The *prefix order* and the *lexicographical order* on $O(t)$ are denoted by $<$ and $<_{lex}$, respectively. Their reflexive closures are denoted by $\leq$ and $\leq_{lex}$, respectively. The minimal element with respect to $\leq_{lex}$ in a subset S of $O(t)$ is denoted by $min_{lex} S$. For a term $t \in T\langle\Omega\rangle(\mathcal{V})$ and an occurrence u of t, t/u and $t[u]$ denote the *subterm of t at u* and the *label of t at u*, respectively. We use $\mathcal{V}(t)$ to denote the set of variables occurring in t. Finally, we define $t[u \leftarrow s]$ as the term t in which we have replaced the subterm at occurrence u by the term s.

Let Ω be a ranked alphabet. An assignment $\varphi : \mathcal{V} \to T\langle\Omega\rangle(\mathcal{V})$, where the set $\{x \mid \varphi(x) \neq x\}$ is finite, is called a $(\mathcal{V}, \Omega)$-*substitution*. The set $\{x \mid \varphi(x) \neq x\}$ is denoted by $\mathcal{D}(\varphi)$ and it is called the *domain of φ*. If $\mathcal{D}(\varphi) = \{x_1, \ldots, x_n\}$, then φ is represented as $[x_1/\varphi(x_1), \ldots, x_n/\varphi(x_n)]$. If $\mathcal{D}(\varphi) = \emptyset$, then φ is denoted by $\varphi_\emptyset$. The set $\bigcup_{x \in \mathcal{D}(\varphi)} \mathcal{V}(\varphi(x))$ is denoted by $\mathcal{I}(\varphi)$ and it is called the set of *variables introduced by φ*. The set of $(\mathcal{V}, \Omega)$-substitutions is denoted by $Sub(\mathcal{V}, \Omega)$. The *composition* of two substitutions φ and ψ, denoted by $\varphi \circ \psi$, is the substitution which is defined by $\psi(\varphi(x))$ for every $x \in \mathcal{V}$. Let $V \subset \mathcal{V}$. The *restriction of φ to V* is denoted by $\varphi|_V$.

The set of all functions from A into B is denoted by $[A \to B]$. If two functions $f, g \in [A \to B]$ are different only for finitely many elements $a_1, \ldots, a_n$ in A and if for every $j \in [n] : g(a_j) = b_j$, then we denote g by $f[a_1/b_1, \ldots, a_n/b_n]$. A function $f : A \to B$ is denoted by $f_\emptyset$, if for every $a \in A : f(a)$ is undefined. If h is a function from A into A, then we write h^k for the $k > 0$-times composition from h with itself, i.e., $h^1 = h$ and $h^{k+1}(x) = h(h^k(x))$ for every $x \in A$.

3 Modular Tree Transducer

In this section we recall the notion of modular tree transducer from [6].

Definition 3.1 An *mt* is a tuple (F, mod, Δ, R), where

- F is a ranked alphabet of function symbols, such that $F^{(0)} = \emptyset$.
- $mod : F \rightarrow \mathbb{N}$ is a mapping.
- Δ is a ranked alphabet of constructors.
- R is the finite set of rewrite rules; for every $f \in F^{(n+1)}$ and $\sigma \in \Delta^{(k)}$ with $n, k \geq 0$, there is exactly one (f, σ)-*rule* in R, i.e., a rule of the form

$$f(\sigma(x_1, \ldots, x_k), y_1, \ldots, y_n) \rightarrow t,$$

where $t \in T\langle F \cup \Delta \rangle(\{x_1, \ldots, x_k\} \cup \{y_1, \ldots, y_n\})$, such that for every function symbol $g \in F$ which occurs in t, the following two conditions hold:
(i) $mod(g) \geq mod(f)$
(ii) if $mod(g) = mod(f)$, then the first argument t_1 of a function call of the form $g(t_1, \ldots, t_m)$ in t is in the set $\{x_1, \ldots, x_k\}$.

An example for an mt is presented in Section 1. Since the mapping *mod* has no influence in our further considerations, we always denote M by (F, Δ, R).

Remark that every mt is a ctn-trs [8], because it is a canonical [6], totally defined [6], and not strictly sub-unifiable term rewriting system which follows from [5] and the definition of the rewrite rules' left hand sides.

In the rest of the paper we assume that the considered mt $M = (F, \Delta, R)$ has the following structure: $F = \{f_1, \ldots, f_r\}$, $\Delta = \{\sigma_1, \ldots, \sigma_\rho\}$, and

$$R = \{f_i(\sigma_j(x_1, \ldots, x_{rank(\sigma_j)}), y_1, \ldots, y_{rank(f_i)-1}) \rightarrow r_{ji} \mid j \in [\rho], i \in [r]\}.$$

Furthermore, we assume that the rewrite rules are enumerated by the bijection $\pi : R \rightarrow [card(R)]$ such that the (f_i, σ_j)-rule has the number $(i-1) \cdot \rho + j$.

4 Deterministic LO-Narrowing Semantics

The deterministic lo-narrowing semantics for the mt M is formalized by the *lo-narrowing function for M* which serves as the base for the correctness proof of the implementation in Section 5. The lo-narrowing function formalizes a depth-first left-to-right traversal over trees which are introduced by the nondeterministic *lo-narrowing relation* that is formalized first.

LO-Narrowing Relation

The sentential forms of the lo-narrowing relation are pairs (t, φ) which consist of a term $t \in T\langle F \cup \Delta \rangle(LV)$ and an (LV, Δ)-substitution φ which is the composition of the applied substitutions along the whole derivation. By allowing only logic variables in sentential forms we omit conflicts with variables occurring in rewrite rules. Because of the particular structure of the rewrite rules' left hand sides, the *lo-narrowing occurrence* is the minimal position of a function symbol f with respect to $\leq_{lex}$, the first argument *rec_arg* of which is in head normal form. In the case that *rec_arg* is a logic variable, every rule for f can be applied. Since the lo-narrowing function has to compute all those different lo-narrowing steps in a fixed order, every step is formalized by a seperate lo-narrowing relation which is indexed by the number of the applied rule. Then the lo-narrowing relation is simply formalized as the union of all indexed lo-narrowing relations.

Definition 4.1 Let $t \in T\langle F \cup \Delta \rangle(LV) \backslash T\langle \Delta \rangle(LV)$ and $\psi \in Sub(LV, \Delta)$.

- The *lo-narrowing occurrence in* t, denoted by $lo-naro(t)$, is the occurrence $min_{lex}\{u \in O(t) \mid t[u] \in F \text{ and } t[u1] \in \Delta \cup LV\}$.
- Let $l \rightarrow r$ be the (f_i, σ_j)-rule in R and $k = \pi(l \rightarrow r)$. The k-lo-narrowing relation $\overset{lo}{\leadsto}_{M,k}$ is defined as follows: For every $t' \in T\langle F \cup \Delta \rangle(LV)$ and $\psi' \in Sub(LV, \Delta)$: $(t, \psi) \overset{lo}{\leadsto}_{M,k} (t', \psi')$, if the following three conditions hold:
  1. $t/lo - naro(t)$ and l are unifiable with most general unifier φ.
  2. $t' = \varphi(t[lo - naro(t) \leftarrow r])$.
  3. $\psi' = \psi \circ (\varphi|_{V(t)})$.
- The *lo-narrowing relation* $\overset{lo}{\leadsto}_M$ is the binary relation $\bigcup_{i \in [card(R)]} \overset{lo}{\leadsto}_{M,i}$.

LO-Narrowing Function

In the case that more than one rule can be applied, the lo-narrowing function applies those rules in the order of their numbers. In particular, the sentential form has to contain some information about the rules which are not applied yet. For this purpose, a sentential form of the lo-narrowing function contains a word of triples. The first two components of a triple form a sentential form of the lo-narrowing relation and the third component includes the numbers of rules which are not applied yet. Triples, the third component of which is the empty set, are not produced by the lo-narrowing function. Hence, we only consider triples from the following set

$$Triple(M) := T\langle F \cup \Delta \rangle(LV) \times Sub(LV, \Delta) \times (\mathcal{P}([card(R)]) \backslash \emptyset).$$

The computation of the third component of a triple is realized by the function

$$app\_rules_M : T\langle F \cup \Delta \rangle(LV) \backslash T\langle \Delta \rangle(LV) \rightarrow \mathcal{P}([card(R)]) \backslash \emptyset$$

$app\_rules_M(t) = \textbf{if } t[lo - naro(t)] = f_i$
$\qquad\qquad \textbf{then if } t[lo - naro(t)1] = \sigma_j \in \Delta \textbf{ then } \{(i - 1) \cdot \rho + j\}$
$\qquad\qquad \textbf{if } t[lo - naro(t)1] \in LV \textbf{ then } \{(i - 1) \cdot \rho + j \mid j \in [\rho]\}.$

Furthermore, a sentential form includes the *set of results* which have been already computed during the depth-first left-to-right traversal over the lo-narrowing tree. A result is a sentential form (t, φ) of the lo-narrowing relation, such that t does not contain any function symbol.

$$Results(M) := \mathcal{P}(T\langle \Delta \rangle(LV) \times Sub(LV, \Delta)).$$

The *initial sentential forms* are computed by the function

$$init\_sf_M : T\langle F \cup \Delta\rangle(LV) \rightarrow Triple(M)^* \times Results(M)$$
$$init\_sf_M(t) = \textbf{if } t \in T\langle\Delta\rangle(LV) \textbf{ then } (\Lambda, \{(t, \varphi_\emptyset)\})$$
$$\textbf{if } t \notin T\langle\Delta\rangle(LV) \textbf{ then } ((t, \varphi_\emptyset, app\_rules_M(t)), \emptyset).$$

In the case that the input term t does not contain any function symbol, the computation of the lo-narrowing function is already finished. Hence, the initial sentential form contains the empty word Λ and the pair $(t, \varphi_\emptyset)$ is the single result. Otherwise, the initial triple is produced and the set of results is empty.

The lo-narrowing function applies the rule in the third component of the rightmost triple with the minimal number. Furthermore, it deletes triples, the third component of which are the empty set. That is the reason, why it has to consider four different cases.

Definition 4.2 The *lo-narrowing function for M* is the function

$$det\_lo\_nar_M : Triple(M)^* \times Results(M) \rightarrow Triple(M)^* \times Results(M)$$

which is defined as follows:
Let $tri \in Triple(M)^*, (t, \varphi, set) \in Triple(M), res \in Results(M), min$ be the minimal element in set, and $(t, \varphi) \overset{lo}{\leadsto}_{M,min} (t', \varphi')$.

$det\_lo\_nar_M(tri\ (t, \varphi, set), res) =$
if $card(set) = 1$
then if $t' \in T\langle\Delta\rangle(LV)$ **then** $(tri, res \cup \{(t', \varphi')\})$
        if $t' \notin T\langle\Delta\rangle(LV)$ **then** $(tri\ (t', \varphi', app\_rules_M(t')), res)$
if $card(set) > 1$
then if $t' \in T\langle\Delta\rangle(LV)$ **then** $(tri\ (t, \varphi, set\backslash\{min\}), res \cup \{(t', \varphi')\})$
        if $t' \notin T\langle\Delta\rangle(LV)$ **then** $(tri\ (t, \varphi, set\backslash\{min\})(t', \varphi', app\_rules_M(t')), res)$

Since the previous definition is very technical, we illustrate it by the following short example.

Example 4.3 The first and second result for the input term $t = add(z_1, z_1)$ are computed by the following three applications of $det\_lo\_nar_{M_1}$ (cf. Figure 1 for the rewrite rules of the mt M_1) to $init\_sf_{M_1}(t) = ((add(z_1, z_1), \varphi_\emptyset, \{3, 4\}), \emptyset)$

$det\_lo\_nar^3_{M_1}((add(z_1, z_1), \varphi_\emptyset, \{3, 4\}), \emptyset)$
$= det\_lo\_nar^2_{M_1}((add(z_1, z_1), \varphi_\emptyset, \{4\}), \{(\alpha, [z_1/\alpha])\})$
$= det\_lo\_nar_{M_1}((\gamma(add(z_2, \gamma(z_2))), [z_1/\gamma(z_2)], \{3, 4\}), \{(\alpha, [z_1/\alpha])\})$
$= ((\gamma(add(z_2, \gamma(z_2))), [z_1/\gamma(z_2)], \{4\}), \{(\alpha, [z_1/\alpha]), (\gamma(\gamma(\alpha)), [z_1/\gamma(\alpha), z_2/\alpha])\})$

5 Implementation of the LO-Narrowing Function

In this section we present the implementation of $det\_lo\_nar_M$ on an abstract machine which is called *graph narrowing machine*. This machine can be considered as an extension of a restricted G-machine [2, 14] by adding mechanisms for logic variables, unification, and backtracking. We start with the introduction of the configurations and instructions of the graph narrowing machine.

Configurations and Instructions of the Graph Narrowing Machine

A configuration of the graph narrowing machine consists of the following components:

- The *program store* is a partial function $ps : PA \rightarrow Instr$ from the *set of program addresses* $PA = \mathbb{N}$ into the *the set of instructions Instr* which is the union of the following sets, where $n, k, i \in \mathbb{N}$:

$$\{FNODE(f, n + 1) \mid f \in F^{(n+1)}\}, \{CNODE(\sigma, k) \mid \sigma \in \Delta^{(k)}\},$$
$$\text{and } \{RECVAR\ i, PAR\ i, VAR\ i, INIT\_RESET\}.$$

  The program store contains a program which realizes the bottom-up creation of the graph representations for the input term $t \in T\langle F \cup \Delta\rangle(LV)$ and right hand sides of rewrite rules in R.

- The *instruction pointer* is an element $ip \in PA \cup \{Gmode, Stop\}$.
  During the graph creation, ip is the program address of the instruction which has to be executed next. If the computation is finished, i.e., the complete set of results is computed, then $ip = Stop$. Otherwise, $ip = Gmode$.

- The *graph* is a partial function $G : Adr \rightarrow GNodes$ from the *set of graph addresses* $Adr = \mathbb{N}$ into the *set of graph nodes* $GNodes = FUN - nodes \cup CON - nodes \cup VAR - nodes$, where the *sets of function nodes, constructor nodes*, and *variable nodes* are defined as follows ($n, k \geq 0$):

  - $FUN - nodes = \{\langle FUN, f, reca, a_1, \ldots, a_n\rangle \mid f \in F^{(n+1)},$
    $$reca, a_1 \ldots, a_n \in Adr\}$$
  - $CON - nodes = \{\langle CON, \sigma, a_1, \ldots, a_k\rangle \mid \sigma \in \Delta^{(k)}, a_1, \ldots, a_k \in Adr\}$
  - $VAR - nodes = \{\langle VAR, a\rangle \mid a \in Adr \cup \{?\}\}$

The graph represents subterms of the first components of the triples in the current sentential form of the lo-narrowing function which still have to be evaluated. In the graph representation, every logic variable is represented by exactly one VAR-node. This is realized by the mechanism of *sharing* [14]. If the second component of such a VAR-node contains a "?", then the corresponding logic variable is not bound. Otherwise, the second component points to the graph representation of the corresponding logic variable's binding.

- The *address stack* is an element $rs \in (Adr \cup CPs)^*$, where the *set of choice points* CPs is the cartesian product $Adr \times \mathbb{N}^4$.
  Every square of the address stack which is not a choice point, refers to the graph representation of a subterm which still has to be evaluated. Those graph representations are evaluated in the order of the references on the address stack from the top to the bottom. Furthermore, the address stack contains choice points which save information that is needed during backtracking, in a similar way as in the WAM [23]. A choice point has the form

$$(adr, alt, sep, ltr, lot),$$

  where

  - adr refers to the term's graph representation which will be evaluated next.
  - alt is the number of the current alternative.
  - sep is the saved environment pointer.
  - ltr is the saved length of the trail.
  - lot is the saved length of the output tape.

- The *stack of substitutions* is an element $ss \in (Adr \cup \{?\})^*$.
  For every logic variable z_i which occurs in the input term t, the i-th square in the stack of substitutions refers to the graph representation of the binding of z_i. Thus, together with the graph and the trail, ss represents the substitution φ in the second component of the rightmost triple in the current sentential form of the lo-narrowing function. Since logic variables which occur in t, are from particular interest, the stack of substitutions is seperated from the trail.
- The *trail* is an element $tr \in Adr^*$.
  It refers to the graph representations of the bindings of bounded variables and it is used for unbinding logic variables during backtracking (cf. [23]).
- The *environment pointer* is an element $ep \in \mathbb{N}$.
  In the case that the address stack contains at least one choice point, the environment pointer points to the position in the address stack which refers to the graph, the copy of which is evaluated just now. Otherwise, it points to the bottom of the address stack.
- The *data stack* is an element $ds \in Adr^*$.
  The data stack is only used for the bottom-up creation of the graph.
- The *output tape* is an element $ot \in (\Delta \cup Adr)^*$.
  The output tape contains the evaluated part of the first component of a result which is evaluated just now. A logic variable is represented on the output tape by the address of the graph representation of its binding.

The *set of configurations* of the graph narrowing machine, denoted by $Conf$, is the cartesian product:

$$[PA \rightarrow Instr] \times (PA \cup \{Gmode, Stop\}) \times (Adr \cup CPs)^* \times [Adr \rightarrow GNodes] \times$$
$$(Adr \cup \{?\})^* \times Adr^* \times \mathbb{N} \times Adr^* \times (\Delta \cup Adr)^*.$$

One configuration is written in the following form

$$(ps, ip, rs, G, ss, tr, ep, ds, ot) \in Conf,$$

where we use the following conventions: The stacks are assumed to grow to the left, where a stack s with m elements is written in the form $s[m] : \ldots : s[1]$. In this case, m is denoted by $length(s)$, and for $l \geq r$, we denote the part $s[l] : s[l-1] : \ldots : s[r]$ of s by $s[l..r]$. The notation $s[i/a]$ means that the value of s at position i changes to a. As usual, the trail and the output tape grow to the right. If we want to separate the n rightmost or leftmost elements $e_1, \ldots, e_n$ from the rest e' of a stack, the trail, or the output tape, we write $e' : e_n : \ldots : e_1$ or $e_1 : \ldots : e_n : e'$, respectively.

Initial Configuration, Translation Schemes, and Graph Creation

In the whole section we assume that the input term $t \in T\langle F \cup \Delta\rangle(LV)$ which includes exactly the logic variables $z_1, \ldots, z_k$, is fixed. Because of lack of space, the definitions of the auxiliary functions which are used for formalizing the implementation, are omitted.

The *initial configuration* of the graph narrowing machine for M and t, denoted by $input(M, t)$, is the following configuration

$$(trans(R, t), 1, \Lambda, G_\emptyset, \underbrace{? : \ldots : ?}_{k-times}, \Lambda, 0, \Lambda, \Lambda).$$

The machine program in the program store is produced by the translation schemes (Figure 2). In the definition of the translation schemes we produce symbolic program addresses and we assume that the produced program is transformed by a loading program into a correct program of the graph narrowing machine. Furthermore, in the initial configuration, the instruction pointer contains the program address 1, where the code for the creation of the graph representation for t starts, the environment pointer is 0, and for every logic variable in t, the stack of substitutions contains a "?". All the other components are empty.

Starting from the initial configuration, the graph representation of t is created by the instructions $FNODE(f, n+1)$, $CNODE(\sigma, k)$, and $VAR\ i$ (Figure 3) which are produced by the translation scheme etr (Figure 2). These instructions work as follows:

- $FNODE(f, n+1)$ produces the graph representation of a function call at a new address $nadr$ by replacing the graph addresses $reca, a_1, \ldots, a_n$ of its $n+1$ arguments on the top of the data stack by $nadr$ and by creating the node $\langle FUN, f, reca, a_1, \ldots, a_n\rangle$ at $nadr$ in G.
- $CNODE(\sigma, k)$ works very similar to $FNODE(f, n+1)$. It replaces the addresses of k successors on the top of the data stack by $nadr$ and it produces a σ-node at $nadr$.
- If $ss[i]$ contains a "?", then $VAR\ i$ produces a $\langle VAR, ?\rangle$-node at a new graph address $nadr$ and it writes $nadr$ into $ss[i]$ and on top of the data stack. Otherwise, i.e., $ss[i]$ contains a graph address adr, $VAR\ i$ only writes adr on top of the data stack.

$$
\begin{array}{ll}
trans(R,t) = & etr(t)\ INIT\_RESET; \\
\text{for every } f_i \in F, \sigma_j \in \Delta: & ca(f_i,\sigma_j): etr(r_{ji})\ INIT\_RESET;
\end{array}
$$

$$
\begin{array}{rl}
etr(f_i(t,t_1,\ldots,t_n)) = & etr(t)\ etr(t_1)\cdots etr(t_n)\ FNODE(f_i, n+1); \\
etr(\sigma_j(t_1,\ldots,t_k)) = & etr(t_1)\cdots etr(t_k)\ CNODE(\sigma_j, k); \\
etr(y_i) = & PAR\ i; \\
etr(x_i) = & RECVAR\ i; \\
etr(z_i) = & VAR\ i;
\end{array}
$$

Fig. 2. Translation schemes.

The semantics of an instruction $inst \in Instr$ is a partial function

$$
C[\![\ inst\ ]\!] : Conf \to Conf.
$$

In its definition in Figure 3 as well as in the definition of the transition function in Figure 4, we underline the components of the configurations which are changed. The creation of new graph addresses is realized by the auxiliary function new which is formalized as a global variable, i.e., $new(k)$ yields k new graph addresses and new is also increased by k. In the initial configuration, new is set to 0.

If the creation of the graph representation of t is finished, then the data stack contains exactly one square which refers to this graph representation. The following instruction $INIT\_RESET$ (Figure 3) which is produced by the translation scheme $trans$ (Figure 2), transfers the reference from the data stack to the address stack.

Transition Function

The one-step semantics of the graph narrowing machine is formalized by the *transition function*

$$
transform : Conf \times Results(M) \to Conf \times Results(M)
$$

which also considers a set of results (cf. Section 4) for being consistent to the lo-narrowing function.

In the case that the instruction pointer contains a program address ip, such that the program store is defined for ip, $transform$ is simply defined by the semantics of the instruction (Figure 3) at ip in the program store, the set of results is not changed, and the address stack rs has one of the following two forms:

1. If the graph representation of t is constructed, then rs is empty.
2. If the graph representation for a rewrite rule's right hand side is constructed, then rs contains at least the top element a which refers to a function call.

Assumptions: $conf = (ps, ip, rs, G, ss, tr, ep, ds, ot)$

$ip \in PA$, one of the following two cases, and let $\{nadr\} = new(1)$ in:

1. $rs = \Lambda$

2. $rs = a : rs'$ and $G(a) \in FUN - nodes$ and $adr = imp\_fc(G, a)$
   and $G(adr) = \langle FUN, g, recad, adr_1, \ldots, adr_m \rangle$ and $rec\_adr = deref(recad, G)$
   and $G(rec\_adr) = \langle CON, \delta, radr_1, \ldots, radr_k \rangle$

$C \llbracket FNODE(f, n+1) \rrbracket (conf) =$ **if** $ds = a_n : \cdots : a_1 : reca : ds'$
   then $(ps, \underline{ip+1}, rs, G[nadr/\langle FUN, f, reca, a_1, \ldots, a_n \rangle], ss, tr, ep, \underline{nadr : ds'}, ot)$

$C \llbracket CNODE(\sigma, k) \rrbracket (conf) =$ **if** $ds = a_k : \cdots : a_1 : ds'$
   then $(ps, \underline{ip+1}, rs, G[nadr/\langle CON, \sigma, a_1, \ldots, a_k \rangle], ss, tr, ep, \underline{nadr : ds'}, ot)$

$C \llbracket VAR\ i \rrbracket (conf) =$
if $ss[i] = "?"$ **then** $(ps, \underline{ip+1}, rs, G[nadr/\langle VAR, ? \rangle], ss[i/nadr], tr, ep, \underline{nadr : ds}, ot)$
if $ss[i] = adr$ **then** $(ps, \underline{ip+1}, rs, G, ss, tr, ep, \underline{adr : ds}, ot)$

$C \llbracket RECVAR\ i \rrbracket (conf) = (ps, \underline{ip+1}, rs, G, ss, tr, ep, \underline{radr_i : ds}, ot)$

$C \llbracket PAR\ i \rrbracket (conf) = (ps, \underline{ip+1}, rs, G, ss, tr, ep, \underline{adr_i : ds}, ot)$

$C \llbracket INIT\_RESET \rrbracket (conf) =$ **if** $ds = gadr$ **then**
if $rs = \Lambda$ **then** $(ps, \underline{Gmode}, \underline{gadr}, G, ss, tr, ep, \underline{\Lambda}, ot)$
if $rs = a : rs'$ **then if** $a = adr$ **then** $(ps, \underline{Gmode}, \underline{gadr : rs'}, G, ss, tr, ep, \underline{\Lambda}, ot)$
   if $a \neq adr$ **then** $(ps, \underline{Gmode}, rs, \underline{G^*}, ss, tr, ep, \underline{\Lambda}, ot)$
   where $G^* = set\_on\_new\_recarg(G, a, adr, gadr)$

Fig. 3. Semantics of the instructions.

In the case that the instruction pointer contains $Gmode$, we distinguish the following five cases, where, in Cases 1 - 3, the graph address adr results from the application of the auxiliary function $deref$ to the graph address a in the topmost square of the address stack. $deref$ realizes the mechanism of dereferencing.

1. If adr refers to a constructor node for σ, then σ is written to the output tape and a is replaced on the address stack by the successors of the σ-node.

2. If adr refers to a $\langle VAR, ? \rangle$-node, then adr is written to the output tape and a is deleted on the address stack.

3. If adr refers to a function node fn, then the evaluation of the represented function call is implemented. For that purpose, the graph address ga of the graph representation of the lo-narrowing occurrence is computed by the auxiliary function imp_fc. Depending on the root $root$ of the recursion argument of the function symbol f at ga and the position of ga with respect to adr, we distinguish the following three cases:

   (a) If $root$ is a constructor σ and $adr = ga$, then the graph representation of the right hand side of the (f, σ)-rule has to be created. For this pur-

$transform((ps, ip, rs, G, ss, tr, ep, ds, ot), res) =$

if $ip \in PA$ and $ps(ip)$ is defined **then**

  $(\mathcal{C}[\![\, ps(ip)\,]\!]\, (ps, ip, rs, G, ss, tr, ep, ds, ot), res)$ (cf. Figure 3)

let $\{nadr\} = new(1)$ in

if $ip = Gmode$ and $rs = a : rs'$ and $adr = deref(a, G)$ **then**

  Case 1: if $G(adr) = \langle CON, \sigma, a_1, \ldots, a_k \rangle$ **then**
       $((ps, Gmode, \underline{a_1 : \ldots : a_k : rs'}, G, ss, tr, ep, ds, \underline{ot : \sigma}), res)$

  Case 2: if $G(adr) = \langle VAR, ? \rangle$ **then**
       $((ps, Gmode, \underline{rs'}, G, ss, tr, ep, ds, \underline{ot : adr}), res)$

     if $G(adr) \in FUN - nodes$ **then**

       if $G(imp\_fc(G, adr)) = \langle FUN, f, reca, a_1, \ldots, a_n \rangle$
          and $reca' = deref(reca, G)$ **then**

          if $G(reca') = \langle CON, \sigma, a_1', \ldots, a_k' \rangle$ **then**

  Case 3a):   if $imp\_fc(G, adr) = adr$ **then**
              $((ps, \underline{ca(f, \sigma)}, rs, G, ss, tr, ep, ds, ot), res)$

  Case 3b):   if $imp\_fc(G, adr) \neq adr$ **then**
              $((ps, \underline{ca(f, \sigma)}, \underline{nadr : rs'}, \underline{copy(G, a, nadr)}, ss, tr, ep, ds, ot), res)$

  Case 3c):   if $G(reca') = \langle VAR, ? \rangle$ **then**
              $((ps, Gmode, \underline{rs^*}, \underline{G^*}, ss, \underline{tr : reca'}, length(rs), ds, ot), res)$
           where $rs^* = nadr : (a, 1, ep, length(tr), length(ot)) : rs'$
              $G^* = replace(copy(G, a, nadr), reca', \sigma_1)$

if $ip = Gmode$ and $rs = (a, i, sep, ltr, lot) : rs'$ and $ep \geq 1$ **then**

       if $ep^* = next\_env(rs, ep)$ **then**

  Case 4a): if $ep^* = 0$ **then**
              $((ps, Gmode, rs, G, ss, tr, \underline{ep^*}, ds, ot), res)$

  Case 4b): if $ep^* > 0$ **then**
              $((ps, Gmode, \underline{nadr : rs}, \underline{copy(G, rs[ep^*], nadr)}, ss, tr, \underline{ep^*}, ds, ot), res)$

if $ip = Gmode$ and $(rs = \Lambda$ or $(rs = (a, i, sep, ltr, lot) : rs'$ and $ep = 0))$ **then**

  Case 5: $(backtrack(ps, ip, rs, G, ss, tr, ep, ds, ot), \underline{res \cup \{(\nu(t), \varphi \circ \nu)\}})$

     where $t = StoT(ot, G, ss, \Lambda)$
        $\varphi = make\_sub(tr, G, ss, \varphi_\emptyset)$
        $\nu = rename((\mathcal{D}(\varphi) \cup \mathcal{I}(\varphi)) \backslash \{z_1, \ldots, z_{length(ss)}\}, length(ss) + 1, \varphi_\emptyset)$

Fig. 4. Transition function $transform$.

pose, the symbolic program address $ca(f, \sigma)$, where the creation starts, is written to the instruction pointer. The code for the creation of the graph representation is produced by the translation scheme etr (Figure 2). This code may consist of the instructions $FNODE(f, n + 1)$, $CNODE(\sigma, k)$, $PAR\ i$, and $RECVAR\ i$, where the latter two instructions work as follows:

- $PAR\ i$ pushes the address of the graph representation of the i-th parameter, i.e., the $i + 1$-th argument, onto the data stack.
- $RECVAR\ i$ pushes the address of the graph representation of the i-th successor of the recursion argument, i.e., the first argument, onto the data stack.

If the creation of the graph representation gr for the right hand side of the (f, σ)-rule is finished, then the data stack contains the address of gr. Next, the instruction $INIT\_RESET$ which is produced by the translation scheme $trans$, has to be executed. In this case, it deletes the topmost address on the address stack. Furthermore, it transfers the address from the data stack to the address stack.

(b) If $root$ is a constructor σ and $adr \neq ga$, then the machine behaves very similar to the previous case with the following differences. Before the graph representation of the right hand side of the (f, σ)-rule is created, the graph at adr is copied by the auxiliary function $copy$ to a new address $nadr$ and a is replaced on the address stack by $nadr$. Of course, this behaviour of the machine is very inefficient, but it is necessary for being consistent to the lo-narrowing function. Furthermore, in this case, the instruction $INIT\_RESET$ replaces the graph at the copy of ga by the produced graph representation of the right hand side. This is realized by the auxiliary function $set\_on\_new\_recarg$.

(c) If $root$ is a $\langle VAR, ? \rangle$-node, then a is replaced on top of the address stack by a choice point which includes the address a. Furthermore, the configuration is initialized for the computation of the first alternative. For this purpose, the graph at address a is copied to a new address $nadr$, $nadr$ is pushed on the address stack, the auxiliary function $replace$ replaces $root$ by a σ_1-node with $\langle VAR, ? \rangle$-nodes as successors, the trail is lengthend by the address of $root$, and the environment pointer points to the new choice point. After this, the configuration has the same form as in one of the two previous cases. Hence, the function call at the leftmost outermost narrowing occurrence is evaluated.

4. If the topmost square of the address stack is a choice point and the environment pointer is greater than 0, then the evaluation of graphs which are refered by a square on top of the address stack is finished. Thus, a graph has to be evaluated which is refered by a square below the topmost choice point. The position of this square in the address stack is computed by the auxiliary function $next\_env$. This function simply decrements the environment pointer ep by one, if the square of the address stack at position $ep - 1$ includes a graph address. Otherwise, i.e., the square of the address stack at

position $ep - 1$ includes a choice point cp, $next\_env$ is called recursively with the saved environment pointer sep in the lowest choice point above cp. The computation of sep is realized by the auxiliary function $sep\_in\_upper\_CP$. Depending on the value ep^* of the function call of $next\_env$, we distinguish the following two cases:

(a) If $ep^* = 0$, then the evaluation of a result of the lo-narrowing function is finished. In order to realize a configuration of the form in Case 5, we simply set ep to 0.

(b) If $ep^* > 0$, then the graph G which is refered by the square of the address stack at position ep^*, has to be evaluated next. For this purpose, G is copied to a new address $nadr$, $nadr$ is pushed on top of the address stack, and ep is set to ep^*.

5. If the address stack is empty or if the topmost square is a choice point and the environment pointer is 0, then the evaluation of a result of the lo-narrowing function is finished. Hence, the result $(\nu(t), \varphi \circ \nu)$ is produced by the auxiliary functions $StoT$ which transforms the output tape into a term t, $make\_sub$ which computes the substitution φ depending on the values of the trail, the stack of substitutions, and the graph, and $rename$ which renames the logic variables by ν to be consistent to the result of the lo-narrowing function. Those auxiliary functions call the auxiliary function $GtoT$ which transforms the graph into the corresponding tree. Furthermore, the computation of the following result is initialized by the auxiliary function $backtrack$ which deletes every square above the topmost choice point for which alternatives are not computed up to now. If such a choice point does not exist, then the complete set of results has been computed. Hence, $backtrack$ writes $Stop$ into the instruction pointer. For resolving bindings of logic variables, $backtrack$ calls the auxiliary function $undo$.

Stop Configuration

The transition function $transform$ is not defined, if the instruction pointer includes $Stop$. In this case, every result of the lo-narrowing function is in the set of results which is computed by the graph narrowing machine. Since such a configuration can be computed, only if the lo-narrowing tree is finite, from the practical point of view, it would be better that the machine stops, whenever a result is computed. But in the framework of this paper, this behaviour would make no sense, because every leaf of the lo-narrowing tree is labeled by a result, and hence, there would be no need to formalize backtracking. Thus, since we want to introduce techniques for verifying implementations which also perform backtracking, we have chosen the presented formalism.

Correctness of the Implementation

After the complete formalization of the implementation we are able to formulate and prove the following theorem which shows the correctness of the implementation.

> **Theorem 5.1** For every modular tree transducer $M = (F, \Delta, R)$, every term $t \in T\langle F \cup \Delta \rangle(LV)$, and every set of results $res \in Results(M)$:
>
> There exist an $i \in \mathbb{N}$ and a word of triples $triples \in Triple(M)^*$, such that
>
> $$det\_lo\_nar^i_M(init\_sf_M(t), \emptyset) = (triples, res), \qquad (1)$$
>
> iff there exist a $j \in \mathbb{N}$ and a configuration $conf \in Conf$, such that
>
> $$transform^j(input(M, t), \emptyset) = (conf, res). \qquad (2)$$

Proof: We show that every single derivation step by the lo-narrowing function can be simulated by a sequence of transition steps of the graph narrowing machine, and vice versa. For this purpose, we introduce the set $imp\_Conf\_Results$ which is a subset of $Conf \times Results(M)$ such that every element corresponds to a sentential form of the lo-narrowing function. Especially, a pair $(conf, res)$ belongs to $imp\_Conf\_Results$ either (1) if it results from an application of Case 5 in Figure 4, or (2) if it results from an application of the $INIT\_RESET$-instruction and Case 5 in Figure 4 is not applied in the sequence of the following transition steps up to the next application of Case 3a) or 3b) in Figure 4. The latter condition is checked by the auxiliary boolean function $ip\_in\_PA\_before\_Case5$.

The set $imp\_Conf\_Results$ is defined as follows:

{ $(conf, res) \in Conf \times Results(M) \mid$
there exists an $i \in \mathbb{N}$: $(conf, res) = transform^i(input(M, t), \emptyset)$ and
the instruction pointer in $conf$ is either $Stop$ or $Gmode$ and
in the latter case, one of the following two conditions holds:
(1) $transform^{i-1}(input(M, t), \emptyset)$ fulfills the conditions in Case 5 in Figure 4
(2) the instruction pointer in $transform^{i-1}(input(M, t), \emptyset)$ is in PA
and $ip\_in\_PA\_before\_Case5(conf, res) = true$ }

Furthermore, as the main point of this proof, we introduce the function

$$conf\_to\_sf : imp\_Conf\_Results \rightarrow Triple(M)^* \times Results(M)$$

$conf\_to\_sf((ps, ip, rs, G, ss, tr, ep, ds, ot), res) =$
if $ip = Stop$ **then** (Λ, res)
if $ip = Gmode$ **then**
$\qquad\qquad (make\_word\_of\_triples(ps, ip, rs, G, ss, tr, ep, ds, ot, \Lambda), res)$

which transforms every pair in $imp\_Conf\_Results$ into the corresponding sentential form of the lo-narrowing function. Thereby, the set of results is simply overtaken. Furthermore, in the case that the instruction pointer is $Stop$, the word of triples in the first component of the sentential form is empty. Otherwise, the word of triples is recursively computed in the rightmost argument of the auxiliary function $make\_word\_of\_triples$ from right to left. Beside this auxiliary function the following auxiliary functions, the definitions of which are also omitted because of lack of space, are needed to formalize the function $conf\_to\_sf$. The triple which corresponds to the configuration in the first argument of $make\_word\_of\_triples$ is computed by the auxiliary function $make\_triple$. Furthermore, in the first argument of $make\_word\_of\_triples$, the auxiliary function $prev\_conf$ computes the configuration which corresponds to the triple that has to be computed next. For this purpose, $prev\_conf$ initializes the arguments of the auxiliary function $make\_prev\_conf$ which simply calls the auxiliary function $backtrack$ and resolves the initialization for the computation of the next alternative.

The first and second component of a triple are constructed by the auxiliary function $make\_triple$ in a very similar way as a new pair is put to the set of results in Case 5 in Figure 4 with the following difference. The first component t can not be computed only by considering the output tape, because t may not be completely evaluated. Hence, t is computed by the auxiliary function $make\_tree$ which transforms the unevaluated graphs into trees that are put together to t by calling the auxiliary function $StoT$. For computing the positions of the references of the unevaluated graphs in the address stack, $make\_tree$ calls the auxiliary function $next\_env$.

Furthermore, the third component of the triple is computed by the auxiliary function $alts$ as follows. In the case that the following step of the lo-narrowing function is deterministic, the value of $alts$ is simply defined by the function app_rules_M (cf. Section 4). Otherwise, $alts$ produces the set of the numbers of all rules which have not been applied up to now by considering the current choice point.

Direction 1 $\Longrightarrow$ *2* of the theorem is shown by proving that every application of $det\_lo\_nar_M$ in (1) can be simulated by a sequence of applications of $transform$. For this purpose, the following claim is used which can be shown by considering a very complex case study:

Claim For every $i \in \mathbb{N}$: If there exist $(trip, res), (trip', res') \in Triples(M)^* \times Results(M)$, such that $det\_lo\_nar_M^i(input(M,t), \emptyset) = (trip, res)$ and

$$det\_lo\_nar_M(trip, res) = (trip', res'),$$

then for every $conf \in Conf$ which fulfills the following two conditions: (i) $(conf, res) \in imp\_Conf\_Results$ and (ii) $conf\_to\_sf(conf, res) = (trip, res)$, there exists a configuration $conf' \in Conf$ which fulfills the following three conditions: (i) $(conf', res') \in imp\_Conf\_Results$, (ii) $conf\_to\_sf(conf', res') = (trip', res')$, and (iii) there exists a $j > 0$ such that

$$transform^j(conf, res) = (conf', res').$$

Direction 2 $\Longrightarrow$ *1* of the theorem is shown by proving that every shortest sequence of applications of $transform$ from one pair in $imp\_Conf\_Results$ to another one corresponds to one application of $det\_lo\_nar_M$ in (1). For this purpose, the following claim is used which can also be shown by considering a very complex case study:

Claim For every pair $(conf, res) \in imp\_Conf\_Results$:
If the set $\{j \mid j > 0 \text{ and } transform^j(conf, res) \in imp\_Conf\_Results\}$ is not empty and i is the minimal element of this set, then

$$conf\_to\_sf(transform^i(conf, res)) = det\_lo\_nar_M(conf\_to\_sf(conf, res)).$$

The proofs of the claims are omitted because of lack of space. □

6 Conclusion

In this paper we have formalized a deterministic graph-based implementation of the lo-narrowing strategy for modular tree transducers and proved its correctness. For this purpose, we have introduced a mechanism for formalizing the implemented depth-first left-to-right traversal over lo-narrowing trees which serves, in our opinion, as a more effective base for proving the correctness of a graph-based implementation than the formalization by *evolving algebras* [10]. Nevertheless, in the context of a stack-based implementation like the *WAM* [23], evolving algebras are an effective mechanism for proving the correctness [4].

Similar to the implementation which is considered in the correctness proof of the *G-machine* in [16], the presented implementation seems not to be very efficient. A lot of storage capacity and time is needed, because subgraphs are copied very frequently and sharing is only used in a very restricted form for being consistent to the lo-narrowing function. But, the introduction of a very efficient implementation was not the aim of the paper. Nevertheless, the efficiency can be increased by integrating efficient *garbage collection methods* and by sharing complete subexpressions. But, for integrating the latter method, we would need a more complex formalism, e.g., *graph rewriting systems* [21], than the lo-narrowing function.

We assume that the presented mechanisms can be extended for proving other deterministic graph-based implementations of narrowing as, e.g., the implementations of *BABEL* [18] in [17, 15]. Especially, in [7] we have extended the presented implementation and the correctness proof to a deterministic implementation of the efficient universal unification algorithm in [8] by adding mechanisms for unification.

References

1. S. Antoy, R. Echahed, and M. Hanus. A needed narrowing strategy. In *POPL'94*, pages 268–279. Portland, 1994.
2. L. Augustsson. A compiler for lazy ML. In *Proceedings of the ACM Symposium on Lisp and Functional Programming*, pages 218–227, 1984.
3. E. Börger, F. Lopez-Fraguas, and M. Rodriguez-Artalejo. A model for mathematical analysis of functional logic languages and their implementations. In *IFIP'94*, *Vol. I*, pages 410–415. IFIP Transactions A-51, 1994. North-Holland.
4. E. Börger and D. Rosenzweig. The WAM - definition and compiler correctness. Technical Report TR-14/92, Dipartimento di Informatica, Universita di Pisa, 1992.

5. R. Echahed. On completeness of narrowing strategies. In *CAAP'88*, pages 89–101. Springer-Verlag, 1988. LNCS 299.
6. J. Engelfriet and H. Vogler. Modular tree transducers. *Theoretical Computer Science*, 78:267–304, 1991.
7. H. Faßbender. *E-Unifikation für primitiv-rekursive Baumfunktionen*. PhD thesis, Universität Ulm, Fakultät für Informatik, März 1995.
8. H. Faßbender and H. Vogler. A universal unification algorithm based on unification-driven leftmost outermost narrowing. *Acta Cybernetica*, 11(3):139–167, 1994.
9. M. Fay. First-order unification in an equational theory. In *Proceeding of the 4th workshop on automated deduction, Austin*, pages 161–167, 1979.
10. Y. Gurevich. Evolving algebras. A tutorial introduction. *Bulletin of the EATCS*, 43:264–284, 1991.
11. W. Hans, R. Loogen, and S. Winkler. On the interaction of lazy evaluation and backtracking. In *PLILP'92*, pages 355–369. Springer-Verlag, 1992. LNCS 631.
12. M. Hanus. The integration of functions into logic programming: From theory to practice. *Journal of Logic Programming*, 19,20:583–628, 1994.
13. J.M. Hullot. Canonical forms and unification. In *Proceedings of the 5th conference on automated deduction*, pages 318–334. Springer-Verlag, 1980. LNCS 87.
14. T. Johnsson. Efficient compilation of lazy evaluation. *SIGPLAN Notices*, 6:58–69, 1984.
15. H. Kuchen, R. Loogen, J.J. Moreno-Navarro, and M. Rodriguez-Artalejo. Graph-based implementation of a functional logic language. In *ESOP'90*, pages 271–290. Springer-Verlag, 1990. LNCS 432.
16. D. Lester. The G-Machine as a representation of stack semantics. In *Functional Programming Languages and Computer Architecture*, pages 46–59. Springer-Verlag, 1987. LNCS 274.
17. J.J. Moreno-Navarro, H. Kuchen, R. Loogen, and M. Rodriguez-Artalejo. Lazy narrowing in a graph machine. In *ALP'90*, pages 298–317. Springer-Verlag, 1990. LNCS 463.
18. J.J. Moreno-Navarro and M. Rodriguez-Artalejo. Logic-programming with functions and predicates: the language BABEL. *Journal of Logic Programming*, 12:191–223, 1992.
19. A. Mück. CAMEL: an extension of the categorical abstract machine to compile functional/logic programs. In *PLILP'92*. Springer-Verlag, 1992. LNCS 631.
20. A. Mück. CAMEL: a verified abstract machine for functional logic programming. In A. Mück, editor, *Proceedings of the Second International Workshop on Functional/ Logic Programming*. University of Munich, Number 9311, 1993.
21. Rinus Plasmeijer and Marko van Eekelen. *Functional Programming and Parallel Graph Rewriting*. International Computer Science Series. Addison Wesley, 1993.
22. U.S. Reddy. Narrowing as the operational semantics of functional languages. In *Symposium on Logic Programming*, pages 138–151. IEEE Comp. Soc. Press, 1985.
23. D.H.D. Warren. An abstract prolog instruction set. Technical Report 309, SRI International, 1983.
24. J.H. You. Enumerating outer narrowing derivations for constructor-based term rewriting systems. *Journal of Symbolic Computation*, 7:319–341, 1989.

A New Calculus for Semantic Matching[*]

Bernd Bütow, Robert Giegerich, Enno Ohlebusch, Stephan Thesing

Technische Fakultät der Universität Bielefeld
Postfach 100131, 33501 Bielefeld, Germany
{berndb|robert|enno|isthesin}@techfak.uni-bielefeld.de

Abstract. In this paper, we present Reverse Restructuring, a new calculus for solving the semantic matching problem. For narrowing, advanced selection rules are commonly seen as an appropriate method to reduce the search space. Our approach to design a special calculus for special goals is another way of reducing the efficiency defects of narrowing. Reverse Restructuring constructs derivations in the reverse direction by guessing terms from which an already known term might be derived. To this end, the rules of the underlying term rewriting system are also applied in the reverse direction, i.e. from right to left. We show the soundness and completeness of this calculus and demonstrate its efficiency for an important class of problems.

1 Introduction

Narrowing is commonly seen as the basis for the amalgamation of logic and functional programming. Narrowing today refers to a whole family of procedures for solving equational goals $s =_? t$ in the equational theory defined by a term rewriting system (TRS) $\mathcal{R}$. The development of different narrowing strategies is driven by the need to improve efficiency by cutting down the search space. Most approaches require certain restrictions on the TRS in order to ensure completeness. A certain endpoint in this line of development is currently marked by the needed narrowing strategy of Antoy et al. [AEH94], where a relatively complex selection rule ensures that no "unnecessary" narrowing step is done. The price paid by this strategy is a severe restriction on the underlying TRSs.

Another line of attacking the efficiency defects of narrowing is the idea to design special calculi for special goals. Of course, this idea makes only sense if the subgoals arising in a computation have the same restricted form. Such an approach was first suggested by Dershowitz et al. [DMS92], introducing the problem of semantic matching: Given an arbitrary term s and a term N in ground normal form, find a substitution σ such that $s\sigma =_\mathcal{R} N$.

At first glance, this does not seem to be a problem worth individual study. It is a common technical device to transform general narrowing goals $s =_? t$ into the form $eq(s, t) =_? true$ by means of the extra rule $eq(X, X) \rightarrow true$.

[*] This work is supported by the DFG-project: "Abstrakte Inferenzmaschine" under Az. Gi 178/1-2

In this sense, all semantic unification problems can be turned into semantic matching problems, and we cannot expect a semantic matching calculus to offer any advantages.

However, the situation changes if we restrict the underlying TRS $\mathcal{R}$ to left-linear or variable-preserving rules. Note that the rule $eq(X, X) \rightarrow true$ violates either restriction. In this setting, Dershowitz et al. [DMS92] gave a calculus for semantic matching which is sound and complete.

The class of problems which can be solved by semantic matching is still quite interesting. In particular, function inversion belongs to this class, i.e. goals of the form $f(X) =_? N$. It has been observed in [Gie90] that compiler code selection can be seen as solving an equation $\delta(X) =_? ilprog$, where $ilprog$ is an intermediate language program, and δ is a derivor mapping target machine programs into intermediate language programs. (We shall elaborate this example in section 4.2.) Current compiler technology [ESL89, Gie90, FHP92, Emm94] solves this problem with an efficiency which narrowing procedures can only dream of.

The overall goal of our work is to improve the efficiency of semantic matching while at the same time improving the flexibility of code selection mechanisms. The present paper is a first step in this direction. Our basic idea is as follows: In solving $f(X) =_? N$, we try to exploit the fact that N is in ground normal form. Rather than guessing instantiations of X that successively construct a rewrite derivation $f(X\sigma) \rightarrow_{\mathcal{R}} N$, we construct this derivation in the reverse direction by guessing terms from which N can be derived. In this paper, we introduce the Reverse Restructuring calculus, show its soundness and completeness for the semantic matching problem, and outline some optimizations. Furthermore, we will provide the Reverse Restructuring computation with some knowledge about the forward computation paths resulting in a further considerable reduction of the search space. More details and full proofs can be found in [BT94].

2 Preliminaries

2.1 Notations

In this section we briefly recall basic notions of term rewriting (see for example [DJ90, Klo92] for more details).

A *signature* is a set Σ of *operators*. Every $f \in \Sigma$ is combined with an *arity* n, $n \geq 0$. Let X, $X \cap \Sigma = \emptyset$, be a countable infinite set of *variables*. We define the set of *terms* $T(\Sigma, X)$ with variables as the smallest set containing X such that $f(t_1, \ldots, t_n) \in T(\Sigma, X)$ if f has arity n and $t_1, \ldots, t_n \in T(\Sigma, X)$. By $\mathcal{V}(t)$, we denote the set of variables occurring in the term t. If $\mathcal{V}(t) = \emptyset$, t is said to be *ground*. An *occurrence* p in a term t is defined as a sequence of natural numbers identifying a subterm in t. The empty sequence, denoted by ϵ, identifies for every term t the whole term itself. For every term $f(t_1, \ldots, t_n)$, the sequence $i.p$ where $i \leq n$ and p is an occurrence in t_i, identifies the subterm of t_i at occurrence p. t/p denotes the *subterm* of t at occurrence p; $t[p \leftarrow t']$ denotes the term obtained from t by replacing t/p by t'. We define the *prefix ordering*

$\leq \subset \mathbb{N}^* \times \mathbb{N}^*$ on occurrences as follows: $u \leq v \Leftrightarrow \exists w : u.w = v$. We write $v > u$ iff $u \leq v$ and $u \neq v$. With $root(t)$, we refer to the outermost symbol, i.e. $root(t) = f$ if $t \equiv f(t_1, \ldots, t_n)$ (the symbol $\equiv$ denotes syntactical equality) and $root(t) = t$ if $t \in X$. A *substitution* is a mapping $\sigma : X \to T(\Sigma, X)$, such that $\mathcal{D}(\sigma) := \{x \in X \mid \sigma(x) \neq x\}$ is a finite set. $\mathcal{D}(\sigma)$ denotes the *domain* of σ. Since $\mathcal{D}(\sigma)$ is finite, we can represent σ as a finite subset of $X \times T(\Sigma, X)$: $\sigma = \{x_1 \leftarrow t_1, \ldots, x_n \leftarrow t_n\}$, where $x_i \neq x_j \Leftrightarrow i \neq j$. In the following we write $x\sigma$ for $\sigma(x)$. $\mathcal{I}(\sigma) := \bigcup_{x \in \mathcal{D}(\sigma)} \mathcal{V}(x\sigma)$ is the set of *variables introduced* by σ. Furthermore, we define the *restriction* of a substitution σ to a set of variables $V \subseteq X$ as $\sigma_{|V} := \{x \leftarrow t \in \sigma \mid x \in V\}$. The extension of substitutions to terms is straightforward, we define $\sigma : T(\Sigma, X) \to T(\Sigma, X)$ by $\sigma(f(t_1, \ldots, t_n)) = f(\sigma(t_1), \ldots, \sigma(t_n))$ for every term $f(t_1, \ldots, t_n)$. For substitutions σ and λ we define $\sigma \leq \lambda$, iff $\exists$ substitution $\theta : \sigma\theta = \lambda$.

A term s *matches* a term t if $s\sigma \equiv t$ for a substitution σ. A term s *unifies* with a term t if $s\sigma \equiv t\sigma$ for a substitution σ. A substitution σ is called *most general unifier (mgu)* of two terms s and t if, for any other unifier σ' of s and t, there exists a substitution γ such that $\sigma' = \sigma\gamma$. If s and t are unifiable the $mgu(s, t)$ exists and is unique up to *variable renaming*.

A rewrite rule is an ordered pair $l \to r$ from $T(\Sigma, X) \times T(\Sigma, X)$ satisfying the additional conditions $\mathcal{V}(l) \supseteq \mathcal{V}(r)$ and $l \notin X$. We speak of a *collapsing* rule if $r \in X$. A term rewriting system (TRS) is a set of rules. A term t *rewrites* in one step to t' at occurrence p (in symbols: $t \to_{l \to r, p, \sigma} t'$) if there is a rule $l \to r \in \mathcal{R}$ and a (smallest) substitution σ with $t/p \equiv l\sigma$ and $t' \equiv t[p \leftarrow r\sigma]$. Sometimes we just write $\to$ or $\to_\mathcal{R}$ instead of $\to_{l \to r, p, \sigma}$. The symbol $\to_\mathcal{R}^n$ denotes a rewriting derivation in n steps; in analogy, we define $\to_\mathcal{R}^{\leq k}$ etc. Finally, $\to_\mathcal{R}^*$ denotes a derivation of zero or more steps. $f \in \Sigma$ is called *defined function* if $root(l) = f$ for the left-hand side of some rule $l \to r$; otherwise it is called *constructor*.

A TRS $\mathcal{R}$ is said to be *terminating* or *noetherian* iff there are no infinite derivations. It is called *confluent* iff $\forall t, t_1, t_2$ with $t \to_\mathcal{R}^* t_1 \wedge t \to_\mathcal{R}^* t_2 \Rightarrow \exists t_3$ such that: $t_1 \to_\mathcal{R}^* t_3, t_2 \to_\mathcal{R}^* t_3$. A TRS is called *canonical* iff it is terminating and confluent. A term which cannot be reduced any more is said to be in *normal form*. In a canonical TRS every term has a unique normal form, which is denoted by $t\downarrow$. Furthermore, a TRS $\mathcal{R}$ is *variable-preserving* if $\forall_{l \to r \in \mathcal{R}} : \mathcal{V}(l) = \mathcal{V}(r)$, i.e. all variables occurring on the left-hand side of a rule also occur on its right-hand side. A term is said to be *linear* if every variable in it occurs only once. If $\forall_{l \to r \in \mathcal{R}} : l$ is a linear term, then $\mathcal{R}$ is called *left-linear*. A rewrite step $t \to_{l \to r, p, \sigma} t'$ is called *innermost* iff there is no rewrite step possible below this position, i.e.: $\forall_{p' > p} : t/p'$ is in normal form. A rewriting derivation is called *innermost* iff every step in it is innermost. In a canonical TRS $\mathcal{R}$ two terms t, t' are defined to be *joinable* (in symbols: $t =_\mathcal{R} t'$) iff $t\downarrow \equiv t'\downarrow$. A *goal* is a pair $(t, t') \in T(\Sigma, X) \times T(\Sigma, X)$ written as $t =_? t'$. An $\mathcal{R}$-*unifier* of this goal is a substitution σ such that: $t\sigma =_\mathcal{R} t'\sigma$. The relation $\leq_\mathcal{R}$ on substitutions is defined analogous to $\leq$.

For substitutions σ, λ, θ we define the following properties: σ is called *idempotent* iff $\forall_{x \in \mathcal{D}(\sigma)} : x\sigma\sigma \equiv x\sigma$; $\sigma = \lambda[V]$ iff $\sigma_{|V} = \lambda_{|V}$, analogously for $=_\mathcal{R}$: $\sigma =_\mathcal{R} \lambda[V]$ iff $\forall_{x \in V} : x\sigma =_\mathcal{R} x\lambda$; $\sigma \leq \lambda[V]$ iff $\exists \theta : \sigma\theta = \lambda[V]$, analogously for $=_\mathcal{R}$: $\sigma \leq_\mathcal{R} \lambda[V]$ iff $\exists \theta : \sigma\theta =_\mathcal{R} \lambda[V]$. σ is called *ground substitution* iff $\forall_{x \in \mathcal{D}(\sigma)} : x\sigma$ is a ground term; it is called *normalized* iff $\forall_{x \in \mathcal{D}(\sigma)} : x\sigma$ is in normal form and it is called *linear* iff $\forall_{x \in \mathcal{D}(\sigma)} : x\sigma$ is a linear term. The following properties of substitutions will be used later:

Lemma 2.1 :

1. *Equivalence ($\equiv$) is preserved under substitutions, i.e.* $s \equiv t \Rightarrow s\sigma \equiv t\sigma \; \forall\sigma$.
2. *If $\mathcal{R}$ is a canonical TRS, then joinability ($=_\mathcal{R}$) is preserved under substitutions, i.e.* $s =_\mathcal{R} t \Rightarrow s\sigma =_\mathcal{R} t\sigma \; \forall\sigma$. *Furthermore, note that $\equiv \subset =_\mathcal{R}$, that is, $s \equiv t \Rightarrow s =_\mathcal{R} t$.*
3. *If $t \to_{l \to r, p, \sigma} t'$ is an innermost step, then σ is a* normalized *substitution.*

Proof. Routine. $\square$

2.2 Semantic Matching

The task of semantic *unification* consists of finding a complete set Θ of $\mathcal{R}$-unifiers for a goal $s =_? t$, that is to say, for every σ with $s\sigma =_\mathcal{R} t\sigma$ there is a $\lambda \in \Theta$ with $s\lambda =_\mathcal{R} t\lambda$ and $\lambda \leq_\mathcal{R} \sigma$. For semantic *matching* we have to find a complete set Θ of $\mathcal{R}$-matchers for a goal $s =_? N$, where N is a term in ground normal form, i.e. for every σ with $s\sigma =_\mathcal{R} N$ there is a $\lambda \in \Theta$ with $s\lambda =_\mathcal{R} N$ and $\lambda \leq_\mathcal{R} \sigma$. In general, semantic matching is as hard as semantic unification, since every goal $s =_? t$ can be transformed into the goal $eq(s,t) =_? true$ using the additional rule $eq(X, X) \to true$.

In [DMS92] it was pointed out that canonical TRSs which are in addition *variable-preserving* or *left-linear* prevent such a transformation. These classes of TRSs are still quite large, so it is reasonable to think about a special calculus which takes advantage of the peculiarities of the semantic matching problem. In [DMS92] the following forward decomposition rules are presented:

Eliminate: $(x =_? s, L_1, \ldots, L_n, \lambda) \rightsquigarrow (L_1\sigma, \ldots, L_n\sigma, \lambda\sigma)$, where $\sigma = \{x \leftarrow s\}$
Decompose: $(f(s_1, \ldots, s_m) =_? f(s'_1, \ldots, s'_m), L_1, \ldots, L_n, \sigma)$
$\rightsquigarrow (s_1 =_? s'_1, \ldots, s_m =_? s'_m, L_1, \ldots, L_n, \sigma)$
Mutate: $(f(s_1, \ldots, s_m) =_? t, L_1, \ldots, L_n, \sigma)$
$\rightsquigarrow (s_1 =_? l_1, \ldots, s_m =_? l_m, r =_? t, \sigma)$,
where $f(l_1, \ldots, l_m) = r$ is a renamed rule in $\mathcal{R}$

This forward decomposition calculus is a complete procedure for solving goals $s =_? N$, where N is in ground normal form, and the underlying TRS is canonical and variable-preserving or left-linear [DMS92].

The above rules are a subset of the transformation rules which were used in [DS87, MMR89, Mit90, Mit94] for semantic unification. However, this approach does not explicitly exploit the fact that the right-hand side of an initial goal has to be in ground normal form. In the next section we will present a calculus which takes more advantage of this since it starts with the result on the right-hand side and tries to find out how the way leads back to the term on the left-hand side.

3 Reverse Restructuring

3.1 Definition of Reverse Restructuring

The following inference rules define the calculus of Reverse Restructuring (RR).

Definition 3.1 : (Reverse Restructuring rules)

Eliminate1: $([x =_? s, L_1, \ldots, L_n], \lambda) \rightleftharpoons ([L_1\sigma, \ldots, L_n\sigma], \lambda\sigma)$
if $s \equiv x$ or $x \notin \mathcal{V}(s)$, where $\sigma = \{x \leftarrow s\}$

Eliminate2: $([s =? x, L_1, \ldots, L_n], \lambda) \rightleftharpoons ([L_1\sigma, \ldots, L_n\sigma], \lambda\sigma)$
if $x \notin \mathcal{V}(s), s \notin X$, where $\sigma = \{x \leftarrow s\}$

Decompose: $([f(s_1, \ldots, s_m) =_? f(s'_1, \ldots, s'_m), L_1, \ldots, L_n], \sigma) \rightleftharpoons$
$([s_1 =_? s'_1, \ldots, s_m =_? s'_m, L_1, \ldots, L_n], \sigma), m \geq 0$

Mutate1: $([s =_? f(s_1, \ldots, s_m), L_1, \ldots, L_n], \sigma) \rightleftharpoons$
$([r_1 =_? s_1, \ldots, r_m =_? s_m, s =_? g(l_1, \ldots, l_q), L_1, \ldots, L_n], \sigma)$,
where $g(l_1, \ldots, l_q) \rightarrow f(r_1, \ldots, r_m)$ is a renamed rule from $\mathcal{R}$
and Eliminate1 does not apply

Mutate2: $([s =_? s', L_1, \ldots, L_n], \sigma) \rightleftharpoons$
$([s =_? g(l_1, \ldots, l_q)\sigma', L_1, \ldots, L_n], \sigma\sigma')$, where $\sigma' = \{x \leftarrow s'\}$,
$g(l_1, \ldots, l_q) \rightarrow x$ is a renamed rule from $\mathcal{R}$ and neither
Eliminate1 nor Eliminate2 does apply.

Note that we assume that the variables in the rewrite rules are always renamed such that the rules contain only fresh variables. Similar to the forward decomposition calculus, the Martelli & Montanari rules – Decompose and Eliminate1/2 [2] – for syntactic matching and unification [MM82] are part of our calculus. The Mutate rules make the difference: Using the Mutate1 rule, the right-hand side of a rule is compared to the right-hand side of the goal. Under the presumption that the goals generated by this comparison are solvable, another goal consisting of the left-hand side of the original goal and the left-hand side of the rule under consideration is added. By means of this transformation a reverse computation, starting with the right-hand side, is established. The Mutate2 rule is necessary for handling collapsing rules. It includes an implicit Eliminate2 step. If in a Decompose step we have $m = 0$, the goal is removed. In this case we speak of a *trivial* Decompose step. The reader may also observe that we deal with a *list* and not with a set of goals. This is because the order of evaluating the equations is strictly sequential from left to right. There is still a high degree of nondeterminism, due to the fact that in general more than one rule can be used for a Mutate1/2 step and that a Decompose and a Mutate rule might be applicable at the same time. However, if an Eliminate1 or an Eliminate2 step can be applied, no other transformation rule is applicable, i.e. *eager variable elimination* is built into our calculus. In order to complete the picture we define the notions *computation* and *result*:

---

[2] In [DMS92] a kind of Eliminate2 rule is implicitly assumed in the left-linear case.

Definition 3.2 : (computation, result) A *computation* of a goal $s =_? s'$ is a sequence of RR steps starting with: $([s =_? s'], \emptyset) \rightleftharpoons (G_1, \sigma_1) \rightleftharpoons \cdots$ A computation $([s =_? s'], \emptyset) \rightleftharpoons {}^*([\,], \sigma)$ is called *successful* and σ is its *result*.

The next example illustrates how RR works in detail. We define the addition on natural numbers which are constructed by 0 and s.

$$R_1 : \quad 0 + X \quad \rightarrow X$$
$$R_2 : \quad s(X) + Y \rightarrow s(X + Y)$$

Example 3.3 : Addition

The goal to solve is $s(0) + U =_? s(s(0))$. Remember that the computation order is not arbitrary: the leftmost goal has to be solved first and the resulting substitution has to be applied to the rest of the goals.

$$([s(0) + U =_? s(s(0))], \emptyset)$$

$\overset{Mutate1_{R_2}}{\rightleftharpoons} \quad ([X + Y =_? s(0), s(0) + U =_? s(X) + Y], \emptyset)$

$\overset{Mutate2_{R_1}}{\rightleftharpoons} \quad ([X + Y =_? 0 + s(0), s(0) + U =_? s(X) + Y], \{X' \leftarrow s(0)\})$

$\overset{Decompose}{\rightleftharpoons} \quad ([X =_? 0, Y =_? s(0), s(0) + U =_? s(X) + Y], \{X' \leftarrow s(0)\})$

$\overset{2 \times Eliminate1}{\rightleftharpoons} \quad ([s(0) + U =_? s(0) + s(0)], \{X' \leftarrow s(0), X \leftarrow 0, Y \leftarrow s(0)\})$

$\overset{Decompose}{\rightleftharpoons} \quad ([s(0) =_? s(0), U =_? s(0)], \{X' \leftarrow s(0), X \leftarrow 0, Y \leftarrow s(0)\})$

$\overset{2 \times Decompose}{\rightleftharpoons} \quad ([U =_? s(0)], \{X' \leftarrow s(0), X \leftarrow 0, Y \leftarrow s(0)\})$

$\overset{Eliminate1}{\rightleftharpoons} \quad ([\,], \{X' \leftarrow s(0), X \leftarrow 0, Y \leftarrow s(0), U \leftarrow s(0)\})$

At first glance, it seems that in comparison to other narrowing derivations the length of a RR computation is much longer; e.g. the goal above can be computed with two innermost narrowing steps. But a single narrowing step is more complicated than a RR step. This is because a subterm of a goal has to be unified with a left-hand side of a rule before the rewrite step can be be carried out. In RR, unification is performed explicitly by a sequence of Eliminate1/2 and Decompose steps. Therefore we will only count the Mutate1/2 steps when the length of a RR computation is compared to the one of a narrowing derivation.

3.2 Soundness

In this section we show that every result computed by RR is an $\mathcal{R}$-unifier.

Theorem 3.4: (soundness of RR) *Let $\mathcal{R}$ be a canonical TRS. If there exists a successful computation of the goal $s =_? s'$ with result σ, then $s\sigma =_{\mathcal{R}} s'\sigma$.*

Proof. The proof is by induction on m, the length of the RR computation:

$m = 1$: Only an Eliminate1/2 or a trivial Decompose step is possible, since only such a step reduces the number of goals. Such a transformation computes a correct solution.

$m = k + 1, k \geq 1$: We split the computation into a first and k subsequent steps: now the first step cannot be an Eliminate1/2 or a trivial Decompose step since this would contradict $m > 1$. We consider the remaining possibilities.

Decompose: We have $([s =_? s'], \emptyset) \rightleftarrows ([s_1 =_? s'_1, \ldots, s_n =_? s'_n], \emptyset)$. Each of these goals can be solved in at most k steps. Therefore we can apply the induction hypothesis from left to right to every goal.
This yields $\forall_{1 \leq i \leq n} : s_i \sigma^1 \cdots \sigma^i =_{\mathcal{R}} s'_i \sigma^1 \cdots \sigma^i$. Consequently, it follows $\forall_{1 \leq i \leq n} :$ $s_i \sigma^1 \cdots \sigma^n =_{\mathcal{R}} s'_i \sigma^1 \cdots \sigma^n$, since joinability is preserved under substitutions (Lemma 2.1). $\sigma := \sigma^1 \cdots \sigma^n$ is the computed result and clearly $s\sigma =_{\mathcal{R}} s'\sigma$.

Mutate1: This means $s' \equiv f(s'_1, \ldots, s'_n)$ and there is a rule $l \rightarrow f(r_1, \ldots, r_n)$ in $\mathcal{R}$, such that $([s =_? s'], \emptyset) \rightleftarrows ([r_1 =_? s'_1, \ldots, r_n =_? s'_n, s =_? l], \emptyset)$. Analogous to the Decompose case we get for the first n goals:
$([r_1 =_? s'_1, \ldots, r_n =_? s'_n], \emptyset) \overset{q_1 \leq k}{\rightleftarrows} ([\,], \sigma^1)$ and thus $f(r_1, \ldots, r_n)\sigma^1 =_{\mathcal{R}} s'\sigma^1$.

For the remaining goal we get $([s\sigma^1 =_? l\sigma^1], \emptyset) \overset{q_2 \leq k}{\rightleftarrows} ([\,], \sigma^2)$. Again the induction hypothesis is applicable which yields $s\sigma^1\sigma^2 =_{\mathcal{R}} l\sigma^1\sigma^2$. From Lemma 2.1 it follows that $f(r_1, \ldots, r_n)\sigma^1\sigma^2 =_{\mathcal{R}} s'\sigma^1\sigma^2$, $l\sigma^1\sigma^2 =_{\mathcal{R}} f(r_1, \ldots, r_n)\sigma^1\sigma^2$. The latter is true, because for every rule $l \rightarrow r \in \mathcal{R}$ it trivially holds that $l =_{\mathcal{R}} r$. We get $s\sigma^1\sigma^2 =_{\mathcal{R}} l\sigma^1\sigma^2 =_{\mathcal{R}} f(r_1, \ldots, r_n)\sigma^1\sigma^2 =_{\mathcal{R}} s'\sigma^1\sigma^2$. Since $\sigma := \sigma^1\sigma^2$ is the result of the computation $([s =_? s'], \emptyset) \overset{k+1}{\rightleftarrows} ([\,], \sigma^1\sigma^2)$ this case is proved. The remaining Mutate2 case is similar. $\square$

Note that soundness does neither depend on left-linearity, variable-preservingness, nor the form of the goal.

3.3 Completeness

Although the completeness proof for left-linear TRSs is more challenging, due to space limitations we present here only the variable-preserving case in detail. The structure of both proofs is similar, which makes it tolerable to restrict the proof for left-linear TRSs to a short sketch of their differences. For a detailed proof see [BT94].

Variable-Preserving TRSs

In this part, we tacitly assume that $\mathcal{R}$ is a canonical and variable-preserving TRS. In Theorem 3.6, we will show that RR is complete w.r.t. to normalized substitutions and goals of the form $s =_? N$, where N is a ground normal form. The simple proof of the next preparatory lemma is omitted.

Lemma 3.5 : *Let s be a term and t be a ground term.*

1. *Every solution σ of the goal $s =_? t$ satisfies $\mathcal{D}(\sigma) \supseteq \mathcal{V}(s)$ and moreover, $\sigma_{|\mathcal{V}(s)}$ is a ground substitution.*
2. *Every right-hand side of a goal which comes into existence during the computation of the goal $s =_? t$ is ground at the time of its evaluation.* $\square$

Theorem 3.6: (Completeness w.r.t. variable-preserving TRSs)
Let s be a term, N a term in ground normal form, and σ a normalized substitution with $\mathcal{D}(\sigma) = \mathcal{V}(s)$. If $s\sigma =_{\mathcal{R}} N$, then there is a successful computation of $s =_? N$ with result $\tilde{\sigma}$ such that $\tilde{\sigma} = \sigma[\mathcal{V}(s)]$.

Proof. For technical reasons, we prove the following more general claim.
Claim: *Let s be a term, t a ground term, and σ a normalized substitution with $\mathcal{D}(\sigma) = \mathcal{V}(s)$. If $s\sigma \to_{\mathcal{R}}^{n} t$ via an innermost derivation, then there is a successful computation of $s =_? t$ with result $\tilde{\sigma}$ such that $\tilde{\sigma} = \sigma[\mathcal{V}(s)]$. Moreover, if no rule is applied at a root position, then we may assume that the successful computation starts with a Decompose step.*

The theorem is a special case of this claim because if $t \equiv N$ is a term in ground normal form, then the existence of an innermost rewriting derivation $s\sigma \to_{\mathcal{R}}^{n} t$ is equivalent to $s\sigma =_{\mathcal{R}} t$. Clearly, σ must be ground, since $\mathcal{R}$ is variable-preserving.

The proof is by induction on n, the length of the rewrite derivation. If $n = 0$, then $s\sigma \equiv t$ yields the syntactic matching problem $s =_? t$ which can be solved by RR. So let $n \geq 1$. Clearly, $s \not\equiv x$ because σ is normalized. So we have $s \equiv f(s_1, \ldots, s_m)$, $m \geq 0$ and $t \equiv g(t_1, \ldots, t_q)$, $q \geq 0$. In the innermost derivation $s\sigma \equiv T_0 \xrightarrow[p_1, R_{i_1}, \sigma_1]{} T_1 \to \cdots \to T_{n-1} \xrightarrow[p_n, R_{i_n}, \sigma_n]{} T_n \equiv t$ every T_j is ground. Let $\mathcal{P}$ be the set of redex positions in the derivation, i.e. $\mathcal{P} := \{p_i \mid 1 \leq i \leq n\}$. We distinguish between: (1) $\epsilon \in \mathcal{P}$ and (2) $\epsilon \notin \mathcal{P}$.

(1) $\epsilon \in \mathcal{P}$: Let $j := \max\{k \mid p_k = \epsilon, 1 \leq k \leq n\}$, so the j-th rewrite step is the last rewrite step at a root position. We have $T_0 \to_{\mathcal{R}}^{\leq n} T_{j-1} \xrightarrow[\epsilon, R_{i_j}, \sigma_j]{} T_j \to_{\mathcal{R}}^{\leq n} T_n$. Thus, if R_{i_j} is the rule $l \to r$, then $T_{j-1} \equiv l\sigma_j$ and $r\sigma_j \equiv T_j$. Note that since the derivation is innermost, all proper subterms of T_{j-1} are in normal form and hence σ_j is normalized. Obviously, σ_j is a ground substitution. Since $l\sigma_j \to r\sigma_j$ was the last rewrite step at a root position and $t = g(t_1, \ldots, t_q)$, we can conclude that either (i) $r \equiv x$ or (ii) $r \equiv g(r_1, \ldots, r_q)$. We further distinguish between these two cases.

(i) $r \equiv x$: First note that in this case $x\sigma_j \equiv r\sigma_j \equiv t$ because σ_j is normalized. Hence $\sigma_j = \{x \leftarrow t\}$. Since $s \notin X$ and $t \notin X$, neither an Eliminate1 nor an Eliminate2 rule is applicable. Thus we can apply a Mutate2 rule to the original goal. This results in the computation $([s =_? t], \emptyset) \overset{Mutate2}{\rightleftarrows} ([s\sigma_j =_? l\sigma_j], \sigma_j)$, where $\sigma_j = \{x \leftarrow t\}$. Since $s\sigma \to_{\mathcal{R}}^{\leq n} l\sigma_j$, the induction hypothesis yields an RR computation for $s \equiv s\sigma_j =_? l\sigma_j$ with result $\tilde{\sigma}^2$ such that $\tilde{\sigma}^2 = \sigma[\mathcal{V}(s)]$. Hence $([s\sigma_j =_? l\sigma_j], \sigma_j) \rightleftarrows^* ([\,], \sigma_j\tilde{\sigma}^2)$ with result $\tilde{\sigma} = \sigma_j\tilde{\sigma}^2$. It is clear that $\tilde{\sigma} = \sigma[\mathcal{V}(s)]$.

(ii) $r \equiv g(r_1, \ldots, r_q)$: Here a Mutate1 step can be applied to $s =_? t$. We obtain $([s =_? t], \emptyset) \overset{Mutate1}{\rightleftarrows} ([r_1 =_? t_1, \ldots r_q =_? t_q, s =_? l], \emptyset)$. Note that in $r\sigma_j \to_{\mathcal{R}}^{\leq n} t$ no rule is applied at a root position. So according to the induction hypothesis, there exists a successful computation

$$([r =_? t], \emptyset) \overset{Decompose}{\rightleftarrows} ([r_1 =_? t_1, \ldots r_q =_? t_q], \emptyset) \rightleftarrows^* ([\,], \tilde{\sigma}^1)$$

such that $\tilde{\sigma}^1 = \sigma_j[\mathcal{V}(r)]$.

Combining these results, we obtain (note $s\tilde\sigma^1 \equiv s$)

$$([s =_? t], \emptyset) \overset{Mutate1}{\rightleftharpoons} ([r_1 =_? t_1, \ldots r_q =_? t_q, s =_? l], \emptyset) \rightleftharpoons^* ([s =_? l\tilde\sigma^1], \tilde\sigma^1)$$
$$\overset{I.H.}{\rightleftharpoons}^* ([\,], \tilde\sigma^1\tilde\sigma^2), \text{ where } \tilde\sigma^2 = \sigma[\mathcal{V}(s)]$$

In the latter computation, the induction hypothesis is applicable because $s\sigma \to_{\mathcal{R}}^{\leq n} l\sigma_j \equiv l\tilde\sigma^1$ (the equality follows from $\mathcal{V}(l) = \mathcal{V}(r)$). Evidently, the result $\tilde\sigma^1\tilde\sigma^2$ satisfies $\tilde\sigma^1\tilde\sigma^2 = \tilde\sigma^2 = \sigma[\mathcal{V}(s)]$.

(2) $\epsilon \notin \mathcal{P}$: In the derivation no rule is applied at a root position. Thus, we have $m = q, f \equiv g$ and hence $s\sigma = f(s_1, \ldots, s_m)\sigma \to_{\mathcal{R}}^n f(t_1, \ldots, t_m) = t$. We proceed by induction on the structure of s. If s were a constant, then this would mean $\epsilon \in \mathcal{P}$, contradicting $\epsilon \notin \mathcal{P}$. Thus f has arity $m \geq 1$. We prove the claim by induction on m. First, we define $\sigma^1 := \sigma_{|\mathcal{V}(s_1)}$ and $\sigma^2 := \sigma_{|\mathcal{V}(s)\backslash\mathcal{V}(s_1)}$. Clearly, $\sigma = \sigma^1\sigma^2 = \sigma^2\sigma^1$. We have the $R\!R$ computation

$$([f(s_1, \ldots, s_m) =_? f(t_1, \ldots, t_m)], \emptyset)$$
$$\overset{Decompose}{\rightleftharpoons} ([s_1 =_? t_1, \ldots, s_m =_? t_m], \emptyset)$$
$$\overset{I.H.}{\rightleftharpoons}^* ([s_2\tilde\sigma^1 =_? t_2, \ldots, s_m\tilde\sigma^1 =_? t_m], \tilde\sigma^1), \text{ where } \tilde\sigma^1 = \sigma^1 = \sigma[\mathcal{V}(s_1)]$$

The latter follows from the structural induction hypothesis and case (1), respectively. Recall that $t_i\tilde\sigma^1 \equiv t_i$ since t_i is ground. Furthermore, $\mathcal{D}(\tilde\sigma^1) \cap \mathcal{D}(\sigma^2) = \emptyset$ and hence $\tilde\sigma^1\sigma^2 = \sigma[\mathcal{V}(s)]$. If $m = 1$, the assertion is proved. Otherwise, we paste together the remaining goals by a new constructor C of arity $m - 1$. Let $s' := C(s_2\tilde\sigma^1, \ldots, s_m\tilde\sigma^1)$. Since $s_i\tilde\sigma^1\sigma^2 \equiv s_i\sigma$ and $s_i\sigma \to_{\mathcal{R}}^{\leq n} t_i$, $i = 2, \ldots, m$, we have the innermost rewrite derivation $s'\sigma^2 \to_{\mathcal{R}}^{\leq n} C(t_2, \ldots, t_m)$, where no rule is applied at a root position. Since σ^2 is normalized, an application of the induction hypothesis on m yields $([s' =_? C(t_2, \ldots, t_m)], \tilde\sigma^1) \overset{I.H.}{\rightleftharpoons}^* ([\,], \tilde\sigma^1\tilde\sigma^2)$ with $\tilde\sigma^2 = \sigma^2[\mathcal{V}(s')]$, where the first step in this $R\!R$ computation is a Decompose step. So the $R\!R$ computation $([s_2\tilde\sigma^1 =_? t_2, \ldots, s_m\tilde\sigma^1 =_? t_m], \tilde\sigma^1) \rightleftharpoons^*([\,], \tilde\sigma^1\tilde\sigma^2)$ exists. Finally, we show $\tilde\sigma^1\tilde\sigma^2 = \sigma[\mathcal{V}(s)]$ by case analysis: If $x \in \mathcal{V}(s_1)$, then $x\tilde\sigma^1\tilde\sigma^2 \equiv x\tilde\sigma^1 \equiv x\sigma$, since $x\tilde\sigma^1$ is ground. If $x \in \mathcal{V}(s) \backslash \mathcal{V}(s_1)$, then $x\tilde\sigma^1\tilde\sigma^2 \equiv x\tilde\sigma^2 \equiv x\sigma^2 \equiv x\sigma$. $\square$

Left-Linear TRSs

Theorem 3.7: (Completeness w.r.t. left-linear TRSs)
Let $\mathcal{R}$ be a canonical, left-linear TRS, s a term, N a term in ground normal form, σ a normalized substitution. If $s\sigma =_{\mathcal{R}} N$, then there exists a computation of $s =_? N$ with result $\tilde\sigma$ such that $\tilde\sigma \leq \sigma[\mathcal{D}(\sigma) \cup \mathcal{V}(s)]$.

Proof. The structure of the proof is the same as the previous one, except that now the generated goals are no more ground at the time of their evaluation. For this reason the bookkeeping of the variables occurring during the computation is more intricate. In order to handle this, we exploit the following two properties: (1) A substitution produced by a $R\!R$ computation does only contain variables from the goal solved plus some new variables, introduced through Mutate1/2 steps. In addition, such a substitution is idempotent. (2) In left-linear TRSs the right-hand sides of a goal stay linear and variable-disjoint during any computation of goals $s =_? N$. Furthermore, the generated substitutions are linear, when restricted to the left-hand side of the goals. The proofs of these two properties are straightforward but also quite lengthy since both are via a complex case analysis. □

4 Efficiency and Optimizations

4.1 Simple Optimizations

The following observations reduce the rule nondeterminism and avoid unnecessary run-time checks, but still preserve completeness.

Lemma 4.1 : (Optimizations)

1. *If a goal is of the form* $c(s_1, \ldots, s_m) =_? c(s'_1, \ldots, s'_m)$, *where c is a constructor, then Decompose and Mutate1/2 rules may be applicable but a computation starting with a Mutate1/2 step cannot be successful.*

2. *If a goal has the form* $s =_? s'$ *and s' is not in normal form, then the Mutate2 rule need not be applied to this goal (this can be shown by strengthening the induction hypothesis in Theorems 3.6 and 3.7).*

3. *If the underlying TRS is variable-preserving then the Eliminate2 rule can be dropped from the calculus since the right-hand side of a goal is always ground at the time of its evaluation (cf. Lemma 3.5). For the same reason, the occur check in the Eliminate1 rule is superfluous.*

4. *If the underlying TRS is left-linear there need not be an occur check in the Eliminate1/2 rules, since it can be shown that a variable never occurs on both sides of a goal. Furthermore, a substitution* $\{x \leftarrow s\}$ *produced by an Eliminate2 step need not be applied to the other goals, since x does not occur in them.* □

Due to the evaluation mechanism of RR there is another simple optimization which prunes needless computations. This is an analogue to the *rejection*-rule as known from innermost narrowing:

Lemma 4.2 : (Constructor conflict) *If in* $([s_1 =_? t_1, \ldots, s_n =_? t_n], \sigma)$ *there is a goal* $s_i =_? t_i$ *where* $root(s_i)$ *is a constructor,* t_i *is not a variable and* $root(s_i) \not\equiv root(t_i)$, *then the RR computation of this goallist fails.*

Proof. Let $s_i =_? t_i$ be a goal where such a conflict appears, and σ_{i-1} the composition of the results of the $i - 1$ previous goals. Since s_i and t_i are no variables the computation of $s_i\sigma_{i-1} =_? t_i\sigma_{i-1}$ cannot proceed by Eliminate1/2. Due to $root(s_i)\sigma_{i-1} \not\equiv root(t_i)\sigma_{i-1}$ a Decompose is also not possible. So only a Mutate1/2 might be possible. Its application would yield a goal $s_i\sigma_{i-1}\sigma' =_? l\sigma'$ with $root(l)$ being a defined function. It follows inductively that on the right-hand side a constructor or variable will never be generated. So this goal is unsolvable. □

4.2 Efficiency of RR

RR has been integrated into a platform for testing different narrowing strategies. Its name is Parametrized Narrowing Machine, or *PaNaMa*, for short.[3] *PaNaMa* allows to declare signatures, TRSs, goals, etc. and computes the narrowing/RR derivations together with the solutions, dead ends, and further informations on the search process. The efficiency results below were obtained using *PaNaMa*. First, we come to an example out of the area of compiler code selection, which was presented in [Emm94].

$$
\begin{aligned}
R_1 : & \quad \delta(add(A, B)) \rightarrow plus(\delta(A), \delta(B)) \\
R_2 : & \quad \delta(mov(A)) \rightarrow \delta(A) \\
R_3 : & \quad \delta(di(A, B)) \rightarrow cont(plus(\delta^{reg}(A), \delta(B))) \\
R_4 : & \quad \delta(c(X)) \rightarrow c(X) \\
R_5 : & \quad \delta^{reg}(A) \rightarrow bb
\end{aligned}
$$

Example 4.3 : Code–Selection

Here add, mov, di denote commands from the target language, i.e. addition, movement of data and addressing with displacement. $plus, cont, bb$ are operands from the intermediate language, denoting addition, memory access and a base register. $c(X)$ makes an expression from a number X. The functions δ and δ^{reg} map target into intermediate program, and are defined via a left-linear canonical TRS. To select machine code for the intermediate program $P = plus(c(1), cont(plus(bb, c(4))))$, we must solve the goal $\delta(X) =_? P$. One out of many possible solutions is $\{X \leftarrow add(c(1), di(Y, c(4)))\}$.

The tree of all narrowing/RR derivations was computed up to a depth of 6 levels. The following table gives an efficiency comparison between different strategies (cf. [Han94] for a survey of current narrowing strategies):

strategy	# solutions found	# branches unfinished	# dead ends
RR	5	13	0
forward decomposition	5	138	92
innermost narrowing	8[4]	98406	66
outermost narrowing	1	742	23
left–right basic narr.	1	15611	389

Note that the term rewriting system is small but not artificial. It scales up to realistic code selector descriptions with hundreds of rules of a similar structure. Current compiler technology provides very efficient code selectors. By adapting

---

[3] The *PaNaMa* system and a Users manual [BT95] are available from the authors.

[4] These solutions are identical (modulo variable renaming) to the solution found by the other narrowing strategies, whereas RR and the forward decomposition calculus find 5 different solutions.

these methods, we expect that Reverse Restructuring can be refined into an efficient computational calculus for problems of this kind.

The next example is more artificial. However, such examples can still provide valuable insights, because they illustrate the situations in which a certain strategy has its strength. The example was given in [AEH94] to relate the *needed narrowing* strategy to others. We now reuse this example to discuss the behaviour of RR.

$$
\begin{array}{llll}
R_1: & f(a) \rightarrow a & R_5: & g(b(X), a) \rightarrow a \\
R_2: & f(b(X)) \rightarrow b(f(X)) & R_6: & g(b(X), b(Y)) \rightarrow c(a) \\
R_3: & f(c(X)) \rightarrow a & R_7: & g(b(X), c(Y)) \rightarrow b(a) \\
R_4: & g(a, X) \rightarrow b(a) & R_8: & g(c(X), Y) \rightarrow b(a)
\end{array}
$$

Example 4.4 : [AEH94]

We try to solve the goal $g(X, f(X)) =_? c(a)$. Innermost and basic narrowing have an infinite search space for this example. Outermost narrowing finds no solution. Needed narrowing finds the most general solution $\sigma = \{X \leftarrow b(X')\}$, trying three derivations one of which succeeds. We computed the search tree up to a depth of 10 levels[5].

The search space is finite with Reverse Restructuring and the forward decomposition calculus, and the computed solution is also the most general one. The narrowing strategies generate infinitely many solutions where each has the form $\sigma_i = \{X \leftarrow b^i(a) \mid i > 0 \in \mathbb{N}\}$.

strategy	# solutions found	# branches unfinished	# dead ends
RR	1	0	3
forward decomposition	1	0	6
innermost narrowing	16	3	2
left-right basic narr.	25	3	4
needed narrowing	1	0	2

With RR the following successful computation is obtained:

$$([g(X, f(X)) =_? c(a)], \emptyset)$$

$\underset{\rightleftharpoons}{Mutate1_{R_6}}$ $\quad ([a =_? a, g(X, f(X)) =_? g(b(X_1), b(Y_1))], \emptyset)$

$\underset{\rightleftharpoons}{2 \times Decompose}$ $\quad ([X =_? b(X_1), f(X) =_? b(Y_1)], \emptyset)$

$\underset{\rightleftharpoons}{Eliminate1}$ $\quad ([f(b(X_1)) =_? b(Y_1)], \{X \leftarrow b(X_1)\})$

$\underset{\rightleftharpoons}{Mutate1_{R_2}}$ $\quad ([f(X_2) =_? Y_1, f(b(X_1)) =_? f(b(X_2))], \{X \leftarrow b(X_1)\})$

$\underset{\rightleftharpoons}{Eliminate2}$ $\quad ([f(b(X_1)) =_? f(b(X_2))], \{X \leftarrow b(X_1), Y_1 \leftarrow f(X_2)\})$

$\underset{\rightleftharpoons}{2 \times Decompose}$ $\quad ([X_1 =_? X_2], \{X \leftarrow b(X_1), Y_1 \leftarrow f(X_2)\})$

$\underset{\rightleftharpoons}{Eliminate1}$ $\quad ([\,], \{X \leftarrow b(X_2), Y_1 \leftarrow f(X_2), X_1 \leftarrow X_2\})$

---

[5] Needed narrowing is not implemented in *PaNaMa*. The result is according to [AEH94].

The coming off well of RR results from the fact that at the beginning of the derivation only R_6 can be applied. The three failing computations chose the rules R_4, R_7 or R_8 instead of R_2 for the second Mutate1 step. These dead ends can be foreseen by using a technique which is introduced next.

4.3 Operator Derivability

In order to further reduce rule nondeterminism, we provide the RR calculus with some knowledge about the forward derivation paths. This is a first approach to combine kind of a forward strategy with a reverse strategy. We will do this by adopting the technique of *operator rewriting* [DS87] or *operator derivability*.

Definition 4.5 : (Operator TRS $\mathcal{R}_{\mathcal{OP}}$) The TRS $\mathcal{R}_{\mathcal{OP}}$ is derived from a TRS $\mathcal{R}$ by the following rules: For each defined function in $\mathcal{R}$, we add a rule $f \to f$ to $\mathcal{R}_{\mathcal{OP}}$. For each rule $g(l_1, \ldots, l_q) \to f(r_1, \ldots, r_m) \in \mathcal{R}$ where f is a defined function, we add a rule $g \to f$ to $\mathcal{R}_{\mathcal{OP}}$.

Definition 4.6 : (Operator derivability) Given a TRS $\mathcal{R}_{\mathcal{OP}}$, we call an operator f derivable from g, if there is a rewriting derivation $g \to^*_{\mathcal{R}_{\mathcal{OP}}} f$.

We now combine the Mutate1 rule with a derivability check, excluding such Mutate1 steps which will never lead to a solution.

Definition 4.7 :

Mutate1':
$([h(s_1, \ldots, s_p) =_? f(s'_1, \ldots, s'_m), L_1, \ldots, L_n], \sigma) \rightleftharpoons$
$([r_1 =_? s'_1, \ldots, r_m =_? s'_m, h(s_1, \ldots, s_p) =_? g(l_1, \ldots, l_q), L_1, \ldots, L_n], \sigma),$
if $g(l_1, \ldots, l_q) \to f(r_1, \ldots, r_m)$ is a renamed rule from $\mathcal{R}$, $h \to^*_{\mathcal{R}_{\mathcal{OP}}} g$,
and *Eliminate1 does not apply*.

In contrast to [DS87], we do not consider collapsing rules for operator derivability, which would add to $\mathcal{R}_{\mathcal{OP}}$ for every collapsing rule as many rules as there are defined functions in $\mathcal{R}$. Of course, we also need not add rules $f \to c$ to $\mathcal{R}_{\mathcal{OP}}$ where c is a constructor, since these rules would never be used in a derivation $h \to^*_{\mathcal{R}_{\mathcal{OP}}} g$, where g is a defined function. Therefore, within the framework of RR, operator derivability becomes a more effective and important optimization than in the context of [DS87].

Theorem 4.8: (completeness of RR w.r.t. operator derivability)
Let $\mathcal{R}$ be a canonical, variable-preserving or left-linear TRS. RR is complete w.r.t. normalized substitutions for goals $s =_? N$, where N is in ground normal form, if Mutate1 is replaced by Mutate1'.

Proof. We prove the claim for variable-preserving TRSs. The left-linear case is analogous.

Recapitulate the proof of Theorem 3.6. There is only one situation (case 1(ii)) where explicitly a Mutate1 step is applied. We show that the prerequisites of a Mutate1' step are also fulfilled. We considered the following innermost rewriting derivation:

$s\sigma \equiv T_0 \xrightarrow[p_1, R_{i_1}, \sigma_1]{} T_1 \to \cdots \to T_{n-1} \xrightarrow[p_n, R_{i_n}, \sigma_n]{} T_n \equiv t$, where $s \equiv f(s_1, \ldots, s_m)$, $m \geq 0$.

For the case that there are rewrite steps at root positions, we defined the j-th step as the last one of those, with $R_{i_j} : l \to r$ as the rule used. We obtained: $s\sigma \to_{\mathcal{R}}^{\leq n} T_{j-1} \equiv l\sigma_j$, $r\sigma_j \equiv T_j \to_{\mathcal{R}}^{\leq n} t$ and concluded that, if R_{i_j} was a non-collapsing rule then the constructed RR computation of $s =_? t$ started with a Mutate1 step using also R_{i_j}. The crucial point is that in $T_0 \to_{\mathcal{R}}^* T_{j-1}$ no collapsing rule is applied at a root position. Such a step $T_i \xrightarrow[\epsilon, l \to x, \sigma_{i+1}]{} T_{i+1}$ would yield $T_{i+1} \equiv x\sigma_{i+1}$. Since the rewrite derivation is innermost, σ_{i+1} is normalized and T_{i+1} is in normal form. This contradicts the further rewrite steps. Because there are no collapsing root rewrite steps in $s\sigma \to_{\mathcal{R}}^{\leq n} T_{j-1} \equiv l\sigma_j$, it follows with the definition of $\mathcal{R}_{OP}$ that $root(s\sigma) \equiv root(s) \to_{\mathcal{R}_{OP}}^* root(l) \equiv root(l\sigma_j)$, which proves our claim. $\Box$

The next example illustrates the usefulness of operator derivability in RR.

$\mathcal{R}:$	$R_1:$	$0 + X \quad \to X$	$\mathcal{R}_{OP}: + \to +$
	$R_2:$	$s(X) + Y \to s(X + Y)$	$* \to *$
	$R_3:$	$0 * X \quad \to 0$	$* \to +$
	$R_4:$	$s(X) * Y \to (X * Y) + Y$	

Example 4.9 : Addition and Multiplication

We add multiplication to our TRS (cf. Example 3.3) and reconsider the beginning of the computation:

$$s(0) + U =_? s(s(0)) \xrightarrow{*} X + Y =_? 0 + s(0),\ 1 + U =_? s(X) + Y$$

At this point only a Decompose step or the application of a Mutate1 step with rule R_4 have to be considered[6]. The path starting with the Mutate1 step leads to no solution. Since $*$ is not derivable from $+$, the Mutate1' rule does not allow this step and so this part of the search tree is pruned.

Now we come back to Example 4.4. There, the computation of the goal $g(X, f(X)) =_? c(a)$ results in the subgoal $f(b(X')) =_? b(Y')$, where Mutate1 steps with the rules R_2, R_4, R_7 and R_8 are possible, though only the selection of rule R_2 leads to a solution. Using the Mutate1' rule, the three failing computations can be avoided: since there are only constructors at the root position of the right-hand sides, the TRS $\mathcal{R}_{OP}$ consists only of the two rules $f \to_{\mathcal{R}_{OP}} f$ and $g \to_{\mathcal{R}_{OP}} g$. Thus g is not derivable from f and consequently, the rules R_4, R_7 and R_8 cannot be used for a Mutate1' step.

There is no overhead from implementing operator derivability since $\mathcal{R}_{OP}$ and its transitive closure[7] can be generated at compile time.

The runtime check of $g \to_{\mathcal{R}_{OP}}^* f$ takes constant time.

---

[6] Since $0 + s(0)$ is not in normal form, a Mutate2 step need not be performed (cf. Lemma 4.1 (2)).

[7] $\mathcal{R}_{OP}$ can be represented as a binary $n \times n$ matrix, where n is the number of defined functions in $\mathcal{R}$. Its transitive closure can be computed using the Floyd-Warshall [Flo62, War62] algorithm in time $\mathcal{O}(n^3)$.

5 Conclusion

RR can be seen as a special calculus that can be used for semantic matching if the underlying canonical TRS is variable-preserving or left-linear. By itself RR is not a sufficient basis for the implementation of a functional logic language. Such a language must also provide, of course, more general calculi for more general goals. While we have indicated that RR can behave very efficiently in certain cases, a compiler or interpreter choosing between different calculi must be based on a deeper analysis of those situations where a particular strategy performs well. Therefore the development of new program analysis methods for functional logic programs will be necessary.

Completeness proofs are always based on a correspondence between (innermost) rewrite derivations and narrowing (or semantic matching) computations. For the forward calculi, this correspondence is based on the same sequence of redex and narrowing positions. In contrast, RR breaks with that simple correspondence (cf. proof of Theorem 3.6), but still specific Mutate1/2 steps can be attributed to certain rewrite steps. Up to now, only the root position of the right-hand side of a goal is considered for a Mutate1/2 step. A further study of the correspondence described above might yield a more refined but still deterministic selection strategy which reduces rule nondeterminism.

References

[AEH94] S. Antoy, R. Echahed, and M. Hanus. A needed narrowing strategy. In *Proceedings of the 21st ACM Symposium on Principles of Programming Languages*, pages 268–278. ACM Press, 1994.

[BT94] B. Bütow and S. Thesing. Reverse restrucuring: Another method of solving algebraic equations. Forschungsberichte der Technischen Fakultät, Abteilung Informationstechnik, Report 94-07, Universität Bielefeld, 1994.

[BT95] B. Bütow and S. Thesing. PaNaMa user's manual. Forschungsberichte der Technischen Fakultät, Abteilung Informationstechnik, Report 95-01, Universität Bielefeld, 1995.

[DJ90] N. Dershowitz and J.-P. Jouannaud. Rewrite systems. In J. van Leuwen, editor, *Handbook of Theoretical Computer Science*, volume B, pages 243–320. Elsevier Science Publisher B.V., 1990.

[DMS92] N. Dershowitz, S. Mitra, and G. Sivakumar. Decidable matching for convergent systems. In *Proceedings of the 11th Conference on automatic deduction (CADE)*, pages 133–146, 1992.

[DS87] N. Dershowitz and G. Sivakumar. Solving goals in equational languages. In S. Kaplan and J.-P. Jouannaud, editors, *Proceedings of the 1st International Workshop on Conditional Term Rewriting Systems*, volume 308 of *Lecture Notes in Computer Science*, pages 45–55, Orsay, France, July 1987.

[Emm94] H. Emmelmann. *Codeselektion mit regulär gesteuerter Termersetzung*. PhD thesis, Universität Karlsruhe, 1994.

[ESL89] H. Emmelmann, F.-W. Schröer, and R. Landwehr. BEG – a generator for efficient back ends. In *Sigplan '89 Conference on Programming Language Design and Implementation*, volume 24, pages 227–237. Sigplan Notices, 1989.

[FHP92] C. Fraser, R. Henry, and T. Proebsting. BURG – fast optimal instruction selection and tree parsing. *ACM Sigplan Notices*, **27**:pages 68–76, 1992.

[Flo62] R.W. Floyd. Algorithm 97. *Communications of the ACM*, 5(6):page 345, 1962.

[Gie90] R. Giegerich. Code selection by inversion of order-sorted derivors. *Theoretical Computer Science*, **73**:pages 177–211, 1990.

[Han94] M. Hanus. The integration of functions into logic programming: From theory to practice. *Journal of Logic Programming*, **19&20**:pages 583–628, 1994.

[Klo92] J.W. Klop. Term rewriting systems. In S. Abramsky, D. Gabbay, and T. Maibaum, editors, *Handbook of Logic in Computer Science*, volume 2, pages 1–116. Oxford University Press, 1992.

[Mit90] S. Mitra. Top-down equation solving and extensions to associative and commutative theories. Masters thesis, Department of Computer and Information Sciences, University of Delaware, 1990.

[Mit94] S. Mitra. *Semantic Unification for Convergent Systems*. PhD thesis, University of Illinois, 1994.

[MM82] A. Martelli and U. Montanari. An efficient unification algorithm. *ACM Transactions on Programming Languages and Systems (TOPLAS)*, 4(2):pages 258–282, 1982.

[MMR89] A. Martelli, C. Moiso, and G.F. Rossi. Lazy unification algorithms for canonical rewrite systems. In H. Ait-Kaci and M. Nivat, editors, *Resolution of Equations in Algebraic Strucutures*, pages 245–274, New York, 1989. Academic Press,.

[War62] S. Warshall. A theorem on boolean matrices. *Journal of the ACM*, 9(1):pages 11–12, 1962.

A Complete Narrowing Calculus for Higher-Order Functional Logic Programming

Koichi Nakahara,[1]* Aart Middeldorp,[2] Tetsuo Ida[2]

[1] Canon Inc.
Shimomaruko, Ohta-ku, Tokyo 146, Japan

[2] Institute of Information Sciences and Electronics
and
Center for Tsukuba Advanced Research Alliance
University of Tsukuba, Tsukuba 305, Japan

Abstract. Using higher-order functions is standard practice in functional programming, but most functional logic programming languages that have been described in the literature lack this feature. The natural way to deal with higher-order functions in the framework of (first-order) term rewriting is through so-called applicative term rewriting systems. In this paper we argue that existing calculi for lazy narrowing either do not apply to applicative systems or handle applicative terms very inefficiently. We propose a new lazy narrowing calculus for applicative term rewriting systems and prove its completeness.

1 Introduction

There is a growing interest in combining the functional and logic programming paradigms in a single language, see Hanus [6] for a recent overview of the field. The underlying computational mechanism of most of these integrated languages is (conditional) narrowing. Examples of such languages include BABEL [9] and K-LEAF [4]. Both BABEL and K-LEAF lack higher-order features. Bosco and Giovannetti [2] extended K-LEAF to the higher-order language IDEAL. The semantics of IDEAL is given by means a translation from IDEAL programs into K-LEAF programs. González-Moreno *et al.* proposed in [5] the language SFL, a higher-order extension of BABEL. The higher-order aspects of these two languages are modeled by means of first-order applicative (conditional) constructor-based term rewriting systems. This means in particular that higher-order unification—like in the higher-order logic programming language λ-PROLOG [11]—is avoided because there are no λ-abstractions around. The following example program is taken from [5]:

$$
\begin{array}{ll}
\text{plus } 0 \, y = y & \text{map } f \, [\,] = [\,] \\
\text{plus } (S \, x) \, y = S \, (\text{plus } x \, y) & \text{map } f \, [x \,|\, y] = [f \, x \,|\, \text{map } f \, y] \\
\text{double } x = \text{plus } x \, x & \text{compose } f \, g \, x = f \, (g \, x)
\end{array}
$$

---

* Most of the work reported in this paper was carried out while the first author was at the University of Tsukuba, Doctoral Program of Engineering.

The functions **map** and **compose** are higher-order. Solving the goal

$$\mathbf{map}\,f\,[\,\mathbf{S\,0},\,\mathbf{0},\,\mathbf{S\,0}\,] = [\,\mathbf{S\,(S\,(S\,0))},\,\mathbf{S\,0},\,\mathbf{S\,(S\,(S\,0))}\,]$$

means finding a substitution for the higher-order variable f such that the value of the left-hand side of the goal equals the right-hand side. One easily verifies that

$$f \mapsto \mathbf{compose\ S\ double}$$

is a solution to the goal, but actually computing such solutions is a different matter. The operational semantics of SFL is a particular kind of conditional narrowing and González-Moreno et al. [5] prove its soundness and completeness with respect to a declarative semantics that is based on applicative algebras over Scott domains.

In this paper we are concerned with lazy narrowing strategies for applicative term rewriting systems. Most lazy narrowing strategies that have been proposed in the literature are defined for constructor-based term rewriting systems, e.g. [1, 10, 14]. An easy but important observation is that while every applicative term rewriting system is a particular kind of term rewriting system, not every applicative constructor-based term rewriting system is a constructor-based term rewriting system. Nevertheless, an applicative orthogonal (constructor-based) term rewriting system is an orthogonal term rewriting system, so lazy narrowing strategies that are defined and proved complete for the latter class can be used as a computation model for higher-order functional logic programming.

We analyze the behaviour of OINC—a simple calculus proposed in [7] which realizes lazy narrowing—for applicative orthogonal term rewriting systems. It turns out that OINC handles applicative terms very inefficiently. We transform OINC into a calculus NCA that deals with applicative terms in an efficient way and we prove the completeness of NCA. We would like to stress that the ideas developed in this paper do not depend on OINC. The only reason for choosing OINC is the simplicity of its inference rules.

This paper is organized as follows. In the next section we introduce applicative term rewriting. In Sect. 3 we recall the calculus OINC. In Sect. 4 we observe that OINC doesn't manipulate applicative term rewriting systems in a very efficient way. The new calculus NCA is defined to overcome this inefficiency. The completeness of NCA is proved in Sect. 5. Section 6 is concerned with a further optimization of our calculus, namely we extend NCA with special inference rules for dealing with strict equality in an efficient way. In Sect. 7 we compare the relative efficiency of NCA and OINC on a small example. We conclude in Sect. 8 with suggestions for future research.

2 Preliminaries

We assume the reader is familiar with the basics of term rewriting. (See [3] and [8] for extensive surveys.) In this preliminary section we recall only some less common definitions and we introduce the notion of applicative term rewriting.

The set of function symbols $\mathcal{F}$ of a term rewriting system (TRS for short) $(\mathcal{F}, \mathcal{R})$ is partitioned into disjoint sets $\mathcal{F}_{\mathcal{D}}$ and $\mathcal{F}_{\mathcal{C}}$ as follows: a function symbol f belongs to $\mathcal{F}_{\mathcal{D}}$ if there is a rewrite rule $l \to r$ in $\mathcal{R}$ such that $l = f(t_1, \ldots, t_n)$ for some terms $t_1, \ldots, t_n$, otherwise $f \in \mathcal{F}_{\mathcal{C}}$. Function symbols in $\mathcal{F}_{\mathcal{C}}$ are called *constructors*, those in $\mathcal{F}_{\mathcal{D}}$ *defined* symbols. A term built from constructors and variables is called a *constructor term*. A *constructor system* (CS for short) is a TRS with the property that the arguments $t_1, \ldots, t_n$ of every left-hand side $f(t_1, \ldots, t_n)$ of a rewrite rule are constructor terms. A left-linear TRS without critical pairs is called *orthogonal*.

We distinguish a binary function symbol $\approx$, written in infix notation. A term of the form $s \approx t$, where neither s nor t contains any occurrences of $\approx$, is called an *equation*. Observe that we do not identify the equations $s \approx t$ and $t \approx s$. A *goal* is a sequence of equations. The *empty* goal is the empty sequence and denoted by $\square$.

In applicative term rewriting we deal with applicative terms. Such terms are built from variables, constants, and a special binary function symbol *application*, which is denoted by juxtaposition of its two arguments. Examples of applicative terms are $(+ (\mathtt{S}\, 0))\, 0$ and $\mathtt{S}\, (x\, 0)$. To distinguish constants from variables in applicative terms we denote the former always in **typewriter** style. Parentheses are omitted under the convention of *association to the left*, which means that missing parentheses are restored by always taking the leftmost possibility, so $(+ (\mathtt{S}\, 0))\, 0$ and $+ (\mathtt{S}\, 0)\, 0$ denote the same term, which is different from $+ \mathtt{S}\, 0\, 0$. The *head-symbol* of an applicative term is the symbol that occurs at the leftmost-innermost position. This symbol is either a constant or a variable. For instance, the head-symbol of $+ (\mathtt{S}\, 0)\, 0$ is $+$ and the head-symbol of $x\, 0$ is x. We assume that every constant f is equipped with a natural number $arity(f)$. Intuitively this number indicates the number of arguments we have to supply in order to evaluate the function or build the data structure. In the following definition we identify a subclass of applicative terms that is used to define applicative term rewriting systems.

Definition 1. A *pattern* is an applicative term t with the property that the head-symbol of every non-variable subterm of t is a constant.

So a pattern is either a variable or a term of the form $f\, t_1 \cdots t_n$ where f is not a variable and $t_1, \ldots, t_n$ are patterns. The term $+ (\mathtt{S}\, 0)\, (+ x)$ is a pattern, but $\mathtt{S}\, (x\, 0)$ isn't. Now we are ready to define applicative term rewriting systems.

Definition 2. An *applicative rewrite rule* is a pair $l \to r$ of applicative terms such that the left-hand side l is a pattern of the form $f\, l_1 \cdots l_n$ with $n = arity(f)$, and $Var(r) \subseteq Var(l)$. An *applicative term rewriting system* ($\mathcal{A}$TRS for short) consists of applicative rewrite rules.

Every $\mathcal{A}$TRS is a TRS. Hence notions defined for TRSs like orthogonality apply to $\mathcal{A}$TRSs. We would like to point out however that the definition of CS doesn't make much sense in the context of $\mathcal{A}$TRSs. The well-known **map** function from functional programming can be specified as the following $\mathcal{A}$TRS:

$$\begin{cases} \mathtt{map}\,f\,\mathtt{nil} & \to \mathtt{nil} \\ \mathtt{map}\,f\,(:x\,y) & \to\, :(f\,x)\,(\mathtt{map}\,f\,y) \end{cases}$$

We have $arity(\mathtt{map}) = 2$. This $\mathcal{A}$TRS is not a CS because the arguments of the two left-hand sides contain (hidden) application symbols, which are in $\mathcal{F}_D$. For example, the arguments of the left-hand side $\mathtt{map}\,f\,\mathtt{nil}$ are $\mathtt{map}\,f$ and $\mathtt{nil}$, not f and $\mathtt{nil}$. Nevertheless, there is a clear separation between constants that define functions ($\mathtt{map}$) and those that build data structures ($\mathtt{nil}$ and $:$). This suggests the following definition.

Definition 3. Let $\mathcal{R}$ be an $\mathcal{A}$TRS. A constant f is said to be *applicatively defined* if it is the head-symbol of the left-hand side of some rewrite rule in $\mathcal{R}$. An *applicative constructor* is a constant that is not defined. We call $\mathcal{R}$ an *applicative constructor system* (*ACS* for short) if the terms $t_1, \ldots, t_n$ in the left-hand side $f\,t_1 \cdots t_n$ of every rewrite rule do not contain defined symbols.

The $\mathcal{A}$TRS defining the $\mathtt{map}$ function is an $\mathcal{A}$CS. We would like to stress that $\mathcal{A}$CSs are not CSs, except in trivial cases. So narrowing strategies that are defined for CSs do not apply to $\mathcal{A}$CSs.

When writing applicative terms we find it convenient to abbreviate $f\,t_1 \cdots t_n$ to $f\,\mathbf{t}_n$. Observe that $\mathbf{t}_n$ is not a term. If $n = 0$ then $f\,\mathbf{t}_n$ denotes the constant f. By the same convention $x\,\mathbf{s}_n\,\mathbf{t}_m$ stands for the term $x\,s_1 \cdots s_n\,t_1 \cdots t_m$. A term of the form $x\,\mathbf{t}_n$ is called a *head-variable* term. In the sequel, when dealing with $\mathcal{A}$TRSs, we usually omit the adjective applicative.

In this paper we consider *untyped* systems since typing does affect neither the design nor the soundness and completeness of our narrowing calculus NCA. However, in Sect. 8 we briefly explain why typing is useful to reduce the search space of NCA.

3 The Outside-In Narrowing Calculus

In this section we recall the outside-in narrowing calculus of [7] and state its completeness.

Definition 4. Let $\mathcal{R}$ be an orthogonal TRS. The *outside-in narrowing calculus*, OINC for short, consists of the following inference rules (E denotes an arbitrary sequence of equations):

[on] *outermost narrowing*

$$\frac{f(s_1, \ldots, s_n) \approx t, E}{s_1 \approx l_1, \ldots, s_n \approx l_n, r \approx t, E}\ \ t \notin \mathcal{V}$$

if there exists a fresh variant $f(l_1, \ldots, l_n) \to r$ of a rewrite rule in $\mathcal{R}$,

[d] *decomposition*

$$\frac{f(s_1, \ldots, s_n) \approx f(t_1, \ldots, t_n), E}{s_1 \approx t_1, \ldots, s_n \approx t_n, E}$$

[v] *variable elimination*

$$\frac{t \approx x, E}{E\theta} \quad \text{and} \quad \frac{x \approx t, E}{E\theta} \quad t \notin \mathcal{V}$$

with $\theta = \{x \mapsto t\}$.

We introduce some useful notations relating to the calculus OINC. If G and G' are the upper and lower goal in the inference rule $[\alpha]$ ($\alpha \in \{\text{on}, \text{d}, \text{v}\}$), we write $G \Rightarrow_{[\alpha]} G'$. This is called an OINC-*step*. The applied rewrite rule or substitution may be supplied as a subscript, that is, we will write things like $G \Rightarrow_{[\text{on}], l \to r} G'$ and $G \Rightarrow_{[\text{v}], \theta} G'$. A finite OINC-*derivation* $G_1 \Rightarrow_{\theta_1} \cdots \Rightarrow_{\theta_{n-1}} G_n$ may be abbreviated to $G_1 \Rightarrow_\theta^* G_n$ with $\theta = \theta_1 \cdots \theta_{n-1}$. A *successful* OINC-derivation ends in the empty goal $\square$. The number of steps in an OINC-derivation $A: G \Rightarrow^* G'$ is denoted by $|A|$. If $|A| \geqslant 1$ then $A_{>1}$ denotes the derivation obtained from A by omitting the first step.

The calculus OINC has been designed with the restriction in mind that initial goals are right-normal.

Definition 5. A goal G is called *right-normal* if the right-hand side t of every equation $s \approx t$ in G is a ground normal form. A goal G' is called *proper* if there exists an OINC-derivation $G \Rightarrow^* G'$ starting from a right-normal goal G.

The restriction to proper goals is motivated from the understanding that functional logic programming languages deal with so-called *strict* equality in order to model non-termination correctly. Every goal consisting of strict equations is right-normal, see Sect. 6.

It is not difficult to show that the term t in a proper goal $E_1, s \approx t, E_2$ has no variables in common with s and E_1. This explains why we don't need the occur-check in the variable elimination rules. It also explains why there is no *imitation* rule in OINC. Finally, the absence of the symmetric outermost narrowing rule

$$\frac{t \approx f(s_1, \ldots, s_n), E}{s_1 \approx l_1, \ldots, s_n \approx l_n, t \approx r, E}$$

is easily explained by the restriction to orthogonal TRSs and proper goals.

Definition 6. Let $\mathcal{R}$ be a TRS and G a goal. A substitution θ is called a *solution* of G if $s\theta =_\mathcal{R} t\theta$ for every equation $s \approx t$ in G.

Ida and Nakahara [7] obtained the following soundness and completeness result.

Theorem 7. *Let $\mathcal{R}$ be an orthogonal TRS and G a right-normal goal. If there exists a successful OINC-derivation $G \Rightarrow_\theta^* \square$ then θ is a solution of G. For every normalizable solution θ of G there exists a successful OINC-derivation $G \Rightarrow_{\theta'}^* \square$ such that $\theta' \leqslant_\mathcal{R} \theta \, [\mathcal{V}ar(G)]$.* $\square$

If the substitution θ in the second part of Theorem 7 is normalized, the subscript $\mathcal{R}$ can be dropped.

4   A Narrowing Calculus for Applicative Systems

Every $\mathcal{A}$TRS is a TRS, hence OINC is complete (in the sense of the second part of Theorem 7) for orthogonal $\mathcal{A}$TRSs. However, OINC doesn't handle applicative terms very efficiently. The problem is that the applicable inference rules and (in the case of [on]) rewrite rules are determined by the outermost symbol of the left-hand side of the leftmost equation of the current goal. In the context of $\mathcal{A}$TRSs this outermost symbol is almost always the binary application symbol, which doesn't carry any useful information for restricting the choice of inference and rewrite rules. Let us consider two examples.

Example 1. The (right-normal) goal $c\,x\,y \approx c\,a\,b$ is solved by the following OINC-derivation:

$$
\begin{aligned}
c\,x\,y \approx c\,a\,b \Rightarrow_{[d]} \quad & c\,x \approx c\,a, y \approx b \\
\Rightarrow_{[d]} \quad & c \approx c, x \approx a, y \approx b \\
\Rightarrow_{[d]} \quad & x \approx a, y \approx b \\
\Rightarrow_{[v],\,\{x \mapsto a\}} \quad & y \approx b \\
\Rightarrow_{[v],\,\{y \mapsto b\}} \quad & \square.
\end{aligned}
$$

Recall that the term $c\,x\,y$ has the (hidden) application symbol as root with $c\,x$ and y as arguments. Likewise the term $c\,a\,b$ consists of the application of $c\,a$ and b. Hence the inference rule [d] decomposes the equation $c\,x\,y \approx c\,a\,b$ into $c\,x \approx c\,a$ and $y \approx b$. We need three decomposition steps before we can bind the variables.

Example 2. Consider the orthogonal $\mathcal{A}$TRS

$$
\mathcal{R} = \begin{cases} \text{inc} & \rightarrow \text{add1} \\ \text{add1}\,x & \rightarrow S\,x \end{cases}
$$

The following OINC-derivation computes the solution $\{x \mapsto 0\}$ of the goal $\text{inc}\,x \approx S\,0$:

$$
\begin{aligned}
\text{inc}\,x \approx S\,0 \Rightarrow_{[on],\,\text{add1}\,x_1 \rightarrow S\,x_1} \quad & \text{inc} \approx \text{add1}, x \approx x_1, S\,x_1 \approx S\,0 \\
\Rightarrow_{[on],\,\text{inc} \rightarrow \text{add1}} \quad & \text{add1} \approx \text{add1}, x \approx x_1, S\,x_1 \approx S\,0 \\
\Rightarrow_{[d]} \quad & x \approx x_1, S\,x_1 \approx S\,0 \\
\Rightarrow_{[v],\,\{x_1 \mapsto x\}} \quad & S\,x \approx S\,0 \\
\Rightarrow_{[d]} \quad & S \approx S, x \approx 0 \\
\Rightarrow_{[d]} \quad & x \approx 0 \\
\Rightarrow_{[v],\,\{x \mapsto 0\}} \quad & \square.
\end{aligned}
$$

It is essential that we choose the (renamed) rewrite rule $\text{add1}\,x_1 \rightarrow S\,x_1$ for the equation $\text{inc}\,x \approx S\,0$ in the first outermost narrowing step. However, we have no way to implement this choice. In order to ensure completeness all rewrite rules whose left-hand side is not a single constant must be used in combination with the outermost narrowing rule.

We overcome the problems mentioned above by looking at the head-symbol rather than the outermost symbol of the left-hand side of the equation under consideration. This is natural since the head-symbol of an applicative term corresponds to the outermost symbol of a functional term. The narrowing calculus defined below implements this idea.

Definition 8. Let $\mathcal{R}$ be an orthogonal $\mathcal{A}$TRS. The calculus NCA—Narrowing Calculus for Applicative TRSs—consists of the following five inference rules:

[ona] *outermost narrowing of applicative terms*

$$\frac{f\,s_n\,t_m \approx t, E}{s_1 \approx u_1, \ldots, s_n \approx u_n, r\,t_m \approx t, E}\ t \notin \mathcal{V}$$

if there exists a fresh variant $f\,u_n \to r$ of a rewrite rule in $\mathcal{R}$,

[onv] *outermost narrowing of head-variable terms*

$$\frac{x\,s_n\,t_m \approx t, E}{(s_1 \approx v_1, \ldots, s_n \approx v_n, r\,t_m \approx t, E)\theta}\ t \notin \mathcal{V}$$

if there exists a fresh variant $f\,u_k\,v_n \to r$ of a rewrite rule in $\mathcal{R}$, $n > 0$, and $\theta = \{x \mapsto f\,u_k\}$,

[da] *decomposition of applicative terms*

$$\frac{f\,s_n \approx f\,t_n, E}{s_1 \approx t_1, \ldots, s_n \approx t_n, E}$$

[dv] *decomposition of head-variable terms*

$$\frac{x\,s_n \approx f\,t_m\,u_n, E}{(s_1 \approx u_1, \ldots, s_n \approx u_n, E)\theta}$$

if $\theta = \{x \mapsto f\,t_m\}$,

[v] *variable elimination*

$$\frac{t \approx x, E}{E\theta}$$

if $\theta = \{x \mapsto t\}$.

Observe that the second variable elimination rule of OINC is subsumed by the inference rule [dv] of NCA. In order to distinguish NCA-derivations from OINC-derivations, we use $\Rrightarrow$ instead of $\Rightarrow$ for the former.

Example 3. The goal $c\,x\,y \approx c\,a\,b$ is solved by the following NCA-derivation:

$$c\,x\,y \approx c\,a\,b \Rrightarrow_{[da]} \qquad x \approx a, y \approx b$$
$$\Rrightarrow_{[dv], \{x \mapsto a\}} y \approx b$$
$$\Rrightarrow_{[dv], \{y \mapsto b\}} \square.$$

With respect to the $\mathcal{ATRS}$ $\mathcal{R}$ of Example 2, the goal $\mathtt{inc}\,x \approx \mathsf{S}\,0$ is solved by the following NCA-derivation:

$$\mathtt{inc}\,x \approx \mathsf{S}\,0 \Rightarrow_{[\mathrm{ona}],\,\mathtt{inc}\to\mathtt{add1}} \quad \mathtt{add1}\,x \approx \mathsf{S}\,0$$
$$\Rightarrow_{[\mathrm{ona}],\,\mathtt{add1}\,x_1\to\mathsf{S}\,x_1} \quad x \approx x_1, \mathsf{S}\,x_1 \approx \mathsf{S}\,0$$
$$\Rightarrow_{[\mathrm{v}],\,\{x_1\mapsto x\}} \quad \mathsf{S}\,x \approx \mathsf{S}\,0$$
$$\Rightarrow_{[\mathrm{da}]} \quad x \approx 0$$
$$\Rightarrow_{[\mathrm{dv}],\,\{x\mapsto 0\}} \quad \square.$$

The inference rule [onv] of NCA is used to bind higher-order logical variables. This is illustrated in the next example.

Example 4. Consider the orthogonal $\mathcal{ATRS}$

$$\mathcal{R} = \begin{cases} \mathtt{plus}\,0\,x & \to x & (1) \\ \mathtt{plus}\,(\mathsf{S}\,x)\,y \to \mathsf{S}\,(\mathtt{plus}\,x\,y) & (2) \\ \mathtt{map}\,f\,\mathtt{nil} & \to \mathtt{nil} & (3) \\ \mathtt{map}\,f\,(x:y) \to (f\,x):(\mathtt{map}\,f\,y) & (4) \end{cases}$$

Here : is a binary constructor, written in infix notation, and $\mathtt{nil}$ is a constant denoting the empty list. We adopt the usual convention of writing $[t_1,\ldots,t_n]$ to denote the list $(t_1:(\cdots(t_n:\mathtt{nil})\cdots))$. The goal $\mathtt{map}\,x\,[\mathsf{S}\,0] \approx [\mathsf{S}\,0]$ is solved by the following NCA-derivation:

$$\mathtt{map}\,x\,[\mathsf{S}\,0] \approx [\mathsf{S}\,0]$$
$$\Rightarrow_{[\mathrm{ona}],\,(4)} \quad x \approx f_1, [\mathsf{S}\,0] \approx (x_1:y_1), (f_1\,x_1):(\mathtt{map}\,f_1\,y_1) \approx [\mathsf{S}\,0]$$
$$\Rightarrow_{[\mathrm{v}],\,\{f_1\mapsto x\}} \quad [\mathsf{S}\,0] \approx (x_1:y_1), (x\,x_1):(\mathtt{map}\,x\,y_1) \approx [\mathsf{S}\,0]$$
$$\Rightarrow_{[\mathrm{da}]} \quad \mathsf{S}\,0 \approx x_1, \mathtt{nil} \approx y_1, (x\,x_1):(\mathtt{map}\,x\,y_1) \approx [\mathsf{S}\,0]$$
$$\Rightarrow_{[\mathrm{v}],\,\{x_1\mapsto \mathsf{S}\,0\}} \quad \mathtt{nil} \approx y_1, (x\,(\mathsf{S}\,0)):(\mathtt{map}\,x\,y_1) \approx [\mathsf{S}\,0]$$
$$\Rightarrow_{[\mathrm{v}],\,\{y_1\mapsto \mathtt{nil}\}} \quad (x\,(\mathsf{S}\,0)):(\mathtt{map}\,x\,\mathtt{nil}) \approx [\mathsf{S}\,0]$$
$$\Rightarrow_{[\mathrm{da}]} \quad x\,(\mathsf{S}\,0) \approx \mathsf{S}\,0, \mathtt{map}\,x\,\mathtt{nil} \approx \mathtt{nil}$$
$$\Rightarrow_{[\mathrm{onv}],\,(1),\,\{x\mapsto \mathtt{plus}\,0\}} \mathsf{S}\,0 \approx y_2, y_2 \approx \mathsf{S}\,0, \mathtt{map}\,(\mathtt{plus}\,0)\,\mathtt{nil} \approx \mathtt{nil}$$
$$\Rightarrow_{[\mathrm{v}],\,\{y_2\mapsto \mathsf{S}\,0\}} \quad \mathsf{S}\,0 \approx \mathsf{S}\,0, \mathtt{map}\,(\mathtt{plus}\,0)\,\mathtt{nil} \approx \mathtt{nil}$$
$$\Rightarrow^{+}_{[\mathrm{da}]} \quad \mathtt{map}\,(\mathtt{plus}\,0)\,\mathtt{nil} \approx \mathtt{nil}$$
$$\Rightarrow_{[\mathrm{ona}],\,(3)} \quad \mathtt{plus}\,0 \approx f_3, \mathtt{nil} \approx \mathtt{nil}, \mathtt{nil} \approx \mathtt{nil}$$
$$\Rightarrow_{[\mathrm{v}],\,\{f_3\mapsto \mathtt{plus}\,0\}} \quad \mathtt{nil} \approx \mathtt{nil}, \mathtt{nil} \approx \mathtt{nil}$$
$$\Rightarrow^{+}_{[\mathrm{da}]} \quad \square.$$

Note that in the $\Rightarrow_{[\mathrm{onv}]}$-step the variable x is bound to the higher-order term $(\mathtt{plus}\,0)$.

Soundness of NCA is expressed in the following theorem.

Theorem 9. *Let $\mathcal{R}$ be an orthogonal TRS and G a right-normal goal. If there exists a successful NCA-derivation $G \Rightarrow^{*}_{\theta} \square$ then θ is a solution of G.*

Proof. Straightforward induction on the length of the successful NCA-derivation $G \Rightarrow^{*}_{\theta} \square$. $\qquad\square$

5 Completeness

In this section we establish the completeness of NCA. The idea behind the proof is straightforward. We show that for every non-empty OINC-derivation $A: G \Rightarrow^+_\theta \square$ there exist an NCA-step $G \Rightarrow_\sigma G'$ and an OINC-derivation $A': G' \Rightarrow^*_{\theta'} \square$ such that $\theta = \sigma\theta'$ and $|A'| < |A|$. Completeness of NCA is then reduced to the completeness of OINC by a routine induction argument. First we define an appropriate notion of descendant for OINC-derivations.

Definition 10. Let $\mathcal{R}$ be an $\mathcal{A}$TRS. In the OINC-derivation

$$E \Rightarrow^* s_1\, s_2 \approx t_1\, t_2, E_1 \Rightarrow_{[\mathrm{d}]} s_1 \approx t_1, s_2 \approx t_2, E_1 \Rightarrow^* E_2$$

the equation $s_1 \approx t_1$ is called an *immediate descendant* of the equation $s_1\, s_2 \approx t_1\, t_2$. In the OINC-derivation

$$E \Rightarrow^* s_1\, s_2 \approx t, E_1 \Rightarrow_{[\mathrm{on}]} s_1 \approx l_1, s_2 \approx l_2, r \approx t, E_1 \Rightarrow^* E_2$$

where $l_1\, l_2 \to r$ is a fresh variant of a rewrite rule in $\mathcal{R}$, the equation $s_1 \approx l_1$ is called an *immediate descendant* of the equation $s_1\, s_2 \approx t$. The notion of immediate descendants generalizes to a notion of *descendant* by reflexivity and transitivity.

In Lemmata 11–16 we observe basic properties of successful OINC-derivations.

Lemma 11. *Let $G = f\, \mathsf{s}_n \approx g\, \mathsf{t}_m, E$ be a goal such that $f \neq g$ or $n \neq m$. In all successful OINC-derivations starting from G the rule $[\mathrm{on}]$ is applied to a descendant of $f\, \mathsf{s}_n \approx g\, \mathsf{t}_m$.*

Proof. Obvious. $\square$

Lemma 12. *Let $G = x\, \mathsf{s}_n \approx g\, \mathsf{t}_m, E$ be a goal such that $m < n$. In all successful OINC-derivations starting from G the rule $[\mathrm{on}]$ is applied to a descendant of $x\, \mathsf{s}_n \approx g\, \mathsf{t}_m$.*

Proof. Obvious. $\square$

Lemma 13. *Let $G = f\, \mathsf{s}_n \approx t, E$ be a goal such that $n < arity(f)$. There exist no successful OINC-derivations starting from G in which the rule $[\mathrm{on}]$ is applied to a descendant of $f\, \mathsf{s}_n \approx t$.*

Proof. We use induction on n. If $n = 0$ then there are no rewrite rules of the form $f \to r$ because $arity(f) > 0$ and hence $[\mathrm{on}]$ is not applicable. Suppose $n > 0$. Let A be an arbitrary successful OINC-derivation starting from G. We distinguish the following three cases:

1. Suppose the inference rule $[\mathrm{d}]$ is applied in the first step of A, so A can be written as

   $$f\, \mathsf{s}_n \approx t, E \Rightarrow_{[\mathrm{d}]} f\, \mathsf{s}_{n-1} \approx t_1, s_n \approx t_2, E \Rightarrow^* \square$$

   with $t = t_1\, t_2$. According to the induction hypothesis the inference rule $[\mathrm{on}]$ is not applied to a descendant of $f\, \mathsf{s}_{n-1} \approx t_1$ in the subderivation $A_{>1}$. Therefore $[\mathrm{on}]$ is not applied to a descendant of $f\, \mathsf{s}_n \approx t$ in A.

2. Suppose the inference rule [on] is applied in the first step of A, so A can be written as

$$f\,\mathbf{s}_n \approx t, E \Rightarrow_{[\text{on}],\, l_1\, l_2 \to r} f\,\mathbf{s}_{n-1} \approx l_1, s_n \approx l_2, r \approx t, E \Rightarrow^* \Box$$

for some rewrite rule $l_1\, l_2 \to r$. We have to show that this is impossible. According to the induction hypothesis the inference rule [on] is not applied to a descendant of $f\,\mathbf{s}_{n-1} \approx l_1$ in the subderivation $A_{>1}$. Because $l_1\, l_2$ is a pattern, l_1 is not a head-variable term, so we may write $l_1 = g\,\mathbf{u}_m$ with $m = arity(g) - 1$. We have either $f \neq g$ or $n - 1 \neq m$. Now Lemma 11 yields the desired contradiction.

3. If the inference rule [v] is applied in the first step of A then by definition there are no descendants of $f\,\mathbf{s}_n \approx t$ left in $A_{>1}$ to which [on] can be applied. Therefore [on] is not applied to a descendant of $f\,\mathbf{s}_n \approx t$ in A.

$\Box$

Lemma 14. *Let $G = f\,\mathbf{s}_n \approx f\,\mathbf{t}_n, E$ be a goal such that $n < arity(f)$. In every successful* OINC-*derivation starting from G the rule [d] is applied to all descendants of $f\,\mathbf{s}_n \approx f\,\mathbf{t}_n$.*

Proof. Easy consequence of Lemma 13. $\Box$

Lemma 15. *Let $G = f\,\mathbf{s}_n \approx t, E$ be a goal such that $n = arity(f)$. If there exists a successful* OINC-*derivation starting from G in which the first step is an application of the rule [on], then the rewrite rule used in this step is of the form $f\,\mathbf{u}_n \to r$.*

Proof. Easy consequence of Lemmata 11 and 13. $\Box$

Lemma 16. *Let $G = f\,\mathbf{s}_n\,\mathbf{t}_m \approx t, E$ be a goal such that $n = arity(f)$ and $m > 0$. If there exists a successful* OINC-*derivation A starting from G in which the rule [on] is applied to a descendant $f\,\mathbf{s}_n\,\mathbf{t}_k \approx t'$ ($1 \leqslant k \leqslant m$) then [on] is applied to the descendant $f\,\mathbf{s}_n \approx t''$ of $f\,\mathbf{s}_n\,\mathbf{t}_m \approx t$.*

Proof. Without loss of generality we may write A as

$$
\begin{aligned}
G \Rightarrow^* \quad & f\,\mathbf{s}_n\,\mathbf{t}_k \approx t', E' \\
\Rightarrow_{[\text{on}],\, g\,\mathbf{u}_\ell \to r} \quad & f\,\mathbf{s}_n\,\mathbf{t}_{k-1} \approx g\,\mathbf{u}_{\ell-1}, t_k \approx u_\ell, r \approx t', E' \\
\Rightarrow^*_{[\text{d}]} \quad & f\,\mathbf{s}_n \approx g\,\mathbf{u}_{\ell-k}, E'' \\
\Rightarrow^*_\theta \quad & \Box
\end{aligned}
$$

where $g\,\mathbf{u}_\ell \to r$ is a fresh variant of a rewrite rule in $\mathcal{R}$. Note that $\ell = n$ if $f = g$. So we have $f \neq g$ or $n \neq \ell - k$. According to Lemmata 11 and 13 the inference rule [on] is applied to the equation $f\,\mathbf{s}_n \approx g\,\mathbf{u}_{\ell-k}$. $\Box$

In Lemmata 17–21 we prove that for certain successful OINC-derivation $A: G \Rightarrow^+_\theta \Box$ there exists an OINC-derivation $A': G' \Rightarrow^*_{\theta'} \Box$ such that $G \Rightarrow_\sigma G'$, $\theta = \sigma\theta'$, and $|A'| < |A|$. In Lemma 22 we show that there are no other cases to consider.

Lemma 17. *Let* $A: s \approx x, E \Rightarrow_\theta^+ \square$ *be an* OINC-*derivation. There exists an* OINC-*derivation* $A': E\sigma \Rightarrow_{\theta'}^* \square$ *with* $\sigma = \{x \mapsto s\}$ *such that* $\theta = \sigma\theta'$ *and* $|A'| < |A|$.

Proof. The first step of A must be an application of the inference rule [v]:

$$A: s \approx x, E \Rightarrow_{[v], \sigma} E\sigma \Rightarrow_{\theta'}^* \square$$

with $\theta = \sigma\theta'$ and $\sigma = \{x \mapsto s\}$. Define $A' = A_{>1}$. We clearly have $|A'| = |A| - 1 < |A|$. $\square$

The initial goals of A and A' in Lemma 17 are connected by a $\Rightarrow_{[v], \sigma}$-step.

Lemma 18. *Let* $A: f\, s_n \approx f\, t_n, E \Rightarrow_\theta^+ \square$ *be an* OINC-*derivation. If* [on] *is never applied to a descendant of* $f\, s_n \approx f\, t_n$ *then there exists an* OINC-*derivation* $A': s_1 \approx t_1, \ldots, s_n \approx t_n, E \Rightarrow_\theta^* \square$ *such that* $|A'| < |A|$.

Proof. By induction on n. If $n = 0$ then we take $A' = A_{>1}$. In this case we clearly have $|A'| < |A|$. Suppose $n > 0$. The first step of A must be an application of [d], so we may write A as

$$f\, s_n \approx f\, t_n, E \Rightarrow_{[d]} f\, s_{n-1} \approx f\, t_{n-1}, s_n \approx t_n, E \Rightarrow_\theta^+ \square.$$

An application of the induction hypothesis to the OINC-derivation $A_{>1}$ yields an OINC-derivation

$$A': s_1 \approx t_1, \ldots, s_{n-1} \approx t_{n-1}, s_n \approx t_n, E \Rightarrow_\theta^* \square$$

with $|A'| < |A_{>1}| = |A| - 1 < |A|$. $\square$

Note that the initial goals of A and A' in Lemma 18 are connected by a $\Rightarrow_{[da]}$-step.

Lemma 19. *Let* $A: x\, s_n \approx f\, t_m\, u_n \Rightarrow_\theta^+ \square$ *be an* OINC-*derivation. If* [on] *is never applied to a descendant of* $x\, s_n \approx f\, t_m\, u_n$ *then there exists an* OINC-*derivation* $A': (s_1 \approx u_1, \ldots, s_n \approx u_n, E)\sigma \Rightarrow_{\theta'}^* \square$ *with* $\sigma = \{x \mapsto f\, t_m\}$ *such that* $\theta = \sigma\theta'$ *and* $|A'| < |A|$.

Proof. Similar to the proof of Lemma 18. $\square$

The initial goals of A and A' in Lemma 19 are connected by a $\Rightarrow_{[dv], \sigma}$-step.

Lemma 20. *Let* $A: f\, s_n\, t_m \approx t \Rightarrow_\theta^+ \square$ *be an* OINC-*derivation with* $n = arity(f)$. *If* [on] *is applied to the descendant* $f\, s_n \approx t'$ *of* $f\, s_n\, t_m \approx t$ *using the rewrite rule* $f\, u_n \to r$ *then there exists an* OINC-*derivation*

$$A': s_1 \approx u_1, \ldots, s_n \approx u_n, r\, t_m \approx t, E \Rightarrow_\theta^* \square$$

such that $|A'| < |A|$.

Proof. We use induction on m. If $m = 0$ then the inference rule [on] with rewrite rule $f\, u_n \to r$ is used in the first step of A. If $n = 0$ then we take $A' = A_{>1}$. If $n > 0$ then we may write A as

$$f\, s_n \approx t, E \Rightarrow_{[\mathrm{on}]} f\, s_{n-1} \approx f\, u_{n-1}, s_n \approx u_n, r \approx t, E \Rightarrow_\theta^+ \square.$$

According to Lemma 14 the inference rule [on] is not applied to descendants of $f\, s_{n-1} \approx f\, u_{n-1}$ in the subderivation $A_{>1}$. Hence we can apply Lemma 18 to $A_{>1}$. This yields an OINC-derivation

$$A': s_1 \approx u_1, \ldots, s_{n-1} \approx u_{n-1}, s_n \approx u_n, r \approx t, E \Rightarrow_\theta^* \square$$

such that $|A'| < |A_{>1}| = |A| - 1 < |A|$. For the induction step, suppose $m > 0$. Let us abbreviate $s_1 \approx u_1, \ldots, s_n \approx u_n$ to E'. We distinguish the following cases:

1. Suppose the inference rule [d] is used in the first step of A. This means that we may write A as $s \approx t, E \Rightarrow_{[\mathrm{d}]} f\, s_n\, t_{m-1} \approx v_1, t_m \approx v_2, E \Rightarrow_\theta^+ \square$ with $t = v_1\, v_2$. An application of the induction hypothesis to the subderivation $A_{>1}$ yields an OINC-derivation

$$B: E', r\, t_{m-1} \approx v_1, t_m \approx v_2, E \Rightarrow_\theta^* \square$$

such that $|B| < |A_{>1}| < |A|$. The OINC-derivation B can be split into

$$B_1: E', r\, t_{m-1} \approx v_1, t_m \approx v_2, E \Rightarrow_{\theta_1}^* (r\, t_{m-1} \approx v_1, t_m \approx v_2, E)\theta_1$$

and

$$B_2: (r\, t_{m-1} \approx v_1, t_m \approx v_2, E)\theta_1 \Rightarrow_{\theta_2}^* \square$$

with $\theta = \theta_1 \theta_2$. The OINC-derivation B_1 can easily be transformed into the OINC-derivation

$$C: E', r\, t_m \approx v_1\, v_2, E \Rightarrow_{\theta_1}^* (r\, t_m \approx v_1\, v_2, E)\theta_1.$$

Because $m > 0$ we can apply the inference rule [d] to the final goal $(r\, t_m \approx v_1\, v_2, E)\theta_1$ of C, yielding the OINC-step

$$D: (r\, t_m \approx v_1\, v_2, E)\theta_1 \Rightarrow_{[\mathrm{d}]} (r\, t_{m-1} \approx v_1, t_m \approx v_2, E)\theta_1.$$

Concatenating the three OINC-derivations C, D, and B_2 yields the desired OINC-derivation

$$A': E', r\, t_m \approx t, E \Rightarrow_\theta^* \square.$$

Note that $|A'| = |C| + |D| + |B_2| = |B| + 1 < |A|$.

2. Suppose the inference rule [on] is used in the first step of A. This means that there exists a fresh variant $v_1\, v_2 \to r'$ of a rewrite rule in $\mathcal{R}$ such that A can be written as

$$s \approx t, E \Rightarrow_{[\mathrm{on}]} f\, s_n\, t_{m-1} \approx v_1, t_m \approx v_2, r' \approx t, E \Rightarrow_\theta^+ \square$$

An application of the induction hypothesis to $A_{>1}$ yields an OINC-derivation

$$B: E', r\, t_{m-1} \approx v_1, t_m \approx v_2, r' \approx t, E \Rightarrow_\theta^* \square$$

such that $|B| < |A_{>1}| < |A|$. The OINC-derivation B can be split into

$$B_1: E', r\,\mathbf{t}_{m-1} \approx v_1, t_m \approx v_2, r' \approx t, E \Rightarrow^*_{\theta_1} (r\,\mathbf{t}_{m-1} \approx v_1, t_m \approx v_2, r' \approx t, E)\theta_1$$

and

$$B_2: (r\,\mathbf{t}_{m-1} \approx v_1, t_m \approx v_2, r' \approx t, E)\theta_1 \Rightarrow^*_{\theta_2} \square$$

with $\theta = \theta_1\theta_2$. We transform B_1 into the OINC-derivation

$$C: E', r\,\mathbf{t}_m \approx t, E \Rightarrow^*_{\theta_1} (r\,\mathbf{t}_m \approx t, E)\theta_1.$$

Next we apply the inference rule [on] to the final goal $(r\,\mathbf{t}_m \approx t, E)\theta_1$ of C, using exactly the same variant $v_1\,v_2 \to r'$. This yields the OINC-step

$$D: (r\,\mathbf{t}_m \approx t, E)\theta_1 \Rightarrow_{[\text{on}]} (r\,\mathbf{t}_{m-1} \approx v_1, t_m \approx v_2, r' \approx t, E)\theta_1.$$

Also in this case the desired OINC-derivation A' is obtained by concatenating C, D, and B_2.

3. It is not possible that the first step of A is an application of the variable elimination rule [v] because then there is no descendant left to which [on] can be applied.

$\square$

Observe that the initial goals of A and A' in Lemma 20 are connected by a $\Rightarrow_{[\text{ona}]}$-step.

Lemma 21. *Let* $A: x\,\mathbf{s}_n\,\mathbf{t}_m \approx t, E \Rightarrow^+_\theta \square$ *be an* OINC-*derivation with* $n > 0$. *If* [on] *is applied to the descendant* $x\,\mathbf{s}_n \approx t'$ *of* $x\,\mathbf{s}_n\,\mathbf{t}_m \approx t$ *using the rewrite rule* $f\,\mathbf{u}_k\,\mathbf{v}_n \to r$ *and* $x\,\mathbf{s}_n \approx t'$ *is the last descendant to which* [on] *is applied then there exists an* OINC-*derivation* $A': (s_1 \approx v_1, \ldots, s_n \approx v_n, r\,\mathbf{t}_m \approx t, E)\sigma \Rightarrow^*_{\theta'} \square$ *with* $\sigma = \{x \mapsto f\,\mathbf{u}_k\}$ *such that* $\theta = \sigma\theta'$ *and* $|A'| < |A|$.

Proof. Similar to the proof of Lemma 20. $\square$

In Lemma 21 the initial goals of A and A' are connected by a $\Rightarrow_{[\text{onv}], \sigma}$-step.

Lemma 22. *Let* G *be a proper goal. For every* OINC-*derivation* $A: G \Rightarrow^+_\theta \square$ *there exist an* NCA-*step* $G \Rightarrow_\sigma G'$ *and an* OINC-*derivation* $A': G' \Rightarrow^*_{\theta'} \square$ *such that* $\theta = \sigma\theta'$ *and* $|A'| < |A|$.

Proof. We have to show that Lemmata 17–21 cover all possible cases. Let G be the goal $s \approx t, E$. The case that t is a variable is covered by Lemma 17, so we may assume that t is not a variable. Because G is proper the right-hand side t is a pattern. Hence t is not a head-variable term. We distinguish two cases.

1. Suppose [on] is not applied to a descendant of $s \approx t$.
   (a) If the head-symbol of s is a constant f then, according to Lemma 11, we must have $s = f\,\mathbf{s}_n$ and $t = f\,\mathbf{t}_n$. This case is covered by Lemma 18.
   (b) If the head-symbol of s is a variable x then, according to Lemma 12, $s = x\,\mathbf{s}_n$ and $t = f\,\mathbf{t}_m\,\mathbf{u}_n$. This case is covered by Lemma 19.

2. Suppose [on] is applied to a descendant of $s \approx t$.

    (a) If the head-symbol of s is a constant f then, according to Lemma 13, we have $s = f\, s_n\, t_m$ with $n = arity(f)$. From Lemma 13 we also infer that [on] is never applied to descendants of the form $f\, s_k \approx t'$ with $k < arity(f)$. Hence the application of [on] is to a descendant of the form $f\, s_n\, t_k \approx t''$ with $0 \leqslant k \leqslant m$. According to Lemma 16 [on] is applied to the descendant $f\, s_n \approx t'''$. Lemma 15 states that the employed rewrite rule is of the form $f\, u_n \to r$. Hence this case is covered by Lemma 20.

    (b) The case that the head-symbol of s is a variable x is covered by Lemma 21, using similar reasoning as in the previous case.

□

The completeness of NCA is an easy consequence of the previous lemma and the completeness of OINC (Theorem 7).

Theorem 23. *Let $\mathcal{R}$ be an orthogonal ATRS and G a right-normal goal. For every normalizable solution θ of G there exists a successful NCA-derivation $G \Rightarrow^{*}_{\theta'} \square$ such that $\theta' \leqslant_{\mathcal{R}} \theta\ [Var(G)]$.*

Proof. According to the second part of Theorem 7 there exists a successful OINC-derivation $A\colon G \Rightarrow^{*}_{\theta'} \square$ such that $\theta' \leqslant_{\mathcal{R}} \theta\ [Var(G)]$. By induction on $|A|$ we show the existence of a successful NCA-derivation $G \Rightarrow^{*}_{\theta'} \square$. The case $|A| = 0$ is trivial. Suppose $|A| > 0$. According to Lemma 22 there exist an NCA-step $G \Rightarrow_{\sigma} G'$ and an OINC-derivation $A'\colon G' \Rightarrow^{*}_{\theta''} \square$ such that $\theta' = \sigma\theta''$ and $|A'| < |A|$. The induction hypothesis yields an successful NCA-derivation $G' \Rightarrow^{*}_{\theta''} \square$. Combining this derivation with the NCA-step $G \Rightarrow_{\sigma} G'$ yields the desired NCA-derivation $G \Rightarrow^{*}_{\theta'} \square$.

□

Inspection of the above proofs reveals that the length of the resulting NCA-derivation $G \Rightarrow^{*} \square$ never exceeds the length of the OINC-derivation $G \Rightarrow^{*} \square$. Moreover, it is not difficult to see that for every application of Lemmata 18–21 in the transformation from $G \Rightarrow^{*} \square$ to $G \Rightarrow^{*} \square$ we gain n steps (corresponding to applications of the inference rule [d] in the OINC-derivation $G \Rightarrow^{*} \square$).

6 Incorporating Strict Equality into NCA

In functional logic programming languages like K-LEAF [4] and BABEL [9] two expressions are considered to be equal if and only if they reduce to the same ground constructor normal form. This so-called *strict equality* is adopted to model non-termination correctly. In the framework of applicative term rewriting, strict equality is realized by *adding* the rewrite rules

$$\begin{cases} c \equiv c & \to \mathbf{true} & \text{if } c \text{ is a nullary constructor,} \\ c\, x_n \equiv c\, y_n \to x_1 \equiv y_1 \wedge \cdots \wedge x_n \equiv y_n & \text{if } c \text{ is an } n\text{-ary constructor with } n > 0, \\ \mathbf{true} \wedge x & \to x \end{cases}$$

to a given $\mathcal{ATRS}$ $\mathcal{R}$, resulting in the $\mathcal{ATRS}$ $\mathcal{R}_s$. Here $\equiv$ denotes strict equality and $\wedge$ is a binary right-associative symbol, written in infix notation, denoting logical conjunction. In our framework a strict equation is an equation of the form $(s \equiv t) \approx \mathbf{true}$, which we abbreviate to $s \approx_s t$. A goal consisting of strict equations is trivially right-normal.

Since $\mathcal{R}_s$ inherits orthogonality from $\mathcal{R}$, we can solve (strict) equations with respect to the calculus NCA and the $\mathcal{ATRS}$ $\mathcal{R}_s$. However, as observed by Ida and Nakahara [7] in the context of OINC, it is much more efficient to add special inference rules for the rewrite rules in $\mathcal{R}_s \backslash \mathcal{R}$. Based on their ideas, we extend NCA in the following definition.

Definition 24. Let $\mathcal{R}$ be an orthogonal $\mathcal{ATRS}$. The calculus NCA$_s$ is obtained by adding the following inference rules to NCA:

[onas]   *outermost narrowing of applicative terms for strict equations*

$$\frac{f\,\mathsf{s}_n\,\mathsf{t}_m \simeq_s t, E}{s_1 \approx u_1, \ldots, s_n \approx u_n, r\,\mathsf{t}_m \approx_s t, E}$$

if there exists a fresh variant $f\,\mathbf{u}_n \to r$ of a rewrite rule in $\mathcal{R}$,

[onvs]   *outermost narrowing of head-variable terms for strict equations*

$$\frac{x\,\mathsf{s}_n\,\mathsf{t}_m \simeq_s t, E}{(s_1 \approx v_1, \ldots, s_n \approx v_n, r\,\mathsf{t}_m \approx_s t, E)\theta}$$

if there exists a fresh variant $f\,\mathbf{u}_k\,\mathbf{v}_n \to r$ of a rewrite rule in $\mathcal{R}$, $n > 0$, and $\theta = \{x \mapsto f\,\mathbf{u}_k\}$,

[das]   *decomposition of applicative terms for strict equations*

$$\frac{c\,\mathsf{s}_n \approx_s c\,\mathsf{t}_n, E}{s_1 \approx_s t_1, \ldots, s_n \approx_s t_n, E}$$

if c is an n-ary constructor symbol,

[imas]   *imitation for strict equations*

$$\frac{x\,\mathsf{s}_n \simeq_s c\,\mathsf{t}_m\,\mathbf{u}_n, E}{(x_1 \approx_s t_1, \ldots, x_m \approx_s t_m, s_1 \approx_s u_1, \ldots, s_n \approx_s u_n, E)\theta}$$

if c is an $(m + n)$-ary constructor symbol and $\theta = \{x \mapsto c\,\mathbf{x}_m\}$ with $\mathbf{x}_m$ fresh variables,

[dvs]   *decomposition of head-variable terms for strict equations*

$$\frac{x\,\mathsf{s}_n \simeq_s y\,\mathsf{t}_m\,\mathbf{u}_n, E}{(x_1 \approx_s t_1, \ldots, x_m \approx_s t_m, s_1 \approx_s u_1, \ldots, s_n \approx_s u_n, E)\theta}$$

if either $x = y$, $m = 0$, and θ is the empty substitution, or $x \neq y$ and $\theta = \{x \mapsto y\,\mathbf{x}_m\}$ with $\mathbf{x}_m$ fresh variables.

Here $s \simeq_s t$ stands for $s \approx_s t$ or $t \approx_s s$.

Observe that the rewrite rules in $\mathcal{R}_s \backslash \mathcal{R}$ for strict equality are no longer needed in NCA$_s$.

Example 5. Let $\mathcal{R} = \{\mathsf{a} \to \mathsf{b}\}$. The following NCA_s-derivation, starting from the goal $G = x\,\mathsf{a}\,\mathsf{a} \approx_s y\,\mathsf{a}$, produces the substitution $\sigma = \{y \mapsto x\,\mathsf{b}\}$:

$$G \Rightarrow_{[\mathrm{dvs}],\,\{y \mapsto x\,x_1\}} \mathsf{a} \approx_s x_1, \mathsf{a} \approx_s \mathsf{a} \Rightarrow_{[\mathrm{onas}],\,\mathsf{a} \to \mathsf{b}} \mathsf{b} \approx_s x_1, \mathsf{a} \approx_s \mathsf{a}$$

$$\Rightarrow_{[\mathrm{imas}],\,\{x_1 \mapsto \mathsf{b}\}} \mathsf{a} \approx_s \mathsf{a} \qquad\qquad \Rightarrow_{[\mathrm{onas}],\,\mathsf{a} \to \mathsf{b}} \mathsf{b} \approx_s \mathsf{a}$$

$$\Rightarrow_{[\mathrm{onas}],\,\mathsf{a} \to \mathsf{b}} \mathsf{b} \approx_s \mathsf{b} \qquad\qquad \Rightarrow_{[\mathrm{das}]} \qquad \Box$$

Note that σ is not a solution of G since the terms $x\,\mathsf{a}\,\mathsf{a}$ and $x\,\mathsf{b}\,\mathsf{a}$ do not reduce to the same ground constructor normal form. However, we would like to stress that σ represents (all) solutions of G in the sense that $\sigma\theta$ is a solution of G for all $\theta = \{x \mapsto c\,t_n \mid c$ is an $(n+2)$-ary constructor and $t_1, \ldots, t_n$ are ground constructor terms$\}$.

Below we state the completeness of NCA_s. The proof, which is essentially the same as the completeness proof of the calculus $\mathrm{S\text{-}OINC}$ studied in [7], is omitted.

Theorem 25. *Let $\mathcal{R}$ be an orthogonal $\mathcal{A}TRS$ and G a right-normal goal. For every normalizable solution θ of G there exists a successful NCA_s-derivation $G \Rightarrow_{\theta'}^* \Box$ such that $\theta' \leqslant_{\mathcal{R}} \theta\ [\mathcal{V}ar(G)]$.* $\Box$

7 Experimental Results

In this section we compare the performance of NCA and OINC on a small example. We have implemented both calculi in Sicstus Prolog 2.1. We solved goals of the form

$$G_n = \mathtt{map}\, f\, [\mathsf{S}^n\, 0,\, \mathsf{S}^{n-1}\, 0,\, \ldots,\, 0] \approx [\mathsf{S}^{2n+1}\, 0,\, \mathsf{S}^{2n-1}\, 0,\, \ldots,\, 0]$$

with respect to the example program in Sect. 1. Here $\mathsf{S}^n\, 0$ denotes the term

$$\underbrace{\mathsf{S}\,(\mathsf{S}\,(\cdots(\mathsf{S}\,0)\cdots))}_{n}.$$

Since for each n there are infinitely many normalized solutions of G_n, we measured the runtime of the two programs to compute the first solution $\{f \mapsto \mathtt{compose}\,\mathsf{S}\,\mathtt{double}\}$. Table 1 shows, for several values of n, these times in milliseconds as well as the length of the resulting successful derivation.

It is apparent that NCA has a much better performance than OINC.

8 Concluding Remarks

We have presented complete narrowing calculi for applicative term rewriting. Applicative term rewriting is a natural first-order framework for dealing with higher-order functions in the setting of functional (logic) programming with lazy semantics. Because of the absence of (λ-)abstraction, applicative term rewriting cannot express all higher-order features. Prehofer [13] describes a full higher-order lazy narrowing calculus.

Table 1. Comparison between OINC and NCA

n	OINC	NCA
1	3800 msec. (73 steps)	120 msec. (42 steps)
2	5448 msec. (136 steps)	179 msec. (77 steps)
3	7401 msec. (210 steps)	250 msec. (118 steps)
4	8799 msec. (295 steps)	305 msec. (165 steps)
5	10769 msec. (391 steps)	380 msec. (218 steps)

Although NCA and NCA$_s$ have been designed to deal efficiently with applicative terms, there remains some room for improvement. In order to ensure completeness of the calculi, the various inference rules have to be applied don't know non-deterministically. In general more than one inference rule is applicable to a given goal. For instance, both [ona] and [da] apply to certain goals. One way to remove this particular non-determinism is by restricting ourselves to $\mathcal{A}$CSs. The restriction to $\mathcal{A}$CSs also enables us to add failure rules which can be used to prune unsuccessful derivations at an early stage. Another source of inefficiency in our calculi is in the inference rules [onv] and [onvs] themselves. In the worst case there are $arity(f) - 1$ different ways to apply the two inference rules with respect to a given rewrite rule $f\,\mathbf{u}_k\,\mathbf{v}_n \to r$. A practical restriction to remove this non-determinism is by adding types to applicative term rewriting systems. In typed systems we can associate a type with every (head-)variable. This implies that we can uniquely determine the number k for the rewrite rule $f\,\mathbf{u}_k\,\mathbf{v}_n \to r$ such that the type of the head-variable x is (unifiable with) the type of the term $f\,\mathbf{u}_k$. For example, consider the $\mathcal{A}$CS $\{\mathtt{plus}\,0\,x \to x\ (1), \mathtt{plus}\,(\mathsf{S}\,x)\,y \to \mathsf{S}\,(\mathtt{plus}\,x\,y)\ (2)\}$ and the goal $z\,0\,(\mathsf{S}\,0) \approx \mathsf{S}\,0$. In the untyped system presented in this paper there are four different ways to apply the inference rule [onv]. Only one of them leads to a successful derivation. One of the three unsuccessful applications of [onv] is

$$z\,0\,(\mathsf{S}\,0) \approx 0 \Rightarrow_{[\mathrm{onv}],\,(2),\,\{z \mapsto \mathtt{plus}\,(\mathsf{S}\,x)\}} 0 \approx y, \mathsf{S}\,(\mathtt{plus}\,x\,y)\,(\mathsf{S}\,0) \approx \mathsf{S}\,0.$$

Notice that the term $\mathsf{S}\,(\mathtt{plus}\,x\,y)\,(\mathsf{S}\,0)$ cannot be typed. In a (polymorphically) typed system we can actually avoid the above [onv] step by letting the type of the variable z be $\mathbf{Nat} \to \mathbf{Nat} \to \mathbf{Nat}$ and observing that the type $\mathbf{Nat} \to \mathbf{Nat}$ of the term $\mathtt{plus}\,(\mathsf{S}\,x)$ isn't unifiable with the type of z. So by type-checking the substitution $\theta = \{x \mapsto f\,\mathbf{u}_k\}$ in [onv] steps we can avoid invalid ones and hence (significantly) reduce the search space.

As a final remark, we emphasize that the basic ideas in this paper do not depend on the calculus OINC. For example, it is only a matter of diligence to extend NCA to a calculus based on the calculus LNC studied in [12]. Because the inference rules of LNC are more complex than those of OINC, the former calculus is complete for arbitrary confluent term rewriting systems and arbitrary initial goals.

Acknowledgements. We thank the referees for their constructive comments on an earlier version of this paper. This work is partially supported by the Grant-in-Aid for Scientific Research (C) 06680300 and the Grant-in-Aid for Encouragement of Young Scientists 06780229 of the Ministry of Education, Science and Culture of Japan.

References

1. S. Antoy, R. Echahed, and M. Hanus, *A Needed Narrowing Strategy*, Proceedings of the 21st ACM Symposium on Principles of Programming Languages, Portland, pp. 268–279, 1994.
2. P.G. Bosco and E. Giovannetti, *IDEAL: An Ideal Deductive Applicative Language*, Proceedings of the IEEE International Symposium on Logic Programming, pp. 89–94, 1986.
3. N. Dershowitz and J.-P. Jouannaud, *Rewrite Systems*, in: Handbook of Theoretical Computer Science, Vol. B (ed. J. van Leeuwen), North-Holland, pp. 243–320, 1990.
4. E. Giovannetti, G. Levi, C. Moiso, and C. Palamidessi, *Kernel-LEAF: A Logic plus Functional Language*, Journal of Computer and System Sciences 42(2), pp. 139–185, 1991.
5. Juan Carlos González-Moreno, M.T. Hortalá-González, and M. Rodríguez-Artalejo, *On the Completeness of Narrowing as the Operational Semantics of Functional Logic Programming*, Proceedings of the 6th Workshop on Computer Science Logic, San Miniato, Lecture Notes in Computer Science 702, pp. 216–230, 1992.
6. M. Hanus, *The Integration of Functions into Logic Programming: From Theory to Practice*, Journal of Logic Programming 19 & 20, pp. 583–628, 1994.
7. T. Ida and K. Nakahara, *Leftmost Outside-In Narrowing Calculi*, report ISE-TR-94-107, University of Tsukuba, 1994. To appear in the Journal of Functional Programming.
8. J.W. Klop, *Term Rewriting Systems*, in: Handbook of Logic in Computer Science, Vol. II (eds. S. Abramsky, D. Gabbay and T. Maibaum), Oxford University Press, pp. 1–116, 1992.
9. J.J. Moreno-Navarro and M. Rodrìguez-Artalejo, *Logic Programming with Functions and Predicates: The Language BABEL*, Journal of Logic Programming 12, pp. 191–223, 1992.
10. J.J. Moreno-Navarro, H. Kuchen, R. Loogen, and M. Rodrìguez-Artalejo, *Lazy Narrowing in a Graph Machine*, Proceedings of the 2nd International Conference on Algebraic and Logic Programming, Nancy, Lecture Notes in Computer Science 463, pp. 298–317, 1990.
11. G. Nadathur and D. Miller, *An Overview of λ-Prolog*, Proceedings of the 5th International Conference on Logic Programming, MIT Press, pp. 810–827, 1988.
12. S. Okui, A. Middeldorp, and T. Ida, *Lazy Narrowing: Strong Completeness and Eager Variable Elimination*, Proceedings of the 20th Colloquium on Trees in Algebra and Programming, Aarhus, Lecture Notes in Computer Science 915, pp. 394–408, 1995.
13. C. Prehofer, *Higher-Order Narrowing*, Proceedings of the 9th IEEE Symposium on Logic in Computer Science, Paris, pp. 507–516, 1994.
14. U.S. Reddy, *Narrowing as the Operational Semantics of Functional Languages*, Proceedings of the IEEE International Symposium on Logic Programming, Boston, pp. 138–151, 1985.

Exploiting Parallelism in Tabled Evaluations[*]

Juliana Freire, Rui Hu, Terrance Swift, David S. Warren

Department of Computer Science
State University of New York at Stony Brook
Stony Brook, NY 11794-4400
Email: {*juliana,ruihu,tswift,warren*}@*cs.sunysb.edu*

Abstract. This paper addresses general issues involved in parallelizing tabled evaluations by introducing a model of shared-memory parallelism which we call *table-parallelism*, and by comparing it to traditional models of parallelizing SLD. A basic architecture for supporting table-parallelism in the framework of the SLG-WAM[14] is also presented, along with an algorithm for detecting termination of subcomputations.

1 Introduction

The deficiencies of SLD resolution are well known, and extended efforts have been made to remedy these deficiencies. For instance, while SLD can be combined efficiently with negation-by-failure in SLDNF, the semantics of SLDNF have proven unacceptable for many purposes, in particular for non-monotonic reasoning. Even without negation SLD is susceptible to infinite loops and redundant subcomputations, making it unacceptable for deductive databases.

The latter deficiency, that of repeating subcomputations, has given rise to several systems which table subcomputations: OLDT [16], SLD-AL [18], and SLG [4, 3] are three tabling methods which have been implemented. At an abstract level, systems which use magic evaluation can also be thought of as tabling systems. Substantiation for this claim stems both from the asymptotic results of [12] and the experimental results of [15]. Tabling is also be relevant for computing the well-founded semantics: besides SLG, well-founded ordered search [13] and the tabulated resolution of [2] also use tabling.

While nearly all of these approaches are sequential, there is a natural parallelism inherent in these evaluation methods which we call *table-parallelism*. At a general level, the idea behind table-parallelism is simple: the table can be thought of as a large structured buffer through which cooperating threads communicate. Parallelization then can take place at each tabled subgoal. If a subgoal is called and is not in the global table, an entry for it is created, and answers are derived for the subgoal using program clause resolution. Otherwise, if this subgoal is a variant of subgoal present in the table, it will consume the answers stored in the table.

* This work was partially supported by CAPES-Brazil and NSF grants CCR-9102159, CCR-9123200, CCR-9404921, CDA-9303181.

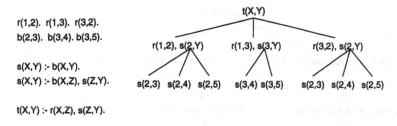

r(1,2). r(1,3). r(3,2).
b(2,3). b(3,4). b(3,5).

s(X,Y) :- b(X,Y).
s(X,Y) :- b(X,Z), s(Z,Y).

t(X,Y) :- r(X,Z), s(Z,Y).

Fig. 1. *Join in Parallel Prolog*

Fig. 1 illustrates how, by reducing redundant subcomputations, tabling can improve performance over an SLD-based resolution method ([1, 15]). An or-parallel (or any sequential) Prolog would have to compute the relation $s(2, Y)$ twice. In a tabling system, if the predicate s were declared as tabled it would be computed only once, and any other subsequent call would simply consult the table. This avoidance of redundant computation has long been recognized as necessary for data-oriented queries. On the other hand, a bottom-up method such as tabling can also introduce an overhead for copying information into and out of tables, making it unsuitable for programs such as list recursion.

We believe that, just as a mixture of SLD and tabled evaluation is arguably most useful for sequential evaluation, a combination of SLD-based and- and or-parallelism and table-parallelism will prove most useful for parallel evaluation. In terms of practical programs, the mixture of SLD and tabling has proven useful for program analysis [7] over flat domains [5], and for the Unification Factoring compiler optimization [6]. While implementation has so far been sequential, both of these algorithms contain bottom-up subcomputations that are amenable to table-parallelism, along with top-down computations amenable to traditional SLD parallelism.

The structure of the paper is as follows:

- We briefly overview general concepts underlying tabling using SLG formalism, and its implementation in the SLG-WAM.
- Using SLG terminology, we introduce the abstract model of table-parallelism and discuss its relation to more traditional and- and or- parallelism, and briefly discuss issues for the analysis of table-parallelism.
- We describe and prove the correctness of a *parallel completion* algorithm which detects termination of subcomputations, allowing a great deal of concurrency between subcomputations.
- In the framework of the SLG-WAM [14] , we present the extensions to tabling operations necessary to implement table-parallelism.

2   A brief overview of SLG

SLG evaluates programs by keeping a table of answers to subgoals, and resolving repeated subgoals against *answer clauses* from the table rather than against

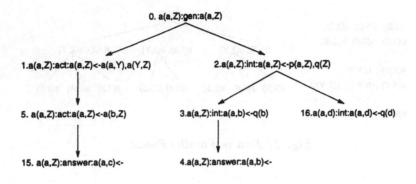

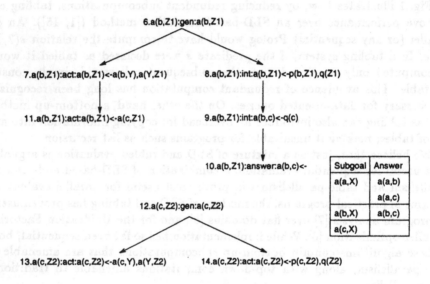

Fig. 2. *SLG forest*

program clauses. By using answer resolution in this manner, rather than repeatedly using program clause resolution as in SLD, SLG avoids looping and thus terminates for programs with finite models [4].

SLG has an efficient partial implementation in the SLG-WAM of XSB[2], which currently is restricted to apply to modularly stratified programs. The SLG-WAM has been shown to compute in-memory deductive database queries about an order of magnitude faster than current set-at-a-time methods, and to compute Prolog queries with little loss in efficiency over the WAM [19].

An SLG evaluation can be modelled by an SLG-forest which grows dynamically as predicates are called. Whenever a tabled subgoal S is called for the first time, it becomes the root of a subtree in the forest. In addition, an entry in the table is created for S, with which answers for S will be associated. If in

---

[2] XSB, along with source code for the sequential SLG-WAM, can be obtained by anonymous ftp from ftp.cs.sunysb.edu.

the course of evaluation variants of S are called, they will be resolved against the answer clauses in the table. Fig. 2 shows the SLG-forest for the program in Example 1. Nodes 0, 6, or 12 — roots of trees which use program clause resolution to produce answers, are called *generator* nodes. Nodes 1, 7 and 13, which consume answers from the table, are called *active* (consumer) nodes. Nodes like 2, 3, 8, 9, and 14 have selected literals which are not tabled; SLD resolution is performed on them instead and they are called *interior* nodes.

Example 1.   :- table a/2.
  p(a,b).   p(a,d).   p(b,c).
  q(b).   q(c).
  a(X,Z):- a(X,Y),a(Y,Z).
  a(X,Z):- p(X,Z),q(Z).
  ?-a(a,Z).

Let us examine elements of the forest in detail. Since new trees are added whenever a call is made to a tabled subgoal that is not a variant of one in the table, the evaluation of a(a,Z) begins by creating a new tree a(a,Z) and resolving the program clause a(X,Z):-a(X,Y),a(Y,Z) against the goal, creating node 1 in Fig. 2 with status *new*. In node 1, since the selected literal a(a,Z) is already tabled, no new tree is created; the node's status is changed to *active* as it appears in Fig. 2. If answers for the subgoal were present, the evaluation would use them for resolution. Instead, since there are no answers, the second clause for a/2 is tried. The program clause a(X,Z):-p(X,Z),q(Z) is resolved against the goal a(a,Z), producing node 2 with status *new*. Since the selected literal for node 2 is not tabled, it will be evaluated using SLD-style program clause resolution. Depth-first resolution of non-tabled nodes continues until node 4 is produced (with status *new*), containing no further subgoals to resolve. The status of node 4 is then changed to *answer*, and the variable binding $\{Z \leftarrow b\}$ forms a solution to the goal a(a,Z). Since the answer is not a variant of any other answer node for a(a,Z), it is inserted into the table and scheduled to be returned to all active nodes whose selected subgoal is a variant of a(a,Z). In our example, the only such active node is node 1. Answer clause resolution thus creates node 5, which in its turn creates other answers. Eventually, the evaluation gives rise to two other tabled subgoals, a(b,Z), and a(c,Z), each of which forms the root of its own tree.

The process of expanding nodes, adding answers and returning answers, continues until all possible resolutions have been made for subgoals, when they are termed *completely evaluated*. Mutually dependent set of subgoals, called a *strongly connected components* or SCCs, can be completed together. Efficient dynamic detection both of subgoals in an SCC and of when an SCC has been completely evaluated is necessary to evaluate programs with negation and to efficiently evaluate definite programs. In a parallel framework, this process of *completion detection* has much in common with termination detection for concurrent systems, and we make use of results from concurrency theory in the correctness proof of our algorithm in Section 4.

In Example 1 there are no mutual dependencies, and so there are three single-ton SCCs {a(a,Z)}, {a(b,Z)}, and {a(c,Z)}. Using a completion algorithm for early detection of completed SCCs, trees for a component can be disposed as soon as they are completely evaluated.

At the engine level, the sequential SLG-WAM [14] makes several changes to the WAM in order to implement SLG. One change, which the parallel model will exploit, is the use of *failure continuations* to return answers to active nodes. Consider again the actions that occur for the answer in node 4, but this time at the engine level. In the sequential SLG-WAM, nodes on the same path share common variables in their environments just as in the WAM: variables are represented as pointers, and the values of these variables obtained by dereferencing the pointers. However, nodes 1 and 4 are not on the same path; in order to return the answer in node 4 to node 1, the engine must backtrack to a common ancestor of node 4 and 1 – i.e. node 0 – and then rebind any variables on the path from node 0 to node 1. Because of the need to backtrack and restore environments to which answers are returned, it is natural to return answers through failure. The choice point stack thus schedules answer clause resolution in addition to its original function of scheduling program clause resolution through WAM **retry** and **trust** instructions.

As will be discussed in detail below, by altering the times at which *answer return choice points* are placed onto the choice point stack, searches can be obtained more suited to a parallel framework. For instance, the sequential SLG-WAM chooses the simple strategy of creating an answer return choice point as soon as a new answer is created, giving it something of a depth-first flavor. A more breadth-first strategy might wait until an evaluation had failed back to the root of its search tree, and then create choice points for all answers that had not previously been returned to active nodes[3].

3 Table-Parallelism

Tabling evaluations in general have four types of operation not found in SLD.

1. *Tabled Subgoal Call*: A check whether a subgoal exists in the table and an insert into the table if not;
2. *New Answer*: A check whether an answer is in the table and an insert to the table if not;
3. *Answer Return*: Consumption of an answer from a table;
4. *Completion*: Determining when an SCC is completely evaluated.

The fundamental idea behind table-parallelism is to parallelize at the *Subgoal Call* operation. (As a practical matter, it may happen that parallelization is declared or inferred useful for only a subset of tabled subgoals). To see how this method differs from more traditional models, consider first that at *Subgoal Call* a tabling system must determine whether a subgoal is new to an evaluation or not: whether to create a generator node for program clause resolution or

---

[3] This is similar to the multistage depth-first strategy proposed for OLDT in [16].

an active node for answer clause resolution. In order to make this decision, the subgoal must be traversed. Indeed, by creating an active node in the first calling environment of a tabled subgoal, and by standardizing apart the variables of the generator node *as the subgoal is traversed for the table check*, a computational model closer to that of the forest in Fig. 2 is achieved[4]. Bindings are not shared between SLG trees in this model, so resolutions for separate trees can be performed in parallel.

While the copying of a subgoal each time a thread is created was a drawback of the Abstract Model [20], the copying does not add a significant overhead for a parallel tabling system over its sequential counterpart. There is every reason to believe that tabling, even with copying, can lead to efficiencies in a parallel context as it does in a sequential, although the decision of when to parallelize through tabling will require further research and experimentation.

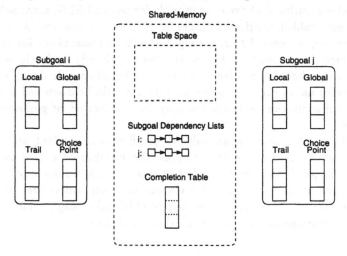

Fig. 3. *Memory Layout*

The algorithms and instructions for parallelizing tabled evaluations in a WAM framework are presented in Sections 4 and 5. Here we provide a brief overview of the main data structures. Memory layout for the parallel SLG-WAM is shown in Fig. 3. Because of the copying of tabled subgoals, issues of environment sharing are avoided (in the simplest model) and the local stack, heap and trail are all private.

Threads[5] for tabled subgoals are rooted in generator nodes and are called *generator* threads: they generate answers and copy them into the table. The originating active node will asynchronously consume answers as they are added to the table, as noted in Section 1. Since communication between the parallel subgoals is through the table, the table is stored in shared-memory.

---

[4] The sequential SLG-WAM shares bindings between the first tabled subgoal and its parent environment, rather than creating an intervening active node.

[5] In order to simplify the discussion in this section, we assume that there is one thread per tabled subgoal.

As mentioned above, the engine must dynamically detect mutually dependent sets of subgoals for completion. Thus, the *completion table* which supports this decision is also kept in shared memory. To present how this data structures are used, we review completion in the simpler model of the sequential SLG-WAM. Evaluations of the sequential SLG-WAM are deterministic, a property which that engine uses heavily. For instance, every time a new answer for a tabled subgoal is generated, it is scheduled on the choice point stack to be returned to the active subgoals that are variants of the tabled predicate. Properties of a depth-first evaluation are also used to detect the completion of an SCC. The first subgoal of an SCC visited by a sequential evaluation is called the *leader* of the SCC. Whenever the system fails over the leader of an SCC, it is provable that each subgoal inside the component is completed and all answers have been generated (details can be found in [3]).

The strategy outlined above is unsuitable for parallel SLG, since active nodes for a particular tabled predicated might be in different threads. As a reflection of this difference, a *Subgoal Dependency List* will be maintained for each tabled subgoal, which keeps track of the active (uncompleted) nodes in a subgoal's tree. While the subgoal dependency list keeps subgoal-specific data, it is kept in a global area for use in the completion algorithm as will be discussed in Section 4. Completed subgoals can be treated simply as a collection of globally available unit clauses.

The SLG-forest grows dynamically as predicates are called and thus, if a thread is created for each tabled subgoal, the number of active threads will depend on the particular program. For programs with a high degree of table-parallelism, the number of active threads can be much larger than the number of actual processors. This problem of how to efficiently assign tabled predicates to processors remains an open issue in table-parallelism.

Table vs. And/Or Parallelism

The use of a copying strategy allows each table-parallel thread to work independently of the others on its own SLG tree. Therefore, there is a great flexibility in the choice of evaluation strategy used to compute answers for the tabled predicate *within* each tree. Furthermore since SLG and SLD predicates can interleave, each tree can consist of a variety of sequential or parallel scheduling strategies. Tabled subgoals could be called — and table-parallelism occur — within a thread spawned by an and- or or- parallel goal; conversely and- or or- parallel execution could be used for SLD predicates within a tree for a tabled subgoal. In this sense, table-parallelism can be considered orthogonal to both and- and or-parallelism.

Because table-parallelism occurs at the selection of new tabled subgoals, it can be invoked by folding subgoals into a (parallel) tabled predicate, thus, table-parallelism can emulate or- and several varieties of and- parallelism. In the case of an or-parallel predicate consisting of N clauses, one need only fold the body of each clause into a unique tabled predicate. If the subgoals are not present in the table, the caller spawns threads for each predicate in turn, failing — as a means of looking for new work — between each. (If the subgoals are present in the table, the consuming process reads the answers out, sequentially). Note that the search space for the program using table-parallelism will be no greater than for the program using or-parallelism, and may terminate in cases where the or-parallel program will not.

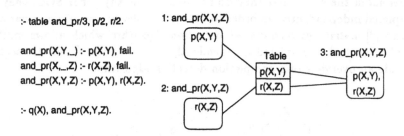

Fig. 4. *And-parallelism in SLG*

Folding can be used to emulate Independent And-Parallelism (see e.g. [10]). Consider the following goal: :- q(X),p(X,Y),r(X,Z). Using independence analysis, we could simulate and-parallelism in parallel SLG, by a program transformation as in Fig. 4, declaring and_pr/3, p/2 and r/2 as tabled. In this case all answers for p/2 and r/2 would be computed in parallel, assuming that they are not already in the table. As they are being computed, they will be fed to the third clause of and_pr/3, where p/2 and r/2 will be active (consumer) nodes. As an aside, we note that since duplicates are eliminated at each tabled predicate, the search space of the SLG and-parallel (table-parallel) program may be less than that of the SLD and-parallel program. As an extension of this example, suppose the goal :-p(X),q(X), and predicates p/1 and q/1 fulfill the conditions of non-strict independence as formulated in [11]. Then the transformation

```
:- table and_pq/1, p/1, q/1.

and_pq(X):- p(X),fail.
and_pq(X):- q(X),fail.
and_pq(X):- p(X),q(X).
```

will execute the goal using table parallelism, given the ability for an instantiated call to q/1 to use the table for q(X).

4 Parallel Completion

Detection of completion is a non-trivial problem in a sequential framework, and the difficulties are compounded for parallel evaluations. While various approaches to completion are possible, implementation of an efficient engine requires a minimum of synchronization.

To understand the synchronization issues that arise when tabling is parallelized, consider Fig. 5 which schematically shows call dependency graphs for two programs, both taken to have 6 tabled predicates, and thus permitting a 6-way parallelism in principle. In Fig. 5a, however, all 6 predicates lie in the same SCC, while in Fig. 5b they lie in three. Accordingly, there will be one concurrent completion operation in Fig. 5a, and three in Fig. 5b. Since in Fig. 5b, the SCCs { q, r, s } and { t, u } do not consume answers from each other, and since they are *maximal* in the sense that they do not depend on any other SCC, they can be completed independently. In order to detect completion, we use a variation of Dijkstra's [8] *distributed termination detection* algorithm which allows multiple maximal SCCs to be completed independently of one another[6]. In this section, we provide an overview of a completion detection algorithm [7].

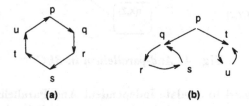

Fig. 5. *Call Dependency Graphs*

Completion detection is split into *leader detection* and *termination detection*. Leader detection produces knowledge of the subgoals in an SCC, and using this knowledge chooses a unique leader. Afterwards, the leader executes the termination detection phase. The completion table, introduced in Section 3, supports both of these functions. The completion table contains a frame for each incomplete tabled subgoal. A frame is added the first time a new subgoal is called, and consists of the following cells.

DFN	Unique Depth-first number
StateFlag	Thread is working
AnswerFlag	Thread has consumed all answers
ColorFlag	Thread may affect the completion of other subgoals
LeaderFlag	Thread is the leader of its SCC

The *StateFlag* indicates whether the subgoal is doing computation or is waiting to be completed. The *StateFlag* can have values *done* (if the subgoal is doing no work), *undone* (if the subgoal still has clauses or answers to resolve) or *unconditionally done* (if the table for the subgoal has been completed). The *AnswerFlag* indicates whether there are answers for the subgoal to process. Conceptually, the *AnswerFlag* can have values *done* (there are no more answers to be resolved) or *undone*, however it is implemented as a pointer into the subgoal dependency list. The *AnswerFlag* for a subgoal S is a disjunction of the next answer pointer for all active nodes in S. A computation needs to maintain the following invariants in order to correctly complete subgoals.

---

[6] The main difference between our algorithm and Dijkstra's is that we must allow leaders to change dynamically to account for the dynamic nature of SCCs.

[7] Full details are available in [9].

Invariant 4.1 (StateFlag) *The StateFlag is done for a given subgoal iff it is performing no answer or program clause resolution (i.e. doing no work).*

Invariant 4.2 (AnswerFlag) *The AnswerFlag is done for a given subgoal iff all active nodes in the SLG tree for the subgoal have resolved all applicable answers in the table.*

We first discuss leader detection and termination detection under the assumptions that these invariants hold, and then indicate how an evaluation can ensure the invariants.

Leader Detection

The first step in the completion of an SCC is to determine a single leader to execute the termination detection. Leader detection is summarized at an abstract level in *Algorithm* 4.1. The leader in the SCC is always chosen to be the subgoal in it with minimal DFN: thus knowledge of the SCC uniquely determines the leader. By searching through shared subgoal dependency lists, each subgoal is able to determine the set of subgoals in its SCC, and by extension whether it is the leader or not. If it depends on a subgoal with smaller DFN, it concludes it is not the leader, and can abort its leader detection. If it depends on a subgoal with greater DFN that is not done with its resolution and leader detection (determined by a *StateFlag* or *AnswerFlag* value of *undone*, in accordance with Invariants 4.1 and 4.2, and by a *LeaderFlag* value of *unknown*, respectively), it waits until this subgoal is *done* and has a *LeaderFlag* set to *false* before it continues its leader detection.

Algorithm 4.1   LEADER DETECTION(Subgoal:S)

        If the S.*StateFlag* is done and S.*AnswerFlag* is *done*
           Mark S.*LeaderFlag* as *unknown*;
           For all subgoals S_{dep} which S depends on
               If S_{dep} has a smaller DFN than S
                  Mark S.*LeaderFlag* as *not leader*;
                  Exit leader detection;
               If either $S_{dep}.AnswerFlag$ or $S_{dep}.StateFlag$ is *undone*
                  Wait until $S_{dep}.AnswerFlag$ is *done* and
                  $S_{dep}.StateFlag$ is *done* or *unconditionally done*;
               If $S_{dep}.LeaderFlag$ is *unknown* or *leader*
                  Wait until $S_{dep}.LeaderFlag$ is *not leader* or
                  until $S_{dep}.StateFlag$ is *unconditionally done*;
           Mark S.*LeaderFlag* as *leader*;

To understand Algorithm 4.1, consider how leader detection would be performed for a subgoal S_{small}. When S_{small} runs out of resolutions and executes *leader detection*, it will conduct a search through subgoal dependency lists to determine the topology of the SCC. As mentioned above, S_{small} can abort its leader detection phase if it finds that it depends on a subgoal with smaller DFN. On the other hand, if S_{small} depends on a subgoal S_{large} with a larger DFN, there are three possible outcomes. First, S_{large} may be in a different SCC than S_{small}, in which case it will complete before S_{small} and effectively remove itself from the subgoal dependency list of S_{small}. Next, S_{small} and S_{large} may be determined to be in the same SCC. And thirdly, S_{small} and S_{large} may initially be found to be in different SCCs, but later a topology change may occur that merges the two SCCs. In the case where S_{small} and S_{large} are in the same SCC, S_{large} knows that it is not the leader of the SCC (it depends on S_{small}), so it writes in the completion table that it has no work and it is not the leader of the SCC. If, on the other hand, S_{large} and S_{small} are initially found to be in different SCCs, and the topology of the SCC changes, termination detection is guaranteed to fail for S_{large}.

Taken together these cases indicate that S_{small} will be prevented from completing its leader detection phase until S_{large} has marked its *StateFlag* as *done* and its *LeaderFlag* as *false*. Thus no subgoal will consider itself a leader until every node that it depends on has executed leader detection and marked itself as *not leader*. Also note that at a given time in a computation, each SCC must have a node with minimal DFN. These observations form the intuition behind the following statement

Proposition 1. Let $\mathcal{E}$ be an SLG evaluation in which Invariants 4.1 and 4.2 hold. By *Algorithm* 4.1, at state S of $\mathcal{E}$, only one leader can be chosen for an SCC. Further, if all subgoals in a particular SCC have their *StateFlag* and *AnswerFlag* values marked as *done*, a leader can be chosen for this SCC.

Algorithm 4.2 TERMINATION DETECTION(Subgoal: S)

    For each subgoal r in S's SCC
        Wait until *r.StateFlag* and *r.AnswerFlag* are *done*;
        If *r.ColorFlag* is black
            Set the *ColorFlags* of all subgoals in S's SCC to white;
            abort to be restarted later;
        Set *StateFlags* of all subgoals $\in$ S's SCC to *unconditionally done*;

Termination Detection

Once a leader is chosen, and the SCC known, a termination detection phase is entered. Termination detection is specified in Algorithm 4.2, whose context in a parallel evaluation will be discussed in Section 5. Whenever a subgoal exhausts all of its program and answer clauses, it is put to sleep after both its *StateFlag* and *AnswerFlag* are set to *done* and it is found not to be the leader. The thread executing this particular subgoal will then switch to other work. (We use the traditional sleeping and waking metaphor in this section for simplicity of presentation). Whenever a new answer is added to a table for any of the active nodes of the tabled subgoal, the subgoal's *AnswerFlag* is effectively set to *undone*. When a subgoal wakes up, it executes the following steps.

1. The subgoal sets its *StateFlag* to *undone*.
2. The subgoal traverses its subgoal dependency list to see if new answers have been returned for its active nodes.
3. If new answers have been returned, the subgoal lays down an *answer return* choice point (presented in Section 5) for each uncompleted subgoal. The subgoal will backtrack through the choice points and continue its normal execution. It iterates steps 2 and 3 while there is work to do.
4. When there is no more work left, the subgoal sets *StateFlag* to *done* (the *AnswerFlag* has been effectively set to *done*) and either starts leader detection, or sleeps if it has already marked itself as *not leader*.

Using Proposition 1, and Invariants 4.1 and 4.2, the following proposition can be proven:

Proposition 2. (Parallel Completion)
Let $\mathcal{E}$ be a computation in which Invariants 4.1 and 4.2 hold. Let L be the leader of an SCC, S, and for each $A \in S$ suppose

$$StateFlag(A) = AnswerFlag(A) = done$$

Then all subgoals in S have been completely evaluated.

The intuition behind Proposition 2 is that if the SCC has been determined precisely and a single leader chosen, if the flag values are as specified and the invariants hold, then all mutually dependent subgoals have performed all answer and program clause resolution. Proposition 2 thus reduces proving correctness of completion to ensuring that the invariants hold.

As an example of how the invariants can fail to hold, consider the following situation illustrated in Fig. 6. Let *root* be the leader of SCC$_{root}$, which is currently being checked for completion. And, suppose SCC$_{root} = \{r_{n-1}, r_{n-2}, \ldots, r_1, r_0 = root\}$. In executing termination detection, *root* will check the status of the flags of each subgoal from r_{n-1} to r_0, in this order. Suppose while *root* is checking the flags for r_i, after finding that all flags for r_l, $l > i$ are *done*, another subgoal r_j, where $0 \le j < i$, generates a new answer and activates a subgoal r_k that has been already checked by *root*. The leader will wrongly assume that r_k is done, and it will incorrectly complete.

Following the model of [8], we address this problem using a *ColorFlag* for each tabled subgoal. The *ColorFlag* indicates that a node in the SCC may have generated new answers after another node was checked as completed by the leader. This flag can be either *black*, if new answers were added, or *white*, otherwise.

Subgoal	DFN	StateFlag	AnswerFlag	ColorFlag
r_{n-1}	n	done	done	white
⋮	⋮	⋮	⋮	⋮
r_k	k+1	done	done	white
⋮	⋮	⋮	⋮	⋮
r_i	i+1	done	done	white
⋮	⋮	⋮	⋮	⋮
r_j	j+1	done	done	black
⋮	⋮	⋮	⋮	⋮
root	1	undone	done	white

check flags →   (wake up)

Fig. 6. *Snapshot of Completion Table*

Initially the *ColorFlags* for all tabled subgoals are set to *white*. When after waking up, a subgoal *r* generates new answers, it will set its own color to *black*. If the leader finds a *black* node, during termination detection, it concludes that nodes that it has already checked might have been awakened, and therefore this procedure is suspended. Later, leader and termination detections will have to be restarted.

With the addition of the *ColorFlag*, Proposition 2 can be strengthened.

Proposition 3. (Termination Detection Algorithm)
Given a set S of mutually dependent tabled subgoals, *Algorithm* 4.2 will mark as complete the subgoals in S iff all subgoals in S are completely evaluated.

The correctness of termination detection depends on the fact that there is only one node, the leader, in each SCC executing this procedure at one time. If multiple nodes in an SCC were allowed to execute termination detection simultaneously, they may interfere with each other in checking and setting the *ColorFlag*.

Given a formalization of a parallel tabled evaluation as a whole, and not just of its completion algorithm, it can be proven that the synchronization required to complete tables causes neither deadlock nor starvation (this formalization is provided in the full version of the paper [9]). We now turn to the description of how the operations of Section 3 can be modified to implement table-parallelism.

5 Implementation Framework

The instruction-level changes needed to parallelize the SLG-WAM are discussed below. Since the changes mainly involve adding concurrency features to tabling operations, and since SLG-WAM instructions generally correspond to primitive tabling operations, a detailed knowledge of the SLG-WAM in *not* required to understand this section. For further elaboration, SLG-WAM instructions for definite programs have been presented in [14].

Changes occur in the following stages of the tabling process:

- In the *Tabled Subgoal Call* operation;
- In the *New Answer* operation, when adding an answer for a tabled subgoal;
- In the *Completion* operation.

Note that the *Answer Return* operation of Section 3 is not affected by parallelization. We discuss each aspect in the following subsections.

Tabled Subgoal Call

As outlined in Section 3, three different kinds of actions can occur when a tabled subgoal is selected. We review the actions which take place in a sequential system. (1) If a tabled subgoal is not already in the table, an entry is added for it and a new frame is created in the completion table. In the case of the SLG-WAM, the subtree rooted in this new subgoal is then explored via program clause resolution. (2) If the subgoal is in the table, but is not completely evaluated, an active node is set up which will use answer clause resolution on the subgoal. This active node cannot be deallocated until the subgoal is known to be completely evaluated. Finally, (3) if the subgoal is in the table and completed, the answers are treated as if they were facts, and the node may be deallocated as soon as it has backtracked through the answers in the table.

In the parallel model, at a call to a new tabled subgoal, the caller must arrange for that subgoal to be scheduled for execution. In addition, an active node is created which will consume answers, and a pointer to this node is added in the subgoal dependency list of the caller. The calling subgoal will then suspend the current call and move on to find other work in its own subtree. On the other hand, if the selected subgoal is in the table and not complete, the subgoal dependency list should be updated and resolution can then begin using answers in the table. Finally, if the subgoal is in the table and is completed, actions are the same in the parallel and sequential models: the answers are treated as facts and neither the completion table nor the subgoal dependency list needs to be updated.

Adding a new SLG subgoal to the table will require a lock to prevent multiple threads from simultaneously adding the same subgoal to the table. In the SLG-WAM the table for each predicate is structured as a tree, so only a subtree needs to be locked allowing a great deal of concurrency for this operation.

New Answer

In principle, parallelism introduces a minimum of change for the operation of adding a new answer to a table. If the answer is not already in the table, the operation will set the subgoal's *ColorFlag* to *black* indicating that other subgoals may have been affected. The *AnswerFlag*, which is a (pointer to a) list of pointers into the table, will be effectively updated as discussed in Section 4. Note that no locking will be necessary when adding answers to a table, since there is a single generating node for each table (though possibly many consuming nodes).

A deeper issue in parallelization is how to schedule return of new answers to existing active subgoals. For the parallel SLG-WAM, we are implementing a strategy somewhat akin to traditional breadth-first searches.

Completion

In a manner similar to traditional semi-naive execution, executions of the COM-PLETION operation in the sequential engine mark the boundaries of an iteration and control the scheduling of answer resolution. At the end of each iteration the COMPLETION instruction checks whether new answers have been added or whether the SCC has instead reached a fixpoint. An important difference however, is that while traditional semi-naive methods use static approximations of SCCs, the completion algorithm *dynamically* determines SCCs. Another difference stems from the fact that the choice point stack controls the scheduling for the engine. The COMPLETION instruction is executed when the engine fails back into a completion choice point frame. This instruction in turn might add *answer return* choice points to the choice point stack. These answers will cause new search space to be explored and, in a terminating program, the engine will eventually backtrack to the completion frame, causing a new iteration or determining a fixpoint.

This breadth-first strategy can be incorporated into parallel execution: whenever a subgoal executes a COMPLETION operation, it returns to its active subgoals any answers added since the last time it checked for completion. If no new answers have been added, rather than determining the fixpoint immediately as in the sequential case, a parallel consuming subgoal must coordinate with others to determine if its active subgoals have indeed been completely evaluated. If the current subgoal is a leader, it checks for the fixpoint as described in Section 4. Otherwise, it must poll at intervals to see whether new answers have been added.

Algorithm 5.1   SCHEDULE ANSWERS(Subgoal: S)
        If S.subgoal_dependency_list is empty
           set S.*StateFlag* to *unconditionally done*;
           deallocate the completion frame for S;
           exit;
        Else
           For each subgoal r_i in S.subgoal_dependency_list
              If (r_i is unconditionally complete
              and there are no unexamined answers)
                Remove r_i from list;
              Else
                If there are unexamined answers for r_i
                    Set up choice points to backtrack through answers;

The parallel COMPLETION operation is specified in *Algorithm* 5.2, and the portion of it which schedules *answer return* is summarized in the procedure

SCHEDULE ANSWERS of *Algorithm* 5.1. Recall that the subgoal dependency list maintains a list of subgoals for each tabled subgoal which are active in that subgoal's tree. In SCHEDULE ANSWERS, a subgoal removes other subgoals from its subgoal dependency list as they become complete. Thus, if the list is empty, the subgoal can mark itself as unconditionally complete. Otherwise, it will run through its subgoal dependency list, removing any subgoals which have become unconditionally complete and scheduling answer resolution otherwise. For each subgoal in the subgoal dependency list, a choice point is set up to backtrack through unresolved answers in the table. The evaluation then returns answers to each subgoal, by failing into these *answer return* choice points.

Algorithm 5.2 COMPLETION(Subgoal: S)

```
        If SCHEDULE ANSWERS exits exit;
        If S depends on a subgoal which is not complete
          If answers have been returned
            S.StateFlag = undone;
            backtrack through new answers;
          Else
            S.StateFlag = done;
            perform LEADER DETECTION(S);
            If S is the leader perform TERMINATION DETECTION(S);
            Else sleep(S);
```

In *Algorithm* 5.2, after a generator subgoal polls for answers (using procedure SCHEDULE ANSWERS) it will set its *StateFlag* to *undone* and continue with its normal, sequential SLG, evaluation. When it runs out of work (or if there was no work to begin with), it resets its *StateFlag* to *done*. If the current subgoal is not the leader of its SCC, it then sleeps. Otherwise, it will start the termination detection described in Section 4.

6 Conclusion

Tabling adds important termination properties to evaluations, whether they are sequential or parallel. In doing so, it may avoid recomputation and reduce the complexity of some programs, while adding an unnecessary overhead to others. In an actual parallel system, then, one would expect some predicates to use table-parallelism, others to use and- or or- parallelism, and still others to be executed sequentially. To attain this level of integration for a parallel engine, it must be possible to detect dynamic SCCs and their completion so that resources may be released and negation performed correctly. Section 4 provided algorithms for ensuring this, and Section 5 indicated how the basic tabling operations must change to implement table-parallelism.

While the development in this paper has taken place within the framework of SLG and the SLG-WAM, we believe that many of the results here are generally applicable to other tabling-style evaluation strategies and engines. For instance, magic set evaluations, which determine SCCs statically, could pick leaders statically, and would not require the leader detection algorithm presented here. However, the termination algorithm, and its associated flags would allow detection of fixpoint of an SCC. Indeed, by keeping in mind the well-known similarities between tabling and magic evaluation (see e.g. [15]) table-parallelism can also be seen as a way to parallelize database queries that may include recursion.

In our implementation framework, only three elements are shared: the table itself, the completion table, and the subgoal dependency list. If the table itself were distributed using techniques as those in [17] and if the completion algorithms were distributed, then a model of distributed table-parallelism could be formulated which would serve as a general mechanism for distributed data queries. Our current efforts regarding table-parallelism are thus twofold: in addition to implementing the shared memory parallelism sketched here, we are also exploring extensions to these algorithms for distributed table-parallelism.

Acknowledgements: We are indebted to Phil Lewis for his contributions to the design of the termination and leader detection algorithms.

References

1. F. Banchilhon and R. Ramakrishnan. An amateur's introduction to recursive query processing strategies. In *Proc. of SIGMOD 1986 Conf.*, pages 16–52. ACM, 1986.

2. R. Bol and L. Degerstedt. Tabulated resolution for the well-founded semantics. In *Proc. ILPS'93 Workshop on Programming with Logic Databases.* MIT Press, 1993.

3. W. Chen, T. Swift, and D.S. Warren. Efficient implementation of general logical queries. *J. Logic Programming*. To Appear.

4. W. Chen and D. S. Warren. Query evaluation under the well-founded semantics. In *Proc. of 12th PODS*, pages 168–179, 1993.

5. M. Codish and B. Demoen. Analysing logic programs using 'prop'-ositional logic programs and a magic wand. In *Proc. of the Int'l Symp. on Logic Programming*, pages 114–130, 1993.

6. S. Dawson, C. R. Ramakrishnan, I. V. Ramakrishnan, K. Sagonas, S. Skiena, T. Swift, and D. S. Warren. Unification factoring for efficient execution of logic programs. In *Proc. of the 22nd Symp. on Principles of Programming Languages.* ACM, 1995.

7. S. Dawson, C.R. Ramakrishnan, and D.S. Warren. Using XSB for abstract interpretation. In *Special Workshop on Abstract Interpretation*, 1995. Eliat, Israel. To Appear.

8. E.W. Dijkstra, W.H.J Feijen, and A.J.M. van Gasteren. Derivation of a termination detection algorithm for distributed computations. *Information Processing Letters*, pages 217 – 219, June 1983.

9. J. Freire, R. Hu, T. Swift, and D.S. Warren. Exploiting parallelism in tabled evaluations. Technical report, SUNY at Stony Brook, 1995. Full version available at http://www.cs.sunysb.edu/~sbprolog.

10. M. Hermenegildo and F. Rossi. On the correctness and efficiency of independent and-parallelism in logic programs. In *N. Amer. Conf. on Logic Programming.*, 1989.
11. M. Hermenegildo and F. Rossi. Non-strict independent and-parallelism. In *Logic Programming: Proc of the 5th Int'l. Conf.*, pages 237–252, 1990.
12. H. Seki. On the power of Alexander templates. In *Proc. of 8th PODS*, pages 150–159. ACM, 1989.
13. P. Stuckey and S. Sudarshan. Well-founded ordered search. In *13th conference on Foundations of Software Technology and Theoretical Computer Science*, pages 161–172, 1993.
14. T. Swift and D. S. Warren. An abstract machine for SLG resolution: definite programs. In *Proceedings of the Symposium on Logic Programming*, pages 633–654, 1994.
15. T. Swift and D. S. Warren. Analysis of sequential SLG evaluation. In *Proceedings of the Symposium on Logic Programming*, pages 219–238, 1994.
16. H. Tamaki and T. Sato. OLDT resolution with tabulation. In *Third Int'l Conf. on Logic Programming*, pages 84–98, 1986.
17. J. D. Ullman. Flux, sorting, and supercomputer organization for AI applications. *Journal of Parallel and Distributed Computing*, 1:133–151, 1984.
18. L. Vieille. Recursive query processing: The power of logic. *Theoretical Computer Science*, 69:1–53, 1989.
19. D.H.D. Warren. An abstract Prolog instruction set. Technical report, SRI, 1983.
20. D.H.D. Warren. Or-parallel models of Prolog. In *Proceedings of the International Conference on Theory and Practice of Software Development.* Springer-Verlag, 1987.

Design and Implementation of Jump Tables for Fast Indexing of Logic Programs*

Steven Dawson

sdawson@cs.sunysb.edu

C.R. Ramakrishnan

cram@cs.sunysb.edu

I.V Ramakrishnan

ram@cs.sunysb.edu

Department of Computer Science
SUNY at Stony Brook
Stony Brook, NY 11794-4400
U.S.A.

Abstract. The principal technique for enhancing the speed of clause resolution in logic programming languages, such as Prolog, is indexing. Given a goal, the primary objective of indexing is to quickly eliminate clauses whose heads do not unify with the goal. Efforts at maximizing the performance of indexing automata have focused almost exclusively on constructing them with small depth, which in turn translates into making fewer transitions. Performance of an automata also critically depends on its ability to make each transition efficiently. This is a problem that has largely been ignored and constitutes the topic of this paper.

Although jump tables ensure that each transition can be done in constant time they are usually very sparse and hence are seldom used in implementations. We describe a novel method to construct dense jump tables for indexing automata. We present implementation results that show that our construction indeed yields jump tables that are dense enough to be practical. We also show that indexing automata that use jump tables based on our method, improve the overall performance of Prolog programs. We also provide experimental evidence that our method is a general technique for compressing transition tables of other finite state automata such as those used in scanners.

1 Introduction

Optimizing clause resolution is a problem of considerable importance for efficient evaluation of resolution-based logic programs. The principal technique used for enhancing the speed of clause resolution steps is indexing. When resolving a goal, a clause becomes applicable if its head unifies with the goal. The sole objective of indexing is to quickly eliminate many clauses that are not applicable. Indexing can yield substantial gains in speed, since it reduces the number of clauses on which unification will be performed and can also avoid the pushing of a choice point. These benefits of indexing have long been recognized, and considerable effort has been made to develop fast and effective indexing techniques for logic programs (see [CRR92, Car87, HM89, PN91, RRW90] for example). The typical

---

* This work was supported in part by NSF grants CCR-9404921, CDA-9303181 and INT-9314412.

approach underlying high performance techniques is to preprocess the clause heads of a logic program into an indexing trie. The essential idea is to partition terms based upon their structure. A tree is formed, and at each point where two terms have different symbols, a separate branch for each term is added. (See Figure 1 below.)

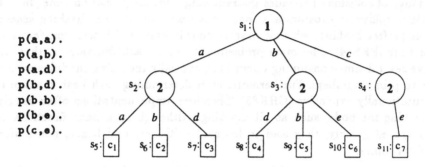

p(a,a).
p(a,b).
p(a,d).
p(b,d).
p(b,b).
p(b,e).
p(c,e).

Fig. 1. Predicate and indexing trie

Observe that the trie is organized as a tree-structured finite-state automaton with the root as the start state, leaves being the final states, and the edges, denoting transitions, representing elementary comparison operations. Each state specifies a position in the goal to be inspected upon reaching that state. The edge labels on the outgoing transitions specify the function symbols expected at that position. A transition is taken if the symbol in the goal at that position matches the label on an outgoing edge. We associate with each final state a set of clauses called its *index* set. On reaching a leaf state, the clauses in its index set are selected for resolution with the goal term. Note that the index set is a superset of the clauses that are applicable. Although our simplistic description above does not discuss variables, several variants to the indexing trie have been proposed for handling them (see [RRW90] for example). Such automata are routinely used to index terms in functional programming [SRR92], automated deduction [McC92] and term rewriting systems [Chr89].

Advances in indexing techniques have resulted in substantial improvement in the execution time of Prolog programs [DRR+95], so much so that indexing time constitutes a significant part of the overall execution time. Thus, devising faster indexing techniques without compromising their effectiveness is a problem of considerable importance for implementing high-performance logic programming systems. Research on fast indexing methods have traditionally focused on techniques for constructing tries of small depth. Indexing based on such tries translates into making fewer state transitions for selecting an index set, thereby improving indexing time. Indexing times can be further improved by executing each transition efficiently. Techniques that facilitate efficient state transitions for indexing have not been well researched so far and constitute the topic of this paper.

Recall that, in order to make a transition from the current state, the corresponding goal symbol in the position specified by the state must match the edge label on one of its outgoing transitions. A simple sequential search of the edge labels for a match is clearly very time consuming. Therefore, Prolog systems, such as SICStus, ALS, and XSB, typically resort to hash-based state transitions. Although hashing reduces the time needed to make a state transition, the possibility of collisions precludes guaranteeing uniform transition time. In order to reduce collisions substantially, one must resort to complex hashing schemes such as perfect hashing, which guarantee no collisions, and hence, constant transition time [FKS84]. The main problem with such hash functions is that they involve several time-consuming operations, typically including modulus with respect to prime numbers. Furthermore, even determining such hash functions is computationally expensive [CHK85]. Therefore, implementations often use simple hashing methods, such as bit masking. Although such hash functions can be computed quickly, they lead to frequent collisions, and hence, non-uniform transition times.

Executing any state transition in constant time can also be guaranteed by directly accessing the next state through jump tables. In this method each function symbol is assigned a unique positive integer as its *id*. Every state is associated with a jump table whose entries either point to next states or are void (denoted by $\perp$). The indexing automaton uses the id assigned to the input symbol inspected in the current state, to directly index into its jump table to make a transition to the next state. Note that it is necessary to store only the segment of the jump table containing valid transitions. If the id of an input symbol indexes into this segment, and the corresponding transition is valid, this transition is taken. Otherwise, indexing failure occurs. For example, in Figure 1 suppose we assign integers $1, 2, 3, 4$, and 5 to the symbols a, b, c, d, and e respectively. The ranges of indices for the jump tables in states s_1, s_2, s_3 and s_4 are $[1..3], [1..4], [2..5]$ and $[5..5]$, respectively, and the jump tables are $[s_2, s_3, s_4], [s_5, s_6, \perp, s_7], [s_9, \perp, s_8, s_{10}]$, and $[s_{11}]$.

Although jump tables facilitate fast transitions, they are seldom used in implementations because they tend to be sparse, *i.e.*, contain many void entries. In the example above the jump tables for states s_2 and s_3 have void entries. For jump tables to become practical, it is critical that they be densely populated. Suppose in the example above we had instead assigned $1, 2, 3, 4$, and 5 to e, d, b, a, and c, respectively. The ranges of jump table indices for states s_1, s_2, s_3, and s_4 would be $[3..5], [2..4], [1..3]$, and $[1..1]$; and the jump tables would be $[s_3, s_2, s_4], [s_7, s_6, s_5], [s_{10}, s_8, s_9]$, and $[s_{11}]$. Note that these jump tables contain no void entries. The problem now is to devise a numbering scheme for symbols that minimizes the total number of void entries in all jump tables. This has remained open, and is addressed in this paper.

Jump tables can be used in conjunction with hashing. This is particularly useful when, for example, the best numbering scheme still yields sparse jump tables for some states, or a state makes transitions on symbols for which numbering is inappropriate (*e.g.*, integers). Thus, jump tables with numbering can be viewed as a complementary scheme to hash tables, and avoids exclusive reliance on hash tables.

Summary of results

1. We show that the problem of assigning numbers to symbols, such that the total number of void entries in jump tables is minimized, is NP-complete (Section 2).
2. We present an efficient heuristic, called the Symbol Numbering Method (SNM), that reduces the number of void entries (Section 3).
3. We provide strong experimental evidence for the practical use of jump tables. The results show that jump tables improve the overall execution time of Prolog programs. Furthermore, the space efficiency of jump tables obtained using SNM is often better (and no worse) than that of hash tables (Section 4).
4. We discuss reasons for the improved space efficiency of jump tables over hash tables and offer ways of further improving the performance of jump tables (Section 5).
5. We argue that SNM is a useful scheme for compressing transition tables of general finite state automata, such as those used in scanners and parsers. In particular, we present experimental results that demonstrate its effectiveness in compressing transition tables (Section 6).

2 Space Minimization of Jump Tables

In this section we formalize the problem of determining the numbering scheme that minimizes the space of jump tables and study its computational complexity.

Notation An indexing trie is a finite state automaton, with S as its set of states, and whose symbols are drawn from an enumerable set $\mathcal{F}$. A transition from a state s on input symbol α to a destination state d is denoted by (α, d). There is a special state $\perp \in S$ with no outgoing transitions, called the *fail state* of the automaton. The set of symbols on which successful transitions can be made from the state $s \in S$, called the *label set* of s, is denoted by λ_s. Let $\tau_s = \{(\alpha_1, s_1), (\alpha_2, s_2), \ldots (\alpha_n, s_n)\}$, where $\alpha_i \in \lambda_s$ and $s_i \in S$ denote the set of outgoing transitions in the state $s \in S$. The set of outgoing transitions is realized as *jump tables*, formalized as follows.

Jump Tables Let $\nu : \mathcal{F} \to N$ be a one-to-one function, called the *numbering* function, that maps each symbol in the alphabet to a distinct natural number. Since $\mathcal{F}$ is enumerable, clearly such a function exists. Let $min_s = \min(\{\nu(x) \mid x \in \lambda_s\})$, and $max_s = \max(\{\nu(x) \mid x \in \lambda_s\})$. The function $jt_s : \{i \in N \mid min_s \leq i \leq max_s\} \to S$, that represents the jump table associated with state s, is defined as:

$$jt_s(\nu(x)) = \begin{cases} s' & \text{whenever } x \in \lambda_s \text{ and } (x, s') \in \tau_s \\ \perp & \text{otherwise} \end{cases}$$

Note that $max_s - min_s + 1$ is the size of the jump table of state s. The number of void elements in the table is $(max_s - min_s + 1) - |\lambda_s|$.

Let s denote the current state of the automaton and y denote the input symbol inspected in the current state. Let $next\_state(s, y)$ be the destination state. Using jump tables, the function $next\_state$ is defined as:

$$next\_state(s, y) = \begin{cases} jt_s(\nu(y)) & \text{if } min_s \leq \nu(y) \leq max_s \\ \perp & \text{otherwise} \end{cases}$$

Note from the above definition that the transition to the destination state can be made in constant time. Given a numbering function ν and an indexing automaton with states S, the total space occupied by jump tables, $Space(\nu, S)$ is given by

$$Space(\nu, S) = \sum_{s \in S} (max_s - min_s + 1)$$

The total size of jump tables, and hence the total number of void entries in the tables, varies with the numbering function. We now show that the problem of minimizing the total size of jump tables is NP-complete.

Complexity of Jump Table Space Minimization The corresponding decision problem of jump table space minimization can be stated as

Given an automaton with states S and a positive integer K, is there a numbering function ν such that $Space(\nu, S) \leq K$?

Membership of the above problem in NP is obvious. By transformation from Optimal Linear Arrangement [GJ79], we show:

Theorem 1 *Jump Table Space Minimization is NP-complete.*

Proof: The NP-complete problem of Optimal Linear Arrangement is stated as follows: Given a graph $G = (V, E)$ and a positive integer M, is there a one-to-one function $f : V \to \{1, 2, \ldots, |V|\}$ such that $\sum_{(u,v) \in E} |f(u) - f(v)| \leq M$?

Given any instance of the above problem, it can be easily transformed into an instance of Jump Table Space Minimization as follows: construct an automaton with states S such that for each edge $(u, v) \in E$ there is a unique state s in S with $\lambda_s = \{u, v\}$. Let $K = |E| + M$. Clearly, the transformation can be done in polynomial time. We now show that a function f exists iff there is a numbering function ν for the automaton with states S satisfying the constraints of the jump table minimization problem.

if: Define $f(u) = |\{x \in V \mid \nu(x) \leq \nu(u)\}|$. Clearly, $f : V \to \{1, 2, \ldots |V|\}$ is one-to-one, and $|f(u) - f(v)| \leq |\nu(u) - \nu(v)|$. Since ν is a solution to the jump table space minimization problem, ν is such that $\sum_{s \in S} (\nu(u) - \nu(v) + 1) \leq K$ where $\{u, v\} = \lambda_s$ and $\nu(u) > \nu(v)$. Hence $\sum_{(u,v) \in E} |\nu(u) - \nu(v)| + 1 \leq K \Rightarrow$ $\sum_{(u,v) \in E} |f(u) - f(v)| \leq \sum_{(u,v) \in E} |\nu(u) - \nu(v)| \leq K - |E| = M$.

only if: Let $g(x) : \mathcal{F} \to \{|V| + 1, |V| + 2, \ldots\}$ be a one-to-one function. Clearly, such a function exists since $\mathcal{F}$ is enumerable. Define $\nu(x) = f(x)$ if $x \in V$ and

$\nu(x) = g(x)$ otherwise. Now, $\nu(x) - \nu(y) = f(x) - f(y)$ for all $x, y \in V$. Since f is a solution to the optimal linear arrangement problem, $\sum_{(u,v)\in E} |f(u) - f(v)| \leq M \Rightarrow \sum_{s\in S}(max_s - min_s) \leq M$ from transformation. Hence, $Space(\nu, S) = \sum_{s\in S}(max_s - min_s + 1) \leq M + |E| = K$. ∎

3 Symbol Numbering Method (SNM)

As shown in the previous section, it is unlikely that there is an efficient algorithm for finding minimum space jump tables. In this section we describe the Symbol Numbering Method, an efficient heuristic for reducing jump table size that, in practice, yields small jump tables. The heuristic is based on two observations. First of all, the space taken by jump tables depends more on the numbers assigned to frequently occurring symbols than on the numbers assigned to other symbols. Secondly, the space tends to depend more on tables having fewer entries than on tables having more entries. This is because larger tables will necessarily have a wider range of numbers, and hence, more flexibility in the numbering of their component symbols. These attributes of the symbols and tables are captured naturally by a bipartite graph representation, called the *index symbol graph* of the program, where the nodes represent symbols and tables, and the edges denote membership of the symbols in the tables.

More formally, let $C = \{\lambda_1, \ldots, \lambda_m\}$ be the collection of label sets of all the states in the indexing automata of all the predicates in the program. Let $A = \{\alpha_1, \ldots, \alpha_n\} = \bigcup_{1\leq i\leq m} \lambda_i$ be the set of all symbols in the automata. The index symbol graph, $G = (V, E)$, is an undirected graph, where $V = \{v_1, \ldots, v_n, v_{n+1}, \ldots, v_{n+m}\}$ and $E = \{(v_i, v_{n+j}) \mid \alpha_i \in \lambda_j\}$. We refer to vertices $v_1, \ldots, v_n$ as *symbol* vertices, and vertices $v_{n+1}, \ldots, v_{n+m}$ as *table* vertices. Note that the jump table for a state has valid entries only in positions corresponding to the symbols in the state's label set. The degree of symbol vertex v_i is equal to the number of label sets in which symbol α_i appears, and the degree of table vertex v_{n+j} is equal to the number of distinct symbols appearing in the label set λ_j.

For example, consider the seven simple predicates in Figure 2a. The indexing automaton for each predicate has one state. The index symbol graph of the program is given in Figure 2b. The symbol vertices are labeled above by the symbols occurring in the predicates. The table vertices are labeled below by the names of the associated predicates. The edges connect each table vertex with the symbol vertices of every symbol that is a member of the corresponding label set. For example, the table vertex labeled "*p6/1*" has three incident edges, connecting it to symbol vertices a, e, and f.

The index symbol graph provides a means of extracting certain "binding" relationships among the symbols. Informally, we say that two symbols are tightly bound to one another if a small difference between their assigned numbers leads to void entries in the jump tables. The larger this difference can become without leading to void entries, the less tightly bound the symbols are considered to be. For example, symbols e and f in Figure 2 are tightly bound, since any difference greater than 1 in their numbers would lead to void entries in the jump table for *p3/1*. On the other hand, symbols *h* and *i* are not as tightly bound, since their numbers could potentially differ by as much as 4 without resulting in any void entries. The heuristic, informally described below, is designed to account for these binding relationships, based on the two observations mentioned at the beginning of this section. The heuristic works in two stages.

In the first stage the index symbol graph is used to determine a rough ordering of the symbols, based on their frequency of occurrence and tightness of binding to other symbols. The result is one or more tree structures, which are traversed in the second stage to perform the actual numbering.

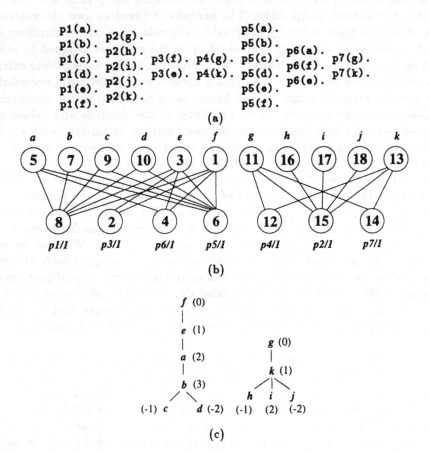

Fig. 2. Seven simple predicates (a), the index symbol graph (b), collapsed spanning trees (c)

Stage 1 The first stage of the heuristic consists of a depth-first search of the index symbol graph, resulting in a spanning tree of each connected component in the graph. Note that, since the index symbol graph is bipartite, in a depth first search we will alternately visit table and symbol vertices. At each step in the depth-first search, the choice of which vertex to visit next is guided by the two observations outlined at the beginning of this section. If the next vertex to be visited is a symbol vertex, then the *maximum degree* symbol vertex is chosen; if the next vertex is a table vertex, then the *minimum degree* table vertex is chosen. The depth-first search begins at the symbol vertex of highest degree, say v_i. Thus, the next vertex to be visited is the table vertex of lowest degree connected to v_i, and we proceed by alternately visiting symbol vertices of highest degree and table vertices of lowest degree. When a symbol vertex is visited, the degree of each table vertex to which it is connected is decreased by one. This reflects the increasing binding among the symbols in the corresponding label sets that remain to be visited.

The index symbol graph in Figure 2b illustrates the first stage of the heuristic. Each vertex is labeled according to the order in which it was visited in the depth-first search. The table vertices are not used in the second stage and are not recorded in the spanning tree. For example, the spanning trees obtained after stage 1 for the index symbol graph in Figure 2b are shown in Figure 2c.

Stage 2 The numbering of each tree (connected component) can be done independently, since the differences between numbers of symbols from distinct connected components have no effect on the size of the jump tables. We begin by assigning the number 0 to the symbol represented by the root. We then perform a preorder traversal of the tree, numbering each symbol as we visit its vertex. Note, however, that the actual numbers assigned may not mirror the preorder traversal. Observe that each chain of vertices between two branch points in the tree is such that a parent symbol occurs in all label sets that contain the child symbol. We number such groups sequentially, and "position" the group so that the distance of the group from the first symbol on its parent chain is minimized. This means that the group will be numbered increasing from the greatest number yet assigned, or decreasing from the least number yet assigned.

Consider the numbering of the left tree in Figure 2c. There is a chain of symbols f, e, a, b, numbered sequentially starting from 0. Next, symbol c is numbered -1, minimizing its difference with f. Finally symbol d is numbered -2, also minimizing its difference with f[2]. The other tree is numbered similarly. It is easily verified that this numbering yields jump tables for this example with no void entries, and hence, no wasted space.

Complexity Let n denote the number of symbols, m the number of tables, and $s = \Sigma_{j=1}^m |\lambda_j|$ the total size of all label sets. The number of vertices in G is $n + m$ and the number of edges is s. Note that $n + m \le 2s$. Thus, the space required for the bipartite graph is $O(s)$. Even if the spanning trees are constructed separately, the space is increased by at most a constant factor.

---

[2] Note that this gives a relative numbering among symbols in one tree. The final, global numbering among all symbols is easily obtained from the relative numbering simply by adding an appropriate constant to the number of each symbol in a tree.

The time required to perform the heuristic is given by the time needed to construct the index symbol graph, plus the time needed to perform the depth-first search, plus the time to traverse each connected component (for numbering). The time required to build the index symbol graph is proportional to its size, $O(s)$. The time to select a vertex to visit is bounded by the maximum degree of any vertex. A symbol may appear in at most m label sets, and any label set may contain at most n symbols. Thus, the maximum degree of any vertex is $O(\max(m, n))$, and the time required for the depth-first search is $O(s \cdot \max(m, n))$. The time required to traverse the trees for numbering is clearly bounded by the depth-first search time. Thus, the overall time for the heuristic is $O(s \cdot \max(m, n))$.

4 Performance

The Symbol Numbering Method has been implemented, and the performance of jump tables obtained using SNM was compared to two hash-based indexing schemes in the XSB system on several benchmark programs[3]. The programs include a sample of the border predicate from the CHAT-80 system, three queries on the Dutch national flag program (dnf), three simple parsers (ll1, ll2, and ll3), a theorem prover (dboyer) and map coloring program (map) from a set of Andorra-I benchmarks, and three queries on a 5000-fact sample of a chemical database (synchem). Each of the programs was compiled using unification factoring [DRR+95], which, with the exception of dboyer, substantially improves the performance of the programs, due in large part to improved indexing.

The two hashing methods tested were modulus with respect to a prime number (mod hash) and bit masking (bit mask hash). Table 1 compares the overall query evaluation times for jump tables to those of the two hashing schemes. Two times are reported for each hashing method. The first is the lowest observed time across a large number of runs with varying memory configurations, and the second is the highest observed time for which the time increase could be attributed to hash collisions[4]. The jump table figures show the execution time (which is nearly constant across all runs) as well as the speedups compared to hashing. Observe that jump table indexing is always as fast as the fastest hashing results and is often substantially faster than the slower hashing results. Mod-based hashing consistently gives the slowest performance (but more consistent performance than bit mask hashing), due to a particularly expensive mod operation on Sparc machines. This is, of course, architecture dependent, but it does indicate that perfect hashing schemes, which rely even more on mod operations, are unlikely to perform well in practice. Bit mask hashing, while offering performance comparable to jump tables in the best case, gives varying results due to (unpredictable) hash collisions.

---

[3] All benchmarks were executed on a Sun SPARCstation 20 running SunOS 5.3, using XSB version 1.4.2. The benchmarks programs are available by ftp from ftp.cs.sunysb.edu in pub/XSB/benchmarks

[4] Hashing in XSB is performed on symbol pointers whose values can vary from one run to another, depending on the memory configurations.

Further light can be shed on these numbers by the considering the execution time of the different indexing instructions. On average, a single mod-based hashing required 2.20 μs, while one bit mask hashing instruction took 0.50 μs. The jump table indexing instruction, which involves two bounds checks followed by a table lookup operation, took 0.62 μs. However, note that any hashing scheme requires some means of verifying the value of the hashed symbol. In the absence of any collision, this verification takes an additional 0.32 μs. Jump table indexing needs no such verification, and hence is faster than either hashing instruction, even in the absence of collisions. The speed of jump table indexing is more apparent in programs where a relatively large portion of the time is spent in indexing (*e.g.*, map).

Program	Mod hash Time Low/High	Bit mask hash Time Low/High	Jump tables		
			Time	Speedup (mod) Low/High	Speedup (bit) Low/High
border	4.29/4.29	3.70/3.70	3.63	1.18/1.18	1.02/1.02
dnf(s1)	4.43/4.43	3.75/5.88	3.68	1.20/1.20	1.02/1.60
dnf(s2)	4.31/4.31	3.66/5.63	3.59	1.20/1.20	1.02/1.57
dnf(s3)	5.05/5.05	4.25/5.00	4.18	1.21/1.21	1.02/1.20
lll1	4.93/6.44	4.24/5.70	4.23	1.17/1.52	1.00/1.35
lll2	5.06/5.06	4.15/5.20	4.13	1.23/1.23	1.00/1.26
lll3	4.24/4.24	3.57/4.40	3.55	1.19/1.19	1.01/1.24
dboyer	4.20/4.20	4.10/4.10	4.03	1.04/1.04	1.02/1.02
map	4.44/5.48	2.58/3.62	2.33	1.91/2.35	1.11/1.55
synchem(alc)	4.49/4.49	3.99/3.99	3.88	1.16/1.16	1.03/1.03
synchem(eth)	4.32/4.32	3.26/3.26	3.21	1.35/1.35	1.02/1.02
synchem(CC)	5.05/5.05	4.65/4.65	4.40	1.15/1.15	1.06/1.06

Table 1. Comparing CPU times for jump table indexing with two hashing schemes

It might be expected that the fast, guaranteed time performance of jump tables must come at some expense in space, and with a naive numbering of the symbols, this might be true. However, our experiments show that, using SNM, even space utilization can be improved. Table 2 compares the space usage and efficiency of jump tables with that of the two hashing methods. Efficiency was computed as the ratio of the minimum space required by any indexing table (*i.e.*, the number of distinct transitions) to the space actually used. For the above benchmarks, jump tables were always at least as space efficient as either hashing method and often gave 100% efficiency. To achieve comparable space performance in the hashing schemes wherever possible, more collisions would necessarily result, giving worse time performance. On the other hand, for hashing to approach the consistent time performance of jump tables would require an even greater sacrifice in space.

It should be pointed out, however, that examples can be constructed for which the space required by even the smallest possible jump table may be too large[5]. Furthermore, due to the presence of symbols that cannot be numbered, such as integers, it may not be possible to compress the jump tables further. In such cases it is reasonable to use a fast hashing method (such as bit masks) as an alternative. For example, in the synchem benchmark, nearly half the tables contained integers and hash-based indexing was used in these cases. The jump tables obtained by SNM for the remaining cases were more than 90% space efficient. The use of jump tables for these cases improved query evaluation times, while increasing the overall space efficiency to over 60%.

Program	Mod hash		Bit mask hash		Jump tables	
	Bytes Low/High	Efficiency Low/High	Bytes Low/High	Efficiency Low/High	Bytes	Efficiency
border	68/68	0.47/0.47	64/64	0.50/0.50	32	1.00
dnf	28/28	0.43/0.43	32/32	0.50/0.50	12	1.00
ll1	56/88	0.27/0.43	32/80	0.30/0.75	24	1.00
ll2	60/108	0.22/0.40	48/64	0.37/0.50	24	1.00
ll3	60/108	0.22/0.40	48/64	0.37/0.50	24	1.00
dboyer	912/1000	0.41/0.45	864/904	0.46/0.48	764	0.54
map	188/188	0.34/0.34	96/144	0.44/0.67	72	0.89
synchem	72792/85640	0.35/0.41	59080/71640	0.42/0.51	48436	0.62

Table 2. Comparing space usage for jump tables with two hashing schemes

Finally, it should be noted that SNM itself, which is done at predicate load time, adds little overhead. In the worst case (synchem), SNM and jump table creation added less than 20% to the (small) load time. In summary, the performance results show that

- jump tables offer guaranteed time performance, and this performance is as good as or better than even the simplest hashing methods;
- using SNM, jump tables need not incur a large space penalty, and may even improve space efficiency;
- jump tables are more practical than perfect hashing schemes – perfect hashing functions are expensive to derive at compile/load time and, since they depend on mod operations, expensive to use at run time.

5 Space Utilization of Jump Tables

Traditional Prolog implementations do not use jump tables, primarily because of possible space degradation. However the experimental results presented in Section 4 show that indexing using jump tables usually outperforms hash-based schemes even in terms of space efficiency. In this section, we characterize the

---

[5] In the worst case, the space taken by jump tables is $O(nt)$, where n is the number of distinct indexing symbols and t is the number of tables.

space utilization of the hashing methods used in well-known Prolog systems, and compare them with the space behavior of jump tables. We identify situations where jump table based schemes perform very well and where degradation is possible. We also suggest ways to prevent such degradation.

5.1 Hash Tables vs. Jump Tables

In a hash-based indexing scheme, the size of the hash table and the number of collisions determine the space needed to represent all the outgoing transitions. In Prolog implementations, the hash function (and hence the size of the hash table) is chosen based on the number of transitions. Note that in order to reduce the possibility of collisions, the size of the table must be considerably larger than the number of transitions. Thus the space efficiency, denoted by e, which is the ratio of number of transitions to the total space needed to represent all the transitions, is bounded for hash tables. The bounds themselves may vary, depending on the hash function used and how collisions are handled.

In ALS Prolog and XSB, which use mod hashing, the hash table size is a prime number greater than the number of transitions. In SICStus Prolog, which uses bit mask hashing, the table size is a power of two, between two and four times the number of transitions[6]. ALS explicitly maintains a set of buckets for each hash entry in a binary tree. In SICStus and XSB, however, the hash table points directly to code. In SICStus, the keys are stored as a part of the hash table, and hence each hash table entry occupies two words. The hash table entry lookup verifies the symbol, and specialized instructions are used to avoid comparing the symbol again in the clause code. Thus part of the job of the initial clause code is taken over by the hash table lookup. In XSB, when there is no collision on a hash entry, the table points to the code for the corresponding clause, which contains unification instructions for all positions in the head of the clause. This code is a part of the try-retry-trust chain that is invoked when the indexed position in the goal is a variable, and hence does not contribute to any additional space overhead. When there is a collision on some entry, a new try-retry-trust chain is created for all the colliding symbols.

Based on the hash functions and the collision management strategy used, we can readily compute the bounds on the space efficiencies of the various hashing schemes. These bounds, as well as those on sequential search and jump table strategies, are given in Figure 3. Observe, from the table, that hash tables guarantee a definite lower bound on space efficiency whereas jump tables offer no such lower limit. On the other hand, the upper bound efficiencies of hash tables are low, while jump tables can achieve 100% efficiency. More interestingly, the performance results in section 4 indicate that, using SNM, jump tables often achieve 100% efficiency. We now discuss the reasons underlying their good space performance, and suggest ways to further improve their space as well as time performance.

---

[6] We use SICStus v2.1 #9 (compact code) for all these comparisons. Larger table sizes are needed to offset the number of collisions incurred by the simpler masking hashing function.

Method	Space Efficiency (e)
Jump Table	$0 < e \leq 1$
Sequential Search	$e = 0.5$
Hashing – XSB	$0.167 \leq e < 1$
– ALS	$0.20 \leq e < 0.25$
– SICStus	$0.125 < e < 0.25$

Fig. 3. Bounds on Space Efficiencies.

5.2 Impact of Type Discipline on Performance of Jump Tables

In the case of well-typed programs, the set of symbols on which transitions are made from any state belong to the same type. Hence, symbols of a given type can be numbered independently of symbols of a different type. This degree of freedom enables SNM to yield jump tables with nearly 100% space efficiency on well-typed programs. It turns out that, although Prolog is an untyped language, programs generally follow an implicit typing discipline. This is the main reason that the jump tables of indexing automata for Prolog programs attain high space efficiencies using SNM.

We can exploit knowledge of types in a program for improving the access time of jump tables even further. For instance, consider a state with the labels a, b and c on its outgoing transitions. If we have type information that whenever this state is reached, the symbol inspected in the input goal belongs to the set $\{a, b, c\}$ (another characteristic of well-typed programs), we can omit the bounds check, and directly index into the jump table. Note that transitions thus made cost no more than a table lookup.

5.3 Inter-module Application of SNM

Our development of SNM was based on complete knowledge of the label sets of all the states in the indexing automata for every predicate in the program. Hence SNM cannot be used at compile time in systems that support separate compilation. However, if all modules are loaded statically, *i.e.*, before the program is executed, SNM can be used at load time. We now naturally extend SNM to systems where modules are loaded dynamically as follows.

When a new module is being loaded, symbols numbered in previously loaded modules cannot easily be renumbered. Hence SNM assigns numbers, distinct from numbers previously used, only to the new symbols in the current module. Due to a loss of type discipline across modules, incremental application of SNM to each load module may yield sparse jump tables. In practice, such loss occurs primarily due to "accidental" synonyms – when the same symbol is used to represent different objects in different modules. Note that accidental synonyms arise only in module systems that are not atom-based, *i.e.*, where all function symbols are drawn from a global space. An atom-based module system that allows the user to intuitively specify the scope of symbols greatly reduces the frequency of this phenomenon, and improves the space efficiency of jump tables with SNM.

6 Generality of SNM

In the previous sections, we have shown the effectiveness of SNM in making jump tables practical for indexing in Prolog programs. We now demonstrate the utility of this technique for compressing transition tables of finite state automata that arise in applications such as parsers and scanners. In these applications, transitions can be stored in a two-dimensional table, indexed by the pair *(state, input symbol)* and transitions can be made with one table lookup operation. However, since transition tables are large in general, a suite of table compression techniques with various time-space tradeoffs have been developed (see [DDH84] for a good introduction). In the following, we first argue why these techniques are not applicable for minimizing jump tables in indexing automata.

6.1 Table Compression Methods

We start with a brief description of compression techniques, currently in widespread use, that guarantee fast constant-access times. These techniques can be broadly classified into *content-based* methods and *structure-based* methods.

Content-based Compression Methods Content based methods exploit the similarity between transitions in different states or on different input symbols. An important compression technique used in scanners, called the *Equivalent Symbols* method (ESM), is based on the observation that many sets of symbols exhibit identical lexical properties. For instance, in many scanners the digits '0', '1', ..., '9' are indistinguishable; *i.e.*, from any state, the destination states on the transitions on '0', ..., '9' are the same. Such symbols are put in one equivalence class, and the transitions on these symbols are replaced by *one transition* labeled by the corresponding equivalence class. The compressed table is represented as a two-dimensional array, indexed by state and equivalence class of symbols.

Two effective content-based techniques for compressing parser tables are *Line Elimination* [Bel74] and *Graph Coloring* [DDH84]. The Line Elimination method is based on the observation that there are many states in an LR parser in which there is only one valid action, regardless of the input symbol. Similarly, there are many input symbols such that the same valid action is performed, regardless of the current state. The rows and columns representing such states and input symbols are removed from the transition table, and the valid actions themselves are stored with the corresponding state or symbol. This process is repeated until the table is irredundant. The Graph Coloring method is based on the following observation: if two states do not perform contradictory valid actions on any input symbol, then their outgoing transitions can be represented by a single row. Note that this method compresses the transition table by reducing the number of states.

Structure-based Compression Methods Structure-based methods compress the tables based only on which positions in the table that represent valid transitions, and not on the transitions themselves. In a commonly used structure-based method, called the *Row Displacement* method (RDM) [AU77], the transition table is mapped to a linear array. Each row in the table is represented by a sequence of consecutive elements in the array such that valid entries of one row do not overlap with the valid entries of another row. The basic idea is to minimize the number of invalid entries in the linear array. All entries in the array are tagged with the corresponding row number, and each access now involves checking the tag to ensure the validity of the transition. The tags are usually maintained in a separate array. Methods such as the RDM are typically used to further compress tables resulting from content-based methods such as ESM.

Applicability to Indexing Automata It appears that the above table compression methods are not applicable for minimizing the jump table space of indexing automata. Firstly, the content-based compression methods exploit the sharing inherent in the automata used by scanners and parsers. But traditional trie-based indexing automata (e.g., path automata and first string automata [RRW90]) are tree structured. This means that no two transitions lead to the same destination state, and hence none of the content based methods apply. Secondly, since the number of possible input symbols is unbounded for indexing automata, fast access schemes for tables compressed using RDM are infeasible.

Even when the indexing automata are structured as DAGs [KS91], the above compression methods cannot be readily applied. Note that the graph coloring method is essentially a state minimization technique and hence yields no further compression on a DAG in which the number of states is already minimized. The line elimination method is applicable only when there is some symbol on which all states make a transition to the same destination state. Clearly, such transitions are not representable without cycles[7] and hence line elimination method is inapplicable. It is not clear if the indexing DAGs exhibit the symmetry needed for ESM to be effective. In any case, we show (in section 6.2 below) that SNM naturally generalizes ESM.

6.2 SNM as a Compression Technique

We now provide experimental evidence of the effectiveness of SNM for compressing transition tables used in scanners. It should be noted that SNM can be applied to tables already compressed using other methods, and that these methods can also be used to compress the tables produced by SNM. This interoperability makes SNM a valuable tool in the suite of compression strategies used in scanners and parsers.

---

[7] Transition tables may exhibit this property after factoring out error entries. However, note that error factoring involves maintaining an independent array that records which positions in the transition table contain valid transitions, and is not suited for making fast transitions.

SNM, as presented thus far in the paper, is a structure-based compression technique. However, it can be easily generalized to become a content-based method as follows. The numbering function was defined as a one-to-one function from the set of symbols to a set of integers. We now generalize the numbering function to be a many-to-one function with the following constraint. Two symbols are mapped to the same integer if and only if they make the same transitions from every state, *i.e.*, they belong to the same equivalence class (as defined by ESM), thus generalizing ESM.

Table 3 compares the space compression obtained by SNM to those obtained by traditional methods on different scanners. All scanners were automatically generated using Flex. The table lists the total space used to represent the transition tables and includes support structures, such as the tag table for RDM and the bounds tables for SNM. The scanners listed in the table were taken from various systems: Cdecl, a system for encoding C and C++ type-declarations; Coral, a deductive database system; Course, a compiler for a Pascal-like language (used in a compilers course); Detex, an utility to remove TeX commands from a text file; Equals, a functional language compiler; Flex, a scanner generator; Postgres, a database management system; Storm, an equational theorem prover; and Web2C, a Web to C translator.

In the table, the space requirements of the uncompressed transition tables are listed under Full Table. Columns ESM and ESM+RDM list the space used to represent the transition tables after applying ESM and ESM followed by RDM respectively. Column SNM lists the space used after applying SNM. The last column, Min Space, lists the minimum space needed to represent the valid transitions, and indicates the sparseness of the transition tables. Each column also lists the percentage reduction in space achieved by the compression technique (compared to the space occupied by the full table).

Program	Full Table	ESM	ESM + RDM	SNM	Min. Space
Cdecl	109568	29140 (73.40%)	23872 (78.21%)	14208 (87.03%)	10724 (90.21%)
Coral	79872	32644 (59.13%)	25944 (67.52%)	8940 (88.81%)	8460 (89.41%)
Course	92160	43268 (53.05%)	36380 (60.53%)	12628 (86.30%)	11676 (87.33%)
Detex	143360	55708 (61.14%)	40628 (71.66%)	19236 (86.58%)	15132 (89.42%)
Equals	154112	57424 (62.74%)	59684 (61.27%)	23288 (84.89%)	18884 (87.75%)
Flex	161280	56288 (65.10%)	21076 (86.93%)	14140 (91.23%)	9168 (94.32%)
Postgres	95232	38024 (60.07%)	38076 (60.02%)	17588 (81.53%)	14016 (85.28%)
Storm	66048	24064 (63.57%)	23220 (64.84%)	12172 (81.57%)	10296 (84.41%)
Web2C	134656	56568 (57.99%)	52728 (60.84%)	29092 (78.47%)	19324 (85.65%)

Table 3. Effectiveness of various scanner table compression methods.

Observe from Table 3 the effectiveness of SNM in reducing space usage. In particular, SNM results in compression factors of over 78% in all the examples, and obtains better compression than any other compression scheme. Furthermore, the compression achieved by SNM is close to the maximum possible.

Recall that compression techniques trade table space for access time. Accessing tables compressed using ESM takes typically 20% longer than accessing full (uncompressed) tables, while compression using ESM+RDM incurs an overhead of 50%. Tables compressed using SNM typically take 65% longer time to access than uncompressed tables. Scanner generators usually compress ESM tables using methods that do not guarantee constant access times. Such methods can achieve compression factors of more than 80% but result in access time overheads of more than 120%. Note that SNM offers a good compromise with 80% space compression and 65% access time overhead. Thus, SNM is a useful compression technique that can be added to the suite of methods currently used to compress scanner tables.

7 Conclusion

Efficient indexing is crucial for high performance logic programming systems. Advanced techniques such as *unification factoring* have resulted in substantial reduction in program execution times. With the increase in proportion of time spent in indexing, it is even more important that the elementary indexing operations themselves are as fast as possible. Jump tables are an effective means of achieving fast, constant-time transitions in indexing automata. In fact, as our experimental results show, indexing based on jump tables is significantly faster than hash-based indexing. Thus, in programs that rely heavily on indexing, the overall execution times can be substantially improved. However, jump tables are seldom used in implementations since they are usually sparse. In this paper, we have addressed the problem of improving the space efficiency of jump tables, thus making the use of jump tables practical.

We devised a scheme that makes jump tables dense, by suitably numbering the symbols. We showed that the problem of determining the numbering scheme that minimizes the total space of jump tables is NP-complete, and presented an efficient heuristic, called the Symbol Numbering Method, to reduce the total space. Experimental results indicate that jump tables improve the overall execution time of Prolog programs, even compared to the fastest hashing method. Furthermore, the space efficiency of jump tables obtained using SNM is often better (and no worse) than that of hash tables. Moreover, SNM is general in the sense that it provides an effective scheme for compressing transition tables of finite state automata, such as scanners and parsers.

Acknowledgements

We thank Mats Carlsson for clarifying several details of the implementation of SICStus Prolog.

References

[AU77]   A.H. Aho and J.D. Ullman. *Principles of Compiler Design.* Addison-Wesley, Reading, Mass., 1977.

[Bel74]   J.R. Bell. A compression method for compiler precedence tables. In *Proc. of the IFIP Congress 74*, pages 359–362, Stockholm, August 1974.

[Car87]   M. Carlsson. Freeze, indexing and other implementation issues in the WAM. In *International Conference on Logic Programming*, pages 40–58, 1987.

[CHK85]   G.V. Cormack, R.N.S. Horspool, and M. Kaiserswerth. Practical perfect hashing. *The Computer Journal*, 28(1):54–58, 1985.

[Chr89]   J. Christian. Fast Knuth-Bendix completion : Summary. In *RTA'89*, pages 551–555. Springer-Verlag LNCS 355, 1989.

[CRR92]   T. Chen, I. V. Ramakrishnan, and R. Ramesh. Multistage indexing algorithms for speeding Prolog execution. In *Joint International Conference/Symposium on Logic Programming*, pages 639–653, 1992.

[DDH84]   P. Denker, K. Dürre, and J. Heuft. Optimization of parser tables for portable compilers. *ACM Transactions on Programming Languages and Systems*, 6(4):546–572, October 1984.

[DRR+95]   S. Dawson, C. R. Ramakrishnan, I. V. Ramakrishnan, K. Sagonas, S. Skiena, T. Swift, and D. S. Warren. Unification factoring for efficient execution of logic programs. In *ACM Symposium on Principles of Programming Languages*, pages 247–258, January 1995.

[FKS84]   L. Fredman, J. Komlós, and E. Szemerédi. Storing a sparse table with O(1) worst case access time. *Journal of the ACM*, 31(3):538–544, July 84.

[GJ79]   M. Garey and D. Johnson. *Computers and Intractibility: A Guide to the Theory of NP-Completness.* W.H. Freeman and Company, 1979.

[HM89]   T. Hickey and S. Mudambi. Global compilation of Prolog. *Journal of Logic Programming*, 7:193–230, 1989.

[KS91]   S. Kliger and E. Shapiro. From decision trees to decision graphs. In *North American Conference on Logic Programming*, pages 97–116, 1991.

[McC92]   W. McCune. Experiments with discrimination-tree indexing and path indexing for term retrieval. *Journal of Automated Reasoning*, 9:147–167, 1992.

[PN91]   D. Palmer and L. Naish. NUA-Prolog: An extension to the WAM for parallel Andorra. In *International Conference on Logic Programming*, pages 429–442, 1991.

[RRW90]   R. Ramesh, I. V. Ramakrishnan, and D. S. Warren. Automata-driven indexing of Prolog clauses. In *ACM Symposium on Principles of Programming Languages*, pages 281–290. ACM Press, 1990.

[SRR92]   R. C. Sekar, I. V. Ramakrishnan, and R. Ramesh. Adaptive pattern matching. In *International Conference on Automata, Languages, and Programming*, number 623 in LNCS, pages 247–260. Springer Verlag, 1992. To appear in *SIAM J. Comp.*

An Abstract Machine for Oz

Michael Mehl, Ralf Scheidhauer, and Christian Schulte

Programming Systems Lab
German Research Center for Artificial Intelligence (DFKI)
Stuhlsatzenhausweg 3, D–66123 Saarbrücken, Germany
{mehl,scheidhr,schulte}@dfki.uni-sb.de

Abstract. Oz is a concurrent constraint language providing for first-class procedures, concurrent objects, and encapsulated search. DFKI Oz is an interactive implementation of Oz competitive in performance with commercial Prolog and Lisp systems. This paper describes AMOZ, the abstract machine underlying DFKI Oz. AMOZ implements rational tree constraints, first-class procedures, local computation spaces for deep guards, and preemptive and fair threads.

1 Introduction

Oz is a concurrent constraint language [20, 19, 17, 6, 22] providing for functional, object-oriented, and constraint programming. It has a simple yet powerful computation model [19, 20], which extends the concurrent constraint model [10, 16] by first-class procedures, deep guards, concurrent state, and encapsulated search.

DFKI Oz [11] is an interactive implementation of Oz based on an incremental compiler and an abstract machine. It features a programming interface based on GNU Emacs, an object-oriented interface to Tcl/Tk, powerful interoperability features, a garbage collector, and support for stand-alone applications. Performance is competitive with commercial Prolog and Lisp systems.

This paper describes an abstract machine, called AMOZ, which covers important aspects of the DFKI Oz abstract machine. AMOZ implements rational tree constraints, first-class procedures, deep guards, and threads, leaving aside mutable state for objects [19], record constraints [21], as well as finite domain constraints and encapsulated search [18].

Constraint Store. By the very idea of concurrent constraint programming, computation emerges from adding constraints to a store. In this paper, we consider constraints over rational trees (as in Prolog II [5]) that enjoy a *variable-centered* normal form: adding constraints results in binding variables. This is utilized in AMOZ: binding variables triggers procedure application, reduction of conditionals, and readiness of threads.

First-class Procedures. Oz provides for first-class procedures typical of modern functional languages (e.g., Haskell [7], Scheme [3], and SML [12]). First-class procedures in Oz support higher-order functional programming [19], concurrent object-oriented programming [6], and encapsulated search [17]. In AMOZ, execution of a procedure definition dynamically creates a procedure (called closure in functional languages) and stores the procedure under a so-called name.

Procedure application is triggered by binding a variable to a name, from which the procedure to be applied is retrieved.

Deep Guards. Deep guards allow any expression in the guard of a conditional. Reduction of a deep guard is done in a local computation space. The main point of discussing deep guards here is to show implementation techniques for local computation spaces. Local computation spaces are needed to encapsulate search, and encapsulation of search is a must in a concurrent and reactive language. It is well known that the problem has not been solved in the Japanese Fifth Generation Project, leaving them with two incompatible language designs: concurrent logic programming and (constraint) logic programming. AKL was the first language that solved this problem, employing a design based on deep guards [8]. Oz, on the other side, employs a higher-order search combinator that uses local computation spaces but does not presuppose deep guards [17].

Threads. Languages like Prolog II and AKL have a single thread of control in which all computations are performed. However, this is insufficient for the fine grained concurrency found in concurrent constraint languages. Since general fairness does not seem practical, Oz provides for multiple threads that are scheduled fairly. In AMOZ threads are lightweight, implemented as multiple stacks of tasks that are scheduled preemptively and fairly.

The design of abstract machines for constraint based languages has been pioneered by the Warren Abstract Machine (WAM) [25, 1]. The implementation of DFKI Oz has been influenced by the AGENTS implementation of the concurrent constraint language AKL [8]. AKL is a deep guard language providing for encapsulated search. However, AKL does not provide for first-class procedures and threads. cc(FD) [23] is a constraint programming language specialized for finite domain constraints. It is a compromise between a flat and a deep guard language in that combinators (i.e., cardinality, disjunction, and implication) can be nested in guards, but procedure applications cannot. As AKL, it does not support first-class procedures and threads.

The paper is organized as follows. Section 2 gives an informal presentation of the computation model, and Sect. 3 gives an example. Section 4 shows unification for rational trees, and Sect. 5 introduces AMOZ. Threads and a limited case of conditional are introduced in Sect. 6. Section 7 extends the abstract machine for local computation spaces. Procedures are introduced in Sect. 8.

2 An Informal Computation Model

This section gives an informal presentation of the computation model underlying the sublanguage of Oz considered in this paper. A full description of Oz's computation model can be found in [20].

The notion of a *computation space* is central to the computation model. A computation space consists of a number of *tasks*[1] connected to a *store*.

---

[1] In other papers on Oz tasks are called actors.

Computation proceeds by reducing tasks with respect to the information contained in the store. A task is reduced as soon as the store contains sufficient information. When a task is reduced new information may be written to the store or new tasks may be created. Tasks are short-lived: they cease to exist once they are reduced. Some tasks may spawn local computation spaces, thus creating a tree of computation spaces. As computation proceeds, new local computation spaces are created and existing spaces are removed or merged with their parent space.

The store consists of a *constraint store* and a *procedure store*. The constraint store contains constraints $x = y$ and $x = f(\overline{y})$ in a normal form. The constraint store grows monotonically. The constraints are interpreted in a fixed first-order structure, called the *universe*. The universe contains rational trees (as in Prolog II [4, 5]), an extension to records is straightforward [21]. Suppose that ϕ is the conjunction of all constraints in the store. We say that the *store entails* a constraint ψ, if $\phi \to \psi$ is valid in the universe. The procedure store contains the bindings of names to procedures (to be explained later).

The tree of computation spaces satisfies the invariant that constraints of a local computation space entail constraints of their parent space ("local spaces know the constraints of global spaces"). A constraint is *imposed* by adding it to the local store and all stores below in the tree of computation spaces. Hence, imposition maintains the invariant on the tree of spaces. A computation space *fails*, if a constraint is imposed such that the constraints in the store become unsatisfiable in the universe. If a computation space fails, all spaces below fail. If a space fails, all its tasks are discarded.

There are two kinds of tasks: *elaborators* and *conditional tasks*. An elaborator is a task that executes an expression. Expressions are:

$$
\begin{aligned}
E, F, G ::= \; & x = y \quad | \quad x = f(\overline{y}) && \textit{constraints} \\
| \; & \textbf{local } x \textbf{ in } E \textbf{ end} && \textit{declaration} \\
| \; & E\,F && \textit{composition} \\
| \; & \textbf{proc } \{x\,\overline{y}\}\, E \textbf{ end} && \textit{procedure definition} \\
| \; & \{x\,\overline{y}\} && \textit{procedure application} \\
| \; & \textbf{if } \overline{x} \textbf{ in } E \textbf{ then } F \textbf{ else } G \textbf{ fi} && \textit{conditional}
\end{aligned}
$$

Elaboration of a *constraint* $x = y$ or $x = f(\overline{y})$ imposes it.

Elaboration of a *declaration* $\textbf{local } x \textbf{ in } E \textbf{ end}$ creates a new variable local to the computation space and an elaborator for E. Within E the new variable is referred to by x. The space is called the *home* of x. Declaration of multiple variables $\textbf{local } x\,\overline{y} \textbf{ in } E \textbf{ end}$ abbreviates $\textbf{local } x \textbf{ in local } \overline{y} \textbf{ in } E \textbf{ end end}$.

Elaboration of a *composition* $E\,F$ creates separate elaborators for E and F.

Elaboration of a *procedure definition* $\textbf{proc } \{x\,\overline{y}\}\, E \textbf{ end}$ chooses a new name a, writes the binding $a : \overline{y}/E$ to the procedure store, and creates an elaborator for the constraint $x = a$. A *name* is a constant in the universe. There are infinitely many different names. Since procedures are associated with new names when they are written to the procedure store, a name cannot refer to more than one procedure.

Elaboration of a *procedure application* $\{x\, y_1 \cdots y_n\}$ waits until there is a name a, such that the constraint store entails $x = a$ and the procedure store contains a binding $a : z_1 \cdots z_n / E$. When this is the case, an elaborator for the expression $E[y_1/z_1, \ldots, y_n/z_n]$, where the formal parameters have been replaced by the actual parameters, is created.

The elaboration of a *conditional* is more involved. We will proceed in two steps. First, we consider the special case **if** $\overline{y}$ **in** $x = f(\overline{y})$ **then** F **else** G **fi**, where variables in $\overline{y}$ are pairwise different ("pattern matching"). This case is especially instructive for AMOZ in Sect. 6. Its elaboration creates a *conditional task*. The conditional task waits until the store either entails $\exists \overline{y}\, x = f(\overline{y})$, in which case an elaborator for **local** $\overline{y}$ **in** $x = f(\overline{y})\, F$ **end** is created, or entails $\neg \exists \overline{y}\, x = f(\overline{y})$, in which case an elaborator for G is created.

The general conditional **if** $\overline{x}$ **in** E **then** F **else** G **fi** subsumes the previous simplified case. Its elaboration creates a conditional task spawning a local computation space. We call the expression $\overline{x}$ **in** E the *guard* of the conditional. A guard is called *deep* if E is not a constraint. The local computation space is created with a store containing the constraints from the parent store and an elaborator for **local** $\overline{x}$ **in** E **end**.

We say that the guard is *entailed* if its associated computation space S is not failed, S has no tasks left, and its parent store entails $\exists \overline{y}\, \phi$, where $\overline{y}$ are the local variables of S and ϕ is the conjunction of constraints of S's store. Due to the monotonic growth of the constraint store, entailment of a guard is a stable property, i.e., it continues to hold when computation proceeds. A conditional task must wait until its guard is either entailed or failed (i.e., its corresponding local computation space is failed). If the guard is failed, the conditional task reduces to an elaborator for the expression G (its **else** constituent). If the guard is entailed, the constraints of the local store are merged with its parent store's constraints. Merging amounts to changing the local variables' home space to the parent space. By this, local variable bindings are made global. Then, the expression F (its **then** constituent) is elaborated.

So far we have not made any assumptions about the order in which tasks are executed. Such assumptions are necessary, however, so that one can write fair and efficient programs. Without such assumptions a single infinite computation, e.g., a data base query server, which is intended to run forever, could lead to starvation of all other computations.

A *thread* is a nonempty sequence of tasks. Each task belongs to exactly one thread. When a computation space is failed, its tasks are discarded, which includes their removal from the threads they reside on.

A thread can *run* by reducing its first task if it is reducible, or otherwise by moving its first task to a newly created thread. Reducing a task on a thread means to reduce the task and replace it with the possibly empty sequence of tasks it has reduced to. The order of replacing tasks is defined as follows. For the task of a composition $E\, F$ the task for E goes before the task for F. For the task of a conditional, the task for the guard goes before the task of the conditional itself.

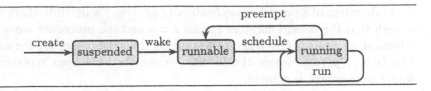

Fig. 1. Different states of threads.

If a thread contains a single not yet reducible task it is called *suspended*, and *runnable* otherwise. Upon creation of a thread it is suspended. If the task of a suspended thread becomes reducible, the thread becomes runnable. We say it is *woken*.

AMOZ is sequential, based on a single worker, where multiple runnable threads are scheduled preemptively and fairly. Only one thread can run at a time, it is called *running*. Making a runnable thread running is called to *schedule* the thread. Figure 1 sketches the handling of threads.

3 An Example: Mapping Lists

A procedure mapping a list Xs to a list Ys by applying a procedure P to all elements of both lists can be written as follows:

```
proc {Map Xs P Ys}
    if Xr X in Xs=c(X Xr) then
        local Y Yr in Ys=c(Y Yr) {P X Y} {Map Xr P Yr} end
    else Xs=nil Ys=nil fi
end
```

Lists are represented as trees $c(t_1 \ c(t_2 \ldots c(t_n \ \text{nil})))$. The procedure is referred to by a variable Map, as to be expected in a language with first-class procedures.

To illustrate the operational semantics of Map, assume that the procedure definition has been elaborated. Now we enter the expression

declare Xs P Ys **in** {Map Xs P Ys}

whose elaboration creates new variables for Xs, P, and Ys and reduces the procedure application {Map Xs P Ys} to a conditional task. The **declare** expression is a variant of the **local** expression whose scope extends to expressions the programmer enters later. The conditional task cannot be reduced since there is no information about the variable Xs in the store.

Now we enter the constraint (every occurrence of '_' creates a fresh variable)

Xs=c(_ c(_ _))

Since Xs=c(_ c(_ _)) entails the constraint in the guard of the conditional, it is reduced with its **then**-part. This imposes the constraint Ys=c(Y Yr), applies P to X and Y, and elaborates the recursive application {Map Xr P Yr}. A new conditional task is created which immediately reduces. Once more a conditional task is

created which this time cannot be reduced. The store now entails Xs=c(_ c(_ _))
and Ys=c(_ c(_ _)). Two elaborators for the application of P have been created,
but cannot reduce, since no definition for P has been elaborated yet. Both have
been moved to newly created threads.

By entering the constraint Xs=c(s(o) c(s(s(o)) nil)) the conditional task is
reduced to its **else**-constituent. Now the store entails Xs=c(s(o) c(s(s(o)) nil))
and Ys=c(_ c(_ nil)). Then we enter a procedure definition for P.

proc {P X Y} if Z in X=s(Z) then Y=Z else Y=o fi end

Both threads where the tasks for the application of P reside on are run, each
creating a conditional task. After their reduction, the store entails:

Xs=c(s(o) c(s(s(o)) nil)) Ys=c(o c(s(o) nil))

Suppose that the definition entered for P would be more involved, e.g., prime
factorization of large integers. In this case, prime factorization of each list element
would proceed in a round-robin fashion.

Threads are created implicitly. However, by using **thread** E **end** as abbrevi-
ation for **local** x **in if in** $x = a$ **then** E **else** $a = a$ **fi** $x = a$ **end**, we can explicitly state that reduction of E must advance fairly.

4   The Constraint Store

This section explains how rational trees are represented in the constraint store
and how they are unified. More details on this can be found in [21].

The constraint store consists of various kinds of *nodes*: tuples, names (explained in Sect. 8), variables, and references. Rational trees are composed of
these nodes. The constraint store is a dynamic memory area, thus nodes must
be explicitly allocated. Nodes are built according to the following definition,
where the code of the abstract machine is presented in a C++-like notation.

```
struct Node {enum {TUPLE, NAME, VAR, REF} tag;
             union {Node *ref;
                    struct {Label label; Node *args[];} tuple;};};
```

As presentation proceeds, the **union** part will be extended to host information
used with names and variables. Representation of nodes in this paper are chosen
with simplicity rather than efficiency in mind. The DFKI Oz asbtract machine
employs tagged pointers instead of a tagged data objects.

Unification of two trees residing in the constraint store works as follows:

```
Bool unify(Node *xin, Node *yin) {
  Node *x = deref(xin); Node *y = deref(yin);
  if (x==y) return True;
  if (x→tag==VAR) { bind(x,y); return True; }
  if (y→tag==VAR) { bind(y,x); return True; }
  if (x→tag==NAME || y→tag==NAME || x→tuple.label≠y→tuple.label)
    return False;
```

```
Node *xargs[] = x→tuple.args; Node *yargs[] = y→tuple.args;
rebind(x,y);
for (int i = 0; i<width(y→tuple.label); i++)
  if (unify(xargs[i],yargs[i]) == False) return False;
return True;
}

Node *deref(Node *n) { return (n→tag==REF) ? deref(n→ref) : n; }
void bind(Node *f, Node *t) { f→tag = REF; f→ref = t; }
void rebind(Node *f, Node *t) { f→tag = REF; f→ref = t; }
```

The function deref follows a chain of references until a non-reference node is reached. The functions bind and rebind make their first arguments into a reference pointing to its second argument. Note that bind and rebind are identical, but as the presentation proceeds they will be enhanced in different ways. After dereferencing both arguments, x and y point to variable, tuple, or name nodes. In case they point to the same node, unification is done. If one of them points to a variable, then bind binds the variable to the other node. If both point to tuples with the same label (which implies the same width, i.e., the same number of subtrees, as well), x is made pointing to y by rebind. Unification continues recursively for all subtrees of the tuples. Otherwise, False is returned. Note that two names can be unified only if they are identical.

In each recursive call the number of tuple and variable nodes in the constraint store is decremented by one. Additionally the invariant holds that chains of references are acyclic. This implies termination of unification. In AMOZ it is important that unification is *variable-centered*: it results in binding variables.

5 Introducing AMOZ

This section introduces AMOZ by presenting compilation and execution of declaration, composition, equation, and tuple construction.

Elaboration is implemented by execution of abstract machine instructions. Elaborating an expression E corresponds to executing the corresponding instructions $C[\![E]\!]$ as given by the compiler. In the following we consider only expressions, that are closed (i.e., without free variables) and renamed apart (i.e., each variable is declared only once).

To compile an expression E, the set V of variables declared in E is computed, where variables declared in procedure definitions are left aside (they are treated in Sect. 8). For each variable $x \in V$ an index $A[\![x]\!]$ is allocated, so that AMOZ can refer to a variable x by its index $A[\![x]\!]$.

AMOZ needs several registers. The *program counter* PC points to the currently executed instruction. The *environment* E is an array mapping the index $A[\![x]\!]$ of variable x to the variable's node in the store: $E[A[\![x]\!]]$. The emulator loop shown in Fig. 2 contains the single instruction ALLOCATE(n). Instructions for an expression E whose set of variables V has n elements are preceded by an instruction ALLOCATE(n) to allocate memory for the variables in E.

```
#define DISPATCH { PC++; goto emulate; }
engine() {
  emulate: switch (*PC) {
              case ALLOCATE(n):
              E = new Node*[n];
              DISPATCH;
              /* further instructions will be filled in here */
            }
  fail: /* handling of failure in unification */
}
```

Fig. 2. The emulator loop of AMOZ.

Execution of **local** x **in** E **end** creates a fresh variable node in the store and writes a reference to it to the environment $E[A[\![x]\!]]$. On the left of the diagram below the instructions obtained by compilation are shown, whereas on the right the implementation of newly introduced instructions is shown.

$C[\![\textbf{local } x \textbf{ in } E \textbf{ end}]\!] \equiv$
```
    CREATE_VAR(A[x])  | case CREATE_VAR(i):
    C[E]              | E[i] = new Node ⟨tag: VAR⟩;
                      | DISPATCH;
```

Composition is compiled into concatenation of the respective instruction sequences: $C[\![E\,F]\!] \equiv C[\![E]\!]C[\![F]\!]$.

An equality constraint is translated to an instruction calling the unification algorithm as presented in Sect. 4.

$C[\![x = y]\!] \equiv$
```
    UNIFY(A[x],A[y])  | case UNIFY(i,j):
                      | if (unify(E[i],E[j])==False) goto fail;
                      | DISPATCH;
```

Tuple construction $x = f(\overline{y})$ proceeds in three steps. First, a node holding the tuple to be constructed is allocated. A reference to it is held in register S. Second, the arguments are constructed and entered to the tuple's node. The last step unifies the constructed tuple with x.

$C[\![x = f(y_1 \ldots y_n)]\!] \equiv$
```
    CREATE_TUPLE(f/n)  | case CREATE_TUPLE(f/n):
    PUT_ARG(1,A[y₁])   | S=new Node ⟨tag:TUPLE,
    ...                |                tuple:⟨label:f/n,
    PUT_ARG(n,A[yₙ])   |                      args:new Node*[n]⟩⟩;
    UNIFY_S(A[x])      | DISPATCH;
                       | case PUT_ARG(i,j):
                       | S→tuple.args[i] = E[j];
                       | DISPATCH;
                       | case UNIFY_S(i):
                       | if (unify(S,E[i])==False) goto fail;
                       | DISPATCH;
```

The compilation scheme presented above is simplified; the integration of optimization techniques known from the WAM [25, 1, 24] like read/write mode unification, and allocation of temporary variables to registers is straightforward and is not detailed.

6 Threads and Matching

This section introduces threads through a restricted form of conditional expression that implements pattern-matching. Conditionals considered herein are of the form **if** $\overline{y}$ **in** $x = f(\overline{y})$ **then** E **else** F **fi**, where the variables in $\overline{y}$ are pairwise distinct. They can be reduced, if and only if the variable x is bound. The case where x is not bound, introduces threads into AMOZ.

Threads in AMOZ are stacks of tasks. They have the type Thread and feature the common operations push, pop, and isEmpty. A *task* TASK(l) points to an abstract machine instruction located at label l. AMOZ is extended by three registers: running (of type Thread) for the currently running thread, runnable (of type ThreadQueue) for the queue of runnable threads, and timeOver for a flag that will be set to TRUE by an external source (e.g., the operating system) after a certain amount of time.

Adding a task to the currently running thread is performed by the PUSH instruction. The RETURN instruction tries to execute the topmost task from the currently running thread. If the currently running thread has no tasks left, another thread is scheduled. Preemption is checked in the PUSH instruction only, since it is the only instruction by which tasks can be added dynamically to a thread.

```
emulate: ...
    case PUSH(l):
      push(running, TASK(l)); PC++;
      if (timeOver) goto preempt; else goto emulate;
    case RETURN:
      goto run;

run:
  if (isEmpty(running)) goto schedule;
  TASK(l) = pop(running); PC = l; goto emulate;
schedule:
  if (isEmpty(runnable)) // terminate AMOZ
  running = dequeue(runnable); goto run;
preempt:
  push(running, TASK(PC)); timeOver=FALSE;
  enqueue(runnable, running); goto schedule;
```

Fig. 3. Extending the emulator loop for threads.

The emulator loop is extended as shown in Fig. 3. Note that the loop deals only with runnable threads, creation and waking of threads is explained below.

We will need in the following that variables are extended such that suspended threads can be attached to them:

struct Node {... {**struct** {ThreadQueue *susp} var; ...} ...};

The special form of conditional compiles as follows:

$C[\![\text{if } y_1 \cdots y_n \text{ in } x = f(y_1 \cdots y_n) \text{ then } E \text{ else } F \text{ fi}]\!] \equiv$

```
    PUSH(L1)              case DELAY(i):
    DELAY(A[x])            S=deref(E[i]);
    MATCH(f/n,Le)          if (S→tag≠VAR) DISPATCH;
    GET_ARG(1,A[y₁])       enqueue(S→var.susp,new Thread ⟨TASK(PC)⟩);
    ...                    goto run;
    GET_ARG(n,A[yₙ])      case MATCH(f/n,le):
    C[E]                   if (S→tag==TUPLE && S→label==f/n) DISPATCH;
    RETURN                 PC=le;
Le: C[F]                   goto emulate;
    RETURN                case GET_ARG(i,j):
L1:                        E[j]=S→tuple.args[i];
                           DISPATCH;
```

The first instruction pushes a task on the running thread, thus fixing where execution proceeds after the conditional. The instruction DELAY checks whether the variable x is bound. In case x is bound, it is matched against the pattern and execution continues with the instructions for either E or F depending on the outcome of the match. Otherwise, the conditional must wait until x is bound. In this case a new thread consisting of the single task to reexecute the DELAY instruction is created. This thread is attached to the node of x. Execution continues by popping the next task from the current thread.

Binding a variable wakes all suspended threads attached to it by adding them to the queue of runnable threads. This is implemented by extending the procedure bind (cf. Sect. 4):

```
void bind(Node *f, Node *t) {
  runnable=concat(runnable,f→susp);
  f→susp=⟨⟩; f→tag=REF; f→ref=t;
}
```

7 Local Computation Spaces

This section introduces local computation spaces to AMOZ, and shows how they are used for implementing conditionals with deep guards.

Local computation spaces are represented in the machine as follows:

```
struct Space {Space *parent;
              NodePair *script[];
              enum {ALIVE, FAILED, ENTAILED} state;
              int counter;
              Instr* entailed, failed;};
```

The parent component points to the space directly above, linking spaces to the tree of spaces. In the *root* space, i.e., the topmost space, it is NULL. AMOZ is equipped with a register curSpace pointing to the current computation space, which is initialized with the root space. A task TASK(s,1) now also carries the space s to which it belongs.

Local constraints are maintained in the script, consisting of pairs of nodes (to be explained later).

The field entailed(failed) points to an instruction where execution proceeds in case the local space is entailed (failed). In our case of a conditional, these fields point to the first instruction of its **then** respectively **else** constituent.

Entailment of computation spaces. A computation space is entailed if no spaces exist below, it has no tasks left, and its local constraints are entailed by its parent's constraints. To check the first two conditions for a space, the field counter counts its tasks and the spaces below. The counter is maintained upon creation of new spaces, failure of spaces, merging of entailed spaces, and upon pushing and popping of tasks to the currently running thread.

A local constraint is entailed if it does not bind any global variables (this is a well known property of rational tree constraints [21], sometimes also referred to as quietness). Binding of variables needs to support entailment checking: the direction of binding becomes important, that is, global variables must not be bound to local variables [21]. This introduces the need to check for locality of variables. Therefore, a variable node in the store contains its home space, which is initialized upon variable creation:

```
struct Node {... struct {ThreadQueue *susp; Space *home;} var; ...};

case CREATE_VAR(i):
  E[i]=new Node ⟨tag:VAR var:⟨susp:⟨⟩ home:curSpace⟩⟩;
  DISPATCH;
```

Checking locality of a variable must take into account that a computation space is merged with its parent's space upon entailment. Testing whether a variable is local to a space is implemented by applying the function isCurrent to the variable's home field.

```
Bool isCurrent(Space *s) {
  return (s==curSpace || s→state==ENTAILED && isCurrent(s→parent));
}
```

In DFKI Oz, the garbage collector shortens parent chains, such that memory used by entailed spaces can be reclaimed.

As in the previous section, upon binding of variables suspended threads need to be woken. DFKI Oz incorporates an important optimization, that only threads below the current space are woken. Maintaining the script will be explained later.

```
void bind(Node *f, Node *t) {
  if (t→tag==VAR && isCurrent(t→home)) swap(f,t);
  if (!isCurrent(f→home)) add(curSpace→script,⟨f,t⟩);
```

```
f→tag=REF; f→ref=t;
wake(f);
if (t→tag==VAR) wake(t);
}


void wake(Node *n) {
runnable=concat(runnable,n→susp);
}
```

Bindings done by the procedure rebind must be done local to a space as well. This can be achieved by doing them only temporarily during unification and undoing them after finishing unify. For sake of brevity we omit the straightforward redefinition of unify and rebind.

Finally, the procedure isEntailed tests whether a space is entailed:

```
Bool isEntailed(Space *s) {
return s→state==ALIVE && s→counter==0 && s→script==⟨⟩
}
```

Maintaining multiple computation spaces. In a space, all constraints from spaces above must be visible. In a sequential implementation this can be achieved by doing variable bindings in place and maintaining a script of globally visible changes, supporting fast access to both local and global bindings. The globally visible changes are bindings of global variables. They are written to the script in the procedure bind. Other schemes for multiple constraint stores are known, e.g., [15, 13, 14].

Suppose that the current space is S_1, and a task in a different space S_2 must be run. All constraints local to spaces between S_1 and the root must be removed, and all constraints between the root and S_2 must be made visible[2]. Removal of constraints is called *leaving*, whereas making constraints visible is called *entering*.

```
void leave(Space *s) {              Bool enter(Space *s) {
if (s==NULL) return;                 if (s==NULL) return True;
leaveSpace(s);                       if (!enter(s→parent)) return False;
leave(s→parent);                     return enterSpace(s);
}                                   }
```

Leaving a single space removes all bindings contained in its script. Entering updates curSpace, and performs unification of all pairs in the script. Note, that it is not sufficient to perform binding, since the left hand side of a script entry may be bound already.

---

[2] DFKI Oz and the AGENTS implementation of AKL use the straightforward optimization not to go up to the root space, but to the closest common ancestor of S_1 and S_2.

```
void leaveSpace(Space *s) {        Bool enterSpace(Space *s) {
 foreach ⟨x,t⟩ in s→script          curSpace=s;
   x→tag = VAR;                      if (s→state==FAILED) return False;
}                                    foreach ⟨x,t⟩ in s→script
                                      if (unify(x,t)==False)
                                       return False;
                                     s→script=⟨⟩; return True;
                                   }
```

Deep guards. Now we consider conditionals with deep guards. Their compilation is as follows:

$C[\![$ **if** $x_1 \cdots x_n$ **in** E **then** F **else** G **fi** $]\!] \equiv$

```
    PUSH(L1)                    case CREATE_SPACE(lt,le):
    CREATE_SPACE(Lt,Le)          Space *s=new Space ⟨state:ALIVE,
    CREATE_VAR(A[[x₁]])                               entailed:lt, failed:le,
    ...                                               script:⟨⟩, counter:0,
    CREATE_VAR(A[[xₙ]])                               parent:curSpace⟩;
    C[[E]]                       curSpace→counter++; curSpace = s;
    CHECK_ENTAILED               DISPATCH;
Lt: C[[F]]                      case CHECK_ENTAILED:
    RETURN                       if (isEntailed(curSpace)) {
Le: C[[G]]                        merge(); goto emulate;
    RETURN                       } else {
L1:                               foreach ⟨x,t⟩ in s→script
                                   enqueue(x→var.susp,newThread(PC));
                                  goto run;
                                 }
```

The instruction CREATE_SPACE links the newly created space to the tree, updates the current space's counter, and enters the created space. By newThread(1) a new thread with one task to execute the instruction at 1 in the current space is created. The instruction CHECK_ENTAILED checks if the space is entailed. In case the space is not yet entailed, new threads are added to the global variables bound in this space. These threads contain the task to reexecute the CHECK_ENTAILED instruction. Otherwise, the function merge merges the current space with its parent space.

```
void merge() {
 curSpace→state=ENTAILED;
 PC=curSpace→entailed;
 curSpace=curSpace→parent;
 curSpace→counter--;
}
```

Running a thread now needs to enter the computation space of the task. When popping a task TASK(s,1), then the space s is entered and its counter is decremented. When this task has been executed (i.e., when reaching run again) entailment is checked, because it could have been the space's last task. AMOZ also needs to handle failure of a space, because failed spaces must be discarded from the tree of spaces. Spaces below a failed space are not marked as failed immediately, instead enter detects them and discards their tasks. The extended emulator loop is shown in Fig. 4.

```
fail:                                   run:
  if (curSpace→status==ALIVE) {           if (isEntailed(curSpace)) {
    curSpace→state = FAILED;                 merge(); goto emulate;
    PC = curSpace→failed;                  }
    curSpace=curSpace→parent;              if (isEmpty(running)) goto schedule;
    curSpace→counter--;                    TASK(s,1)=pop(running);
    goto emulate;                          s→counter--;
  }                                        leave(curSpace);
  goto run;                                if (enter(s)==False) goto fail;
                                           PC=1;
                                           goto emulate;
```

Fig. 4. Extending the emulator loop for local computation spaces.

The cost of pushing and popping tasks of the form TASK(s,1) can be reduced by having two kinds of tasks TASK_S(s) and TASK_L(1). The latter will simply jump to the instruction at label 1, where the (costly) former will enter space s and maintain counter as explained above. Now tasks of kind TASK_S(curSpace) need to be pushed just before curSpace is left.

Pattern matching conditionals as shown in the previous section can be used as an optimization to conditionals with deep guards. This makes indexing techniques applicable as known for Prolog [24], Concurrent Logic Programming languages [9], and CC languages [2]. Further techniques for optimization of flat guards including composition of equations and arithmetic tests have been integrated into DFKI Oz.

8 First-class Procedures

This section introduces procedure definition and application to AMOZ. Nodes carrying the tag NAME are extended to support binding to procedures. Procedures resemble closures known from functional programming languages.

```
struct Node {... struct {Instr *lb; int arity; Node *free[];} proc;};
```

A procedure definition is compiled to instructions creating the procedure and instructions for the body of the procedure ($z_1, \cdots, z_k$ denote the *free* variables of local $y_1 \cdots y_n$ in E end):

```
C[[proc {x y₁ ··· yₙ} E end]] ≡
```

$$C[\![\text{proc}\,\{x\,y_1\cdots y_n\}\,E\,\text{end}]\!] \equiv$$

`PROCDEF(Lb,n,k)`	`case PROCDEF(l,n,k):`
`UNIFY_S(A[[x]])`	`S=new Node ⟨tag:NAME, proc:⟨lb:l,arity:n,`
`MOVE_C(A[[z₁]],1)`	`                           free:new Node*[k]⟩⟩;`
`···`	`DISPATCH;`
`MOVE_C(A[[zₖ]],k)`	`case MOVE_C(i,j):`
`JUMP(L1)`	`S→proc.free[j]=E[i];`
`Lb: B[[y₁ ··· yₙ, E]]`	`DISPATCH;`
`L1:`	`case JUMP(l):`
	`PC=l; goto emulate;`

The PROCDEF instruction creates a new NAME node. The free field is filled by MOVE_C instructions. Free variables are addressed by a new register F within the body of a procedure similarly to how other variables are addressed by E. Thus we allow access to F in instructions like UNIFY.

$B[\![y_1\cdots y_n, E]\!]$ compiles the body: it allocates an environment of size $k = m+n$, where m is the number of declared variables within E (cf. Sect. 5), and n is the number of arguments. Parameters, which are passed in *argument registers* A[1] to A[n] as in the WAM, are first saved into the environment. Then the compiler creates code for the body.

$$B[\![y_1\cdots y_n, E]\!] \equiv$$

`ALLOCATE(k)`	`case MOVE_E(i,j):`
`MOVE_E(1,A[[y₁]])`	`E[j]=A[i];`
`···`	`DISPATCH;`
`MOVE_E(n,A[[yₙ]])`	
`C[[E]]`	
`RETURN`	

An application $\{x\,y_1\cdots y_n\}$ checks whether x is bound to a name, moves $y_1\cdots y_n$ into A[1] to A[n], pushes the instruction following the application on the currently running thread, sets register F to point to x's procedure, and jumps to the body of the procedure. Since an ALLOCATE instruction in the procedure's body will change E we have to modify the task data structure to also contain the environment E and free variable F registers.

$$C[\![\{x\,y_1\cdots y_n\}]\!] \equiv$$

`PUSH(L1)`	`case PUSH(l):`
`DELAY(A[[x]])`	`push(running,TASK(curSpace,l,E,F));`
`MOVE_A(A[[y₁]],1)`	`DISPATCH;`
`···`	`case MOVE_A(i,j):`
`MOVE_A(A[[yₙ]],n)`	`A[j]=E[i];`
`APPLY(A[[x]],n)`	`DISPATCH;`
`L1:`	`case APPLY(i,n):`
	`S=deref(E[i]);`
	`if (S→tag≠NAME || S→proc.arity≠n) goto fail;`
	`F=S→proc.free;`
	`PC=S→proc.lb; goto emulate;`

In Fig. 5 we show the compilation of a procedure P, which unifies its argument with the variable Z.

```
                               ALLOCATE(2)
                               CREATE_VAR(E[0])    % Z → E[0]
                               CREATE_VAR(E[1])    % P → E[1]
                               CREATE_TUPLE(a/0)
        local Z P in           UNIFY_S(E[0])       % Z=a
          Z=a                  PROCDEF(Lp,1,1)     % Z is free
          proc {P X}           UNIFY_S(E[1])
             X=Z               MOVE_C(E[0],0)      % Z→F[0]
          end                  JUMP(Le)
        end              Lp: ALLOCATE(1)
                               MOVE_E(1,E[0])      % X → E[0]
                               UNIFY(E[0],F[0])    % X=Z
                               RETURN
                         Le: RETURN
```

Fig. 5. Example code for a procedure definition.

The scheme presented above imposes an overhead to the handling of procedures compared to the first order case as exemplified by Prolog. Therefore the DFKI Oz compiler performs some important optimizations to eliminate these extra costs, as we will describe now.

The ALLOCATE instruction takes memory from a heap (and not from a stack as Prolog does) and there is no DEALLOCATE instruction at the end of a procedure. This is due to concurrency: when the end of a procedure is reached, there may be suspended threads referring to the environment. DFKI Oz does the following: the compiler inserts a DEALLOCATE instruction, and environments are allocated from a free-list. Every environment has a flag which is set if there exist suspended threads. The DEALLOCATE instruction checks this flag, and frees the environment only if the flag is not set.

The execution of the instructions for a procedure definition is quite costly due to procedure creation. The instructions are only executed once per procedure, whereas the procedure itself can be applied many times. To reduce the cost of a procedure application the instructions PUSH, DELAY and APPLY can be collapsed into one instruction. Additionally, the compiler tries to determine by static analysis whether x in $\{x\, y_1 \cdots y_n\}$ is bound to a procedure with arity n. Experience shows that it succeeds in most cases, especially in all cases where procedures are used as in Prolog: the compiler replaces the PUSH, DELAY, APPLY sequence by a special instruction FASTAPPLY taking as argument the instruction address for the procedure definition of x. This eliminates the extra costs of the general scheme.

Acknowledgements

We would like to thank Tobias Müller, Konstantin Popov, and Gert Smolka for helping us with the design and/or implementation of the abstract machine for Oz. We are grateful to the Programming Systems Group at SICS for sharing their experiences in implementing SICStus Prolog and AKL with us. Martin Henz, Martin Müller, Tobias Müller, Joachim Niehren, Gert Smolka, Ralf Treinen, Peter Van Roy, and the anonymous referees provided helpful comments on this paper. The research reported in this paper has been supported by the Bundesminister für Bildung, Wissenschaft, Forschung und Technologie (FTZ-ITW-9105), the Esprit Project ACCLAIM (PE 7195), and the Esprit Working Group CCL (EP 6028).

References

1. Hassan Aït-Kaci. *Warren's Abstract Machine: A Tutorial Reconstruction.* Logic Programming Series. The MIT Press, Cambridge, MA, 1991.
2. Per Brand. A decision graph algorithm for CCP languages. In Leon Sterling, editor, *Proceedings of the 1995 International Conference on Logic Programming*, pages 433–448, Kanagawa, Japan, June 1995. The MIT Press.
3. William Clinger and Jonathan Rees. The Revised[4] Report on the Algorithmic Language Scheme. *LISP Pointers*, IV(3):1–55, July-September 1991.
4. Alain Colmerauer. Prolog and infinite trees. In K.L. Clark and S.-A. Tärnlund, editors, *Logic Programming*, pages 153–172. Academic Press, 1982.
5. Alain Colmerauer. Equations and inequations on finite and infinite trees. In *Proceedings of the 2nd International Conference on Fifth Generation Computer Systems*, pages 85–99, 1984.
6. Martin Henz, Gert Smolka, and Jörg Würtz. Object-oriented concurrent constraint programming in Oz. In Vijay Saraswat and Pascal Van Hentenryck, editors, *Principles and Practice of Constraint Programming*, chapter 2, pages 27–48. The MIT Press, Cambridge, MA, 1995.
7. Paul Hudak, Philip Wadler, et al. Report on the programming language Haskell. Technical Report YALEU/DCS/RR/777, Yale University, 1990.
8. Sverker Janson. *AKL - A Multiparadigm Programming Language.* Dissertation, SICS Swedish Institute of Computer Science, Uppsala University 1994, SICS Box 1263, S-164 28 Kista, Sweden, 1994.
9. Shmuel Kliger and Ehud Shapiro. From decision trees to decision graphs. In Saumya Debray and Manuel Hermenegildo, editors, *North American Conference on Logic Programming*, pages 97–116, Austin, TX, 1990. The MIT Press.
10. Michael J. Maher. Logic semantics for a class of committed-choice programs. In Jean-Louis Lassez, editor, *Logic Programming, Proceedings of the Fourth International Conference*, pages 858–876, Melbourne, 1987. The MIT Press.
11. Michael Mehl, Tobias Müller, Konstantin Popov, and Ralf Scheidhauer. DFKI Oz user's manual. DFKI Oz documentation series, German Research Center for Artificial Intelligence (DFKI), Stuhlsatzenhausweg 3, D-66123 Saarbrücken, Germany, 1994.
12. Robin Milner, Mads Tofte, and Robert Harper. *Definition of Standard ML.* The MIT Press, Cambridge, MA, 1990.

13. Johan Montelius and Khayri A. M. Ali. An And/Or-parallel implementation of AKL. *New Generation Computing*, 13–14, August 1995.
14. Andreas Podelski and Gert Smolka. Situated simplification. In *Proceedings of the First International Conference on Principles and Practice of Constraint Programming*, LNCS, Marseille, France, September 1995. Springer-Verlag. To appear.
15. Andreas Podelski and Peter Van Roy. The beauty and beast algorithm: quasi-linear incremental tests of entailment and disentailment over trees. In Maurice Bruynooghe, editor, *Logic Programming: Proceedings of the 1994 International Symposium*, pages 359–374, Ithaca, NY, November 1994. The MIT Press.
16. Vijay A. Saraswat and Martin Rinard. Concurrent constraint programming. In *Proceedings of the 7th Annual ACM Symposium on Principles of Programming Languages*, pages 232–245, San Francisco, CA, January 1990. ACM Press.
17. Christian Schulte and Gert Smolka. Encapsulated search in higher-order concurrent constraint programming. In Maurice Bruynooghe, editor, *Logic Programming: Proceedings of the 1994 International Symposium*, pages 505–520, Ithaca, NY, November 1994. The MIT Press.
18. Christian Schulte, Gert Smolka, and Jörg Würtz. Encapsulated search and constraint programming in Oz. In Alan H. Borning, editor, *Second Workshop on Principles and Practice of Constraint Programming*, Lecture Notes in Computer Science, vol. 874, pages 134–150, Orcas Island, WA, May 1994. Springer-Verlag.
19. Gert Smolka. A foundation for higher-order concurrent constraint programming. In Jean-Pierre Jouannaud, editor, *1st International Conference on Constraints in Computational Logics*, Lecture Notes in Computer Science, vol. 845, pages 50–72, München, Germany, September 1994. Springer-Verlag.
20. Gert Smolka. The definition of Kernel Oz. In Andreas Podelski, editor, *Constraints: Basics and Trends*, Lecture Notes in Computer Science, vol. 910, pages 251–292. Springer-Verlag, 1995.
21. Gert Smolka and Ralf Treinen. Records for logic programming. *Journal of Logic Programming*, 18(3):229–258, April 1994.
22. Gert Smolka and Ralf Treinen (ed.). DFKI Oz documentation series. Deutsches Forschungszentrum für Künstliche Intelligenz, Stuhlsatzenhausweg 3, D–66123 Saarbrücken, Germany, 1994.
23. Pascal Van Hentenryck, Vijay Saraswat, and Yves Deville. Design, implementation and evaluation of the constraint language cc(FD). In Andreas Podelski, editor, *Constraint Programming: Basics and Trends*, Lecture Notes in Computer Science, vol. 910, pages 293–316. Springer-Verlag, 1995.
24. Peter Van Roy. 1983-1993: The wonder years of sequential Prolog implementation. *Journal of Logic Programming*, 19/20:385–441, May/July 1994.
25. David H. D. Warren. An abstract Prolog instruction set. Technical Note 309, SRI International, Artificial Intelligence Center, Menlo Park, CA, October 1983.

Remark

The DFKI Oz system and papers of authors from the Programming Systems Lab at DFKI are available through WWW at http://ps-www.dfki.uni-sb.de/ or through anonymous ftp from ps-ftp.dfki.uni-sb.de.

Uniform PERs and Comportment Analysis

Alan Mycroft[1] and Kirsten Lackner Solberg[2]

[1] Computer Laboratory, Cambridge University,
New Museums Site, Pembroke Street, Cambridge CB2 3QG, United Kingdom
Alan.Mycroft@cl.cam.ac.uk

[2] Dept. of Math. and Computer Science, Odense University, Denmark and
DAIMI, Aarhus University, Denmark
kls@daimi.aau.dk

Abstract. Hunt showed that the notion of PER-based strictness properties subsumed the incomparable notions of ideal- and projection-based properties on non-lifted value spaces. We extend Hunt's idea so it can encompass other *comportment* properties (such as totality) by separating the information and fixpoint orderings. We then define a class of *uniform* PERs (in the sense that they treat non-bottom ground elements, e.g. integers, identically) and then show how these can be hereditarily defined on function spaces. These (or subsets thereof) can be taken as a space of abstract values for abstract interpretation in the style of Nielson.

1 Introduction

Strictness properties are properties of functions between domains; in principle they are intended to capture the notion of how the function reacts to changes in the definedness of its arguments rather than changes between incomparable values of its arguments. A comparison to the notion of partial differentiation may prove fruitful. Since many properties conforming to this notion (e.g. totality) exclude $\perp$, we use the word *comportment property* as coined by Cousot and Cousot [4] to avoid abusing the epithet "strictness property" by allowing it to encompass totality. Here comportment is considered within a denotational formalism.

This work started as an experiment to discover the expressive power of a natural notion of uniform PER as a basis for describing comportment properties. It transpires that the framework of "higher order abstract interpretation" in [4] describes a similar set of comportment properties after being specialised as in the second part of that work. The interest is that PERs in the traditional abstract interpretation framework can express similar properties as simple properties in the higher-order framework. We express no preference between these routes; the aim was to understand better PER-based properties. Moreover we do not discuss implementation complexity.

For many years there were two forms of strictness property, the ideal-based form and the projection-based form [8, 2, 12, 6, 5, 4]. Cousot and Cousot [3] consider a Galois connection view of strictness properties.

Given domains D and E and a continuous function $f : D \to E$ (the denotational semantics of a program function) these properties can be summarised as follows.

Let I range over the *ideals* of D (non-empty Scott-closed sets) and J over those of E. The ideal-based strictness properties are those of the form

$$W_{I,J}^{\text{ideal}}(f) \Leftrightarrow f(I) \subseteq J.$$

Similarly, let α range over projections on D (continuous, idempotent functions such that $\alpha(x) \sqsubseteq x$) and β over those on E. The projection-based strictness properties are those of the form

$$W_{\alpha,\beta}^{\text{proj}}(f) \Leftrightarrow \beta \circ f = \beta \circ f \circ \alpha.$$

Hunt [6] observed that PER-based properties generalise both the above forms. (Actually Hunt only considered strict and inductive PER-properties which suffice for his generalisation but we relax this so as to be able to encompass as many comportment properties as possible.)

A PER P on D is a relation on D which is symmetric and transitive. It is inductive if it contains limits of chains when seen as a subset of $D \times D$. Such a PER P defines a property W_P^{PER} on D given by its diagonal

$$W_P^{\text{PER}}(d) \Leftrightarrow d \in |P| \quad \text{where} \quad |P| = \{d \mid dPd\}.$$

Hunt essentially defines the PER-properties of functions in two stages, first defining the basic PER-properties and then the PER-properties as the conjunctive completion (meet completion) of these. Letting P range over (a class of, see below) PERs of D and Q over (a class of) PERs of E, the basic PER-properties appear as

$$W_{P,Q}^{\text{basic-PER}}(f) \Leftrightarrow (\forall x, y \in D)(xPy \Rightarrow (fx)Q(fy)).$$

Conjunctions (intersections) of these give PER-properties.

The reader familiar with the presentations of abstract interpretation in the style of Nielson or Burn, Hankin and Abramsky *et al.* will here note an absence of specifying *how* the properties at higher-order are related to those at lower-order (or first-order). Indeed, Cousot and Cousot argue that the framework of abstract interpretation should leave this unspecified; only particular examples instantiating the framework should do this (see e.g. the two parts of [4]). In general the argument is that the framework should allow one to select representations for program properties at each type independently; all that is required is that function properties, ranged over by C, should be in Galois correspondence $\gamma : C \to \mathcal{P}(D \to E)$ where $\mathcal{P}(\cdot)$ is the power-*set* construct. This yields properties

$$W_C^{\text{Galois}}(f) \Leftrightarrow f \in \gamma(C).$$

This paper leans towards the framework propounded by Nielson [9]. Here the framework requires that the properties on higher types are related to those on lower types (albeit possibly in different ways for different uses of a given type). This framework is tighter in that is forbids the parasitic possibility that, e.g., the space of properties for a type depends uncomputably on its type structure. On the other hand the attempt to forbid 'unreasonable' interpretations in the framework generally requires the framework to be repeatedly enhanced when desirable interpretations are found not to fit.

One can, perhaps over-simplistically, observe that the former view approach is platonic in that it prescribes a framework and then searches for instances which explain or yield various analyses (these instances appear always to have been decidable). Similarly the effect of the latter approach has generally been constructivist: abstract spaces have been crafted which explain exactly the range of phenomena at hand; abstract spaces for higher types are expected to be derivable from those at lower types (cf. intuitionistic implication). The old arguments transfer beautifully: the Platonist argues that the constructivist is doomed to extend his constructions each time the world demands (and never models it perfectly); the constructivist chides the generality of the Platonist for allowing parasitic solutions which have no application or physical reality.

The present construction of *uniform* PERs, both at ground type and hereditarily at higher types, is designed to capture as many comportment properties as possible (hopefully all) in a constructive manner. It was developed concurrently with [4] and attempts to capture comportments from the constructive, as opposed to platonic, viewpoint. Interestingly the two approaches differ (ours yields four extra properties at type Int → Int). These are accounted for by the fact that uniform PERs naturally separate the notion of a function being total from it being constant; the application in [4] does not. Certainly, it is possible to imagine a variant of each which models the other.

The range of comportment properties expressible by uniform PERs include not only strictness properties but also totality properties. These latter are also captured by the annotated type system of Solberg *et al.* [10]. Note also the independent work by Benton [1] and Jensen [7] on expressing strictness properties in logical form.

1.1 Overview

In Section 2 we recall the definition of PERs then define the subset and Egli-Milner orderings on PERs.

Section 3 defines the notion of uniform PER on the integers. These PERs are uniform in the sense that they treat all the integers identically (as in [5]). We observe that some of the uniform PERs on the integers are *not* strict. Far from being a problem these non-strict PERs describe the property of being a non-bottom value (or at higher types being a total function) in contrast to most work on ideal- or projection-based program analysis (subject to the unsurprising need to use the Egli-Milner ordering on PERs which we define). The empty PER also appears as a uniform PER but is not proscribed because of its possible use to represent some sort of dead code. Next we form following Hunt the PERs on the function space and again we observe that some of these are not strict.

In Section 4 we take a closer look at the uniform PERs on $\mathbf{Z}_{\perp} \to \mathbf{Z}_{\perp}$. It appears that we are able to express all the properties that are expressible with ideal-based [5, 8], projection-based [12], and rule-based [1, 10] program analyses. Some new properties also emerge, e.g. the property of being constant on the integers, the property of being non-bottom on the integers, the property of being constant and non-bottom on the integers, the property of being non-bottom on the integers *or* the bottom-function. These include the comportment properties of [4].

For each property there exists an optimisation that can be applied to the code implementing an expression with that property. So again we can see the advantage of also considering the non-strict PERs. Moreover for each property there exists a function which has that property as the best/most exact description of the expression.

The next step is to define the language in Section 5, its standard semantics, and a non-standard semantics for program analysis. We define the semantics and the program analysis as two different interpretations of the language (as in [6]). Following Hunt's proof of correctness of the analysis we show that the standard and non-standard semantics of terms satisfy a ternary logical relation which relates pairs of standard values with an abstract value.

The final step is to define the abstract-denotations of the constants, e.g. *if*, *fix*, +, such that their standard-denotation is related to the abstract-denotation. The problems here is mostly for *fix*. Not all the PERs we are dealing with are strict therefore we cannot just follow Hunt using a union operator to form the fixpoint. (Starting with the least PER in the subset ordering (the empty PER) and applying the functional to it would result in the empty PER and hence the fixpoint would be the empty PER.) Following ideas dating from at least [8] we start with the least strict PER and a form of the Egli-Milner ordering which we define on PERs.

2 Formalism

We start by defining the notion of a PER on a set and then consider the possible orderings on PERs when the set has order-structure:

2.1 PERs on domains

Recall that a PER on a set S is a relation on S that is *symmetric* and *transitive*. Both the domain and range of the PER is equal to the diagonal-part of P, defined by $|P| = \{x \mid (x, x) \in P\}$. For a given set A of PERs, the *properties* associated with A are the set of diagonals of members of A. PERs can be ordered in at least two different ways: the subset ordering and the Egli-Milner ordering.

Definition 1 *The subset ordering*
The PER P is less than or equal to the PER Q, written $P \leq Q$, if $P \subseteq Q$.

2.2 The Egli-Milner ordering

The subset ordering does not take into account the structure of the domain on which the PERs are built. The least PER in the subset ordering is the empty PER. The result of applying a functional to the empty PER is the empty PER, therefore the fixpoint of a functional is the empty PER. What we want to do is to start with the least, strict, PER. But starting there may not yield a chain under the subset ordering. The obvious choice is to use the Egli-Milner ordering.

Let D be a cpo. We define the Egli-Milner ordering on the space of PERs over D by treating the PERs as subsets of $D \times D$.

Definition 2 *The Egli-Milner ordering on PERs over D*
Let P and Q be PERs over D. We define $P \sqsubseteq_{EM}^{PER} Q$ if $P \sqsubseteq Q$ when considered as subsets of $D \times D$, i.e.

$$(\forall p \in P \exists q \in Q : p \sqsubseteq_{D \times D} q) \wedge (\forall q \in Q \exists p \in P : p \sqsubseteq_{D \times D} q).$$

□

We also define a PER P being strict and downwards closed by inheritance from $D \times D$:

Definition 3 *Strict and downwards closed PERs*
A PER P on D is *strict* if $(\perp_D, \perp_D) \in P$; it is *downwards closed* P if it is downwards closed when seen as a subset of $D \times D$, i.e.

$$((d_1, d_2) \sqsubseteq (d_3, d_4) \wedge (d_3, d_4) \in P) \Rightarrow (d_1, d_2) \in P$$

□

For strict PERs (i.e. Hunt's work) we would expect the subset ordering to coincide with the Egli-Milner ordering:

Lemma 4 *Egli-Milner ordering on strict and downwards closed PERs*
For all strict PERs P on D and for all downwards closed PER Q on D:

$$P \sqsubseteq_{EM}^{PER} Q \Leftrightarrow P \leq Q$$

Proof We assume P is a strict PER on D and that Q is a downwards closed PER on D.
First we assume $P \sqsubseteq_{EM}^{PER} Q$ and show $P \leq Q$. We assume $(x, y) \in P$ then from Definition 2 part one we get that there exits $(x', y') \in Q$ such that $(x, y) \leq (x', y')$. Since Q is downwards closed it can be the case that $x' = x$ and $y' = y$. Therefore we have that P is a subset of Q.
Next assume $P \leq Q$ and show $P \sqsubseteq_{EM}^{PER} Q$. Since P is a subset of Q then for all $(x, y) \in P$ we have $(x, y) \in Q$ such that $(x, y) \leq (x, y)$ which is the first part of Definition 2.
We have $(\perp_D, \perp_D) \in P$ since P is strict. Therefore for all $(x, y) \in Q$ we have $(\perp_D, \perp_D) \leq (x, y)$, which is the second part of Definition 2. ■

3 Uniform PERs on Types

Although the intuition is to define a class of uniform PERs associated with a *domain* it turns out to be more natural to define the classes of uniform PERs associated with (the standard interpretation of) a *type*. (For the purposes of abstract interpretation representations of these PERs can be used as abstract values and it is too restrictive to insist that (accidentally) isomorphic domains described by different types should have identical sets of abstract values). We will continue to refer to (e.g.) "the set of uniform PERs on $\mathbf{Z}_\perp$" when this reads better.

We start with a set Σ, of types, ranged over by σ and τ:

$$\sigma ::= \text{Int} \mid \sigma \to \sigma \mid \sigma \times \sigma$$

For each type σ there is a standard domain D_σ^S:

$$D_\text{Int}^S = \mathbf{Z}_\perp \qquad D_{\sigma \to \tau}^S = D_\sigma^S \to D_\tau^S \qquad D_{\sigma \times \tau}^S = D_\sigma^S \times D_\tau^S$$

where $\to$ is continuous function space. We now define a PER on a type:

Definition 5
Let σ be a type, then P is a PER on σ if P is a PER on D_σ^S. □

For each type σ we now define a finite set of *uniform PERs*, $\mathcal{U}(\sigma)$, consisting of certain PERs on σ. A uniform PER on the integers is a PER on the standard domain of the integers which treats all the integers in the same way. The reason for only looking at uniform PERs is that in a comportment analysis we are typically interested in knowing whether an expression evaluates to an integer or undefined value—not about it being any particular integer. The uniform PERs on the integers are:

Definition 6
A PER P on Int is *uniform* if, whenever π is a permutation on $\mathbf{Z}_\perp$ $(= D_\text{Int}^S)$ leaving $\perp$ unchanged, then $\forall x, y \in \mathbf{Z}_\perp : (x, y) \in P \Leftrightarrow (\pi x, \pi y) \in P$. □

The set of uniform PERs on the integers, which we call $\mathcal{U}(\text{Int})$, thus contains the following 7 elements:

- $1A = \emptyset$
- $2A = \{(x, x) \mid x \in \mathbf{Z}\}$
- $3A = \mathbf{Z} \times \mathbf{Z}$
- $1B = \{(\perp, \perp)\}$
- $2B = \{(x, x) \mid x \in \mathbf{Z}_\perp\}$
- $3B = \mathbf{Z} \times \mathbf{Z} \cup \{(\perp, \perp)\}$
- $4 = \mathbf{Z}_\perp \times \mathbf{Z}_\perp$

and they are related by the subset ordering and the Egli-Milner ordering as in Fig. 1. For the strict PERs $\{1B, 2B, 3B, 4\}$ we observe that the Egli-Milner ordering coincides with the subset ordering.

Note that $\mathcal{U}(\text{Int})$ is intersection and union closed: intersection closedness is desirable for abstract interpretation as it ensures each (standard) value has a best abstract approximation. Union closedness ensures that no more information than necessary is lost on a merge resulting from *if then else*.

We define the *uniform properties* $\mathcal{UP}(\sigma)$ associated with σ in the obvious way:

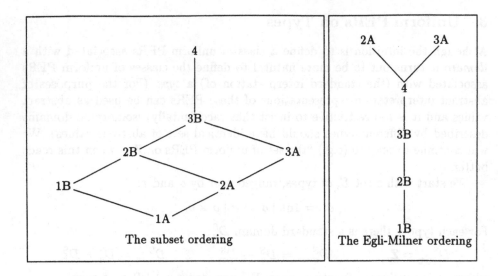

Fig. 1. The subset ordering and the Egli-Milner ordering on Int

Definition 7
To the set $\mathcal{U}(\sigma)$ of uniform PERs there is associated the set $\mathcal{UP}(\sigma)$ of properties:

$$\mathcal{UP}(\sigma) = \{|P| \mid P \in \mathcal{U}(\sigma)\}$$

□

The uniform properties of Int are as one might expect: $\mathcal{UP}(\text{Int}) = \{\emptyset, \{\bot\}, \mathbf{Z}, \mathbf{Z}_\bot\}$. These correspond exactly to the four values in [4], viz. $\emptyset$, $\bot$, $\not\!\bot$, $\top$. Note further that omitting PERs 2A and 2B from our definition of uniformity leads to the same property space on Int; moreover doing so seems to lead to the Cousots' comportment property space on Int → Int.

Starting from $\mathcal{U}(\text{Int})$ we will define uniform PERs on compound types. First we recall constructions which derive PERs at compound types from PERs of their components:

Definition 8
Given a PER P on σ and a PER Q on τ we can construct a PER $P\boxed{\rightarrow}Q$ on $\sigma \rightarrow \tau$ as:

$$P\boxed{\rightarrow}Q = \{(f,g) \in D^S_{\sigma\rightarrow\tau} \times D^S_{\sigma\rightarrow\tau} \mid (\forall(a,b) \in P)(fa,gb) \in Q\}$$

and a PER $P\boxed{\times}Q$ on $\sigma \times \tau$ as:

$$P\boxed{\times}Q = \{((a,c),(b,d)) \in D^S_{\sigma\times\tau} \times D^S_{\sigma\times\tau} \mid (a,b) \in P, (c,d) \in Q\}$$

(Note these are particular PERs, not spaces of PERs.)

□

Now we define the set of *uniform* PERs on compound types inductively. Doing so is facilitated by defining them mutually with a set of *basic* PERs.

Definition 9 *The uniform PERs on compound types*
Given types σ and τ and their associated sets $\mathcal{U}(\sigma)$ and $\mathcal{U}(\tau)$ of uniform PERs, we define, at type $\sigma \to \tau$, the set of basic PERs, $\mathcal{B}(\sigma \to \tau)$ as:

$$\mathcal{B}(\sigma \to \tau) = \{P\,\boxed{\to}\,Q \mid P \in \mathcal{U}(\sigma), Q \in \mathcal{U}(\tau)\}$$

and the set of uniform PERs, $\mathcal{U}(\sigma \to \tau)$ as:

$$\mathcal{U}(\sigma \to \tau) = \{\cap S \mid S \subseteq \mathcal{B}(\sigma \to \tau)\}.$$

Similarly we define, at type $\sigma \times \tau$, the set of basic PERs, $\mathcal{B}(\sigma \times \tau)$, as:

$$\mathcal{B}(\sigma \times \tau) = \{P\,\boxed{\times}\,Q \mid P \in \mathcal{U}(\sigma), Q \in \mathcal{U}(\tau)\}$$

and the set of uniform PERs, $\mathcal{U}(\sigma \times \tau)$, as:

$$\mathcal{U}(\sigma \times \tau) = \{\cup S \mid S \subseteq \mathcal{B}(\sigma \times \tau)\}.$$

$\square$

In general $\mathcal{B}(\sigma \to \tau)$ is not intersection closed; hence the definition of $\mathcal{U}(\sigma \to \tau)$ as its intersection-closure. Note that $\mathcal{U}(\sigma \to \tau)$ is not union closed. We resist the temptation to form the union-closure at higher types not from any polemic grounds but from a desire to preserve the "abstract function spaces are function spaces on abstract values" property found in [2, 5, 6].

Fact 10
Let σ be a type and A and B be members of $\mathcal{U}(\sigma)$ then: $A \cap B \in \mathcal{U}(\sigma)$ but in general $A \cup B \notin \mathcal{U}(\sigma)$. $\square$

Section 4 considers the uniform PERs and properties on Int $\to$ Int and the need for intersection closure in some detail.

For Observation 12 below we need the following lemma:

Lemma 11 *Intersections and unions*
For PERs $P_1, P_2, \ldots P_n$ on σ and one PER Q on τ:

$$(P_1 \cup P_2 \cup \cdots \cup P_n)\,\boxed{\to}\,Q = (P_1\,\boxed{\to}\,Q) \cap (P_2\,\boxed{\to}\,Q) \cap \cdots \cap (P_n\,\boxed{\to}\,Q)$$

Proof Let

$$P_{\mathrm{L}} = (P_1 \cup P_2 \cup \cdots \cup P_n)\,\boxed{\to}\,Q$$

and

$$P_{\mathrm{R}} = (P_1\,\boxed{\to}\,Q) \cap (P_2\,\boxed{\to}\,Q) \cap \cdots \cap (P_n\,\boxed{\to}\,Q)$$

We show $P_L = P_R$ by the following chain of equivalences.

$$(f, f') \in P_R \Leftrightarrow \forall i : [(d, d') \in P_i \Rightarrow ((fd, f'd') \in Q)]$$
$$\Leftrightarrow (\forall i : (d, d') \in P_i) \Rightarrow ((fd, f'd') \in Q)$$
$$\Leftrightarrow ((d, d') \in \cup P_i) \Rightarrow ((fd, f'd') \in Q)$$
$$\Leftrightarrow (f, f') \in P_L$$

Observation 12

We have

$$\mathcal{U}(\sigma_1 \times \sigma_2 \to \tau)$$
$$= \{\cap S \mid S \subseteq \mathcal{B}(\sigma_1 \times \sigma_2 \to \tau)\}$$
$$= \{\cap S \mid S \subseteq \{P \boxed{\to} Q \mid P \in \mathcal{U}(\sigma_1 \times \sigma_2), Q \in \mathcal{U}(\tau)\}\}$$
$$= \{\cap S \mid S \subseteq \{P \boxed{\to} Q \mid P \in \{\cup S_1 \mid S_1 \subseteq \mathcal{B}(\sigma_1 \times \sigma_2)\}, Q \in \mathcal{U}(\tau)\}\}$$

Consider the uniform PERs on $\sigma_1 \times \sigma_2 \to \tau$ but with the restriction that the product PERs involved are basic PERs. That is look at:

$$\{\cap S \mid S \subseteq \{P \boxed{\to} Q \mid P \in \mathcal{B}(\sigma_1 \times \sigma_2), Q \in \mathcal{U}(\tau)\}\} = P_B$$

Now since we have $\mathcal{B}(\sigma) \subseteq \mathcal{U}(\sigma)$ we arrive at $\mathcal{U}(\sigma_1 \times \sigma_2 \to \tau) \supseteq P_B$. Showing $\mathcal{U}(\sigma_1 \times \sigma_2 \to \tau) \subseteq P_B$ completes the proof, so consider a PER in $\mathcal{U}(\sigma_1 \times \sigma_2 \to \tau)$ in the form:

$$(P_{11} \cdots \cup P_{1n} \boxed{\to} Q_1) \cap \cdots \cap (P_{k1} \cup \cdots \cup P_{kn} \boxed{\to} Q_k)$$

where $P_{ij} \in \mathcal{B}(\sigma_1 \times \sigma_2)$ and $Q_i \in \mathcal{U}(\tau)$. By applying Lemma 11 we get

$$(P_{11} \boxed{\to} Q_1) \cap \cdots \cap (P_{1n} \boxed{\to} Q_1) \cap \cdots \cap (P_{k1} \boxed{\to} Q_k) \cap \cdots \cap (P_{kn} \boxed{\to} Q_k)$$

So we have that the PER is in P_B and therefore $\mathcal{U}(\sigma_1 \times \sigma_2 \to \tau) \subseteq P_B$. $\quad\square$

Following Observation 12, in this paper we restrict the types to the form

$$\sigma ::= \text{Int} \mid \sigma \times \cdots \times \sigma \to \sigma$$

(This has the sole effect of forbidding products in function results and is done to avoid arguments concerning appropriate properties for product values.) Indeed, we could have achieved much the same effect by treating such restricted use of product types as shorthand for curried functions. However the present formalism is more natural and enables us to discuss properties of products. Note that because of Observation 12, it suffices to consider $\mathcal{B}(\sigma \times \tau)$ instead of $\mathcal{U}(\sigma \times \tau)$ since the restriction means that all products appear to the left of function types.

4 Examples in Int → Int

As an example let us take a look at the basic and uniform on Int → Int. PERs on Int → Int are subsets of the standard domain of (Int → Int) × (Int → Int). Given two uniform PERs P and Q on Int a *basic* PER on the function space Int → Int can be constructed as:

$$P \boxed{\rightarrow} Q = \{(f,g) \in (\mathbf{Z}_\perp \rightarrow \mathbf{Z}_\perp) \times (\mathbf{Z}_\perp \rightarrow \mathbf{Z}_\perp) \mid \forall(a,b) \in P \Rightarrow (fa, gb) \in Q\}$$

The set of all basic PERs on the function space Int → Int is:

$$\mathcal{B}(\text{Int} \rightarrow \text{Int}) = \{P \boxed{\rightarrow} Q \mid P, Q \in \mathcal{U}(\text{Int})\}$$

The set of all uniform PERs on the function space $\mathbf{Z}_\perp \rightarrow \mathbf{Z}_\perp$ is:

$$\mathcal{U}(\text{Int} \rightarrow \text{Int}) = \{\cap S \mid S \subseteq \mathcal{B}(\text{Int} \rightarrow \text{Int})\}$$

The reason why it is not sufficient only to use the basic PERs is that some expressions lack a least and best PER. Consider the term $\lambda x.\text{if } x = \perp \text{ then } \perp \text{ else } 42$; we have that

$$(\lambda x.\text{if } x = \perp \text{ then } \perp \text{ else } 42) \in |1B \boxed{\rightarrow} 1B|$$

and

$$(\lambda x.\text{if } x = \perp \text{ then } \perp \text{ else } 42) \in |3A \boxed{\rightarrow} 2A|.$$

However, the greatest lower bound in the subset ordering of PERs of the form $P \boxed{\rightarrow} Q$ of the two PERs 1B $\boxed{\rightarrow}$ 1B and 3A $\boxed{\rightarrow}$ 2A is the PER 3B $\boxed{\rightarrow}$ 1A which is empty. One of the missing PERs *is* the greatest lower bound in the subset ordering of 1B $\boxed{\rightarrow}$ 1B and 3A $\boxed{\rightarrow}$ 2A.

The properties, $\mathcal{UP}(\text{Int} \rightarrow \text{Int})$, on Int → Int are:

- empty $= \emptyset$
- strcon $= \{f \in \mathbf{Z}_\perp \rightarrow \mathbf{Z}_\perp \mid (\forall a, b \in \mathbf{Z} : fa = fb \in \mathbf{Z}) \wedge (f\perp = \perp)\}$
- ide $= \{f \in \mathbf{Z}_\perp \rightarrow \mathbf{Z}_\perp \mid (\forall a \in \mathbf{Z} : fa \in \mathbf{Z}) \wedge (f\perp = \perp)\}$
- div $= \{f \in \mathbf{Z}_\perp \rightarrow \mathbf{Z}_\perp \mid \forall a \in \mathbf{Z}_\perp : fa = \perp\}$
- strcon|div $= \{f \in \mathbf{Z}_\perp \rightarrow \mathbf{Z}_\perp \mid (\forall a, b \in \mathbf{Z} : fa = fb \in \mathbf{Z}_\perp) \wedge (f\perp = \perp)\}$
- ide|div $= \{f \in \mathbf{Z}_\perp \rightarrow \mathbf{Z}_\perp \mid (\forall a \in \mathbf{Z} : fa \in \mathbf{Z}) \wedge f\perp = \perp\} \cup \text{div}$
- strict $= \{f \in \mathbf{Z}_\perp \rightarrow \mathbf{Z}_\perp \mid f\perp = \perp\}$
- cgt $= \{f \in \mathbf{Z}_\perp \rightarrow \mathbf{Z}_\perp \mid \forall a, b \in \mathbf{Z}_\perp : fa = fb \in \mathbf{Z}\}$
- totcon $= \{f \in \mathbf{Z}_\perp \rightarrow \mathbf{Z}_\perp \mid \forall a, b \in \mathbf{Z} : fa = fb \in \mathbf{Z}\}$
- tot $= \{f \in \mathbf{Z}_\perp \rightarrow \mathbf{Z}_\perp \mid \forall a \in \mathbf{Z} : fa \in \mathbf{Z}\}$
- cgt|div $= \{f \in \mathbf{Z}_\perp \rightarrow \mathbf{Z}_\perp \mid \forall a, b \in \mathbf{Z}_\perp : fa = fb \in \mathbf{Z}_\perp\}$
- totcon|div $= \{f \in \mathbf{Z}_\perp \rightarrow \mathbf{Z}_\perp \mid \forall a, b \in \mathbf{Z} : fa = fb \in \mathbf{Z}_\perp\}$
- tot|div $= \{f \in \mathbf{Z}_\perp \rightarrow \mathbf{Z}_\perp \mid \forall a \in \mathbf{Z} : fa \in \mathbf{Z}\} \cup \text{div}$
- all $= \{f \in \mathbf{Z}_\perp \rightarrow \mathbf{Z}_\perp\}$

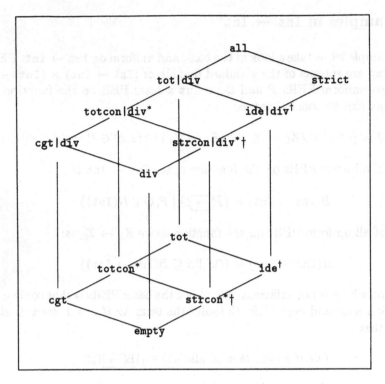

Fig. 2. The Subset ordering on $\mathcal{UP}(\text{Int} \to \text{Int})$

and they are related by the subset ordering as in Fig. 2 and by the Egli-Milner ordering as in Fig. 3.

In the figures, names have been chosen to match [4] except that 'cgt' is used for 'convergent' leaving 'con' to indicate 'constant'. Annotations: † indicates a PER added by conjunctive completion; * indicates a PER not in [4]. For PERs in [4] the '|' character in the name indicates it is added by disjunctive completion. Note that [4] has an error in that ide|div ⊆ tot|div is omitted from the Hasse diagram.

We see that subset ordering coincides with the Egli-Milner ordering for the strict PERs (div, cgt|div, strcon|div, totcon|div, ide|div, tot|div, strict, and all). Also note that some of the PER-properties are Egli-Milner equivalent, viz.

$$\text{totcon|div} \equiv_{\text{EM}} \text{cgt|div}$$
$$\text{strict} \equiv_{\text{EM}} \text{ide|div}$$
$$\text{all} \equiv_{\text{EM}} \text{tot|div}$$

This is also left unresolved in [4].

One question to be asked is: "are all these properties useful?". A property is useful if an optimisation is possible only when it holds (or, indirectly, if it

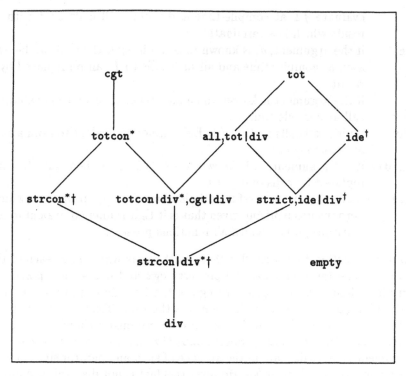

Fig. 3. The Egli-Milner ordering on $\mathcal{UP}(\text{Int} \to \text{Int})$

is needed to avoid information loss when calculating such a property—cf. joint strictness). Let f be an expression with property:

empty	falsity property—which no function possesses.
strcon	if the argument, a, is known to be an integer then $f0$ can be evaluated at compile-time and all these calls to f can be replaced by the result and if the argument is known not to terminate, then we can replace such calls with the result $\bot$.
ide	if the argument is known not to terminate, then we can replace the calls with the result $\bot$ and in the case where we know that the argument, a, is an integer then fa can be evaluated at compile-time.
div	replace all calls to f by the result $\bot$.
strcon|div	if the argument is known not to terminate, then we can replace the calls with the result $\bot$ and if the argument, a, is known to be an integer then all such calls can be replaced by calls of e.g. $f0$.
ide|div	Egli-Milner equivalent to its convex closure strict compared to which there do not seem to be additional optimisations.
strict	evaluate argument before call.

cgt	evaluate $f\bot$ at compile-time and replace all calls to f with that result which is a terminating value.
totcon	if the argument, a, is known to be an integer then $f0$ can be evaluated at compile-time and all such calls to f can be replaced by the result.
tot	if the argument is known to be an integer then we can evaluate the call at compile-time.
cgt|div	replace all calls to f by $f\bot$; Egli-Milner equivalent to (and a subset of) totcon|div.
totcon|div	if the argument, a, is known to be an integer then calls fa can be replaced by calls of e.g. $f0$.
tot|div	there appear no useful optimisations for this property (which is perhaps unsurprising given that is it Egli-Milner equivalent to all).
all	truth property—which all functions possess.

The next question to ask is: "Do there exist terms which have each of these properties?". The term $\lambda x.4$ has the property cgt and it does not possess any property less than cgt. The property cgt is the best description of the term. Similarly $\lambda x.\bot$ has the property div which is the best. Temporarily suppose e' is the term $true \oplus false$ where $\oplus$ denotes non-deterministic choice; then we can construct a term with the best property cgt\|div viz. *if e' then $\lambda x.4$ else $\lambda x.\bot$*. This argument is more delicate in the absence of such an operator since any such e' must reduce to *true, false* or *bot*. However this fact is not discernible uniformly (for any analysis method there are undetectable tautologies) and hence for any analysis method there is such a term with best property cgt\|div. The term *fix$\lambda f.\lambda x.$if $x=0$ then 1 else $f(x-1)$* has the property strcon as the best one. Now we are able to construct terms for the remainder as we did for the property cgt\|div.

5 Implementation use

We assume a simple typed functional language, whose object types coincide with (and are modelled by) the meta-language types above. Its syntax is

$$e = x \mid c \mid ee' \mid \lambda x.e'$$

where x ranges over variables and c over constants. Its semantics is given in terms of type-indexed semantic functions $\mathcal{E}_\sigma^I$ by

$$\mathcal{E}_\sigma^I[\![x]\!]\rho = \rho x$$
$$\mathcal{E}_\sigma^I[\![c]\!]\rho = c^I$$
$$\mathcal{E}_\sigma^I[\![ee']\!]\rho = \mathbf{app}_{\sigma\to\tau}^I(\mathcal{E}_{\sigma\to\tau}^I[\![e]\!]\rho)(\mathcal{E}_\sigma^I[\![e']\!]\rho)$$
$$\mathcal{E}_{\sigma\to\tau}^I[\![\lambda x.e']\!]\rho = \mathbf{lam}_{\sigma\to\tau}^I(\lambda d.\mathcal{E}_\tau^I[\![e']\!]\rho[d/x])$$

where c^I (including $\mathbf{lam}_\sigma^I$ and $\mathbf{app}_\sigma^I$) are given by an interpretation which also specifies the interpretation of types as below.

For each type σ we have beside the standard domains D_σ^S an abstract domain D_σ^A. We will take D_σ^A to be the set of names of uniform PERs on σ; e.g. $D_\sigma^A = \mathcal{N}(\sigma)$. The names of the uniform PERs are defined as follows:

$$\mathcal{N}(\mathtt{Int}) = \{1A, 2A, 3A, 1B, 2B, 3B, 4\}$$

$$\mathcal{N}(\sigma \to \tau) = \{[n_1, m_1; \ldots; n_k, m_k] \mid n_i \in \mathcal{N}(\sigma), m_i \in \mathcal{N}(\tau), k = |\mathcal{N}(\sigma)|\}$$

in one name all the n_i's are distinct

$$\mathcal{N}(\sigma \times \tau) = \{(n, m) \mid n \in \mathcal{N}(\sigma), m \in \mathcal{N}(\tau)\}$$

Names on the function space $\sigma \to \tau$ can be seen as a graph of a function from names on σ to names on τ. For each type there is a concretisation function γ_σ from the set of names of PERs on the type, $\mathcal{N}(\sigma)$, to the set of uniform PERs on the type, $\mathcal{U}(\sigma)$. (Hunt refers to this type-indexed family of maps as *the logical concretisation map*.)

$$\gamma_\sigma : \mathcal{N}(\sigma) \to \mathcal{U}(\sigma)$$

$$\gamma_{\mathtt{Int}}(1A) = \emptyset \qquad\qquad \gamma_{\mathtt{Int}}(1B) = \{(\bot, \bot)\}$$
$$\gamma_{\mathtt{Int}}(2A) = \{(x, x) \mid x \in \mathbf{Z}\} \qquad \gamma_{\mathtt{Int}}(2B) = \{(x, x) \mid x \in \mathbf{Z}_\bot\}$$
$$\gamma_{\mathtt{Int}}(3A) = \mathbf{Z} \times \mathbf{Z} \qquad\qquad \gamma_{\mathtt{Int}}(3B) = \mathbf{Z} \times \mathbf{Z} \cup \{(\bot, \bot)\}$$
$$\gamma_{\mathtt{Int}}(4) = \mathbf{Z}_\bot \times \mathbf{Z}_\bot$$
$$\gamma_{\sigma \to \tau}([n_1, m_1; \ldots; n_k, m_k]) = (\gamma_\sigma n_1 \boxed{\to} \gamma_\tau m_1) \cap \ldots \cap (\gamma_\sigma n_k \boxed{\to} \gamma_\tau m_k)$$
$$\gamma_{\sigma \times \tau}((n, m)) = \gamma_\sigma n \boxed{\times} \gamma_\tau m$$

When using this representation (of functions by graphs) it is important to recall Hunt's observation that the γ are not injective—two (extensionally) different functions between properties may describe the same PER on the function space.

The Egli-Milner ordering on names of uniform PERs is inherited from the Egli-Milner ordering on the uniform PERs:

Definition 13

Let n and m be in $\mathcal{N}(\sigma)$ then $n \sqsubseteq_{EM}^{\mathcal{N}} m$ if $\gamma_\sigma(n) \sqsubseteq_{EM}^{PER} \gamma_\sigma(m)$; $\sqcup_{EM}^{\mathcal{N}}$ is similarly inherited (and well-defined as $\mathcal{N}(\sigma)$ is finite and closed over lubs). $\qquad\square$

Standard and PER name-based interpretations are given in Fig. 4. The application and projection functions are monotonic and continuous:

Lemma 14

The two families of functions $\mathtt{app}_{\sigma \to \tau}^S$ and $\mathtt{app}_{\sigma \to \tau}^A$ are monotonic with respect to ordering on the standard domain, $\leq$, and the Egli-Milner ordering on names of PERs, $\sqsubseteq_{EM}^{\mathcal{N}}$, respectively.

Proof From domain theory we know that $\mathtt{app}_\sigma^S$ is monotonic. We prove that $\mathtt{app}_\sigma^A$ is monotonic by induction on the type σ. For details see Solberg's PhD [11]. $\blacksquare$

Lemma 15

The functions π_1^S, π_1^A, π_2^S, and π_2^A are monotonic with respect to ordering on the standard domain, $\leq$, and the Egli-Milner ordering on name of PERs, $\sqsubseteq_{EM}^{\mathcal{N}}$, respectively.

$$\pi_1^S : D_{\sigma\times\tau}^S \to D_\sigma^S \qquad\qquad \pi_1^A : D_{\sigma\times\tau}^A \to D_\sigma^A$$
$$\pi_1^S((n,m)) = n \qquad\qquad\qquad \pi_1^A((n,m)) = n$$

$$\pi_2^S : D_{\sigma\times\tau}^S \to D_\tau^S \qquad\qquad \pi_2^A : D_{\sigma\times\tau}^A \to D_\tau^A$$
$$\pi_2^S((n,m)) = m \qquad\qquad\qquad \pi_2^A((n,m)) = m$$

$$\mathbf{app}_{\sigma\to\tau}^S : D_{\sigma\to\tau}^S \to D_\sigma^S \to D_\tau^S \qquad \mathbf{app}_{\sigma\to\tau}^A : D_{\sigma\to\tau}^A \to D_\sigma^A \to D_\tau^A$$
$$\mathbf{app}_{\sigma\to\tau}^S\ fx = fx \qquad\qquad\qquad \mathbf{app}_{\sigma\to\tau}^A\ [n_1, m_1; \ldots; n_k, m_k]n_i = m_i$$

$$\mathbf{lam}_{\sigma\to\tau}^S : (D_\sigma^S \to D_\tau^S) \to D_{\sigma\to\tau}^S \qquad \mathbf{lam}_{\sigma\to\tau}^A : (D_\sigma^A \to D_\tau^A) \to D_{\sigma\to\tau}^A$$
$$\mathbf{lam}_{\sigma\to\tau}^S\ f = f \qquad\qquad\qquad \mathbf{lam}_{\sigma\to\tau}^A\ f = [n_1, fn_1; \ldots; n_k, fn_k],$$
$$\text{where } (n_i \in D_\sigma^A) \wedge$$
$$(k = |D_\sigma^A|)$$

$$\mathbf{fix}_\sigma^S : D_{\sigma\to\sigma}^S \to D_\sigma^S \qquad\qquad \mathbf{fix}_\sigma^A : D_{\sigma\to\sigma}^A \to D_\sigma^A$$
$$\mathbf{fix}_\sigma^S\ f \quad = \sqcup \{d_i\} \qquad\qquad \mathbf{fix}_\sigma^A\ n \quad = \sqcup_{EM}^{\mathcal{N}} \{d_i\}$$
$$d_0 = \bot_{D_\sigma^S} \qquad\qquad\qquad d_0 = \bot_{D_\sigma^A}^{EM\mathcal{N}}$$
$$d_{i+1} = \mathbf{app}_{\sigma\to\sigma}^S\ fd_i \qquad\qquad d_{i+1} = \mathbf{app}_{\sigma\to\sigma}^A\ nd_i$$

Fig. 4. Standard and non-standard interpretations for $\mathcal{E}_\sigma^{\mathcal{I}}$

Proof From domain theory we know that π_1^S and π_2^S are monotonic. We prove that π_1^A and π_2^A are monotonic by induction on the type $\sigma \times \tau$. For details see Solberg's PhD [11]. ∎

Lemma 16
The functions $\mathbf{app}_\sigma^S$, $\mathbf{app}_\sigma^A$, π_1^S, π_1^A, π_2^S, and π_2^A are continuous with respect to ordering on the standard domain, $\leq$, and the Egli-Milner ordering on name of PERs, $\sqsubseteq_{EM}^{\mathcal{N}}$, respectively.

Proof It is a consequence of Lemma 14 and 15 and by recalling that monotonic functions from a space of finite height are continuous. ∎

5.1 Ternary logical relations

Now we can define the relations by:

$$(d, d') \in \gamma_\sigma(a) \Leftrightarrow (d, d')\mathcal{R}_\sigma a$$

Definition 17 *Ternary Logical Relations*
A type-indexed family of relations $(\mathcal{R}_\sigma)$ is logical (terminology conventionally abused by writing "$\mathcal{R}$ is a logical relation") if for all σ and τ:

$$(f, f')\mathcal{R}_{\sigma\to\tau}h \Leftrightarrow (\forall d, d' \in D_\sigma^S, \forall a \in D_\sigma^A)$$
$$(d, d')\mathcal{R}_\sigma a \Leftrightarrow (\mathbf{app}_{\sigma\to\tau}^S\ fd, \mathbf{app}_{\sigma\to\tau}^S\ f'd')\mathcal{R}_\tau(\mathbf{app}_{\sigma\to\tau}^A\ ha)$$
$$(p_1, p_2)\mathcal{R}_{\sigma\times\tau}p \Leftrightarrow (\pi_1^S(p_1), \pi_1^S(p_2))\mathcal{R}_\sigma\pi_1^A(p) \wedge (\pi_2^S(p_1), \pi_2^S(p_2))\mathcal{R}_\tau\pi_2^A(p)$$

Proposition 18
The relation $\mathcal{R}$ is logical.

Proof We prove that $\mathcal{R}$ is a logical relation by induction on the type. For details see Solberg's PhD [11].

Proposition 19
The relation $\mathcal{R}$ is inductive with respect to ordering on the standard domain, $\leq$, and the Egli-Milner ordering on name of PERs, $\sqsubseteq^{\mathcal{N}}_{\text{EM}}$, respectively.

Proof We prove that the relation $\mathcal{R}$ is inductive by induction on the type. For details see Solberg's PhD [11]. ∎

The standard fixpoint and the abstract fixpoint of related values are related:

Proposition 20
Let $(f, f)\mathcal{R}_{\sigma \to \sigma}h$ then $(\text{fix}^{S}_{\sigma}f, \text{fix}^{S}_{\sigma}f)\mathcal{R}_{\sigma}(\text{fix}^{A}_{\sigma}h)$

Proof We assume $(f, f)\mathcal{R}_{\sigma \to \sigma}h$. First we show that $(d^{S}_{i}, d^{S}_{i})\mathcal{R}_{\sigma}d^{A}_{i}$ holds for all i by induction on i. Next from Lemma 14 we know that app^{S} and app^{A} are monotonic therefore are $\{d^{S}_{i}\}$ and $\{d^{A}_{i}\}$ chains and since $\mathcal{R}$ is inductive (Proposition 19) we have $(\sqcup_{i}\{d^{S}_{i}\}, \sqcup_{i}\{d^{S}_{i}\})\mathcal{R}_{\sigma}(\sqcup^{\mathcal{N}}_{\text{EM}}\{d^{A}_{i}\})$ as required.

Case $i = 0$: We need to show $(\perp_{D^{S}_{\sigma}}, \perp_{D^{S}_{\sigma}})\mathcal{R}_{\sigma}\perp^{\text{EM}\mathcal{N}}_{D^{A}_{\sigma}}$ which is trivially implied by strictness of $\gamma_{\sigma}(\perp^{\text{EM}\mathcal{N}}_{D^{A}_{\sigma}})$.

Case $i + 1$: Assume $(d^{S}_{i}, d^{S}_{i})\mathcal{R}_{\sigma}d^{A}_{i}$ and show $(d^{S}_{i+1}, d^{S}_{i+1})\mathcal{R}_{\sigma}d^{A}_{i+1}$: we have $d^{S}_{i+1} = \text{app}^{S}_{\sigma} f d^{S}_{i}$ and $d^{A}_{i+1} = \text{app}^{A}_{\sigma} h d^{A}_{i}$. Now, since $\mathcal{R}$ is logical (Proposition 18), we have $(\text{app}^{S}_{\sigma} f d^{S}_{i}, \text{app}^{S}_{\sigma} f d^{S}_{i})\mathcal{R}_{\sigma}(\text{app}^{A}_{\sigma} h d^{A}_{i})$ as required.

Now the soundness and completeness of the analysis is:

Theorem 21
Let ρ be a standard environment and let δ be an abstract environment. Suppose for all constants c of type τ we have $(c^{S}, c^{S})\mathcal{R}_{\tau}c^{A}$ then for all σ and e we have

$$(\rho, \rho)\mathcal{R}\delta \Rightarrow (\mathcal{E}^{S}_{\sigma}[\![e]\!]\rho, \mathcal{E}^{S}_{\sigma}[\![e]\!]\rho)\mathcal{R}_{\sigma}(\mathcal{E}^{S}_{\sigma}[\![e]\!]\delta)$$

Proof We prove the Theorem by induction on the type. For details see Solberg's PhD [11]. ∎

Example 22
We calculate

$$\begin{aligned}
\mathcal{E}^{A}_{\text{Int}\to\text{Int}}[\![\lambda x.4]\!]\rho &= \text{lam}^{A}_{\text{Int}\to\text{Int}} (\lambda d.\mathcal{E}^{A}_{\text{Int}}[\![4]\!]\rho[d/x]) \\
&= \text{lam}^{A}_{\text{Int}\to\text{Int}} (\lambda d.4^{A}) \\
&= \text{lam}^{A}_{\text{Int}\to\text{Int}} (\lambda d.2A) \\
&= [1A, 2A; 2A, 2A; 3A, 2A; 1B, 2A; 2B, 2A; 3B, 2A; 4, 2A]
\end{aligned}$$

Now

$$\lambda x.4 \in \mathcal{UP}(\gamma_{\text{Int}\to\text{Int}}[1A, 2A; 2A, 2A; 3A, 2A; 1B, 2A; 2B, 2A; 3B, 2A; 4, 2A])$$
$$= \text{cgt}$$

as required. We also have

$$\mathcal{E}^A_{\text{Int}\to\text{Int}}[\![\textit{fix}(\lambda x.4)]\!]\rho =$$
$$\text{fix}^A_{\text{Int}\to\text{Int}}[1A, 1B; 2A, 1B; 3A, 1B; 1B, 1B; 2B, 1B; 3B, 1B; 4, 1B]$$

and

$$d_0 = 1B$$
$$d_1 = \text{app}^A_{\text{Int}\to\text{Int}} \ [1A, 2A; 2A, 2A; 3A, 2A; 1B, 2A; 2B, 2A; 3B, 2A; 4, 2A]1B$$
$$= 2A$$
$$d_2 = \text{app}^A_{\text{Int}\to\text{Int}} \ [1A, 2A; 2A, 2A; 3A, 2A; 1B, 2A; 2B, 2A; 3B, 2A; 4, 2A]2A$$
$$= 2A$$

and hence

$$\mathcal{E}^A_{\text{Int}\to\text{Int}}[\![\textit{fix}(\lambda x.4)]\!]\rho = 2A$$

We calculate

$$\mathcal{E}^A_{\text{Int}\to\text{Int}}[\![\lambda x.\bot]\!]\rho = \text{lam}^A_{\text{Int}\to\text{Int}} \ (\lambda d.\mathcal{E}^A_{\text{Int}}[\![\bot]\!]\rho[d/x])$$
$$= \text{lam}^A_{\text{Int}\to\text{Int}} \ (\lambda d.\bot^A)$$
$$= \text{lam}^A_{\text{Int}\to\text{Int}} \ (\lambda d.1B)$$
$$= [1A, 1B; 2A, 1B; 3A, 1B; 1B, 1B; 2B, 1B; 3B, 1B; 4, 1B]$$

and hence

$$\lambda x.\bot \in \mathcal{UP}(\gamma_{\text{Int}\to\text{Int}}[1A, 1B; 2A, 1B; 3A, 1B; 1B, 1B; 2B, 1B; 3B, 1B; 4, 1B])$$
$$= \text{div}$$

as required. We also have

$$\mathcal{E}^A_{\text{Int}\to\text{Int}}[\![\textit{fix}(\lambda x.\bot)]\!]\rho =$$
$$\text{fix}^A_{\text{Int}\to\text{Int}}[1A, 1B; 2A, 1B; 3A, 1B; 1B, 1B; 2B, 1B; 3B, 1B; 4, 1B]$$

and

$$d_0 = 1B$$
$$d_1 = \text{app}^A_{\text{Int}\to\text{Int}} \ [1A, 1B; 2A, 1B; 3A, 1B; 1B, 1B; 2B, 1B; 3B, 1B; 4, 1B]1B$$
$$= 1B$$
$$d_2 = \text{app}^A_{\text{Int}\to\text{Int}} \ [1A, 1B; 2A, 1B; 3A, 1B; 1B, 1B; 2B, 1B; 3B, 1B; 4, 1B]1B$$
$$= 1B$$

and hence $\mathcal{E}^A_{\text{Int}\to\text{Int}}[\![\textit{fix}(\lambda x.\bot)]\!]\rho = 1B$.

6 Conclusion

We have defined the notion of uniform PERs on the integers and by following the framework of Hunt [6] we have lifted this notion to higher types in one particular way. In the work of Hunt all the PERs were strict. Here we have encountered PERs which are not strict; since they are useful too, we had to handle the fixpoint iteration in another way. We defined the Egli-Milner ordering on PERs and used it for the fixpoint iteration.

More work is need to clarify why our approach yields four extra properties at type Int $\rightarrow$ Int compared to the approach in [4]. The significance of some of the properties being Egli-Milner equivalent (both in this work and [4]) also needs more fuller investigation.

We now briefly compare this work with related papers. Firstly we show that the strictness properties of Burn, Hankin and Abramsky [2] naturally embed in the uniform PER properties presented here. To be precise, let $\mathcal{U}'(\text{Int}) = \{1\text{B} = \{(\perp, \perp)\}, 4 = \mathbf{Z}_\perp \times \mathbf{Z}_\perp\}$ and $\mathcal{U}'(\sigma \rightarrow \tau)$ be defined inductively in the same manner as $\mathcal{U}(\sigma \rightarrow \tau)$. Then for all types σ we have $\mathcal{U}'(\sigma) \subseteq \mathcal{U}(\sigma)$ and moreover $\mathcal{U}'(\sigma)$ is isomorphic to $BHA(\sigma)$. Here $BHA(\sigma)$ is the set of strictness properties for type σ defined by Burn, Hankin and Abramsky:

$$BHA(\text{Int}) = 2 = \{0, 1\}$$
$$BHA(\sigma \rightarrow \tau) = BHA(\sigma) \rightarrow BHA(\tau)$$

Note that this embedding can also be seen as selecting uniform PER representatives of the uniform ideals of [5]. It also occurs as an abstraction of the comportment interpretation of [4] which abstracts our uniform PER interpretation.

We also compare with the properties based on annotated types described by Solberg et al. [10]. All the properties of that work can be modelled by the uniform PERs except for two: the property of knowing that a function does not have a WHNF and the property of knowing that a function does have a WHNF. This is not a lack of the uniform PER notion, but due to the fact that we have used non-lifted function spaces in which $\lambda x.\perp = \perp$ and hence WHNF properties are inexpressible.

Acknowledgements

Thanks to Radhia and Patrick Cousot for engaging arguments on the nature of abstract interpretation. Sebastian Hunt explained further about PERs in abstract interpretation. Thanks to Nick Benton, Hanne Riis Nielson, Flemming Nielson and Andrew Pitts for interesting discussions. We wish also to thank the PLILP referees for their comments which have aided the presentation. This work is supported by EU ESPRIT BRA 8130 "LOMAPS" and has been facilitated by Danish support for one author to visit Cambridge and by French support for the other to spend sabbatical leave at Ecole Polytechnique.

References

1. Nick Benton. *Strictness Analysis of Functional Programs*. PhD thesis, University of Cambridge, 1993. Available as Computer Laboratory Technical Report No. 309.
2. Geoffrey L. Burn, Chris Hankin and Samson Abramsky. Strictness analysis for higher-order functions. *Science of Computer Programming*, 7:249–278, 1986.
3. Patrick Cousot and Radhia Cousot. Galois connections based abstract interpretation for strictness analysis. In D. Bjørner, M. Broy and I. V. Pottosin, editors, *Proceedings of the International Conference on Formal Methods in Programming and their Applications*, LNCS 735, pages 98–127, 1993.
   [Currently available via http://www.ens.fr/~cousot/]
4. Patrick Cousot and Radhia Cousot. Higher-order abstract interpretation (and application to comportment analysis generalizing strictness, termination, projection and PER analysis of functional languages), invited paper. In *Proceedings of the 1994 International Conference on Computer Languages, ICCL'94*, pages 95–112, IEEE Computer Society Press, 1994.
   [Currently available via http://www.ens.fr/~cousot/]
5. Christine Ernoult and Alan Mycroft. Uniform ideals and strictness analysis. In *ICALP 91, LNCS 510*, 1991.
   [Currently available via http://www.cl.cam.ac.uk/users/am]
6. Sebastian Hunt. *Abstract Interpretation of Functional Languages: From Theory to Practice*. PhD thesis, Imperial College, London, 1991.
7. Thomas P. Jensen. *Abstract Interpretation in Logical Form*. PhD thesis, Imperial College, London, 1992.
8. Alan Mycroft. *Abstract Interpretation and Optimising Transformation for Applicative programs*. PhD thesis, University of Edinburgh, 1981.
9. Flemming Nielson. Two-level semantics and abstract interpretation. *Theoretical Computer Science*, 69:117–242, 1989.
10. Kirsten Lackner Solberg, Hanne Riis Nielson and Flemming Nielson. Strictness and totality analysis. In *Proceedings of Static Analysis Symposium, LNCS 864*, 1994.
11. Kirsten Lackner Solberg. *Annotated Type Systems for Program Analysis*. PhD thesis, Odense University, to appear.
12. Philip Wadler and John Hughes. Projections for strictness analysis. In *Proceedings of Functional Programming Languages and Computer Architectures '87, LNCS 274*, 1987.

Uniqueness Type Inference

Erik Barendsen   Sjaak Smetsers

University of Nijmegen, Computing Science Institute,
Toernooiveld 1, 6525 ED Nijmegen, The Netherlands
email: erikb@cs.kun.nl, sjaak@cs.kun.nl

Abstract. We extend the Uniqueness Type System with uniqueness inference. Using a notion of principal uniqueness variant, the type system is shown to be effective in the sense that a uniqueness variant of a given term (uniqueness type) can be determined automatically. The presented algorithm serves as a basis for type checking in the language Clean. At many stages it is used by some concrete implementation.

1   Introduction

Standard functional programming languages are unable to deal with operations with side effects. Except for admitting a these operations (such as the manipulation) one relies on typical referential transparency since these involve functions performing computations, or mutations) the state of an inner object?

In fact, various notions of these have been thought up as solutions for all of this up, one of these is the restriction of destructive operations to such objects that occur only once. If formulated as a type system, this idea is related to linear logic is to Girard et al, 1992. Troelstra (1992)). In this logic, assumptions are regarded as resources. Some 'linear' resources can be used exactly once others (non-linear resources) can be discussed, or used more than once. Wadler (1990) describes a linear type system for Pascal in which e.g. files are considered as linear resources. Unrestricted use is modelled by the resources' duplication operation from one to logue. (This logic without will be referred to as bare linear logic)

Uniqueness typing is related to linear typing in that it also uses an occurrence counting mechanism. The type system, however, is more involved than linear typing. In fact, uniqueness typing is a combination of pure linear typing (yielding 'unique' objects) and the limited typing (for 'non unique' objects) without restrictions on their number of occurrences), connected by a subtyping mechanism: non unique objects should not be required with the (duplicatewise unique) copies of linear resources, modelled by the linear logic.

As most implementations of functional languages use graph-like structures, we focus on graph rewriting as our model of computation. The objects in our typing system are graphs that are related to type-the roots in term graph rewrite systems (TGRS, see Barendregt et al, 1987)). Moreover usage of a subobject

Supported by NWO and STW.

Uniqueness Type Inference

Erik Barendsen   Sjaak Smetsers*

University of Nijmegen, Computing Science Institute
Toernooiveld 1, 6525 ED Nijmegen, The Netherlands
e-mail **erikb@cs.kun.nl**, **sjakie@cs.kun.nl**

Abstract. We extend the Uniqueness Type System with *uniqueness polymorphism*. Using a notion of 'principal uniqueness variants' the type system is shown to be effective in the sense that a uniqueness variant of a given conventional type can be determined automatically. The presented algorithm serves as a basis for type checking in the language *Clean*. We illustrate the system by some concrete examples.

1   Introduction

Traditional functional programming languages are unable to deal with operations with side effects. Indeed, by admitting these operations (such as file manipulations) one risks the loss of referential transparency since these involve changing ('destructively updating' or 'mutating') the state of an input object.

In recent years, various proposals have been brought up as solutions to this shortcoming. One of these is the restriction of destructive operations to input objects that occur only once. If formulated as a type system, this idea is related to *linear logic* (see Girard et al. (1989), Troelstra (1992)). In this logic, assumptions are regarded as *resources*. Some ('linear') resources can be used exactly once; others ('unlimited' resources) can be discarded, or used more than once. Wadler (1990) describes a *linear type system* for λ-calculus in which, e.g., files are considered as linear resources. Unlimited use is modelled by the 'resource duplication' operator ! from linear logic. (The logic without ! will be referred to as *pure* linear logic.)

Uniqueness typing is related to linear typing in that it also uses an occurrence counting mechanism. The type system, however, is more involved than linear typing. In fact, uniqueness typing is a combination of pure linear typing (yielding 'unique' objects) and traditional typing (for 'non-unique' objects without restrictions on their number of occurrences), connected by a subtyping mechanism. Non-unique objects should not be confused with the (duplicatewise unique) copies of linear resources produced by ! in linear logic.

As most implementations of functional languages use graph-like structures, we focus on graph rewriting as our model of computation. The *objects* in our typing system are *graphs* that are subject to rewrite rules in *term graph rewrite systems* (TGRS's, see Barendregt et al. (1987)). Multiple usage of a subobject

---

* Supported by NWO and STW.

corresponds to multiple *references* to a particular node. Our graphs may contain arbitrary sharing, including cyclic structures.

The *types* are annotated versions of the types in a traditional type system (referred to as 'conventional typing') which is present in most functional programming languages.

Conventional typing in TGRS's can be defined in terms of local requirements for type assignment to nodes in a graph. In an application of a function **F**, say, these constraints relate the type of the **F** node and the type of its argument nodes to the symbol type of **F**. The symbol types are declared in a so-called *type environment*. Note that the non-inductive nature of graphs prevents a more traditional (inductive) natural deduction formulation of typing.

Uniqueness typing formulates additional requirements on the reference structure of the graphs in question. In the types of functions, it can now be specified that a given argument should be unique (by the annotation •). In any application, the concrete function argument should have reference count 1, so the function has indeed 'private' access to its argument. This is used for destructive operations: e.g., **WriteChar** : (Char$^\times$, File$^\bullet$) $\rightarrowtail$ File$^\bullet$. Here, $\times$ denotes 'non-unique'. Besides enabling the use of destructive updates, uniqueness typing can help to improve storage management: if it is clear from a rewrite rule that a certain function argument is discarded, and it is specified as unique, then it will certainly become obsolete after rewriting. Re-usage of its space can therefore be anticipated. This leads to so-called *compile-time garbage collection*.

The development of the uniqueness type system (introduced in Barendsen and Smetsers (1993a), see also (1993b)) has been guided by the following objectives.

First of all, it is natural to require that the type system is *consistent* in the sense that 'typed programs cannot go wrong'. This means that typability is preserved under evaluation, also known as the *subject reduction property*, which is proved to hold in the uniqueness type system in Barendsen and Smetsers (1993a).

Second, for practical use the type system should be *effective*: suitable types (if they exist) should be computable. For conventional typing this has been shown in Barendsen and Smetsers (1993a). The present paper deals with the decidability problem for the uniqueness type system.

A side condition (also mentioned by Guzmán and Hudak (1990)) was that the system must be *easy to use* and to reason about; expressing destructive functions should be natural. The presented type system fits to the well-known Hindley-Milner approach to type inference. A programmer writing traditional functional programs should not have to worry about uniqueness. Indeed, programs that are type correct in the Hindley-Milner sense remain so, by considering all types as non-unique. On the other hand, programmers should have maximal freedom to exploit uniqueness typing. There are no essential restrictions from a pragmatic point of view: arbitrary uniqueness variants of data structures and higher-order functions are allowed. For example, one can define functions taking destructive operations as arguments and/or deliver them as a result. It is even possible to introduce new data structures containing such destructive operations, cf. Achten et al. (1993).

The research presented here served as a basis for the incorporation of uniqueness typing in the lazy functional language *Clean*, see Plasmeijer and van Eekelen (1995). The most important application of uniqueness types so far is the implementation of an efficient high-level library for writing arbitrary interactive applications (see Achten et al. (1993) and Achten and Plasmeijer (1995)). Section 7 contains some examples written in *Clean* showing the merits of the type system.

Structure of this paper

We start with a brief introduction to graph rewriting. Section 3 summarizes earlier results on conventional typing. Section 4 contains a short informal explanation of the uniqueness type system. This section is based on Barendsen and Smetsers (1993a). In Section 5, the system is extended with 'uniqueness polymorphism'. A type inference procedure is presented in Section 6. We give some examples in the programming language *Clean* in Section 7, and conclude with a discussion of related work (Section 8).

2 Graph Rewriting

Term graph rewrite systems have been introduced in Barendregt et al. (1987). This section summarizes some basic concepts which are provided by Barendsen and Smetsers (1992); see also (1994).

We will consider finite directed graphs in which each node is labelled with a *symbol*. Each symbol **S**, say, has a fixed arity determining the number of outgoing edges (the *arguments*) of any node labelled with **S**. Rewrite rules specify transformations of graphs. Each rewrite rule is represented by a special graph containing two roots. These roots determine the left-hand side (the *pattern*) and the right-hand side (the *result*) of the rule. The following picture shows the effect of rewriting a graph according to the rule $\mathbf{Add}(\mathbf{Succ}(x), y) \rightarrow \mathbf{Succ}(\mathbf{Add}(x, y))$.

3 Type Assignment for Graphs

The work on *conventional typing*, reported in Barendsen and Smetsers (1993a) and (1993b), has been inspired by van Bakel et al. (1992) who consider TRS's.

Our types are built up from type variables, the standard type constructor $\rightarrow$, and other type constructors (like List) that can be introduced via *algebraic type specifications*. An example of such a specification is

$$\text{List}(\alpha) = \textbf{Nil} \mid \textbf{Cons}(\alpha, \text{List}(\alpha)),$$

which specifies the type constructor List, and moreover the data constructors **Nil** and **Cons** (for building list objects and defining functions by pattern matching).

The symbols of graph rewrite systems are supplied with a type by a *type environment* $\mathcal{E}$. Such an environment contains declarations of the form

$$\textbf{F} : (\sigma_1, \ldots, \sigma_k) \rightarrowtail \tau,$$

where k is the arity of $\textbf{F}$. The part of $\mathcal{E}$ corresponding to the data constructors is determined by the algebraic specifications; e.g. for lists one has

$$\textbf{Nil} : \text{List}(\alpha), \qquad \textbf{Cons} : (\alpha, \text{List}(\alpha)) \rightarrowtail \text{List}(\alpha).$$

Due to the separation of specifications (rewrite rules, algebraic types) from applications one needs an *instantiation mechanism* to deal with different occurrences of symbols. E.g., for the identity function symbol $\textbf{I}$ (with obvious rewrite rule), a type environment would normally contain $\textbf{I} : \alpha \rightarrowtail \alpha$. This type is considered as schematic: at an applicative occurrence of $\textbf{I}$ one can conclude $\textbf{I} : \tau \rightarrowtail \tau$ for each τ.

The use of higher-order functions in TGRS's (in which each symbol has a fixed arity) is made possible by adding a *Currying* procedure. Our results can be applied to ordinary rewrite systems and purely applicative ones (with application as only function), as well as to hybrid variants (such as used in *Clean*). We will not go into this here.

The notion of typing cannot be defined inductively, but is specified in terms of local requirements for a type assignment to nodes.

Formally, a *typing* of a graph g is a function $\mathcal{T}$ assigning a type to each node of g, such that for any application

say with $\mathcal{E}$ containing

$$\textbf{F} : (\sigma_1, \ldots, \sigma_k) \rightarrowtail \tau,$$

there is an substitution $*$ such that

$$\mathcal{T}(n_1) = \sigma_1^*, \quad \ldots, \quad \mathcal{T}(n_k) = \sigma_k^*,$$
$$\mathcal{T}(n) = \tau^*.$$

A typing (for a graph) is called *principal* if all other typings can be obtained via instantiation. The following result is proved in Barendsen and Smetsers (1993a).

Principal Typing Theorem 1. *There is a recursive function* $\text{Type} = \text{Type}^{\mathcal{E}}$ *such that*

$$g \text{ is } \mathcal{E}\text{-typable} \Rightarrow \text{Type}(g) \text{ is a principal } \mathcal{E}\text{-typing for } g;$$
$$g \text{ is not } \mathcal{E}\text{-typable} \Rightarrow \text{Type}(g) = \mathsf{fail}.$$

4 Introduction to Uniqueness Typing

This section gives a short informal overview. Uniqueness typing combines conventional typing and a reference count bookkeeping. All conventional types are now annotated with *uniqueness attributes*, which refer to the reference structure of graph objects. The attributes are • (for 'unique') and × (for 'non-unique').

A symbol type declaration

$$\mathbf{F} : (\sigma^{\bullet}, \tau^{\times}) \rightarrowtail \cdots$$

means that, apart from the conventional typing requirements, **F**-applications are restricted to those in which the first argument is unique, i.e. has reference count 1. There are no reference requirements for the second argument. In the same way, uniqueness of result types is specified. In a type correct application

where **F**'s argument type and **G**'s result type match, the • attribute ensures that the graph remains type-correct while **F**'s arguments is subject to reduction: the reference count of **F**'s argument will remain 1. Naturally, the structure of the rewrite rules for **F** and **G** should respect the corresponding environment types.

Sometimes, uniqueness is not required. If $F : \sigma^{\times} \rightarrowtail \cdots$ then still $\mathbf{F}(\mathbf{G}(\cdots))$ is type correct. This is expressed in a subtype relation, such that roughly $\sigma^{\bullet} \leq \sigma^{\times}$. Offering a non-unique argument if a function requires a unique one fails: $\sigma^{\times} \not\leq \sigma^{\bullet}$. In an application, an argument is non-unique if either its root symbol has a non-unique result type, or its reference count is greater than 1. The latter case is covered by a *correction* mechanism which converts each unique type into a non-unique variant.

In short, the *offered type* (by an argument) is determined by the *result type* of its topmost symbol and the reference count of that argument. The application is type correct if the argument's offered type (i.e. the type after correction) is a subtype of the requested type involved. The subtype relation is defined in terms of the ordering • $\leq$ × on attributes.

Pattern matching is an essential aspect of term graph rewriting, causing a function to have access to arguments via *data paths* instead of a single reference. This 'deeper access' is taken into account by the uniqueness types of data constructors: the result of such a constructor is taken to be unique whenever one of its arguments is. As a result, any uniqueness assumption on a 'deeper' function argument is translated into uniqueness requirements for the complete data path leading to that argument. This propagation requirement for data constructors is expressed in a requirement for type constructors using the relation $\leq$. For example, the types $\mathsf{List}^u(\mathsf{Int}^v)$ with $u \leq v$ are all uniqueness variants of lists of natural numbers. This indeed excludes $\mathsf{List}^\times(\mathsf{Int}^\bullet)$ (a list with unique elements but a non-unique spine).

In general, propagation requirements for algebraic type constructors are determined by analyzing the specification of these, see Barendsen and Smetsers (1993a). For the function type constructor $\rightarrow$ there are no propagation assumptions (it has no standard term constructors).

The way references are counted can be refined. Sometimes, multiple access to a unique argument is harmless if one knows that only one of the references will be present at the moment of evaluation. This is explained in Barendsen and Smetsers (1993a). Indeed, if one evaluates the conditional

If c Then t Else e

in the standard way, the condition is evaluated first (so access from **F** to **N** is not unique), and subsequently one of the alternatives is chosen and evaluated. Depending on this choice, either the reference of **G** to **N** or that of **H** to **N** is left and unique.

Our uniqueness typing system is parametric in the reference counting mechanism. We thus abstract from any particular (straightforward or refined) reference analysis: we assume that a classification of references into 'unique' and 'non-unique' is provided, indicated by a *marking* with labels $\odot$ and $\otimes$ (for unique and non-unique references respectively). The type correction leading to the offered type (denoted below by $\uparrow \cdots$) along a specific reference depends on the reference label. For example, $\uparrow \mathsf{Int}^\bullet$ for $\odot$-references, $\uparrow \mathsf{Int}^\times$ for $\otimes$-references if the involved environment result type is $\mathsf{Int}^\bullet$. In case of $\mathsf{Int}^\times$, both offered versions are $\mathsf{Int}^\times$. We illustrate this refined reference count by an example:

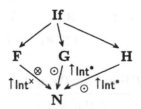

5   Uniqueness Typing

Formally, a uniqueness type is a pair $S = \langle \sigma, A \rangle$ consisting of a conventional type σ and an attribute function A, which assigns a uniqueness attribute to each node in the tree representation of σ. We will usually denote the attributes as superscripts in the conventional type, for example $\text{List}^{\bullet}(\text{Int}^{\times})$. By $|S|$ we denote the underlying conventional type σ.

Uniqueness Polymorphism

To denote families of uniqueness types in a compact way, and to develop a notion of *most general attribution* we introduce uniqueness *schemes*. We extend the attribute set with *attribute variables* $a, b, \ldots$, and use *coercion statements* to indicate restrictions on and relations between attribute variables.

For example, the type of the list constructor **Cons** is formulated as

$$\textbf{Cons} : (\alpha^a, \text{List}^b(\alpha^a)) \rightarrowtail \text{List}^b(\alpha^a) \qquad [b \leq a].$$

In general, a symbol type declaration contains a *set* of coercion statements. An *instantiation* of a symbol type consists of an attribute assignment (mapping attribute variables to attributes) which does not conflict with the coercion statements involved, and an attributed instantiation of the conventional variables. For example, to obtain a spine-unique list one instantiates b with $\bullet$. Lists of unique integers are obtained by choosing $a := \bullet$, and subsequently $\alpha^{\bullet} := \text{Int}^{\bullet}$. The coercion statement $b \leq a$ forces b to be instantiated with $\bullet$.

A uniqueness type assignment to a graph now consists of a conventional typing together with an attribution and a finite set of coercion statements (a so-called *coercion environment*). In order to express the validity of such typings we parameterize the subtyping relation $\leq$ on types with a coercion environment.

Coercion environments

Let Γ be a coercion environment, and let u, v be uniqueness attributes. We say that $u \leq v$ is *derivable* from Γ (notation $\Gamma \vdash u \leq v$) if $\Gamma \vdash u \leq v$ can be produced by the axioms

$$\Gamma \vdash u \leq v \qquad \text{if } (u \leq v) \in \Gamma,$$
$$\Gamma \vdash u \leq u, \qquad \Gamma \vdash u \leq \times, \qquad \Gamma \vdash \bullet \leq u$$

and rule

$$\frac{\Gamma \vdash u \leq v \quad \Gamma \vdash v \leq w}{\Gamma \vdash u \leq w}.$$

This denotation is extended to finite sets of coercion statements: $\Gamma \vdash \Gamma'$ if $\Gamma \vdash u \leq v$ for each $(u \leq v) \in \Gamma'$. By $u = v$ we denote the pair $u \leq v, v \leq u$. We say that Γ is *consistent* if $\Gamma \nvdash \times \leq \bullet$.

An attribute substitution $\diamond$ is said to *respect* Γ if $\Gamma^{\diamond}$ (defined in the obvious way) is consistent.

As has been mentioned before, the relation $\leq$ is used to define the subtype relation: the validity of $\langle \sigma, A \rangle \leq \langle \sigma, B \rangle$ in Γ depends subtypewise on the validity of $\Gamma \vdash A(p) \leq B(p)$ and/or $\Gamma \vdash B(p) \leq A(p)$ (with p a position in the syntax tree of σ). The directions of these positionwise inequalities depend on the covariance and contravariance of the type constructors. One has, for example,

$$b \leq a \quad \vdash \quad \mathsf{Int}^a \xrightarrow{} \mathsf{Bool}^c \leq \mathsf{Int}^b \xrightarrow{} \mathsf{Bool}^x.$$

The attributes of corresponding occurrences of the type constructor $\to$ (in the left-hand and the right-hand side of an inequality) should be identical; see Barendsen and Smetsers (1993a) for a motivation. The same is required (to ensure substitutivity of the subtyping relation) for variables. E.g.,

$$\Gamma \vdash \mathsf{Int}^u \xrightarrow{v} \beta^w \leq \mathsf{Int}^{u'} \xrightarrow{v'} \beta^{w'} \quad \text{iff} \quad \Gamma \vdash u' \leq u, v = v', w = w'.$$

The relation between node type (of a function argument) and requested type (of a function), along $\odot$ and $\otimes$ references respectively, can be fully expressed in terms of coercion statements. This avoids an explicit type correction operation yielding the intermediate 'offered type' ($\uparrow \cdots$ in Section 4) in question. This is done by defining

$$\Gamma \vdash S \leq^{\odot} T \Leftrightarrow \Gamma \vdash S \leq T,$$
$$\Gamma \vdash S \leq^{\otimes} T \Leftrightarrow \Gamma \vdash S \leq T, \times \leq \mathrm{attribute}(T).$$

Here attribute(T) denotes the topmost attribute of T.

Expressing propagation

For any type S, the set of coercion statements expressing algebraic propagation constraints ('internal consistency') is denoted by $\mathrm{Alg}(S)$. For example,

$$\mathrm{Alg}(\mathsf{List}^u(\mathsf{Int}^v \xrightarrow{w} \mathsf{List}^y(\mathsf{Char}^z))) = [u \leq w, y \leq z].$$

Observe that the constructor $\to$ is propagating in neither of its arguments.

A *uniqueness type environment* consists of symbol type declarations of the form

$$\mathbf{F} : (S_1, \ldots, S_k) \rightarrowtail T \quad [\Gamma]$$

where Γ consists of attribute restrictions including internal consistency, i.e.

$$\Gamma \vdash \mathrm{Alg}(S_1), \ldots, \mathrm{Alg}(S_k), \mathrm{Alg}(T).$$

For example (see also Section 7)

$$\mathbf{Append} : (\mathsf{List}^b(\alpha^a), \mathsf{List}^c(\alpha^a)) \rightarrowtail \mathsf{List}^d(\alpha^a) \quad [b \leq a, c \leq a, d \leq a, c \leq d].$$

Uniqueness typing of graphs

Let $\mathcal{E}$ be a uniqueness type environment. An $\mathcal{E}$-typing for a graph g is a pair $\langle \mathcal{U}, \Gamma \rangle$, where $\mathcal{U}$ assigns a Γ-consistent uniqueness type to each node in g, such that for each application

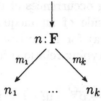

(where $m_1, \ldots, m_k$ are the markings of the references), say with $\mathcal{E}$ containing

$$\mathbf{F} : (S_1, \ldots, S_k) \rightarrowtail T \quad [\Gamma_{\mathbf{F}}],$$

the following holds. There is an attribute assignment $\diamond$ respecting $\Gamma_{\mathbf{F}}$ and a consistent substitution $*$ on the resulting type $(S_1^\diamond, \ldots, S_k^\diamond) \rightarrowtail T^\diamond$ such that

$$\Gamma \vdash \Gamma_{\mathbf{F}}^\diamond,$$
$$\Gamma \vdash \mathcal{U}(n_1) \leq^{m_1} (S_1^\diamond)^*, \quad \ldots, \quad \Gamma \vdash \mathcal{U}(n_k) \leq^{m_k} (S_k^\diamond)^*,$$
$$\Gamma \vdash \mathcal{U}(n) = (T^\diamond)^*.$$

It it easy to show that uniqueness typing is substitutive: if $\langle \mathcal{U}, \Gamma \rangle$ is a typing for g, and $\diamond$ is an attribute assignment respecting Γ, then $\langle \mathcal{U}^\diamond, \Gamma^\diamond \rangle$ is also a typing for g. This substitutivity also holds with respect to instantiation of conventional variables; see Barendsen and Smetsers (1993a).

6 Type Inference

Below, we will show that uniqueness types can be determined in an effective way. In fact, the typing algorithm described at the end of this section has been incorporated in the Nijmegen *Clean* compiler.

We proceed in two steps. First we show how to determine an attribute environment (if it exists) that completes a uniqueness type assignment to a uniqueness typing. Note that determining such an environment is indeed the essential step, since all restrictions on attributes can be expressed by coercion statements: a type $\mathsf{Int}^\bullet$ is equivalent to Int^a in any environment with $a \leq \bullet$.

Suppose we are given (in a fixed environment $\mathcal{E}$) a uniqueness type assignment $\mathcal{U}$ for g, such that its attribute-free version $|\mathcal{U}|$ is a conventional typing for g. Then we can decide whether $\mathcal{U}$ can be extended to a uniqueness typing for g. Moreover, if the answer is positive we can effectively calculate a suitable environment Γ containing the 'minimal requirements' on the attributes in $\mathcal{U}$.

Theorem 2. *Let $\mathcal{E}$, g and $\mathcal{U}$ be given, such that $|\mathcal{U}|$ is a (conventional) $\mathcal{E}$-typing for g. Then there is a coercion environment $\Gamma = \Gamma^{\mathcal{E}}(g, \mathcal{U})$ such that*

$$\Gamma^{\mathcal{E}}(g, \mathcal{U}) \text{ is consistent} \Rightarrow \langle \mathcal{U}, \Gamma^{\mathcal{E}}(g, \mathcal{U}) \rangle \text{ is an } \mathcal{E}\text{-uniqueness typing for } g;$$
$$\Gamma^{\mathcal{E}}(g, \mathcal{U}) \text{ is inconsistent} \Rightarrow \text{ there is no } \Gamma \text{ such that } \langle \mathcal{U}, \Gamma \rangle \text{ is an }$$
$$\mathcal{E}\text{-uniqueness typing for } g.$$

Moreover if $\Gamma^{\mathcal{E}}(g, \mathcal{U})$ is consistent, then for any Γ' one has

$$\langle \mathcal{U}, \Gamma' \rangle \text{ is an } \mathcal{E}\text{-uniqueness typing for } g \Rightarrow \Gamma' \vdash \Gamma^{\mathcal{E}}(g, \mathcal{U}).$$

Proof (Sketch). For each node n in g (as displayed above), set

$$\Gamma_n^{\mathcal{E}}(g, \mathcal{U}) = \text{Alg}(\mathcal{U}(n))$$
$$\cup\, \Gamma_{\mathbf{F}}$$
$$\cup\, \text{Sub}^{m_1}(\mathcal{U}(n_1), S_1) \cup \cdots \cup \text{Sub}^{m_k}(\mathcal{U}(n_k), S_k)$$
$$\cup\, \text{Eq}(\mathcal{U}(n), T).$$

Here, $\text{Sub}^m(U, V)$ is the set of coercion statements expressing that U is $\leq^m$-related to an instance of V. This set depends on the attributes of U and V in the part of the syntax trees up to the variable nodes in V. Moreover $\text{Eq}(U, V)$ contains subtypewise attribute equalities $u = v$ related to the same parts of the trees. Finally we set

$$\Gamma^{\mathcal{E}}(g, \mathcal{U}) = \bigcup_{n \in g} \Gamma_n^{\mathcal{E}}(g, \mathcal{U}). \quad \square$$

Theorem 3. *Consistency of coercion environments is a decidable property.*

Proof. Let Γ be a coercion environment. We define the 'unification' operation U, yielding an attribute substitution or fail, by recursion, as follows.

$$
\begin{aligned}
U(\Gamma) &= \text{fail} & &\text{if } \Gamma \text{ contains } (\times \leq \bullet); \text{ otherwise:} \\
&= \text{id (the identity)} & &\text{if } \Gamma \text{ does not contain } a \leq \bullet \text{ or } \times \leq a; \\
&= U(\Gamma[a := \bullet]) \circ [a := \bullet] & &\text{if } \Gamma \text{ contains } (a \leq \bullet); \text{ otherwise:} \\
&= U(\Gamma[a := \times]) \circ [a := \times] & &\text{if } \Gamma \text{ contains } (\times \leq a).
\end{aligned}
$$

This procedure terminates (note that the number of attribute variables decreases with each recursion step). Then Γ is consistent if and only if $U(\Gamma) \neq \text{fail}$. $\square$

Definition 4. Let $\mathcal{T}$ be a typing for g.
   (i) A *g-attribution* of $\mathcal{T}$ is a uniqueness typing $\langle \mathcal{U}, \Gamma \rangle$ for g such that $|\mathcal{U}| = \mathcal{T}$.
   (ii) Such an attribution is called *principal* if for any g-attribution $\langle \mathcal{U}', \Gamma' \rangle$ there exists an attribute substitution $\diamond$ with

$$\mathcal{U}' = \mathcal{U}^{\diamond} \quad \text{and} \quad \Gamma' \vdash \Gamma^{\diamond}.$$

Principal Attribution Theorem 5. *There is a computable function* Attr $=$ Attr$^{\mathcal{E}}$ *such that for each* $|\mathcal{E}|$-*typing for* g *the following holds. If* $\mathcal{T}$ *has a* g-*attribution, then* Attr$^{\mathcal{E}}(g, \mathcal{T})$ *is a principal attribution; otherwise* Attr$^{\mathcal{E}}(g, \mathcal{T}) =$ fail.

Proof. Given $\mathcal{T}$ and g, construct an attributed version $\mathcal{U}_0$ of $\mathcal{T}$ by giving each subtype a fresh attribute variable. Set $\Gamma_0 = \Gamma^{\mathcal{E}}(g, \mathcal{U}_0)$. Apply Theorem 3; if Γ_0 is consistent, set Attr$^{\mathcal{E}}(g, \mathcal{T}) = \langle \mathcal{U}_0, \Gamma_0 \rangle$; otherwise set Attr$^{\mathcal{E}}(g, \mathcal{T}) =$ fail. Principality follows from Theorem 2. $\square$

Canonical Representations

For a clear presentation of uniqueness types, it is convenient to simplify uniqueness typings $\langle \mathcal{U}, \Gamma \rangle$ which are produced by the above procedure. This yields an 'equivalent' $\langle \mathcal{U}', \Gamma' \rangle$ by moving implicit attribute assignments (such as $a \leq \bullet$) from Γ to $\mathcal{U}$, and removing redundant information from Γ.

Given $\langle \mathcal{U}, \Gamma \rangle$, we first compute $\diamond = U(\Gamma)$ from the proof of Theorem 3. Now we simplify $\Gamma^{\diamond}$ and extend the substitution $\diamond$ by

(1) removing redundant statements $\bullet \leq \bullet$, $\bullet \leq \times$, $\times \leq \times$, $a \leq \times$, $\bullet \leq a$;
(2) removing cycles $a \leq a_1, a_1 \leq a_2, \ldots, a_{n-1} \leq a_n$, replacing $a_1, \ldots, a_n$ by a in the remaining inequalities, and adding $[a_1 := a, \ldots, a_n := a]$ to the attribute substitution;
(3) removing inequalities $a \leq b$ that follow by transitivity, i.e. in the presence of $a \leq c, c \leq b$ for some c.

Say this procedure yields $\Gamma', \diamond'$. Then $\langle \mathcal{U}^{\diamond'}, \Gamma' \rangle$ is a simplified version of $\langle \mathcal{U}, \Gamma \rangle$.

Uniqueness Type Inference in Clean

In *Clean*, type checking is concerned with the determination of a suitable environment type for each function symbol, such that all program parts are well-typed. Since recursive specifications are allowed, determination of such an environment is undecidable in general. In fact, the *Clean* compiler adopts the Hindley-Milner approach towards recursion: in the definition of, say, **F**, all occurrences of **F** are typed with **F**'s environment type (i.e., without instantiation). Indirect recursion is treated similarly. This enables us to determine a conventional environment type for **F**, and for the rewrite rule for **F**, using the Principal Type Theorem for conventional types.

Having obtained an adequate conventional typing for **F**, we proceed as follows. We construct a fresh attribution of **F**'s environment type, which determines the types of the arguments (the left-hand side of the rule). Using the Principal Attribution Theorem, we compute the most general attribution of the right-hand side typing. Together with the requirement that the root type of the right-hand side should be a subtype of **F**'s result type (see Barendsen and Smetsers (1993b)), this gives a correct coercion environment for the initial attribution. The simplification procedure is then applied to obtain the final typing. See Section 7 for an example.

7 Examples

In this section we will give some examples illustrating the use of uniqueness typing in practice. All programs are written in *Clean* which has a Haskell-like syntax. The syntax for types will be self-explanatory. We will occasionally omit the superscript × and leave the algebraic consistency requirements implicit.

The first example shows how a function can be defined that copies the contents of a file to another file using the following I/O primitives.

$$\text{FWriteC} : (\text{Char}, \text{File}^\bullet) \rightarrowtail \text{File}^\bullet$$
$$\text{FReadC} : \text{File} \rightarrowtail (\text{Bool}, \text{Char}, \text{File})$$

FWriteC requires a unique input file which is updated destructively. The output is also unique, to prepare for further manipulations. FReadC is a non-destructive function. Its output consists of three parts: a boolean indicating success or failure of the read operation, the character that has been read from the input file, and the file that results from the input file after deleting the first character. The function FCopy is defined straightforwardly.

$$
\begin{aligned}
\text{FCopy src dst} \quad &= \text{If ok} \\
&\qquad (\text{FCopy nsrc (FWriteC char dst)}) \\
&\qquad \text{dst} \\
&\text{where (ok, char, nsrc)} = \text{FReadC src}
\end{aligned}
$$

To illustrate the type inference algorithm at the end of Section 6, we will show how this type can be derived. This is done best by considering the graph representation of the above rule. We apply explicit selector functions to obtain ok, char and nsrc. The initial attributed conventional typing $\mathcal{U}_0$ (see the proof of the Principal Attribution Theorem 5) is given below. We only display the relevant parts. For the sake of readability, some attribute equalities (that will result from Eq) have already been taken into account by choosing equal attributes.

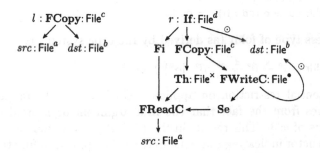

Note that both references to the node *dst* can be marked ⊙; see the explanation at the end of Section 4. The typing procedure gives the following attribute restrictions.

If	$c \leq d, b \leq d$
FCopy	$\times \leq a, \bullet \leq b$
FWrite	$b \leq \bullet$
FRead	$a \leq \times$
right-to-left coercion	$d \leq c$

Unification gives the attribute substitution $[b := \bullet, a := \times]$ (showing consistency of the restrictions). Simplification gives the final environment type

$$\mathbf{FCopy} : (\mathsf{File}^{\times}, \mathsf{File}^{\bullet}) \rightarrowtail \mathsf{File}^{c}.$$

In the second example, a function is given to append one list to another. To examine the effect on its uniqueness typing, we specify two distinct versions of append.

```
append [] li       = li
append (hd : tl) li = hd : append tl li
```

The type derived by the type system is

$$\mathsf{append} : ([\alpha^a]^b, [\alpha^a]^c) \rightarrowtail [\alpha^a]^d \qquad [c \leq d]$$

The coercion statement specifies that spine uniqueness of the result list $(d = \bullet)$ depends only on the spine attribute c of the second argument.

As mentioned in Wadler (1990), storage use can be optimized whenever the first list argument is spine unique. In contrast to Wadler (1990), the recursive call of append cannot be evaluated immediately, since (a part of) the result of this function is likely to be discarded by a context in which append is applied. However, if it is guaranteed that the spine of the list is always evaluated in any context, a smart compiler might produce an efficient loop version of append in which no auxiliary storage at all is needed.

An alternative definition of append can be obtained by using the well-known function foldr:

```
foldr op zero []        = zero
foldr op zero (hd : tl)  = op hd (foldr op zero tl)
```

The uniqueness type of foldr (as derived by the type system) is

$$\mathsf{foldr} : (\alpha^a \xrightarrow{\times} \beta^b \xrightarrow{d} \beta^b, \beta^b, [\alpha^a]^c) \rightarrowtail \beta^b$$

The only essential restriction on applications of foldr w.r.t. uniqueness of arguments comes from the fact that the first argument op is used twice in the right-hand size of foldr. This results in the (default) non-unique attribute at the first $\rightarrow$ constructor indicating that foldr may not be applied to functions to which only one reference is allowed (see Barendsen and Smetsers (1993a) for further explanation).

Now one can define append as follows.

```
append l1 l2 = foldr (:) l2 l1
```

The derived uniqueness type of append is the same as the type derived for the previous version. The equality of both types (in particular with respect to uniqueness information) suggests an equivalent operational behaviour: the results are constructed in essentially the same way, namely by copying the first list and redirecting the final tail pointer to the second list (which is the most obvious way to append two lists). This shows that the derived uniqueness properties give some additional information on append. E.g., if one accidently exchanges both list arguments in the application of foldr, i.e.

```
append l1 l2 = foldr (:) l1 l2
```

the conventional type remains the same; the uniqueness type, however, becomes different. In general, one can use uniqueness typing not only to produce more efficient code but also for a partial correctness check of programs.

The final example has been taken from Guzmán and Hudak (1990). It shows a function mark for marking a graph, e.g. to avoid that nodes of a graph are visited more than once during its traversal. The definition of mark as it can be found in Guzmán and Hudak (1990) is as follows.

```
mark g loc =
    if g == empty || g [loc,0] = True
    then g
    else let* left = g [loc,1]
             right = g [loc, 2]
             g' = update! g [loc,0] True
         in mark (mark g' left) right
```

Guzmán and Hudak (1990) infer the following type for mark (where ws stands for 'write access' and r for 'read access').

$$\text{mark} : \text{Array}(\iota, \text{Int}) \xrightarrow{ws} \iota \xrightarrow{r} \text{Array}(\iota, \text{Int}),$$

implying that all updates can be performed destructively.

In the remainder of this section, we will derive a version of mark in *Clean* with similar uniqueness properties. This will be done in three steps. The first version we consider is obtained by a more or less direct translation of the original mark function. To represent graphs, we introduce the following auxiliary synonym types.

```
:: Graph       :== Array Node
:: Node        :== (Marking, Arity, Arguments)
:: Marking     :== Bool
:: Arity       :== Int
:: Arguments   :== Array Int
```

The definition of mark in *Clean* is as follows.

```
mark g loc | marking
  = g
  = m_args new_g arity args
  where
  (marking,arity,args) = lookup g loc
  new_g = update g loc (True,arity,args)

  m_args g 0 args   = g
  m_args g n args   = m_args g (lookup args n) (n - 1) args
```

For the sake of clarity, we prefer to denote array selection as an application of a function called lookup. Now suppose lookup and update are typed as follows.

$$\text{lookup} : (\text{Array}^b(\alpha^a), \text{Int}) \rightarrowtail \alpha^a,$$
$$\text{update} : (\text{Array}^\bullet(\alpha^a), \text{Int}, \alpha^a) \rightarrowtail \text{Array}^\bullet(\alpha^a).$$

Then the function mark will be rejected by the type system. This is due to the fact that g is used twice in the right-hand side of mark. Consequently, both references to g are $\otimes$-marked, thus preventing unique usage of g.

Note that in the original version the lookup and update operations are evaluated in a certain order (depending on the special semantics of the let * construct). One can also enforce sequentialisation by defining a slightly modified version of lookup

$$\text{lookup} : (\text{Array}^b(\alpha), \text{Int}) \rightarrowtail (\alpha, \text{Array}^b(\alpha))$$

Besides the selected element, lookup yields the array from which selection took place as a result. This array can be used for further manipulations (similar to the files in the results of the file operations mentioned in the FCopy example). Obviously, the elements of the array are no longer unique.

The second definition of mark, including its uniqueness type, is

```
mark : (Array^b(α), Int) ↦ Array^b(α)
mark g loc | marking
  = g
  = m_args upd_g arity args
  where
  (marking,arity,args) = node
  (node,look_g)   = lookup g loc
  upd_g           = update look_g loc (True,arity,args)
```

The final version of mark that even leaves the uniqueness of nodes intact can be obtained by using a new array operation, called exchange.

$$\text{exchange} : (\text{Array}^b(\alpha^a), \text{Int}, \alpha^a) \rightarrowtail (\alpha^a, \text{Array}^b(\alpha^a))$$

This operation is based on the observation that one can safely select an element from an array without destroying uniqueness properties of this array if a new (unique) element is inserted at the place of the selected one. Note that this is essentially what the uniqueness type of exchange indicates.

This leads to the following definition for mark

```
mark : (Array^b(α^a), Int) ⟼ Array^b(α^a)
mark g loc | marking
  = g
  = m_args new_g arity args
  where
  (marking,arity,args) = node
  (node,exc_g)         = exchange g loc empty_node
  (empty, new_g        = exchange look_g loc (True,arity,args)
```

First note that the resulting graph is fully unique whenever the input graph is also fully unique. Moreover, the storage cell containing the constructor of the node being marked becomes garbage. Obviously one can recycle this cell immediately for building the new marked node instead of returning it to free storage. In fact, one can even reuse most of its old contents; the only part that has to be updated is the mark field. Lastly, it requires only a simple analysis to derive that the node exchanged with the selected node is returned by the second call of exchange. Hence, it suffices to allocate only one such node during program execution.

8 Conclusions and Related Work

We have developed a very powerful type system to characterize reference structures in graphs, in order to express uniqueness constraints of function arguments. In the present paper, polymorphism has been extended to uniqueness attributes. The resulting system turns out to be decidable.

The system has been implemented as part of the *Clean* compiler. The applicability is not restricted to languages whose *syntax* is based on graph rewriting. In fact, the type system is also suited for other functional languages, since most of them use graph-like structures in their *implementations*.

The weighted reference count analysis has been inspired by Guzmán and Hudak (1990). This paper addresses the mutability problem using a 'single threaded polymorphic lambda calculus' ($poly-\lambda_{st}$). It uses the operational semantics of lambda-graph reduction of Wadsworth (1971). In our paper the analysis is performed in the formalism of graph rewriting, which is obviously more direct.

In Wadler (1990), a type system including linear types is developed. The paper also uses Wadsworth's lambda reduction. The relation between linear and uniqueness typing can be summarized as follows.

$$\text{uniqueness typing} = \text{pure linear typing}$$
$$+ \text{ conventional typing}$$
$$+ \text{ subtyping}$$
$$+ \text{ strategy-aware reference analysis.}$$

Based on the idea of uniqueness typing, Jacobs (1993) developed a *logical* system explicitly mixing conventional and linear constructive logic.

In Sastry et al. (1993), the update problem is addressed by determining a safe evaluation order via abstract interpretation. It would be interesting to investigate how our strategy-dependent reference analysis can be refined using their techniques. This would lead to a more liberal notion of typing.

References

Achten, P.M. and M.J. Plasmeijer (1995). The ins and outs of Clean I/O, *Journal of Functional Programming* 5, pp. 81–110.

Achten, P.M., J.H.G. van Groningen and M.J. Plasmeijer (1993). High level specification of I/O in functional languages, *in:* J. Launchbury and P. Sansom (eds.), *Proceedings of the International Workshop on Functional Languages*, Ayr, Scotland, 6–8 July 1992, Workshops in Computing, Springer-Verlag, Berlin, pp. 1–17.

van Bakel, S.J., J.E.W. Smetsers and S. Brock (1992). Partial type assignment in left-linear term rewriting systems, *in:* J.C. Raoult (ed.), *Proceedings of the 17th Colloqium on Trees and Algebra in Programming (CAAP'92)*, Rennes, France, Lecture Notes in Computer Science 581, Springer-Verlag, Berlin, pp. 300–322.

Barendregt, H.P., M.C.J.D. van Eekelen, J.R.W. Glauert, J.R. Kennaway, M.J. Plasmeijer and M.R. Sleep (1987). Term graph reduction, *in:* J.W. de Bakker, A.J. Nijman and P.C. Treleaven (eds.), *Proceedings of the Conference on Parallel Architectures and Languages Europe (PARLE)* II, Eindhoven, The Netherlands, Lecture Notes in Computer Science 259, Springer-Verlag, Berlin, pp. 141–158.

Barendsen, E. and J.E.W. Smetsers (1992). Graph rewriting and copying, *Technical Report 92-20*, Computing Science Institute, University of Nijmegen.

Barendsen, E. and J.E.W. Smetsers (1993a). Conventional and uniqueness typing in graph rewrite systems, *Technical Report CSI-R9328*, Computing Science Institute, University of Nijmegen.

Barendsen, E. and J.E.W. Smetsers (1993b). Conventional and uniqueness typing in graph rewrite systems (extended abstract), *in:* R.K. Shyamasundar (ed.), *Proceedings of the 13th Conference on Foundations of Software Technology and Theoretical Computer Science*, Bombay, India, Lecture Notes in Computer Science 761, Springer-Verlag, Berlin, pp. 41–51.

Barendsen, E. and J.E.W. Smetsers (1994). Extending graph rewriting with copying, *in:* H.J. Schneider and H. Ehrig (eds.), *Graph Transformations in Computer Science, International Workshop*, Dagstuhl Castle, Germany, Lecture Notes in Computer Science 776, Springer-Verlag, Berlin, pp. 51–70.

Girard, J.-Y., Y. Lafont and P. Taylor (1989). *Proofs and Types*, Cambridge Tracts in Theoretical Computer Science 7, Cambridge University Press.

Guzmán, J.C. and P. Hudak (1990). Single-threaded polymorphic lambda calculus, *Proceedings of the 5th Annual Symposium on Logic in Computer Science*, Philadelphia, IEEE Computer Society Press, pp. 333–343.

Jacobs, B.P.F. (1993). Personal communication.

Plasmeijer, M.J. and M.C.J.D. van Eekelen (1995). Concurrent Clean 1.0 language report, Computing Science Institute, University of Nijmegen, To appear.

Sastry, W., A.V.S. Clinger and Z.M. Ariola (1993). Order-of-evaluation analysis for destructive updates in strict functional languages with flat aggregates, *Proceedings of the Conference on Functional Languages and Computer Architectures (FPCA)*, Copenhagen, Denmark, ACM Press, pp. 266–276.

Troelstra, A.S. (1992). *Lectures on Linear Logic*, CSLI Lecture Notes 29, CSLI, Stanford.

Wadler, P. (1990). Linear types can change the world!, *Proceedings of the Working Conference on Programming Concepts and Methods*, Israel, North-Holland, Amsterdam, pp. 385–407.

Wadsworth, C.P. (1971). *Semantics and Pragmatics of the Lambda Calculus*, Dissertation, Oxford University.

Modes of Comprehension:
Mode Analysis of Arrays
and Array Comprehensions

B.C. Massey   E. Tick

Dept. of Computer Science
University of Oregon
Eugene, OR 97403, USA
bart,tick@cs.uoregon.edu
Phone: +1-503-346-4436

Abstract. A scheme is presented to enable the mode analysis of concurrent logic programs manipulating arrays containing both ground and non-ground elements. To do this we leverage constraint-propagation mode analysis techniques. The key ideas are to restrict multiple assignments only to variables at the *leaves of paths*, and to extend the language family with *array comprehensions*. The result is a language not significantly different than generic committed-choice languages, which can be *safely* mode analyzed, producing useful (not overly conservative) information, even for programs that assign to unbound array elements. An implementation of the analysis is presented.

1   Introduction

"Given for one instant an intelligence which could comprehend all the forces
by which nature is animated and the respective positions of the beings which
compose it, if moreover this intelligence were vast enough to submit these
data to analysis...to it nothing would be uncertain..."
Pierre Simon de Laplace
Oeuvres, vol VII, Theorie Analytique des Probabilites

Modes in logic programs are a restricted form of types that specify whether a logic variable (or some subterm thereof) is produced or consumed by a procedure. For example, in a pure logic program, a well-known feature of the list concatenation procedure are its two permissible modes: either two input lists are consumed, producing their concatenation, or a concatenated list is consumed, producing pairs of output lists (that if concatenated equal the input list). For such *nondeterminate* logic programs, there are numerous methods for statically determining modes [1, 2, 3, 10, 13] and several optimizations afforded by the mode information, notably converting backtracking programs into functional programs, and specializing unification operators into matches and assignments, and thereby making them execute faster.

Mode information has also been shown to be quite useful in the efficient compilation of *indeterminate* logic programming languages, i.e., committed-choice programs [6, 8, 14, 17, 20]. Mode information allows demand-driven execution [9, 19] and static partitioning of a concurrent logic program into threads of higher granularity for more efficient multiprocessor execution [5, 8]. Mode information is useful not only for compiler optimization but also for static bug detection.

All the previous schemes and implementations (ours included) deal with the problem of aliasing through array indices, which is a well-known stumbling block in many types of static analysis, by restricting array elements to ground terms. Allowing arbitrary expressions for both array elements and indices makes determining aliases undecidable; the restriction of array elements to ground terms makes safe mode analysis possible. In other words: use writable array elements at your own risk! The problem is more pervasive than it might sound offhand — a program with even a single use of non-ground arrays cannot be safely analyzed with currently available technologies.

This paper describes a novel concurrent logic programming language extension and corresponding mode analysis extension that allow safe static mode analysis that is not overly conservative for programs that use arrays with unbound elements. The objective of the paper is to describe the problem and our solution, giving examples of the technique. The method has been implemented as an extension to a previous mode analyzer that we have built [14, 16].

This paper is organized as follows. Section 2 defines the array moding problem and gives a series of increasingly flexible solutions. Section 3 discusses array comprehensions, a technique borrowed from functional languages, which allow constraint-based mode analysis to be exploited. Section 4 gives some examples of the techniques described. Sections 5 and 6 formalize the extensions to the constraint-based mode analysis technique and describe an implementation. Conclusions and future work are summarized in Section 7.

2 Arrays and Constraint-Based Mode Analysis

We first describe the standard functional-style operations upon arrays:[1]

- An array is created by
  array_ini(N,M), where N is the number of elements, and M is the resulting array. M is initialized to contain unbound variables (in contrast to KL1).
- An array element may be referenced by
  array_sel(M,I,E), where M is the array, I is the element index, and E is the Ith element of M.
- An array may be updated by
  array_upd(M,I,E,M'), where M is the array, I the index, E the new element, and M' is a copy of M except that the Ith element of M is replaced by E in M'.

---

[1] We adopt the names from Sastry *et al.* [12], although similar builtins can be found in languages such as KL1, Strand, Parlog, etc. Many languages, such as KL1, instead provide a array_sel_upd/5 primitive which acts like a combination of array_sel/3 and array_upd/4.

Before we discuss the semantics of these operations, let us review the mode analysis technique we plan to exploit. Ueda and Morita [20] proposed a mode analysis scheme based on the representation of procedure paths and their relationships as rooted graphs ("rational trees"). Unification over rational trees combines the mode information obtainable from the various procedures. For example, in a procedure that manipulates a list data stream, we might know that the mode of the *car* of the list (that is the current message) is the same mode as the *cadr* (second message), *caddr* (third message), etc. This potentially infinite set of "paths" is represented as a regular graph. Furthermore, a caller of this procedure may constrain the *car* to be input mode. By unifying the caller and callee path graphs, modes can be propagated. The analysis is restricted to "moded" flat committed-choice logic programs. These are programs in which the mode of each path in a program is constant, rather than a function of the occurrences of the path. As a consequence of this restriction, such programs can have only a single producer of any variable (although multiple consumers are allowed). These are not regarded as major restrictions, since most non-moded flat committed-choice logic programs may be transformed (by the programmer) to moded form in a straightforward fashion [20].

Ueda and Morita define the subterm s "derived" via a *path* p within a term t (written $p(t) \vdash s$) as follows: p derives s in t iff for some predicate f and some functors $a, b, \ldots$ the subterm denoted by descending into t along the sequence $\{<f, i>, <a, j>, <b, k>, \ldots\}$ (where $<f, i>$ is the ith argument of the functor f) is s. A path thus corresponds to a descent through the structure of some object being passed as an argument to a function call. f is referred to as the "principal functor" of p. A program is "moded" if the modes of all possible paths in the program are consistent, where each path may have one of two modes: **in** or **out**. For a formal definition, see [17]. For example the *cadr* of the first argument of procedure q has input mode. We represent this using the notation $m(\{<q, 1>, <., 2>, <., 1>\}) = $ **in**. The notation m/p is used to represent the modes of all subtrees emanating from ("below") p. Specifically, $m/p = $ **in** is shorthand for $\forall s \ m(p \cdot s) = $ **in** (and similarly for **out**), and $m/p = m/q$ is shorthand for $\forall s \ m(p \cdot s) = m(q \cdot s)$ (and similarly for $\neq$).

The scheme described by Ueda and Morita has been implemented and tested in various incarnations [6, 14, 16, 17] so we want to extend these techniques to solve the array aliasing/moding problem. We will go into further details of the mode analysis in Section 5, but for now, let us discuss language restrictions.

To illustrate the sort of aliasing problems that can arise in the mode analysis of programs containing arrays, consider the procedure:

```
bad :-
  array_ini( 1, M ),
  array_sel( M, 0, E1 ),
  array_sel( M, 0, E2 ),
  E1 = 3,
  E2 = 4.
```

The second assignment will cause mode analysis to fail, i.e., the procedure will be determined to be "nonmoded." The key point is that static detection of aliases is undecidable if we allow *arbitrary expressions* as indices, unless we restrict the language somehow.

One language restriction which solves the aliasing problem is to keep all array elements fully ground. This works because of the single-assignment property of the languages, i.e., in ground terms aliasing is not an issue. Mode-analysis savants are aware of this solution. For example, KL1 initializes array elements with integers, and if the programmer ensures they stay ground, then mode analysis could legitimately reject programs which make assignments to any element [18]. This restriction actually covers quite a number of useful programs; however, our goal is to relieve the programmer of any burden and to cover the more general case of arrays with non-ground terms.

Our solution is to *forbid multiple assignments to an element of an array along any given path "below" that array*. In the previous example, the array M is a flat array of a single element, so it is trivially disallowed. Consider the procedure:

```
good :-
    array_ini( 1, M ),
    array_sel( M, 0, E1 ),
    array_sel( M, 0, E2 ),
    E1 = f( A, B ),
    bind( E2 ).

bind( f( A, B ) ) :- A = 1, B = 2.
```

The three assignments bind variables at paths rooted in **array_sel/3**:

$$\{<\text{array\_sel}, 1>, <*>\}$$
$$\{<\text{array\_sel}, 1>, <*>, <f, 1>\}$$
$$\{<\text{array\_sel}, 1>, <*>, <f, 2>\}$$

Here $<*>$ is a placeholder representing the dereference of the array itself.[2] The key point is that each of the three assignments lands at a unique location in the tree of paths, and thus mode analysis should deduce that the program is fully moded, which it is. To make this happen, we extend the mode analysis rules to include a simple reduction rule for **array_sel/3**: let $p = \{<\text{array\_sel}, 1>, <*>\}$, and $q = \{<\text{array\_sel}, 3>\}$, then $m/p \neq m/q$.[3] In other words, we alias all array elements of a given array together, regardless of their index — the placeholder $<*>$ represents an arbitrary array element. The details of this and other extensions are discussed in Section 5.

However, this scheme will disallow the following moded procedure:

```
bad :-
    array_ini( 2, M ),
    array_sel( M, 0, E1 ),
```

---

[2] There is nothing corresponding to $<*>$ in Ueda and Morita's path definitions: its lack of a name reminds us that it represents an array or vector and its lack of an index reminds us that all indices are collapsed into one.

[3] The inequality here may look odd. It derives from the fact that tell unification argument must have opposite modes: What comes in from one argument must go out on the other argument. The same argument holds for the array M and its element E.

```
      array_sel( M, 1, E2 ),
      E1 = 3,
      E2 = 4.
```

This is because we collapse indices and treat **E1** and **E2** as aliases. Thus the mode analysis is overly-conservative, and reports a mode conflict. This example may seem tolerable if we can still assign to unique paths, but actually this simple example is representative of a very serious impracticality: there is no practical way to assign a top-level value to more than one element of a multi-element array! The top-level indices are collapsed (as are all similar internal functors). So what we have is a safe, but far-too-conservative scheme.

We need some coherent way to *override* the restrictions we placed on assignment, allowing us to initialize whole arrays in some fashion digestible by the mode analyzer. Essentially we need to ensure that the indices used to bind a given path are mutually exclusive. This is one purpose of "array comprehensions" in functional languages, e.g., [11, 4]. An array comprehension typically is something like (fictional syntax):

```
      int A[20] {0,1: A[i] = 1;
          2..19: A[i] = A[i-2] + A[i-1]}
```

This example would create an array of 20 Fibonacci numbers. In the next section we expand upon the concept of array comprehensions and show how it meshes with the mode analysis framework previously given.

3 Array Comprehensions

We extend our language to support "array comprehensions" (AC): *procedures designed to concurrently bind given paths in a given array at more than one index*. Since the indices are provided by the implementation rather than the programmer, and are thus guaranteed to be disjoint, if any binding is guaranteed to be safe, all are. Our AC syntax will be identical to that of normal procedures, except for a special guard **array_csel/4**. Presence of such a guard in a procedure indicates that the procedure is an AC.

The semantics are those of concurrent invocation: when an AC is invoked on an array **M**, multiple invocations of the procedure are spawned, each with a unique index **I** of **M**.[4] A formal semantics is given in Section 6. Informally:

- **array_csel(M0,M,I,E)** is the aliasing AC guard where **M0** is the array, **M** is a read-only copy of **M0**,[5] **I** is the element index, and **E** is the Ith element of **M**. The array is input and the latter two arguments are produced by the guard for consumption in the body of the AC. The array **M0** may be further

---

[4] Note that this is just semantics: there are a number of possible ways in which the concurrency might be limited in an implementation, for efficiency reasons — this is a topic of future research. We are unaware of any solution to date of the analogous efficiency problem in non-strict functional language implementations of array comprehensions.

[5] For programmer convenience, **M0** and **M** would be combined and the compiler would differentiate between the two.

instantiated only by binding E: performing a **array_sel/3** on M in the AC body produces a read-only alias into M.

- A procedure containing clauses with **array_csel/4** guards is an *array comprehension*. An array comprehension is automatically invoked multiple times per call, one invocation being produced for each unique element of the array argument given to **array_csel/4**. If an invocation matches more than one clause in the AC, a clause will be chosen nondeterministically in the usual fashion. All arguments of an AC, other than the array, must be input.[6]

Thus array comprehensions allow a variable at a given path within (or "below") an array to be bound *at more than one index, as long as the indices are part of the same AC*. This is the fundamental contribution of this paper. Extension of the mode analysis system to handle arrays is reasonably straightforward, as discussed in Section 5.

4 Some Examples

As our first example, consider a hashed symbol table, sketched in Figure 1 (auxiliary procedures **hash/3** and **sym_match/3** are not supplied). The initialization procedure **init_symtab/2** uses the only array comprehension of the program, because symbol table entries are normally fully ground and thus can be manipulated using array functions only. Given an array consisting of unbound variables, **init_symtab/2** calls **null_table/1** to bind them all to nil. Procedure **null_table/1** can be reused for arbitrary-sized tables, and does not need to reference the element index in its array comprehension. The modes of **init_symtab/2** are

$$m(\{< \text{init\_symtab}, 1 >\}) = \textbf{in}$$
$$m(\{< \text{init\_symtab}, 2 >, < \text{symtab}, 1 >\}) = \textbf{out}$$
$$m(\{< \text{init\_symtab}, 2 >, < \text{symtab}, 2 >\}) = \textbf{out}$$
$$m(\{< \text{init\_symtab}, 2 >, < \text{symtab}, 2 >, < * >\}) = \textbf{out}$$

Note that the modes for **null_table/1** are

$$m(\{< \text{null\_table}, 1 >\}) = \textbf{in}$$
$$m(\{< \text{null\_table}, 1 >, < * >\}) = \textbf{out}$$

Since the array is initialized by **null_table/1**, the caller can safely read its elements.

Because the calls to **array_sel/3** in **insert_sym/3** and **lookup_sym/3** can be mode analyzed, we are guaranteed to get a mode error if **Chain** or any of its aliases are ever assigned to. This is precisely Ueda and Morita's "well-moded programs don't go wrong" condition [20]; we will get a static error rather than the runtime error which would otherwise result.

As a second example (Figure 2), we present a more complex array comprehen-

---

[6] Because in general an AC is invoked multiple times, allowing output variables would lead to multiple bindings for the output variables.

```
init_symtab( NBuckets, SymTab ) :-
  array_ini( NBuckets, Table ),
  null_table( Table ),
  SymTab = symtab( NBuckets, Table ).

null_table( Table ) :-
  array_csel( Table, _, _, E ) |
  E = [].

insert_sym( symtab( NBuckets, Table ), Sym, SymTab_p ) :-
  hash( Sym, NBuckets, Hash ),
  array_sel( Table, Hash, Chain ),
  Chain_p = [ Sym | Chain ],
  array_upd( Table, Hash, Chain_p, Table_p ),
  SymTab_p = symtab( NBuckets, Table_p ).

lookup_sym( symtab( NBuckets, Table ), Name, Sym ) :-
  hash( Sym, NBuckets, Hash ),
  array_sel( Table, Hash, Chain ),
  follow_chain( Chain, Name, Sym ).

follow_chain( [], _, Sym ) :-
  Sym = nomatch.
follow_chain( [ Sym | Syms ], Name, Sym_p ) :-
  sym_match( Sym, Name, B ),
  follow_chain_1( B, [ Sym | Syms ], Name, Sym_p ).

follow_chain_1( true, [ Sym | _ ], _, Sym_p ) :-
  Sym_p = Sym.
follow_chain_1( false, [ _ | Chain ], Name, Sym_p ) :-
  follow_chain( Chain, Name, Sym_p ).
```

Fig. 1. Hashed Symbol Table Example

sion to compute the Fibonacci sequence. Already, we see important advantages of ACs over chains of **array_upd/4** operations. First, the AC will update all elements in parallel rather than sequentially if possible (although in this example only the first two elements can be updated in parallel). Second, even if multiple references to the array are held, the array may be updated in place rather than by copying.

Our final example (Figure 3) illustrates binding of nested arrays, showing the generality of the technique. The program **conv(Ints,Base,V)** converts an array of positive integers **Ints** into an array **V** of radix **Base** numbers, by sequentially computing residues. A single integer **Int** and its conversion is represented by the term:

```
item( Int, { D0-R0, D1-R1, ..., D31-R31 } )
```

```
fib( M ) :-
  array_ini( 20, M ),
  fib_array( M ).

fib_array( M0 ) :-
  array_csel( M0, M, I, E ),
  I >= 2 |
  I1 := I - 1,
  I2 := I - 2,
  array_sel( M, I1, V1 ),
  array_sel( M, I2, V2 ),
  E := V1 + V2.

fib_array( M0 ) :-
  array_csel( M0, M, I, E ),
  I < 2 |
  E = 1.
```

Fig. 2. Fibonacci Example

where Dn is the nth radix **Base** digit and Rn is the residue at the nth position. An array of **item/2** terms is initialized by **conv/3** and **init_all/2**. Number conversion occurs in **trans_all/2** and **trans_int/3** with nested comprehensions. The innermost comprehension is sequentialized by the dependency incurred by the **array_sel/3** goal. The process computing the nth digit must wait for the residue of the $(n-1)^{st}$ digit to be computed.

5 Mode Analysis of ACs

A critical first step in mode analysis of array functions and comprehensions is to note that different arrays may have different modes. Thus, for instance, one cannot blithely unify *all* **array_sel/3** goals together! Though we cannot decide precisely which array is denoted by a variable occurrence, we can safely approximate this notion. Thus, in the spirit of constraint-propagation mode analysis, we will initially assume that all array functions refer to *unique* arrays, and find relationships based on shared variable occurrences. This is similar to what is done in our current mode analyzers [14] to deal with the unification operator =/2.

To accomplish this, we first separate all the occurrences of array functions, by assigning a unique integer subscript to each static occurrence of **array_ini/2**, **array_sel/3**, and **array_upd/4** in the program. We then apply mode analysis rules (Figure 4) giving the relationship between the arguments of these functions.

Figure 5 graphically illustrates the rules using notation introduced in [17]. In brief, constraint graphs are layered networks consisting of structure nodes

```
conv( NInts, Ints, Base, V ) :-
    array_ini( NInts, V ),
    init_all( Ints, V ),
    trans_all( Base, V ).

init_all( Ints, V0 ) :-
    array_csel( V0, V, I, E ) |
    array_sel( Ints, I, N ),
    array_ini( 32, D ),
    E = item( N, D ).

trans_all( Base, V0 ) :-
    array_csel( V0, V, _, E ),
    E = item( N, D ) |
    trans_int( N, Base, D ).

trans_int( N, Base, D0 ) :-
    array_csel( D0, D, I, E ),
    I =:= 0 |
    Digit := N mod Base,
    Residue := ( N - Digit ) / Base,
    E = pair( Digit, Residue ).

trans_int( N, Base, D0 ) :-
    array_csel( D0, D, I, E ),
    I > 0 |
    I1 := I - 1,
    array_sel( D, I1, RL ),
    get_residue( RL, Residue_p ),
    Digit := Residue_p mod Base,
    Residue := ( Residue_p - Digit ) / Base,
    E = pair( Digit, Residue ).

get_residue( pair( _, Res0 ), Res ) :-
    Res = Res0.
```

Fig. 3. Radix Conversion Example

(squares) and variable nodes (circles). For example, **array_sel/3** has three children corresponding to its three arguments. The first child is a variable M leading to an array placeholder, leading to a variable E. Variable E also has a parent extending from the third argument of **array_sel/3** because the abstract meaning of the function (in the mode domain) is to collapse all top-level elements in the array into one element. Each variable node in a graph is labeled with its potential modes (one per input arc, since different input arcs can "see" a variable through different moded "glasses"). For instance, variable E in **array_sel/3** has possible modes [in,out] or [out,in], enforcing the constraint that inputs in the

$$m(\{<\text{array\_ini}_i/2,1>\}) = \textbf{in}$$
$$m(\{<\text{array\_ini}_i/2,2>\}) = \textbf{out}$$
$$m/\{<\text{array\_ini}_i/2,2>, <*>\} = \textbf{in}$$

$$m(\{<\text{array\_sel}_i/3,1>\}) = \textbf{in}$$
$$m/\{<\text{array\_sel}_i/3,1>, <*>\} \neq m/\{<\text{array\_sel}_i/3,3>\}$$
$$m(\{<\text{array\_sel}_i/3,2>\}) = \textbf{in}$$

$$m(\{<\text{array\_upd}_i/4,1>\}) = \textbf{in}$$
$$m/\{<\text{array\_upd}_i/4,1>\} \neq m/\{<\text{array\_upd}_i/4,4>\}$$
$$m/\{<\text{array\_upd}_i/4,3>\} \neq m/\{<\text{array\_upd}_i/4,4>, <*>\}$$
$$m(\{<\text{array\_upd}_i/4,2>\}) = \textbf{in}$$

Fig. 4. Mode Analysis Rules For Array Functions

first argument are seen as outputs in the third argument and vice versa. In fact, from these two parent paths, all graph nodes below E (represented as shaded subtree) are viewed as having opposite modes, effectively implementing the tell unification "M[*] = E."

Each occurrence of a variable corresponds to a particular path from a root to that variable's node. The occurrence's mode is considered inverted if a traversal from its root to the node contains an odd number of **out** modes. An example is variable T in **array_ini/2** which has mode **in**. Further details of tell unification and the graph reduction semantics can be found in [17]. Which representation of the moding rules, axiomatic or graphical, is most understandable is a matter of taste. We wish to point out that the graph segments defining the array functions and comprehensions are built from previously defined primitives.

The rules for moding an AC are similar in spirit to the rules for the simpler cases: the key is the correct moding of **array_csel/4**. The array must be read-only in the body of the AC, in order to ensure that the array is updated only through the fourth argument of **array_csel/4**. For instance if there is a body goal in **null_table** in the previous example which would alter Table$_1$, a mode conflict would be forced. On the other hand, the caller of the AC must have called the AC with the idea that the AC would bind some paths in the array. Thus we propagate the modes of the array elements through (**Table**), but allow the AC body to access the array in a read-only fashion through the second argument to **array_csel/4**. An example of this is the use of D in the body of **trans_int** in Figure 3.

The mode analysis rules for the AC function **array_csel/4** are given in Figure 6. Note that *all* arguments to an AC are treated as input by the callee.

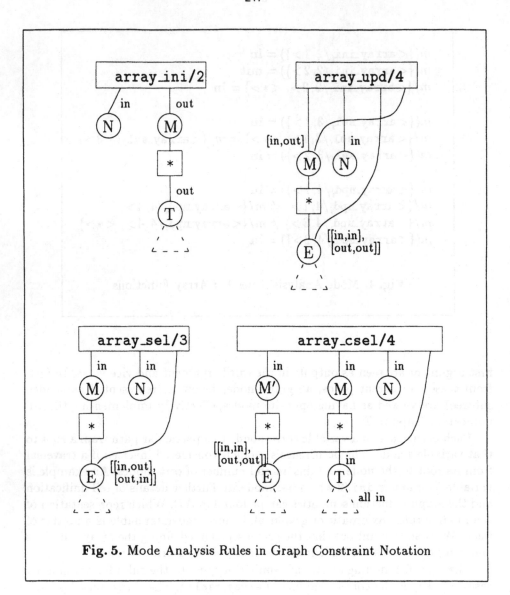

Fig. 5. Mode Analysis Rules in Graph Constraint Notation

$$m(\{<\text{array\_csel}_i/4, 1>\}) = \text{in}$$
$$m/\{<\text{array\_csel}_i/4, 1>, \ <*>\} = m/\{<\text{array\_csel}_i/4, 4>\}$$
$$m/\{<\text{array\_csel}_i/4, 2>\} = \text{in}$$
$$m(\{<\text{array\_csel}_i/4, 3>\}) = \text{in}$$

Fig. 6. Mode Analysis Rules For Array Comprehensions

This is because multiple invocations of the AC will be created, and thus if more than one invocation tries to bind some path beginning at an argument to the AC, some assignment will fail.

A prototype implementation of these rules was added to one of our current mode analyzers. This mode analyzer [16] operates upon a finite set of paths, attempting to construct this set in such a way that the modes of all variable occurences of the target program are uniquely and correctly determined. The original mode analyzer consists of 4487 lines of KL1 (including comments); 3362 lines of application-specific source code plus 1125 lines of KL1 devoted to general-purpose data structures such as sets and lists. The modified analyzer consists of 4021 lines of application-specific KL1. Of the 659 lines of new code, 427 lines comprise a module giving the new mode analysis rules for array functions and comprehensions, and the remaining 232 lines represent changes to the original code necessary to integrate this module. There is no appreciable difference in the runtime performance of the analyzer (see [16] for the performance of the original system).

6 Semantics of Array Operations

To illustrate the operational semantics of the array operations, we now give a reference implementation of array functions and array comprehensions in KL1, representing arrays as lists. There is no reason to believe that the transformation perserves modeness, nor is there any reason to be concerned if it does not. Simply, modeness is not at issue for the metalanguage used to define the semantics, even if that metalanguage is KL1 itself. Actually, the reference language used is KL1 with the exception that names beginning with $ are not visible by the programmer. This prevents the programmer from observing the internals of the particular representation used in the reference implementation.

To simplify the following presentation, we assume two restrictions upon the use of ACs. First, only one **array_csel/4** guard is allowed per clause. Second, the **array_csel/4** guard for each clause of a given AC must have the "same" first argument, i.e., if some clause of a procedure g has an AC guard for an array at some argument path p, the AC guards of all clauses of g must be for the array argument at p. The first restriction can be checked during mode analysis and the second can be checked syntactically.

We first give a reference implementation of the "ordinary" array functions, in Figure 7. These involve straightforward translations of indexing to recursion.

Next, we consider the semantics of array comprehensions. Let p/k be a procedure of arity k whose clause heads have been flattened and consistently renamed, and each of whose clauses contain an array comprehension guard **array_csel(MO,M, I,E)**. Our strategy is to rewrite p to remove the array comprehension as follows:

1. Construct a new procedure $\$p/(k+2)$ by prepending the arguments I and E of the **array_csel/4** guard to the head of each clause of p, and then replacing the **array_csel/4** guard in each clause with the guard "**MO = M.**"

```
array_ini( K, M ) :-
  K > 0 |
  K1 := K - 1,
  array_ini( K1, $array(L) ),
  M = $array( [ _ | L ] ).
array_ini( 0, M ) :-
  true |
  M = $array( [] ).

array_sel( $array( [ _ | Es ] ), I, X ) :-
  I > 0 |
  I1 := I - 1,
  array_sel( $array( Es ), I1, X ).
array_sel( $array( [ E | _ ] ), 0, X ) :-
  true |
  X = E.

array_upd( $array( [ E | Es ] ), I, X, M ) :-
  I > 0 |
  I1 := I - 1,
  array_upd( $array( Es ), I1, X, $array( L ) ),
  M = $array( [ E | L ] ).
array_upd( $array( [ _ | Es ] ), 0, X, M ) :-
  true |
  M = $array( [ X | Es ] ).
```

Fig. 7. Semantics of "Ordinary" Array Functions

2. Let $length/1 be a special guard which returns the length of a array. (It
   is easy to see how to rewrite the program in terms of the reference implementation to remove this guard if necessary.) Construct a new procedure
   $p'/(k + 1) as follows. Let the second and succeeding arguments of $p' be
   the same as those of p, and be denoted by ellipsis. Then p' is given by

```
$p'( $I, ... ) :-
  $I < $length( M0 ) |
  array_sel( M0, $I, $E ),
  $p'( $I, $E, ... ),
  $I1 := $I + 1,
  $p( $I1, ... ).

$p( $I, ... ) :-
  $I >= $length( M0 ) |
  true.
```

3. Replace all calls p(...) with calls $p(0,...).

Some notes about the generality of the scheme presented in this paper. Our
underlying mode analysis allows variables to remain unwritten. An important

implication of this is that a program may terminate with incomplete data structures. This "zero-or-one writer" principle is slightly different than the "guaranteed one writer" principle discussed by Ueda and Morita [20]. Conceivably, the underlying mode analysis used in this paper can be applied to nondeterminate logic programs; nothing in our treatment of array comprehensions precludes this possibility.

7 Conclusions

We have described a new technique for permitting safe mode analysis of concurrent logic programs that use arrays with both ground and non-ground elements. The implications of this work are that a broader class of programs can thus be optimized by mode analysis. We believe this is the first practical, *implemented* solution offered for this problem, particularly because it leverages proven mode analysis technology that is efficient.

We extend the concurrent logic language family to include builtin array predicates, both guards and body goals, that allow array comprehensions and standard array updates and assignments. Thus algorithms that use fully-ground arrays can be programmed in the usual manner and are permissible under our mode analysis. In addition, programs that use non-fully-ground arrays can be programmed using our extensions, and if safe, will be accepted by our mode analysis.

We have implemented the array rules within one of our mode analyzers: future work entails implementations within our other constraint-based analyzers [14, 16], and implementing comprehensions, via concurrent spawning, within the Monaco system [15, 7]. It would be very difficult, in terms of static analysis, to implement an alternative to concurrent spawning of ACs. Although, sequential spawning may result in higher granularity tasks and therefore improved performance. We believe that the concurrent scheme uses no more process resources than existing vector implementations in concurrent logic languages such as KL1. Finally, we may need to further extend our basic scheme as we gain experience in its use, if we identify inflexibilities which make programming with arrays more difficult than we currently expect.

Acknowledgements

Bart Massey was supported by a grant by the Institute for New Generation Computer Technology (ICOT). Evan Tick was supported by an NSF Presidential Young Investigator award, with matching funds from Sequent Computer Systems Inc. We thank Kazunori Ueda and the anonymous referees who made insightful comments and suggestions on several drafts of this paper.

References

1. M. Bruynooghe and G. Janssens. An Instance of Abstract Interpretation Integrating Type and Mode Inference. In *International Conference and Symposium on Logic Programming*, pages 669–683. University of Washington, MIT Press, August 1988. Extended version in Journal of Logic Programming, 1994.

2. S. K. Debray. Static Inference of Modes and Data Dependencies in Logic Programs. *ACM Transactions on Programming Languages and Systems*, 11(3):418–450, July 1989.

3. S. K. Debray and D. S. Warren. Automatic Mode Inference for Prolog Programs. *Journal of Logic Programming*, 5(3):207–229, September 1988.

4. P. Hudak, S. Peyton-Jones, and P. Wadler. Report on Programming Language Haskell: A Non-Strict, Purely Functional Language, Version 1.2. *ACM SIGPLAN Notices*, 27(5):1–164, May 1992.

5. A. King and P. Soper. Schedule Analysis of Concurrent Logic Programs. In *Joint International Conference and Symposium on Logic Programming*, pages 478–492. Washington D.C., MIT Press, November 1992.

6. M. Koshimura and R. Hasegawa. A Mode Analyzer for FGHC Programs in a Model Generation Theorem Prover. In *Proceedings of the 47th Annual Convention IPS Japan*, 1993. In Japanese.

7. J. Larson, B. Massey, and E. Tick. Super Monaco: Its Portable and Efficient Parallel Runtime System. In *EURO-PAR*, Stockholm, August 1995.

8. B. C. Massey and E. Tick. Sequentialization of Parallel Logic Programs with Mode Analysis. In *International Conference on Logic Programming and Automated Reasoning*, number 698 in Lecture Notes in Artificial Intelligence, pages 205–216, St. Petersburg, July 1993. Springer-Verlag.

9. B. C. Massey and E. Tick. Demand-Driven Execution of Concurrent Logic Programs. In *International Conference on Parallel Architectures and Compilation Techniques*, pages 215–224, Montreal, August 1994. North-Holland.

10. C. S. Mellish. Some Global Optimizations for a Prolog Compiler. *Journal of Logic Programming*, 2(1):43–66, April 1985.

11. R. S. Nikhil. Id (Version 90.0) Reference Manual. Technical Report CSG Memo 284-a, MIT Laboratory for Computer Science, 545 Technology Square, Cambridge, MA 02139, USA, July 1990.

12. A. V. S. Sastry, W. Clinger, and Z. Ariola. Order-of-Evaluation Analysis for Destructive Updates in Strict Functional Languages with Flat Aggregates. In *Conference on Functional Programming Languages and Computer Architecture*, pages 266–275. Copenhagen, ACM Press, June 1993.

13. R. Sundararajan. *Data Flow and Control Flow Analysis of Logic Programs*. PhD thesis, Department of Computer Science, University of Oregon, 1994. Also available as Technical Report CIS-TR-94-08.

14. E. Tick. Practical Static Mode Analyses of Concurrent Logic Languages. In *International Conference on Parallel Architectures and Compilation Techniques*, pages 205–214, Montreal, August 1994. North-Holland.

15. E. Tick and C. Banerjee. Performance Evaluation of Monaco Compiler and Runtime Kernel. In *International Conference on Logic Programming*, pages 757–773. Budapest, MIT Press, June 1993.

16. E. Tick and M. Koshimura. Static Mode Analyses of Concurrent Logic Languages. *Journal of Programming Language Design and Implementation*, 1995. Accepted. Also available as University of Oregon Technical Report CIS-TR-94-06.

222

17. E. Tick, B. C. Massey, F. Rakoczi, and P. Tulayathun. Concurrent Logic Programs *a la Mode*. In E. Tick and G. Succi, editors, *Implementations of Logic Programming Systems*, pages 239–244. Kluwer Academic Publishers, 1994.
18. K. Ueda. Report on the Optimization of Concurrent Logic Language Implementations, March 1994. In Japanese.
19. K. Ueda and M. Morita. Message-Oriented Parallel Implementation of Moded Flat GHC. In *International Conference on Fifth Generation Computer Systems*, pages 799–808, Tokyo, June 1992. ICOT.
20. K. Ueda and M. Morita. Moded Flat GHC and Its Message-Oriented Implementation Technique. *New Generation Computing*, 13(1):3–43, 1994.

Better Consumers for Deforestation (Extended Abstract)

Wei-Ngan CHIN and Siau-Cheng KHOO

Dept of Information Systems & Computer Science
National University of Singapore

Abstract. We describe a novel approach to achieve better deforestation by pre-processing a program before subjecting it to actual deforestation. In particular, our approach performs some simple syntactic analyses, and proactively specialises all functions in a program into *treeless consumer form*. This simplifies greatly the task of deforestation in the later stage. The transformation we use is based on the well-understood fold/unfold strategy with generalisation on terms. We ensure the termination of the transformation. Compared to other existing semantics-based approaches, our syntactic-based approach is considerable simpler, but surprisingly powerful.

1 Introduction

In recent years, there has been considerable interest in high-level optimisations for functional programs that are based on automatic fold/unfold transformation techniques. One well-known transformation technique, called deforestation [Wad88], is able to eliminate unnecessary intermediate data structures. A nice characteristics of deforestation is that its termination can be guaranteed by restricting the class of functions to be transformed to those with the *pure treeless* property.

The main result of deforestation can be stated as follows: given an arbitrary expression consisting of only pure treeless functions, it is always possible to transform it to an equivalent pure treeless expression.

Though elegant, the pure treeless grammar form is quite restrictive. To alleviate this problem, a producer-consumer model on functional expressions was proposed in [Chi92] to help guide the extension of deforestation (called *fusion*) to both first-order and higher-order programs.

Given an arbitrary pair of nested calls: $f(g(x))$, this model regards g as a *producer* of intermediate term, and f as its *consumer*. In other words, each function is regarded as a potential consumer of data (through its input parameters) if it is used as an outer call, and a potential producer of data (through its results) if it is used as an inner call. Based on this view, a terminating fusion transformation can be guaranteed for each nested composition whose inner call satisfies a *safe* producer form (see [Chi92]), and its outer call satisfies a *safe* consumer form (see Section 2.1). Such an approach applies its syntactic restrictions more selectively and is thus an improvement over Wadler's pure deforestation.[1]

Lately, better deforestation has been achieved via more powerful semantics-based analyses. Such analyses can identify more safe producers and more safe consumers. One example is the fairly sophisticated grammar analysis proposed in [Sør94].

We would like to suggest a different way to improve deforestation. We advocate the use of *pre-processing techniques* to help obtain more safe consumers and safe producers. This has the advantage that we can continue to rely on the use of simple syntactic analyses, without risking non-terminating transformation.

In this paper, we focus on a pre-processor for obtaining more safe consumers. (A corresponding pre-processor to help obtain more safe producers is the subject of a future work.) Our pre-processor is essentially a *program specialisation* transformation algorithm that is fairly aggressive in obtaining better consumers. Our main contribution is a set of extremely simple syntactic analyses which can help guarantee the transformer's termination.

Section 2 introduces a simple first-order functional language and ends with a description of the expected input and output of our pre-processor (specialisation algorithm). Section 3 proposes a subclass of functions which are guaranteed to behave as safe consumers for deforestation. This sub-class of functions, which includes the pure-treeless form, has an off-line *non-increasing* property for its parameter variables. This subclass is further enlarged in Section 4 by using an on-line analysis, based on an *in-situ cyclic increasing* property, to selectively generalise sub-terms which could cause infinite (non-terminating) program specialisation. With the help of the on-line analysis, we can convert each unsafe consumer to an equivalent safe consumer using our pre-processor.

Due to space constraint, we omit the proofs to most of the claims made in this paper. Instead, the interested readers can refer to [CK95] for details.

2   A Simple Language

Consider a simple first-order language based on the call-by-name (non-strict) semantics.

---

[1] An alternative automatic approach to deforestation has been suggested by [GLPJ93, SF93] which relies on a set of 'canned' transformation theorems for programs expressed via certain higher-order (or categorical) combinators (e.g. foldr/build functions or catamorphism). Compared to fold/unfold approach, this approach is more efficient (takes fewer transformation steps) but is slightly more restricted in its class of transformable programs.

Definition 1: *Functions and Expressions*
Each function is defined by a set of non-overlapping equations of the form.

$$f(p_{11}, \ldots, p_{1n}) = t_1 ;$$
$$\vdots \qquad \qquad \vdots$$
$$f(p_{m1}, \ldots, p_{mn}) = t_m ;$$

The LHS constructor patterns are defined by the grammar

$$p ::= v \mid C(p_1, \ldots, p_n)$$

and the RHS expressions are defined by grammar

$$t ::= v \mid C(t_1, \ldots, t_n) \mid f(t_1, \ldots, t_n) \mid let \ (v_i, t_i)_{i=0}^{n} \ in \ t$$

where $n \geq 0$

A function definition can be written more succinctly using the alternative notation:

$$f \stackrel{def}{=} ?\{(p_{i1}, \ldots, p_{in}) \Rightarrow t_i\}_{i=1}^{m}$$

The expressions of our simple language include variables (v), constructors (C), functions (f) and *let* constructs. The *let* construct is non-recursive and shall be used mainly to facilitate generalisations. Note that we use the tuple constructor, $(t_1, \ldots, t_n)$, as an instance of the more general data constructor, $C(t_1, \ldots, t_n)$.

We shall consider transformation over a set of mutual-recursive functions in a program. A set of mutual-recursive functions, $M = \{f_1, .., f_m\}$, is collectively known as an M-set of functions. Each function f_i in M can also be written as f_i^M while f^M denotes an arbitrary function belonging to the M-set.

A program thus consists of a sequence of mutual-recursive functions, $[M1, .., Mr]$, that are arranged in a *bottom-up order* such that if a function call of f^{Mx} exists in the RHS of the equations of f^{My}, then it must be the case that $x \leq y$.

We introduce a special context notation with multiple holes.

Definition 2: *Context with Multiple Holes*
A *hole*, $\#m$, is a special variable labelled with a number, m.

A *context*, $\hat{e}\langle\rangle^F$, is an expression with a finite number of holes, defined by the grammar:

$$\hat{e}\langle\rangle^F ::= v \mid \#m \mid C(\hat{e_1}\langle\rangle^F .. \hat{e_n}\langle\rangle^F) \mid f(\hat{e_1}\langle\rangle^F .. \hat{e_n}\langle\rangle^F)$$
$$\mid let \ (v_i, \hat{e_i}\langle\rangle^F)_{i=1}^{n} \ in \ \hat{e}\langle\rangle^F ;$$

such that $f \notin F$

Note that the context notation is parameterised by a set of functions, F, which must not appear in the context itself.

Each expression e can be decomposed into a context $\hat{e}\langle\rangle^F$ and a sequence of sub-terms $[t_1, \ldots, t_n]$ using the notation $\hat{e}\langle t_1, \ldots, t_n\rangle^F$. This notation is equivalent to:

$$[t_1/\#1, .., t_n/\#n](\hat{e}\langle\rangle^F)$$

which stands for the substitutions of sub-terms, $t_1, \ldots, t_n$, into their respective holes, $\#1, .., \#n$, for context $\hat{e}\langle\rangle^F$.

We shall abbreviate a list of terms $t_1, \ldots, t_n$ by $\vec{t}$, so that a function call $f(t_1, \ldots, t_n)$ can be abbreviated as $f(\vec{t})$.

We next define the class of functions which shall be supplied as input to our pre-processor, called *constructor-based* (or CB) functions:

Definition 3: *CB-equations & CB-functions*

An equation of the function f_d^M (from an M-set) has the form:

$$f_d^M(\vec{p}_i) = \hat{e}\langle f_{i_1}^M(\vec{ti}_1), .., f_{i_m}^M(\vec{ti}_m)\rangle^M ;$$

This equation is said to be a constructor-based function if it satisfies the following properties:

1. Each of its mutual-recursive calls, $f_{i_1}^M(\vec{ti}_1), .., f_{i_m}^M(\vec{ti}_m)$, has only constructor terms as their arguments, $\vec{ti}_1, .., \vec{ti}_m$. A constructor term is an expression built from only constructors and variables.

2. Each parameter pattern from $\vec{p}_i$ (on the LHS) and argument patterns $\vec{ti}_1, .., \vec{ti}_m$ of its mutual recursive calls (on the RHS) are *linear* constructor terms. A constructor term is *linear* if each variable in it occurs only once.

Correspondingly, a function f_d^M is said to be a CB-function if all the equations of the M-set are in the CB-form.

Note that any program can be transformed into an equivalent one satisfying 1 and 2 above as follows:

- Every nested function call inside the arguments of mutual-recursive calls can simply be abstracted using a *let* construct. For example, the equation below:

  $f_1 \, C_1(v_1, v_2) = \hat{e}_a\langle f_1 \, C_1(f_2(v_1), v_2)\rangle ;$

  can be converted to the following CB-equation :

  $f_1 \, C_1(v_1, v_2) = \hat{e}_a\langle let \; (w_1, f_2(v_1)) \; in \; f_1 \, C_1(w_1, v_2)\rangle ;$

- Each non-linear constructor-based argument can be made linear by abstracting (using *let*) every variable with repeated occurrence(s).

2.1 Treeless Consumers

In [Chi92], it was suggested that each parameter of a function is a safe consumer if its corresponding argument is a variable in all its recursive calls. This property was also known as the *non-accumulating* parameter criterion. If all the parameters of a given mutual-recursive equation have this property, we shall refer to the equation as a *treeless consumer*. More formally:

Definition 4: *Treeless Consumer*

An equation of the function f_d^M (from the M-set) is said to be a *treeless consumer*, if it has the form:

$$f_d^M(\vec{p}_i) = \hat{e}\langle f_{i_1}^M(\vec{v}_1), .., f_{i_m}^M(\vec{v}_m)\rangle^M ;$$

Correspondingly, a function f_d^M is said to be a treeless consumer function if all the equations of the M-set are in treeless consumer form.

Notice that each recursive call from $f_{i_1}^M(\vec{v}_1), .., f_{i_m}^M(\vec{v}_m)$ only has variables as its arguments. Our goal is then to convert each function in a program into treeless consumer. Of course, this can be easily achieved via abstracting the call arguments. However, doing so does not help eliminate intermediate data structures during deforestation. In the rest of the paper, we will demonstrate how treeless consumers can be obtained more sensibly via our specialisation algorithm.

3 Safe Non-Increasing Consumers

In this section, we present a classification of consumers, and consider a sub-class of functions which can be classified as safe consumers of intermediate data terms. In addition, these functions share the common property that they all can be transformed into treeless consumers using our specialisation algorithm. We also present an informal specification of the algorithm.

3.1 Consumer Classification

Each M-set of CB-functions can be viewed as a consumer at three different levels. At the highest level, the whole function is viewed as a consumer of intermediate data term. At the mid-level, each equation (instead of each function) is viewed as a consumer of those terms that it matches. At the lowest level, each parameter variable (of the LHS pattern of an equation) is viewed as a consumer for the corresponding sub-term matched by its equation.

Hence, when we speak of consumers we could be referring to either *parameter variables*, *equations* or the *functions* themselves. Unless there is ambiguity, we shall not distinguish between these consumers.

Before we classify the consumers, let us define the *depth* of a variable v in a constructor term p, as follows:

Definition 5: *Depth of Variable*
The depth of a variable v in a constructor term p can be measured by $depth(p, v)$ where:
$$depth(C(p_1, \ldots, p_m), v) = 1 + max\{depth(p_1, v), .., depth(p_m, v)\}$$
$$\text{if } (v \sqsubseteq C(p_1, \ldots, p_m)) ;$$
$$depth(C(p_1, \ldots, p_m), v) = -1 \text{ if } not(v \sqsubseteq C(p_1, \ldots, p_m)) ;$$
$$depth(v_j, v) = 0 \text{ if } v_j \equiv v ;$$
$$depth(v_j, v) = -1 \text{ if } v_j \not\equiv v ;$$
Note that $(t_1 \sqsubseteq t_2)$ is a predicate to test if t_1 is a sub-term of t_2, while the proper sub-term relationship is defined by $(t_1 \sqsubset t_2) = (t_1 \sqsubseteq t_2) \wedge (t_1 \not\equiv t_2)$.

We classify the different consumer behaviours of parameter variables, as follows:

Definition 6: *Consumer Classification*

Consider a CB-equation:

$$f_d^M(\vec{p}_i) = \hat{e}\langle f_{i_1}^M(\vec{t}_1), .., f_{im}^M(\vec{t}_m)\rangle^M \; ;$$

Suppose $v_1, .., v_n$ are parameter variables from $\vec{p}_i$. With respect to a particular recursive call, $f_{i_k}^M(\vec{t}_k)$, the parameter variable v_j from $\{v_1, .., v_n\}$ can be classified as either:

1. **decreasing** iff $(v_j \sqsubseteq (\vec{t}_k)) \wedge depth((\vec{t}_k), v_j) < depth((\vec{p}_i), v_j)$
2. **increasing** iff $(v_j \sqsubseteq (\vec{t}_k)) \wedge depth((\vec{t}_k), v_j) > depth((\vec{p}_i), v_j)$
3. **stagnating** iff $(v_j \sqsubseteq (\vec{t}_k)) \wedge depth((\vec{t}_k), v_j) = depth((\vec{p}_i), v_j)$
4. **vanishing** iff $not(v_j \sqsubseteq (\vec{t}_k))$

Informally, these properties refer to what happen to each sub-term of a CB-function call when it undergoes a one-step unfold via a particular recursive call of the equation. If every sub-term appears in *decreasing* depth, then the transformed call gets progressively smaller. If every sub-term appears at *stagnating* depth, then the transformed call effectively remains unchanged in size. Alternatively, if every sub-term *vanishes* from the previous call, then the transformed call will simply become one of a finite number of different RHS recursive calls. These three properties are desirable for the termination of program transformers, and are collectively known as *non-increasing* properties.

We now define non-increasing consumers, as follows.

Definition 7: *Non-Increasing Consumers*

A parameter variable in an equation is said to be *non-increasing* if it is either *decreasing*, *stagnating* or *vanishing* for each recursive call of the equation.

A CB-equation is said to be *non-increasing* if all its parameter variables (from its LHS pattern) are *non-increasing*.

A CB-function, f_i^M, is said to be *non-increasing* if all the equations of its mutual-recursive set, $\{f_1^M, .., f_m^M\}$, are *non-increasing*.

The basic idea of using non-increasing property to ensure termination of a specialisation process is already present (in various forms) in a number of past work. In their Elimination Procedure for logic programs, Proietti & Pettorossi [PP91] made use of non-ascending clauses to ensure terminating transformation. Similarly, Chin required parameters to be non-accumulating for safe fusion transformation [Chi92], while Holst [Hol91] analysed for non-increasing static parameters to ensure a terminating partial evaluator. All these previous proposals have relied on off-line static analyses which make use of conservative approximations.

Our present proposals are inspired from these previous works and can be regarded as improvements over them. This is because, in addition to a syntactic (off-line) analysis, we also make direct use of program specialisation and on-line analyses to obtain better consumer.

3.2 Types of Matching Calls and Equations

Before we present our first result on safe specialisation over non-increasing CB-functions, we briefly describe our specialisation algorithm which turns these CB-functions into treeless consumers. Because the termination of this specialisation relies on finding a matching function call, we need to define the key concepts of matching for function calls. But first, let's provide syntactic notation for constructor terms/contexts, which are made up exclusively from variables and constructors.

Definition 8: *Constructor Terms, Contexts & Holes*
A constructor term (or pattern), p, is an expression of the form:

$$p ::= v \mid C(p_1, \ldots, p_n) \; ;$$

A *constructor context*, $p\langle\rangle$, is an expression context with a finite number of holes, as defined by grammar:

$$p\langle\rangle ::= \#n \mid C(p_1\langle\rangle, .., p_n\langle\rangle) \; ;$$

Definition 9: *Call Matching*
A function call $f(\vec{a})$ is said to *unify* (*match*) with a LHS call pattern $f(\vec{p})$ if:

$$(\vec{p}) = s\langle s'_1, \ldots, s'_m, v_{m+1}, \ldots, v_n \rangle \wedge (\vec{a}) = s\langle v'_1, \ldots, v'_m, t_{m+1}, \ldots, t_n \rangle$$

Note that $\vec{p} \, [v'_1/s'_1, .., v'_m/s'_m] \, [t_{m+1}/v_{m+1}, .., t_n/v_n] = \vec{a}$
where $[t_{m+1}/v_{m+1}, .., t_n/v_n]$ is an *input substitution* of sub-terms $t_{m+1}, \ldots, t_n$ into variables $v_{m+1}, \ldots, v_n$ of pattern $\vec{p}$, while $[v'_1/s'_1, .., v'_m/s'_m]$ is an *output substitution* which replaces the sub-patterns $s'_1, \ldots, s'_m$ of $\vec{p}$ by variables $v'_1, \ldots, v'_m$.

A call $f(\vec{a})$ is said to *fully-match* a call pattern $f(\vec{p})$ if it *unifies* with the call pattern, yielding an empty output substitution [] (with $m = 0$).

A call $f(\vec{a})$ is said to *exactly-match* a call pattern $f(\vec{p})$ if it matches the call pattern, yielding an input substitution that is simply a variable renaming.

3.3 Pre-Processing Strategy

To make a function a better consumer, we specialise its body; in particular, those relevant function calls in its body will be subject to *fold/unfold* transformation. The general specialisation algorithm involved is called $\mathcal{T}^F$; it is the usual fold/unfold strategy with the appropriate termination and matching criteria imposed, as will be described below. In the latter sections, we extend this strategy to include *generalisation*.

With respect to a set of functions F (whose calls are to be unfolded), we briefly describe how an expression is being transformed under $\mathcal{T}^F$. A more complete definition of $\mathcal{T}^F$ can be found in [CK95].

We partition the set of rules into two subsets: rules for transforming a function call (the (u)-rules) and rules for transforming other language constructs (the (v)-rules), as shown in Figure 1.

Procedure 1: *Informal Rules for Define/Unfold/Fold Transformation*

(u) $\mathcal{T}^F[\![f(\vec{a})]\!]$

    (u^t) if $f(\vec{a})$ satisfies termination criteria

        $\Rightarrow f(\vec{a})$

    (u^f) if $f(\vec{a})$ exactly matches a specialisation point

        $\Rightarrow f'(\vec{v'})$

        where $\vec{v'}$ are free variables of a, and

            f' is the new function associated with that specialisation point

    (u^u) otherwise

        $\Rightarrow f_{\text{NEW}}(\vec{v'})$

        where $\vec{v'}$ are free variables of $\vec{a}$,

        $f_{\text{NEW}} \overset{\text{def}}{=} ?\{\vec{p'}_j \Rightarrow \mathcal{T}^F[\![t'_j]\!]\}_{j\in J}$, and

        $?\{\vec{p'}_j \Rightarrow t'_j\}_{j\in J}$ is the instantiation of f's body with argument $\vec{a}$

(v^1) $\mathcal{T}^F[\![v]\!] \Rightarrow v$

(v^2) $\mathcal{T}^F[\![C(t_1,\ldots,t_n)]\!] \Rightarrow C(\mathcal{T}^F[\![t_1]\!],\ldots,\mathcal{T}^F[\![t_n]\!])$

(v^3) $\mathcal{T}^F[\![\text{let } (v_i,t_i)_{i=1}^n \text{ in } t]\!] \Rightarrow \text{let } (v_i,\mathcal{T}^F[\![t_i]\!])_{i=1}^n \text{ in } \mathcal{T}^F[\![t]\!]$

Fig. 1. Informal Specification of The Specialisation Algorithm

The (v)-rules are self-explanatory. Rule (u^t) specifies the criteria for termination of the transformation. Specifically, the transformation is terminated when either:

1. the pertaining call cannot be unfolded because the argument does not unify with any of the LHS pattern of the callee's equations, or
2. the call pattern is in treeless form, and hence not profitable to specialise, or
3. the call does not belong to the set of functions, F, of interest.

Actual call unfolding is specified by rule (u^u). When the argument a does not fully matches any one of the LHS patterns of f's equation, call unfolding leads to the selection of more than one of the RHS of the callee's equations (via instantiation). We define a new function f_{NEW} to capture all the selected RHS's, and replace the original call by a call to f_{NEW}. Transformation is then propagated to all the RHS's of f_{NEW} definition such that all calls to functions in F existing in these RHS's will be treated by the (u)-rules. This specialisation process will continue as long as some calls to functions in F are yet to be specialised. The entire transformation process can therefore be viewed as a *tree-of-derivations*[2]

---

[2] This is similar to the SLD trees from logic programming (see [GS94]).

of calls to functions in F, where the nodes are calls to functions in F (with the root being the first relevant function call to be transformed), and a directed link connecting two nodes in the tree if transforming the call at the source node directly leads (omitting the (v)-rules) to the transformation of the call at the sink node.

Introducing a new function whenever a call is unfolded has its advantage. We only allow instantiation of parameters to take place within new function definitions. Since these are always in strict context, the non-strict semantics of the program is preserved during specialisation (see [RFJ89]). Furthermore, giving a new (function) name to each function call unfolded enables the definition of specialisation points for folding, as discussed below.

To further improve the termination property of the specialisation, especially when dealing with recursive functions, *folding* is introduced. This is specified by rule (u^f). A *specialisation point* is created, and kept, whenever a new function is introduced by rule (u^u)-rule. It contains information about the current call pattern and the new function name. A future call that *exactly matches* the call pattern of a specialisation point will not get unfolded, instead, it is simply replaced by a call to the corresponding new function associated with the specialisation point.

Sequence of Derivation A particular path in the tree of derivations is called a *sequence-of-derivations*. In particular, we call a sequence of transformations an *n-step derivation* if the sequence consists of n (u)-links. An n-step derivation can be expressed diagrammatically as follows:

$$f_{i_0}(\vec{a}_0) \overset{(u^u)}{\leadsto} f_{i_1}(\vec{a}_1) \overset{(u^u)}{\leadsto} \dots \overset{(u^u)}{\leadsto} f_{i_n}(\vec{a}_n)$$

Labels containing the transformation rule used are attached to the links in the sequences. Other helpful information, such as the name of the equation used to obtain the call at the sink node, can be included in the labels if necessary. On the other hand, we shall omit the labels attached to a link when it is clear from the context.

As we mentioned in the beginning of this section, the power of this algorithm lies in its ability to produce treeless consumers. This fact is made formal later (Proposition 3).

3.4 Safe Specialisation for Non-Increasing Consumers

We return to the case of non-increasing consumers (Definition 7), and present an elementary result which states that all non-increasing CB-functions can be transformed into safe consumers of intermediate constructor terms, as follows:

Theorem 1 (First Safe Specialisation Result). *Consider a non-increasing CB-function f_k^M. For any list of constructor terms $\vec{c}_0$, the specialisation via applying $\mathcal{T}^M[f_k^M(\vec{c}_0)]$ always terminates.*

As an example, consider the following function *path* which detects a path:

$$path(Nil) \qquad\qquad\qquad = \textit{True} \; ;$$
$$path(Cons(a,Nil)) \qquad\quad = \textit{True} \; ;$$
$$path(Cons(a,Cons(b,ys))) \quad = arc(a,b) \;\&\&\; path(Cons(b,ys))$$

where $arc(a,b) = \textit{True}$ if there is a direct arc between nodes a and b in a graph. *path* is a non-increasing function. Its function call $path(Cons(b,ys))$ in the third equation can therefore be safely specialised via the algorithm $\mathcal{T}$. This results in a tree-of-derivation comprising two sequences, each of which is obtained from an instantiation of variable ys:

$\mathcal{T}[\![path(Cons(b,ys))]\!] \qquad$ (replaced by $path'(b,ys)$)

(1) $\overset{(u^u),[Nil/ys]}{\rightsquigarrow} \qquad$ (terminated)

(2) $\overset{(u^u),[Cons(b1,ys1)/ys]}{\rightsquigarrow} \mathcal{T}[\![path(Cons(b1,ys1))]\!] \overset{(u')}{\rightsquigarrow} \qquad$ (folded to $path'(b1,ys1)$)

As a result, application of $\mathcal{T}$ produces the following treeless functions:

$$path(Nil) \qquad\qquad\qquad = \textit{True} \; ;$$
$$path(Cons(a,Nil)) \qquad\quad = \textit{True} \; ;$$
$$path(Cons(a,Cons(b,ys))) \quad = arc(a,b) \;\&\&\; path'(b,ys) \; ;$$
$$path'(b,Nil) \qquad\qquad\quad = \textit{True};$$
$$path'(b,Cons(b1,ys1)) \qquad = arc(b,b1) \;\&\&\; path'(b1,ys1) \; ;$$

The above safe specialisation result is most similar to Proietti & Pettorossi's work based on non-ascending clauses for logic programs[PP91]. Our approach concentrates on the consumers of each set of mutual-recursive functions individually. In contrast, Proietti & Pettorossi's work simply considered all mutual-recursive definitions altogether. This could cause slightly worse result because variables at increased depths of auxiliary (as opposed to mutual-recursive) predicates (*c.f.* calls) will be over-generalised, even though they do not contribute towards non-terminating transformation.

4 Towards Better Consumers

In order to extend our work to include increasing CB-functions, we need to scrutinise the non-termination behaviour of call-specialisation for increasing functions. We identify two groups of CB-*equations*, namely:

- non-cyclic equations (includes *terminating* and *indirect* equations),
- cyclic equations.

With a proper understanding of their termination/non-termination characteristics, we shall later present suitable on-line specialisation techniques to help obtain better consumers.

Grouping of CB-equations can be obtained by forming an *equation call graph* of a program. Declaratively, an equation call graph consists of a set of nodes and a set of directed arcs (links) joining two nodes in the graph. A node corresponds to an equation in the program. A link directed from node E_i to node E_j exists in the graph if and only if there is a call in the RHS of equation E_i which can unify with the LHS of equation E_j. A node is called a *leaf* if it does not have an outgoing link.

Using the equation call graph, we partition a program into sets of mutually-recursive equations. We can then classify our equations, and corresponding recursive calls, as follows.

Definition 10: *Terminating, Indirect and Cyclic Equations*
Consider an equation call graph for a M-set of CB-equations:

Each set of equations whose corresponding nodes are linked together in circles are *cyclic* equations.

Each equation whose corresponding node is a leaf is a *terminating* equation. Also, each equation whose corresponding node has all its links pointing only to the nodes of *terminating* equations is a *terminating* equation.

The rest of the equations are called *indirect* equations. Their corresponding nodes are not in any cycle, but have at least one path (of links) leading to the corresponding node of a cyclic equation.

Correspondingly, a function call which matches a cyclic/ terminating/ indirect equation is a cyclic/ terminating/ indirect call.

4.1 Non-Cyclic Equations

Terminating and indirect equations are *non-cyclic* (non-recursive) equations of the recursive CB-functions. They do not contribute towards non-termination when unfolded. This is because whenever a *terminating* or *indirect* function call is encountered during specialisation, we can show that the sequence of unfolds involving these equations is always finite.

Theorem 2 (Second Safe Specialisation Result). *Given a M-set of CB-functions which are defined using either non-cyclic or non-increasing equations, or both. For any list of constructor terms $\vec{c}_0$, the specialisation via applying $T^M[\![f^M(\vec{c}_0)]\!]$ always terminates.*

As an example, the following CB-function is considered a safe consumer despite having two *increasing* indirect equations.
$$g1(C1(v)) = \hat{e}_a\langle g1(C2(C1(v)))\rangle ;$$
$$g1(C2(v)) = \hat{e}_b\langle g1(C3(C2(v)))\rangle ;$$
$$g1(C3(v)) = \hat{e}_c\langle g1(C4(v))\rangle ;$$
$$g1(C4(v)) = \hat{e}_d\langle g1(C3(v))\rangle ;$$

To show the effect of our specialisation, we apply T to the call $g1(C2(C1(v)))$ appearing in the first equation:

$$\underbrace{T[\![g1(C2(C1(v)))]\!]}_{\text{replaced by } g21(v)} \begin{aligned} &\rightsquigarrow T[\![g1(C3(C2(C1(v))))]\!] \text{ replaced by } g22(v) \\ &\rightsquigarrow T[\![g1(C4(C2(C1(v))))]\!] \text{ replaced by } g23(v) \\ &\rightsquigarrow T[\![g1(C3(C2(C1(v))))]\!] \text{ folded to } g22(v) \end{aligned}$$

This changes the first equation to the following transformed equivalence.

$$g1(C1(v)) = \acute{e}_a\langle g21(v)\rangle \ ;$$
$$g21(v) \quad = \acute{e}_b\langle g22(v)\rangle \ ;$$
$$g22(v) \quad = \acute{e}_c\langle g23(v)\rangle \ ;$$
$$g23(v) \quad = \acute{e}_d\langle g22(v)\rangle \ ;$$

Notice that the new functions introduced are already consumers in treeless form. The rest of the function calls in the other equations of *g1* can be treated likewise.

4.2 Safe Derivation and Treeless Consumers

At this juncture, the reader may have noticed the impact of our transformation towards building functions that are consumers in treeless form. In this section, we show that if the application of of $\mathcal{T}^F$ does terminate, the equations/functions created are indeed *safe* consumers; furthermore, these consumers are either *treeless*, or can be made into one by further applying $\mathcal{T}$. Thus, deforestation can be effectively performed on the transformed program.

To simplify our presentation, we shall, for the moment, consider as the input language a special class of functions, called *linear-recursive constructor-based* (or LRC) functions:

Definition 11: *LRC-equations & LRC-functions*
Consider a mutual-recursive M-set of CB-functions.
An equation of the function f_d^M (from the M-set) is said to be an LRC-equation if it is of the form:

$$f_d^M(\overrightarrow{p}_i) \quad = \hat{e}\langle f_e^M(\overrightarrow{a}_i)\rangle^M \ ;$$

such that it is linear recursive, *ie.*, at most one M-recursive call in its RHS.

A function f_d^M is said to be an LRC-function if all the equations of the M-set are in the LRC-form.

The results obtained from LRC-functions carry over to CB-functions as well. However, dealing with LRC-functions enables us to focus more on sequence-of-derivation when we discuss the specialisation process, rather than tree-of-derivation, which is usually needed when we consider CB-functions.

Working on LRC-functions, we first define the notion of *safe derivation*, and demonstrate that the equations/functions obtained from a safe derivation are indeed safe consumer. It then becomes obvious that these equations/functions are treeless consumers.

Definition 12: *Safe Derivations*
A sequence-of-derivation is said to be *safe* if it is of finite length and its last function call either is a terminating call or exactly matches some intermediate function call in the sequence. A tree-of-derivation is said to be *safe* if every path of the tree is a *safe* sequence-of-derivation.

Each *safe* derivation (tree or sequence) formed by applying $\mathcal{T}^M$ corresponds to a set of safe equations (either non-cyclic or non-increasing) that can be extracted from the specialisation, as proposed below.

Proposition 3. (Safe Equations from Safe Derivations)
*Each safe sequence-of-derivation (or tree-of-derivation) corresponds to a set
of equations which are either non-cyclic or non-increasing.*

Proof. Consider a sequence-of-derivation of the form:

$$f_{d1}^{M} s_1\langle v1_1, \ldots, v1_{m1}\rangle \rightsquigarrow f_{d2}^{M} s_2\langle v2_1, \ldots, v2_{m1}\rangle \rightsquigarrow \ldots$$
$$\ldots \rightsquigarrow f_{dn}^{M} s_n\langle vn_1, \ldots, vn_{mn}\rangle \rightsquigarrow f_{d(n+1)}^{M} s_{n+1}\langle t_1, \ldots, t_{m(n+1)}\rangle$$

From the specialisation algorithm, $\mathcal{T}^M$, this derivation introduces the following new equations:

$$f_{new1}(v1_1, \ldots, v1_{m1}) = f_{d1}^{M} s_1\langle v1_1, \ldots, v1_{m1}\rangle ;$$
$$f_{new2}(v2_1, \ldots, v2_{m2}) = f_{d2}^{M} s_2\langle v2_1, \ldots, v2_{m2}\rangle ;$$
$$\vdots \qquad\qquad \vdots$$
$$f_{newn}(vn_1, \ldots, vn_{mn}) = f_{d(n+1)}^{M} s_{n+1}\langle t_1, \ldots, t_{m(n+1)}\rangle ;$$

The above equations were also transformed by $\mathcal{T}^M$ (through an unfold followed by define step) to:

$$f_{new1}(..) \quad = e\hat{\imath}_1\langle f_{new2}(v2_1, \ldots, v2_{m2})\rangle^M ;$$
$$f_{new2}(..) \quad = e\hat{\imath}_2\langle f_{new3}(v3_1, \ldots, v3_{m3})\rangle^M ;$$
$$\vdots \qquad\qquad \vdots$$
$$f_{new(n-1)}(..) = e\hat{}_{n-1}\langle f_{newn}(vn_1, \ldots, vn_{mn})\rangle^M ;$$
$$f_{newn}(..) \quad = e\hat{}_n\langle f_{d(n+1)}^{M} s_{n+1}\langle t_1, \ldots, t_{m(n+1)}\rangle\rangle^M ;$$

Regard $\{f_{new1}, .., f_{newn}\}$ as belonging to the same mutual-recursive M-set of functions. Then, each equation of $f_{new1}, .., f_{new(n-1)}$ is a non-increasing equation. There are two cases to consider for the equation of f_{newn}:

- **Case 1:** If $f_{dn}^{M} s_{n+1}\langle t_1, \ldots, t_{m(n+1)}\rangle$ is a terminating function call, then the equation of f_{newn} is a terminating equation, and thus non-cyclic.
- **Case 2:** If $f_{dn}^{M} s_{n+1}\langle t_1, \ldots, t_{m(n+1)}\rangle$ exactly matches a previous function call $f_{dj}^{M} s_j\langle vj_1, \ldots, vj_{mj}\rangle$, then a fold-step by $\mathcal{T}^M$ would yield:

$$f_{newn}(..) \quad = e\hat{}_n\langle f_{newj}(v'_1, \ldots, v'_{mj})\rangle^M ;$$

where $(v'_1, \ldots, v'_{mj})$ are free variables of $s_{n+1}\langle t_1, \ldots, t_{m(n+1)}\rangle$. Such an equation is non-increasing.

Hence, only non-cyclic or non-increasing equations are obtained from safe sequence-of-derivations. The above result carries over to safe tree-of-derivations. □

In the proof above, we observe in the RHS of the new equations created, the function calls are in treeless form, except for the last equation, f_{newn}, which may satisfies Case 2 above. In this case, we further notice that the function call can be unfolded away. Thus, the resulting equations from a safe derivation are treeless consumers. The result again carries over to *safe tree-of-derivations*.

4.3 Cyclic Equations

Specialising calls which match cyclic equations could possibly loop in such a manner that those part of the call arguments beyond certain depth will not be consumed. For instance, given a set of cyclic equations for the form:

$$f_{d1}^M \; s_1\langle v1_1, \ldots, v1_{m1}\rangle = \hat{e_1}\langle f_{d2}^M \; s_2\langle t1_1, \ldots, t1_{m2}\rangle\rangle^M \; ;$$
$$\vdots \qquad\qquad \vdots$$
$$f_{di}^M \; s_i\langle vi_1, \ldots, vi_{mi}\rangle \;\; = \hat{e_2}\langle f_{d(i+1)}^M \; s_{i+1}\langle ti_1, \ldots, ti_{m(i+1)}\rangle\rangle^M \; ;$$
$$\vdots \qquad\qquad \vdots$$
$$f_{dn}^M \; s_n\langle vn_1, \ldots, vn_{mn}\rangle = \hat{e_n}\langle f_{d1}^M \; s_1\langle tn_1, \ldots, tn_{m1}\rangle\rangle^M \; ;$$

Specialising a call, $f_{di}^M(s_i\langle..\rangle)$, will lead to the following cyclic derivation sequence:

$$f_{di}^M \; s_i\langle..\rangle \rightsquigarrow f_{di+1}^M \; s_{i+1}\langle..\rangle \rightsquigarrow .. \rightsquigarrow f_{dn}^M \; s_n\langle..\rangle \rightsquigarrow f_{d1}^M \; s_1\langle..\rangle \rightsquigarrow f_{di}^M \; s_i\langle..\rangle$$

Such a cyclic derivation leading back to the same call pattern shall be referred to as an *endo-derivation*. We shall abbreviate a composite derivation involving one or more steps as:

$$f_{di}^M \; s_i\langle..\rangle \overset{*}{\rightsquigarrow} f_{di}^M \; s_i\langle..\rangle$$

From Theorem 2, we note that cyclic equations must be non-increasing before their functions are considered to be safe consumers. Otherwise, they could result in infinitely many different calls of unbounded size via their recursive loops. We consider how to handle *increasing* cyclic equations next.

4.4 In-situ Increasing Terms & Their Generalisation

In [Hol91], Holst introduced the notion of *in-situ increasing parameters* to characterise the problem of infinitely many different calls of unbounded size. A variant of such notion is defined as follows:

Definition 13: *In-Situ Increasing Cyclic Term/Variable*
Consider an endo-derivation for a function call which matches a cyclic equation with a LHS call pattern $f_i^M \; s_i\langle s'_1, .., s'_m, v_{m+1}, .., v_n\rangle$.
$$f_i^M \; s_i\langle v_1, .., v_m, v_{m+1}, .., v_n\rangle \overset{*}{\rightsquigarrow} f_i^M \; s_i\langle v'_1, .., v'_m, t_{m+1}, .., t_n\rangle$$
The term t_k of call $f \; s_i\langle v'_1, .., v'_m, t_{m+1}, .., t_n\rangle$ is an *in-situ increasing cyclic term* if $(v_k \sqsubset t_k)$.

Correspondingly, a variable v_k of a cyclic equation with LHS pattern $f_i^M \; s_i\langle s'_1, .., s'_m, v_{m+1}, .., v_n\rangle$ is an *in-situ increasing cyclic* variable if one of its derivation sequence (shown above) results in an in-situ increasing cyclic term t_k in the same position as the parameter variable.

Notice that this definition is applicable to arbitrary argument, p_k, instead of variables, v_k, by simply substituting $[p_k/v_k]$.

During the specialisation of cyclic call, in-situ increasing cyclic terms can grow unboundedly. Therefore, *they should be generalised to prevent non-termination*.

We enhance the specialisation algorithm to include generalisation of any in-situ increasing term. This is described informally below (please refer to [CK95] for formal specification):

(u) $\mathcal{T}^F[f(\vec{a})]$

⋮

(u^g) if $f(\vec{a})$ contains in-situ increasing terms $t_1, .., t_n$

$\Rightarrow \mathcal{T}^F[\text{let } (v_i, t_i)_{i=1}^n \text{ in } f(\vec{a'})]$

where $(\vec{a}) = (\vec{a'})[t_1/v_1, \ldots, t_n/v_n]$

⋮

Generalisation should be attempted after the current function call fails to be treated by both rules (u^t) and (u^f), but before it is unfolded (rule (u^u)).

Our method differs from Holst's in three aspects: Firstly, in-situ increment is detected at the level of pattern terms/variables, instead of at the parameter level (Holst's in-situ increasing *parameters*); secondly, our method relies on pattern-matching equations, instead of functions; thirdly, we use on-line instead of off-line generalisation techniques. There are a number of reasons for having these differences. Firstly, pattern-matching equations are more appropriate for call-by-name languages because they help highlight the *strict* parameters of our functions. Secondly, pattern-matching equations allow derivation/analysis to be applied more accurately at the equation level (instead of function level). Lastly, their use make it easier for derivation techniques to permit decreasing variables to cancel out prior increases in variables' depth; this is not achievable by Holst's abstract interpretation technique since an increase followed by a decrease would have been approximated to an increase.

By preventing unbounded growth of the terms in the derivation sequences, we can convert each increasing cyclic equation to a corresponding set of non-increasing equations. Such a procedure is outlined next.

Procedure 2: *Cyclic Derivation and Generalisation*
Consider an increasing cyclic equation:

$f_d^M(\vec{p_i})$ $= \hat{e}\langle f_e^M(\vec{a_i})\rangle^M$;

Deriving a tree-of-derivation on the above equation (using $\mathcal{T}^M$ with generalisation) results in sequences of the form:

$$f_d^M(\vec{p_i}) \rightsquigarrow f_e^M(\vec{a_i}) \rightsquigarrow f_{e1}^M(\vec{a_{i_1}}) \rightsquigarrow .. \rightsquigarrow f_{e(n-1)}^M(\vec{a_{i_{n-1}}}) \rightsquigarrow f_{en}^M(\vec{a_{in}})$$

In each of these derivation sequences, we always generalise in-situ increasing cyclic terms. Each derivation sequence is stopped when it encounters a call $(f_{en}^M(\vec{a_{in}}))$ which

- matches a terminating equation, or
- exactly-matches (modulo variable renaming) a previous call $(f_{em}^M(\vec{a_{im}}))$ in the sequence where $m < n$

If the above procedure stops, each of the derivation sequences is finite: either the last call in the sequence matches a terminating equation, or the sequence is an endo-derivation (because it exactly-matches a previously occurring term). As a result, they represent safe derivation sequences from which it is possible to extract a set of terminating and/or non-increasing equations.

Again, the above procedure for converting each unsafe cyclic equation to its safe equivalence can be applied to arbitrary CB-equation, we now present our final result for CB-functions, as follows:

Theorem 4 (Third Safe Specialisation Result). *Given a M-set of CB-functions which are defined using only:*

- *cyclic equations, and/or*
- *terminating and indirect equations.*

For any list constructor terms $\vec{c}_0$, the specialisation via applying $\mathcal{T}^M[\![f^M(\vec{c}_0)]\!]$ (with rule u^g) always terminates, after we have converted all increasing cyclic equations to corresponding sets of non-increasing equations.

To illustrate the above result, consider a set of cyclic equations:

$$g4(C1(x), y, z) = \hat{e}_a\langle g4(x, z, C3(y))\rangle \ ;$$
$$g4(C2(x), y, z) = \hat{e}_b\langle g4(x, y, z)\rangle \ ;$$

The first cyclic equation is increasing but can be transformed via the following tree-of-derivation. Note the use of indentation to illustrate branching.

$$g4(x, z, C3(y))$$

$$\overset{(u^u)}{[x/C1(x1)]} \ g4(x1, C3(y), C3(z)) \overset{(u^g)}{[C3(y)/w1, C3(z)/w2]} \ g4(x1, w1, w2) \overset{(u')}{\leadsto}$$

$$\overset{(u^u)}{[x/C2(x2)]} \ g4(x2, z, C3(y)) \overset{(u')}{\leadsto}$$

The resulting tree-of-derivation is non-increasing. As a result, the following set of safe non-increasing equations are obtained.

$$g4(C1(x), y, z) = \hat{e}_a\langle g4a(x, y, z)\rangle \ ;$$
$$g4(C2(x), y, z) = \hat{e}_b\langle g4(x, y, z)\rangle \ ;$$
$$g4a(C1(x1), y, z) = \hat{e}_a\langle let \ (w1, C3(y)) \ (w2, C3(z)) \ in \ g4(x1, w1, w2)\rangle \ ;$$
$$g4a(C2(x2), y, z) = \hat{e}_b\langle g4a(x2, y, z)\rangle \ ;$$

5 Conclusion and Related Work

This main theme of this paper is on the use of a specialisation algorithm to help obtain more safe consumers for deforestation. The net effect of the algorithm is the specialisation of functions with constructor terms as arguments in order to obtain equivalent functions which are treeless consumers. Such a transformation actually has a very similar outcome as that of Thiemann's technique for eliminating repeated pattern-matching by partial evaluation [Thi93]. Our novelty here is the use of a set of very simple but powerful syntactic analyses to ensure termination, and its subsequent use as a pre-processor for deforestation.

Initially, a simple off-line analysis, based on *non-increasing* consumers, was proposed to help identify functions which could act as safe consumers of constructor terms during specialisation. Later, we obtained better consumers by classifying *equations* into either *terminating*, *indirect*, or *cyclic*. Classification by equations instead of functions enables us to capture more accurately the recursive behaviour of the functions. We show that *terminating* and *indirect* equations can be *increasing* without causing non-termination in specialisation, while *cyclic* equations have to be *non-increasing*. A new on-line analysis, based on in-situ increasing cyclic criterion, was also proposed to help transform each increasing cyclic equation to an equivalent set of non-increasing equations. This step helps obtain more safe consumers from unsafe ones.

Compared to the more sophisticated semantics-based approaches (such as [Hol91, Sør94]), our syntactic-based analyses are considerably simpler, but surprisingly powerful. The resulting treeless consumers can be used to help achieve better deforestation. Future work will address pre-processing techniques for better producer functions.

A final remark. Our modular approach to developing transformers is also useful for extending the present result to higher-order language. In [CD94], we proposed a transformation algorithm which is able to transform *most* higher-order expressions to equivalent first-order form. This transformer has been used as a pre-processor to help extend the deforestation algorithm to higher-order programs. Similarly, it is possible to allow the specialisation algorithm to take its input from the higher-order removal transformer, and hence provide a simple extension of this result to the higher-order case.

6 Acknowledgements

We thank Morten Sørensen and Peter Thiemann for detailed comments to an earlier draft of this paper. We also thank Simon Peyton-Jones as well as the referees of PLILP for their insightful comments. The support of NUS research grants, RP920614 & RP930611, are acknowledged.

References

[CD94] Wei-Ngan Chin and John Darlington. A higher-order removal method. *Submitted for Publication*, September 1994.

[Chi92] Wei-Ngan Chin. Safe fusion of functional expressions. In *7th ACM Lisp and Functional Programming Conference*, pages 11–20, San Francisco, California, June 1992.

[CK95] Wei-Ngan Chin and Siau-Cheng Khoo. Better consumers for program specialisation. Technical report, Dept of IS/CS, NUS, August 1995.

240

[GLPJ93] A. Gill, J. Launchbury, and S. Peyton-Jones. A short-cut to deforestation. In *6th ACM Conference on Functional Programming Languages and Computer Architecture*, Copenhagen, Denmark, June 1993.

[GS94] R. Glück and Morten H. Sørensen. Partial deduction and driving are equivalent. In *PLILP*, Madrid, Spain, (Lect. Notes Comput. Sc., Berlin Heidelberg New York: Springer, 1994.

[Hol91] Carsten Kehler Holst. Finiteness analysis. In *5th ACM Conference on Functional Programming Languages and Computer Architecture*, pages 473–495, Cambridge, Massachusetts, August 1991.

[PP91] M. Proietti and A. Pettorossi. Unfolding - definition - folding, in this order for avoiding unnecessary variables in logic programs. In *Proceedings of PLILP*, Passau, Germany, (Lect. Notes Comput. Sc., vol 528, pp. 347–258) Berlin Heidelberg New York: Springer, 1991.

[RFJ89] C. Runciman, M. Firth, and N. Jagger. Transformation in a non-strict language : An approach to instantiation. In *Glasgow Functional Programming Workshop*, August 1989.

[SF93] T. Sheard and L. Fegaras. A fold for all seasons. In *6th ACM Conference on Functional Programming Languages and Computer Architecture*, Copenhagen, Denmark, June 1993.

[Sør94] Morten H. Sørensen. A grammar-based data-flow analysis to stop deforestation. In *Colloquium on Trees and Algebra in Programming (CAAP) LNCS 787*, Edinburgh, April 1994.

[Thi93] Peter Thiemann. Avoiding repeated tests in pattern-matching. In *3rd International Workshop on Static Analysis*, Padova, Italy, (Lect. Notes Comput. Sc., vol 724, pp. 141–152) Berlin Heidelberg New York: Springer, 1993.

[Wad88] Phil Wadler. Deforestation: Transforming programs to eliminate trees. In *European Symposium on Programming*, pages 344–358, Nancy, France, March 1988.

Efficient Compile-Time Garbage Collection for Arbitrary Data Structures

Markus Mohnen

Lehrstuhl für Informatik II, Aachen University of Technology
Ahornstraße 55, 52056 Aachen, Germany
email: markusm@zeus.informatik.rwth-aachen.de

Abstract. This paper describes a *compile-time garbage collection* (ct-gc) method in the setting of a first-order functional language with data structures. The aim is to obtain information on positions in a program where certain heap cells will become obsolete during execution. Therefore we develop an abstract interpretation for the detection of *inheritance information* which allows us to detect whether the heap cells of an argument will be propagated to the result of a function call. The abstract interpretation itself is independent of the evaluation strategy of the underlying language. However, since the actual deallocations take place after termination of functions, the information obtained by the abstract interpretation can only be applied in an eager context, which may be detected using strictness analysis in a lazy language. In order to increase efficiency we show how the number of recomputations can be decreased by using only parts of the abstract domains. The worst case time complexity is essentially quadratic in the size of the program. We illustrate the method developed in this paper with several examples and we demonstrate how to use the results in an eager implementation. Correctness of the analysis is considered, using a modified denotational semantics as reference point. Experimental results demonstrate the practicability of our method. A main goal of our work is to keep both the run-time and the compile-time overhead as small as possible.

1 Introduction

One of the advantages of modern functional languages is the ability to work with dynamic data structures without the necessity to control memory allocations explicitly. On the other hand, this prevents a programmer from expressing the reusability of heap cells of intermediate data structures. However, since the available memory is restricted in concrete computers, there must be some means of deallocating obsolete memory cells. But well-known garbage collection mechanisms make no (or only little) use of the special structure of the underlying functional programs.

We present an *abstract interpretation* which exploits this special structure. The underlying observation is that data cells which were created specially for a certain function call become obsolete, if they are not inherited to the *function result*, i.e. if they are not reachable from the top cell of the result. In contrast to other ctgc approaches (like for instance [JM89]) we do not try to reuse obsolete cells as soon as possible, i.e. during function execution, because this would cause severe changes in the abstract machine and some runtime overhead. Therefore, application of our technique is only possible in 'eager' situations, i.e. either in an eager language or at positions in lazy programs which were detected as eager by a strictness analysis.

Furthermore, we show that our method can be implemented *very efficiently* by using test arguments instead of the full domains. The worst case complexity for general programs is essentially $O(n^2)$ where n is the number of function definitions of the program. Especially for realistic programs we will show that the computation of the fixpoint can be done almost in *linear time*.

We will present experimental results which prove that the memory consumption is dramatically improved. Moreover, in combination with traditional garbage collection (gc), even the run-time is improved, which is due to less frequent gc cycles.

We finish this introduction by an outline of the contents of the paper. The next section gives a more detailed intuitive description of our method. Section 3 formalises this intuition and emphasises the problem of arbitrary data structures and their finite representation. After that, we give a brief discussion of the correctness of our method and explain how to use the results of the abstract interpretation in an eager implementation. The next section contains a discussion of efficiency issues. We show how to decrease the time spent for computing the fixpoint and give experimental results. The paper concludes with a comparison with related work and some prospects for further research.

2 Intuitive Description

We want to address the problem in the setting of a first-order functional language with constructors. For the moment, we will assume eager evaluation, although the abstract interpretation itself is also applicable to programs of a lazy language. A constructor on the right hand side on a function definition causes the allocation of a corresponding heap cell, if it is encountered during the execution. We assume *boxed* representation of basic types, i.e. the heap cell representing a constructor contains just pointers to other heap cells, which may represent constructors or basic values. The loosening of some of these restrictions will be discussed in Sec. 8.

The general idea of the optimisation can be summarised in the following way: if there is a subterm $f(g(t))$ on the right hand side of a function definition and we know that f does *not* inherit parts of the result of the evaluation of $g(t)$, then deallocate all cells belonging to these parts which were allocated during the execution of $g(t)$. The deallocation takes place when the call of f terminates. In an eager implementation this can easily be done, since the termination point of a function is fixed.

In order to illustrate which kind of information we want to derive, we consider the following example:

```
datatype ListOfInt ::= Nil | Cons(int,ListOfInt)
append : ListOfInt x ListOfInt -> ListOfInt
append(Nil, L) := L
```

```
append(Cons(a,L1), L2) := Cons(a,append(L1,L2))
```

An obvious observation is that append sets up copies of all Cons cells contained in the first argument, but *not* of the int values stored there. We say that append *inherits* the second level of its first argument and all levels of its second argument, but not the first level of its first argument. The term *level* corresponds to the definition of the underlying type. For instance, the type

```
datatype ListOfList ::= LNil | LCons(ListOfInt, ListOfList)
```

has three levels: one for the list built up from LNil and LCons, one for the ListOfInt constructors and one for the actual int entries.

The basic idea of the abstract interpretation, is to represent the (*infinite*) set with elements of *variable* size which corresponds to a type by means of a *finite* domain of elements with a fixed size, where each element has one component for each level of the type. This component is a binary value, where the value 1 is used to indicate that the corresponding level may contain a cell which is part of one of the parameters of the function call. The value 0 guarantees that there is no such cell, i.e. all cells are new.

In the above example we choose $\mathfrak{A}_{\texttt{ListOfInt}} = \{0, 1\}^2$ as abstract domain for lists over integers. The components represent all constructors resp. all entries in a list.

The corresponding abstract functions are essentially built from the abstract constructors. The formal parameters of a constructor can be divided into two classes: those which are of the same type as the constructor and those which are not. E.g. the constructor Cons for lists has the second argument in the first class and the first argument in the second. The abstract interpretation of a constructor is performed in two stages:

1. All parameters of class two are put (at appropriate positions) into a new value. All other positions in this value are set to 0. These are the binary values indicating the inheritance of the constructors and the positions for parameters of class two of other constructors for this sort. This step copes with the creation of a new heap cell.
2. The maximum (wrt. the order within the abstract domain) of this value and all parameters of class one is returned as result. In this way the information already contained is those parameters in preserved.

If one of those parameters contains a nonzero entry at the position for the constructors, then the result will also. This captures our intuition, that if at least one cell of a list is inherited, we approximate that all cells are inherited. In our example, the constructors are abstractly interpreted by:

$$A[\![\texttt{Nil}]\!] : \mathfrak{A}_{\texttt{ListOfInt}} \quad A[\![\texttt{Cons}]\!] : \mathfrak{A}_{\texttt{int}} \times \mathfrak{A}_{\texttt{ListOfInt}} \to \mathfrak{A}_{\texttt{ListOfInt}}$$
$$A[\![\texttt{Nil}]\!] = (0,0) \quad A[\![\texttt{Cons}]\!](a,l) = (0,a) \sqcup l$$

In Sec. 6 we will show that we do not need to determine the values of the abstract functions for all possible argument combinations. Instead, we will use a single *test argument* for each level of each argument to determine whether this particular level has an impact on the result, i.e. if there are elements from this level, which are part of the result. The test arguments are those with exactly one nonzero entry at a position representing this level. The result can only contain nonzero entries if a nonzero entry in the initial test argument is propagated. For example, the abstract values of append for all possible test arguments are:

$$A[\![append]\!]((0,0),(0,1)) = (0,1) \quad A[\![append]\!]((0,0),(1,0)) = (1,0)$$
$$A[\![append]\!]((0,1),(0,0)) = (0,1) \quad A[\![append]\!]((1,0),(0,0)) = \underline{(0,0)}$$

We can see that the abstract value of append for the test argument $((1,0),(0,0))$ is $(0,0)$. This means that no Cons or Nil of the first level of the first argument is inherited to the result of append

In order to demonstrate the intended use, we expand our example from above by definitions for the functions filterle, filtergt and quicksort:

```
filterle : ListOfInt x int -> ListOfInt
filterle(Nil,b) := Nil
filterle(Cons(a,L),b) := if a<=b then Cons(a,filterle(L,b))
                                  else filterle(L,b)
filtergt (* analogous *)
quicksort : ListOfInt -> ListOfInt
quicksort(Nil) := Nil
quicksort(Cons(a,L)) := append(quicksort(filterle(L,a)),
                               Cons(a,quicksort(filterge(L,a))))
```

We can infer that all these functions do not inherit the constructors of their ListOfInt–typed argument. During the execution of a call to quicksort with a parameter not equal to Nil, assuming left-to-right eager evaluation, the expression filterle(L,a) is evaluated first, creating an intermediate data structure which is fed into a recursive call of quicksort. All constructors of the intermediate list are copied by quicksort, which means that we can deallocate everything which was allocated for the intermediate list *after* the termination of quicksort. Accordingly, every intermediate result can be deallocated after the surrounding function call has terminated, except for the outermost append (this will maybe be deallocated on the level of the calling function) and the second quicksort (because append does not copy its second argument).

3 Abstract Interpretation

In this section we will formalise what we have described informally in the previous section. First, the abstract syntax of our language will be presented. In order to simulate pattern matching, a set of selector and test functions will be be associated with each constructor. Thereafter, we will discuss how to choose the abstract domains, where we focus on the problems arising due to arbitrary data structures. The abstract interpretation is basically determined by the way the constructors and selectors are interpreted w.r.t. the abstract domains.

3.1 Abstract Syntax

Let $S = BS \cup CS$ be the set of *sorts*, where BS and CS are finite, disjoint sets of *(basic) sorts*, containing at least a sort bool $\in BS$, and *constructed sorts*. Correspondingly, we assume that there are disjoint families of *variables* $X = (X^s \mid\mid s \in S)$, *defined functions* $DF = (DF^{s_1,\ldots,s_n \to s} \mid\mid n \in \mathbb{N}_0, s_1,\ldots,s_n, s \in$

S), *basic functions* $\Omega = (\Omega^{bs_1,\dots,bs_n \to bs} \| n \in \mathbb{N}_0,\ bs_1,\dots,bs_n,\ bs \in BS)$ and *constructors* $C = (C^{s_1,\dots,s_n \to cs} \| n \in \mathbb{N}_0,\ s_1,\dots,s_n \in S$ and $cs \in CS)$. From the constructors we can derive

1. the family of *selectors* $CSel := (CSel^{cs \to s} \| cs \in CS$ and $s \in S)$ where the set of selectors of type $cs \to s$ is defined by

$$CSel^{cs \to s} := \{c_{sel}^j \| \exists c \in C^{s_1,\dots,s_n \to cs},\ 1 \le j \le n\colon s_j = s\}$$

2. the family of *constructor tests* $CTest := (CTest^{cs \to \text{bool}} \| cs \in CS)$ where the set of constructor tests of type $cs \to \text{bool}$ is defined by

$$CTest^{cs \to \text{bool}} := \{\text{is-}c \| \exists c \in C^{s_1,\dots,s_n \to cs}\}$$

We use these auxiliary functions to simulate pattern matching. We define the family of all *expressions* over $X, DF, \Omega, C, CSel$ and $CTest$ by $E := (E^s \| s \in S)$, where the sets E^s are defined by:

1. $X^s \subseteq E^s$
2. $f \in (\Omega \cup C \cup CSel \cup CTest)^{s_1,\dots,s_n \to s}$ and $e_i \in E^{s_i} \implies f(e_1,\dots,e_n) \in E^s$
3. $F \in DF^{s_1,\dots,s_n \to s}$ and $e_i \in E^{s_i} \implies F(e_1,\dots,e_n) \in E^s$
4. $e \in E^{\text{bool}}, e_1, e_2 \in E^s \implies$ if e then e_1 else e_2 fi $\in E^s$

A *program* is a finite set of *definitions* (one for each defined function)

$$F(x_1,\dots,x_n) := e$$

with $F \in DF^{t_1,\dots,t_n \to t}$, $x_i \in X^{t_i}$ and $e \in E^t$ with variables $\{x_1,\dots,x_n\}$.

The intended declarative semantics is the usual fixpoint semantics. The data representation we have in mind is a graph reduction machine where the graph is represented by means of a heap. If a constructor is evaluated during the execution, a heap cell representing this constructor is created on the heap. Arguments are already represented in the heap, and so references to the arguments are simply copied into the cell. Note, that we assume a *boxed* representation of basic values. Accordingly, variables always refer to heap cells.

3.2 Abstract Domains

Assume that we have sets V_{bs} for each basic sort $bs \in BS$. Denotational semantics uses sets V_{cs}, which are fixpoints of the equations:

$$V_{cs} = \bigcup_{c \in C^{s_1,\dots,s_n \to cs}} \{c(v_1,\dots v_n) \| v_i \in V_{s_i}\} \qquad \forall cs \in CS$$

This reflects the free interpretation of constructors. In the strict case, these sets consist of the finite terms. Infinite terms are added to obtain the sets for the non-strict semantics. But, of course, as we want to do an abstract interpretation we have to guarantee termination, so we need at least domains with only finite ascending chains instead. Furthermore, since abstract values with variable size can only be implemented inefficiently, we prefer finite domains, where values can be represented with a fixed size. The first (obvious) step is to replace V_{bs} by $I_{bs} := \{0, 1\}$, since our interest is focused on whether parts of the arguments are inherited to the result, and not on the actual value of the result. Later on, we will use test values with a single '1' for a particular level of an argument. If we can observe that this '1' propagates through the computation, we have detected (possible) inheritance w.r.t. this level. Additionally, we add a value to each constructor, indicating its inheritance behaviour. This yields sets I_{cs} in a similar way as fixpoints of equations.

We still have the problem of infinity, since we have essentially only reduced, for instance, the set of lists over integers to the set of lists over $\{0, 1\}$. One possibility for obtaining finite sets, we call it the *horizontal* approach, is to build the elements of the abstract domain by gratuitously restricting the *height* of the elements of I_s, for instance only the list of length 7 or less. This approach is used in [JM89]. The obvious drawback is the selection of the threshold, since there is no possibility to derive a good choice from the program itself.

We propose a different approach, which has a more *vertical* appearance. If I_s has n levels, we define $A_s \simeq \{0, 1\}^n$ and we map an element of I_s to an element of A_s by taking the maximum of all entries of level i as entry for the i-th

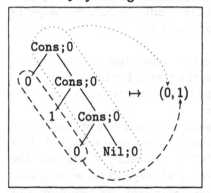

component (see Fig. 1). There is an obvious simplification of the domains, which could be applied here. Instead of $\{0, 1\}^n$ we could choose $\{0, \ldots, n\}$, with a value k representing the fact that levels k and above are not inherited. This approach is taken in [Hug92] and [PG92]. The underlying observation is that it is not possible to define a functional program which inherits a certain constructor c and simultaneously does not inherit a constructor d below c. This is due to the fact that functional programs can not change parameters stored in existing heap cells for con-

Fig. 1.: $I_{\mathrm{ListOfInt}}$ to $A_{\mathrm{ListOfInt}}$

structors. But there is a drawback: this approach works for list structures, but *not* for the general case. We will come back to this point later.

Our first attempt to define the abstract sets starts with setting $A_{bs} := \{0, 1\}$ for all basic sorts $bs \in BS$. We then can define the sets A_{cs} for the constructed sorts as the fixpoints of

$$A_{cs} = \{0, 1\} \times \prod_{c \in C^{s_1, \ldots, s_n \to cs}} \prod_{\substack{1 \le i \le n \\ s_i \ne cs}} A_{s_i} \qquad \forall cs \in CS$$

In order to evaluate the first Cartesian product $\prod$ we need an order on all constructors of target sort cs. The second product filters out all direct recursiveness since we want to combine all constructors of cs in one level. If we apply this to our example, we can compute that $A_{\mathrm{ListOfInt}} = \{0, 1\}^2$. Note that we assume that a $\prod$ over an empty index creates a set, which is a neutral element for subsequent Cartesian products. Therefore, these parts do not contribute to the resulting domain. E.g. the constructor Nil has no influence.

But there is a drawback: it is possible that we have a constructed sort cs,

where we get $|A_{cs}| = \infty$. A minimal example for this consists of the data defin-
itions:

datatype T1 ::= c1 T2   datatype T2 ::= c2 T1

which are valid definitions in Miranda or Haskell. The corresponding sets are
empty for strict semantics and both have exactly one element of infinite length
for non-strict semantics:

$$V_{T1} = \{c1(c2(c1(\cdots)))\} \quad \text{and} \quad V_{T2} = \{c2(c1(c2(\cdots)))\}$$

Accordingly, we get the abstract set equations:

$$A_{T1} = \{0,1\} \times A_{T2} \quad \text{and} \quad A_{T2} = \{0,1\} \times A_{T1}$$

which have the unique solution $A_{T1} = A_{T2} = \{0,1\}^\omega$, the set of all infinite
sequences of binary numbers.

Of course, sets of this kind are not useful for abstract interpretation, because
in order to guarantee termination of the abstract interpretation we need sets
with only finite ascending chains under an appropriate order.

Therefore, we must enhance our notion. The problem of our first attempt is
that *direct recursion* and *indirect recursion* are handled in different ways. In
the above example, it would be convenient to choose $A_{T1} = A_{T2} = \{0,1\}^1$: one
level for all occurrences of c1 and c2.

The indirect recursion, or, to be more precise, the fact that the constructed
sorts need not form a proper hierarchy is also the reason for the failure of the
simplified approach. It assumes that the abstract levels of the type can be ordered
in a single chain, which is not always possible

The next definition will enable us to express indirect recursion of constructed
sorts: Let $cs_1, cs_2 \in CS$. We say that cs_1 *depends on* cs_2 ($cs_1 \leftarrow cs_2$) iff there
exists a constructor $c \in C^{s_1,\dots,s_{i-1},cs_2,s_{i+1},\dots,s_n \to cs_1}$. As usual, $\leftarrow^*$ denotes the
transitive and reflexive closure of $\leftarrow$. If we have $cs_1 \leftarrow^* cs_2$ and $cs_2 \leftarrow^* cs_1$, then
we say that cs_1 and cs_2 are *mutually recursive dependent*. By definition, this
is an equivalence relation, and we denote the equivalence class of $cs \in CS$ by

$$[cs] := \{cs' \in CS \,||\, cs' \leftarrow^* cs \text{ and } cs \leftarrow^* cs'\}$$

Additionally, we define $[bs] := \{bs\}$ and $A_{[bs]} := \{0,1\}$ for all basic sorts $bs \in BS$.
Again, we can obtain sets $A_{[cs]}$ by the equations:

$$A_{[cs]} = \{0,1\} \times \prod_{cs' \in [cs]} \prod_{c \in C^{s_1,\dots,s_n \to cs'}} \prod_{\substack{1 \le i \le n \\ [s_i] \ne [cs]}} A_{[s_i]}$$

These are suitable sets for abstract interpretation, since we have the following
lemma:

Lemma 1. $|A_{[cs]}| < \infty$ *for all* $cs \in CS$.

If we use this definition for the example, we get $[T1] = [T2] = \{T1, T2\}$ and
$A_{[T1]} = \{0,1\}$. Additionally, this notion is compatible with our first attempt, if
we define $A_{cs} := A_{[cs]}$.

Now we are in the position to define our abstract domains. Note that, for
each $s \in CS \cup BS$, A_s can be "flattened" to a set $\{0,1\}^n$ for some $n \in \mathbb{N}$.

Thus we can define a partial order $\leq_s$ by "bitwise" comparison. The resulting structure $\mathfrak{A}_s := \langle A_s, \leq_s \rangle$ is isomorphic to $\langle \mathfrak{P}(\{1,\ldots,n\}), \subseteq \rangle$, which means that we have a complete lattice with bottom element $\bot_s = (0,\ldots,0)$, top element $\top_s = (1,\ldots,1)$, and lub-operator which is a 'bitwise or'

$$(a_1,\ldots,a_n) \sqcup_s (b_1,\ldots,b_n) = (a_1 \vee b_1,\ldots,a_n \vee b_n)$$

Figure 2 shows the abstract domains for our example types ListOfInt and ListOfList, and the relation between ListOfList and $\mathfrak{A}_{\text{ListOfList}}$.

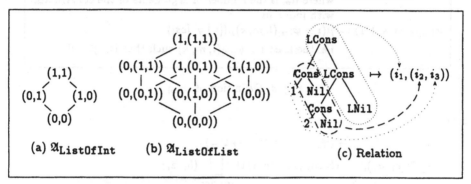

(a) $\mathfrak{A}_{\text{ListOfInt}}$ (b) $\mathfrak{A}_{\text{ListOfList}}$ (c) Relation

Fig. 2. Abstract domains for ListOfInt and ListOfList

3.3 Interpretation of Constructors

In order to formalise which parts of an abstract value are affected by a *particular constructor (or selector)* of a *particular type* of an equivalence class $[cs]$, we need arbitrary, but fixed orderings on all constructed types of class $[cs]$ and on all constructors of a target type in $cs' \in [cs]$. More formally, we assume that $[cs] = \{cs_1,\ldots,cs_{l_{cs}}\}$ and $C^{cs} := (C^{s_1,\ldots,s_n \to cs} \parallel n \in \mathbb{N}_0, s_1,\ldots,s_n \in S) = \{c_1,\ldots,c_{m_{cs}}\}$ for all $cs \in CS$. Implicitly, we have already used these orderings in the definition of $A_{[cs]}$.

In order to locate those parameters which are not directly recursive, we need the following auxiliary functions. Let $c \in C^{s_1,\ldots,s_n \to cs}$ be such that cs has index l in the order of the sorts and c has index m in the order of constructors. We define $\varphi_{[cs]} : \mathbb{N}_0^3 \to \mathbb{N}_0$ (see Fig. 3) such that

$\varphi_{[cs]}(l,m,k) = j$ iff t_k is the non-recursive parameter of c which has the position j in the second component of the abstract values of $A_{[cs]}$.

These preparations are sufficient to give a definition for the abstract meaning of selectors:

$$A : CSel^{cs \to s} \to (\mathfrak{A}_{cs} \to \mathfrak{A}_s)$$

$$A[\![c_{s \bullet l}^k]\!]((b,\bar{a})) := \begin{cases} (b,\bar{a}) & \text{if } [cs] = [s_k] \\ \text{proj}_{\varphi_{[cs]}(l,m,k)}(\bar{a}) & \text{otherwise} \end{cases}$$

if $c \in C^{s_1,\ldots,s_n \to cs}$ such that cs has index l in the order of the sorts and c has index m in the order of constructors

With this definition we now can easily define the abstract meaning of construc-

$$\varphi_{[cs]}(0, m, k) := 0$$
$$\varphi_{[cs]}(l+1, 0, k) := \varphi_{[cs]}(l, m_l, n'_l)$$

where m_l is the number of constructors of the sort with index l, and n'_l the number of arguments of the last constructor of this sort

$$\varphi_{[cs]}(l, m+1, 0) := \varphi_{[cs]}(l, m, n_m))$$

where n_m is the number of arguments of the constructor with index m

$$\varphi_{[cs]}(l, m, k+1) := \mu(i > \varphi_{[cs]}(l, m, k)).([t_i] \neq [cs])$$

i.e. the least $i > \varphi_{[cs]}(l, m, k)$ such that $[t_i] \neq [cs]$

Fig. 3. Definition of $\varphi_{[cs]}$

tors:

$$A : C^{s_1, \ldots, s_n \to cs} \to (\mathfrak{A}_{s_1} \times \cdots \times \mathfrak{A}_{s_n} \to \mathfrak{A}_{cs})$$
$$A[\![c]\!]((b_1, \bar{a}_1), \ldots, (b_n, \bar{a}_n)) := (b, \bar{a}) \sqcup \bigsqcup_{[s_i]=[cs]} (b_i, \bar{a}_i)$$

if $c \in C^{s_1, \ldots, s_n \to cs}$ such that cs has index l in the order of the sorts and c has index m in the order of constructors and a is a minimal value such that for all $1 \leq k \leq n$ holds: $A[\![c_{sel}^k]\!]((b, \bar{a})) = (b_{\varphi_{[cs]}(l, m, k)}, \bar{a}_{\varphi_{[cs]}(l, m, k)})$

Firstly, a new vector $(b, \bar{a})$ is created by placing those arguments, which do not represent references to structures of the same class, into a vector filled with 0. Secondly, the information on references to the same class, which is contained in the remaining arguments are taken into account by building the maximum.

Obviously, the only way to set an entry to 1 is that there is a 1 in the parameters of the abstract function.

3.4 The Complete Abstract Interpretation

We are now able to extend the interpretation of constructors to the interpretation of a program. Firstly, we define the abstract meaning of each basic function by:

$$A : \Omega^{bs_1, \ldots, bs_n \to bs} \to (\mathfrak{A}_{bs_1} \times \cdots \times \mathfrak{A}_{bs_n} \to \mathfrak{A}_{bs})$$
$$A[\![f]\!](a_1, \ldots a_n) := 0$$

This is reasonable since we can assume that the result of a basic function is always represented in a newly created heap cell. Similarly, we can define the abstract meaning of the constructor test functions:

$$A : CTest^{cs \to \text{bool}} \to (\mathfrak{A}_{cs} \to \mathfrak{A}_{\text{bool}}) \qquad A[\![\text{is-}c]\!](a) := 0$$

The definition of the abstract semantic of an expression is as usual; let

$$\text{Env}_X := (X^s \to \mathfrak{A}_s \mid\mid s \in S)$$

be the family of *variable assignments* and

$$\text{Env}_{DF} := (DF^{s_1, \ldots, s_n \to s} \to (\mathfrak{A}_{s_1} \times \cdots \times \mathfrak{A}_{s_n} \to \mathfrak{A}_s) \mid\mid n \in \mathbb{N}, s_1, \ldots, s_n, s \in S)$$

be the family of *function assignments*. The meaning of an expression is inductively defined by:

$$A : E^s \times \mathrm{Env}_X \times \mathrm{Env}_{DF} \to \mathfrak{A}_s \quad \forall s \in S$$

$$A[\![x]\!](\chi, \varrho) := \chi(x) \quad \text{if } x \in X \quad A[\![f(e_1, \ldots, e_n)]\!](\chi, \varrho) := A[\![f]\!](A[\![e_1]\!], \ldots, A[\![e_n]\!])$$

$$\text{if } e_i \in E^{s_i}, f \in (\Omega \cup C \cup CSel \cup CTest)^{s_1, \ldots, s_n \to s}$$

$$A[\![F(e_1, \ldots, e_n)]\!](\chi, \varrho) := \varrho(F)(A[\![e_1]\!], \ldots, A[\![e_n]\!])$$

$$\text{if } e_i \in E^{s_i}, F \in DF^{s_1, \ldots, s_n \to s}$$

$$A[\![\text{if } e \text{ then } e_1 \text{ else } e_2 \text{ fi} \in E^s]\!](\chi, \varrho) := A[\![e_1]\!](\chi, \varrho) \sqcup A[\![e_2]\!](\chi, \varrho)$$

$$\text{if } e \in E^{bool}, e_1, e_2 \in E^s$$

Given a program $P = \{F_i(x_{i,1}, \ldots, x_{i,n_i}) := e_i \parallel 1 \leq i \leq p\}$ with $F_i \in DF^{s_{i,1}, \ldots, s_{i,n_i} \to s_i}$, $x_{i,j} \in X^{s_{i,j}}$ and $e_i \in E^{s_i}$ with free variables $x_{1,i}, \ldots, x_{i,n_i}$ we define the semantics of each F_i as a function

$$A[\![F_i]\!] : \mathfrak{A}_{s_{i,1}} \times \cdots \times \mathfrak{A}_{s_{i,n_i}} \to \mathfrak{A}_{s_i}$$

These functions can be computed as the least fixpoint of the equations

$$A[\![F_i]\!](a_{i,1}, \ldots, a_{i,n_i}) = A[\![e_i]\!]([x_{i,j}/a_{i,j}], [F_k/A[\![F_k]\!]])$$

Since the underlying domains do not have infinitely ascending chains and all abstract functions are monotonous, we can compute this fixpoint according to the Theorem of Knaster and Tarski.

4 Correctness

Due to the restrictions in space we can only give an outline of the correctness proof. The justification of the abstract interpretation A is done w.r.t. a modified non-strict denotational semantics. It is sufficient to proof the correctness for a non-strict semantics only, since the property of inheritance is only of interest if the function terminates. Since a function which terminates for strict semantics yields the same value with non-strict semantics, we can directly reuse the theorem.

In order to formalise that the result of a function contains parts of the arguments, we must be able to distinguish between components of the original parameters and copies of these components. This is not possible within the domains of the usual denotational semantics $I[\![.]\!]$, which are the sets of finite and infinite partial terms $V^s_{\perp,\infty}$. Therefore the domains $V^s_{\perp,\infty}$ are annotated with binary values at each node. This leads to new domains $\hat{V}^s_{\perp,\infty}$, and we can define a modified semantics $\bar{I}[\![.]\!]$ on these domains, which actually does not use the annotations, except that all data created by the modified semantics is annotated with 0. Therefore it is easy to show that the original semantics is equivalent to the modified one, if only arbitrary annotations are added.

If we annotate parts of the input with '1' we can observe whether this tag is propagated to the result of the function. This makes it possible to distinguish between original input and copies of it. We relate the annotated domains $V^s_{\perp,\infty}$

with the abstract domains $\mathfrak{A}_s$ via a family of functions

$$\text{abst} = (\text{abst}^s : V^s_{\bot,\infty} \to \mathfrak{A}_s \parallel \forall s \in S)$$

These functions create a component of the abstract value by collecting all corresponding annotations in the annotated value and combines them via the maximum. It is easy to extent these functions to functions $\text{abst}^{\text{env}\chi}$ and $\text{abst}^{\text{env}DF}$ which map annotated environments to abstract environments.

For all expressions $e \in E^s$ and all environments $\bar\chi$, $\bar\varphi$ we can prove that

$$\text{abst}^s(\bar{\text{I}}[\![e]\!](\bar\chi, \bar\varphi)) \leq_s A[\![e]\!](\text{abst}^{\text{env}\chi}(\bar\chi), \text{abst}^{\text{env}DF}(\bar\varphi))$$

Especially, if we have $A[\![e]\!](\chi, \varphi) = \bot_s$ for some abstract environments χ and φ, we know that there can not be corresponding annotated environments $\bar\chi$ and $\bar\varphi$ such that $\bar{\text{I}}[\![e]\!](\bar\chi, \bar\varphi)$ contains a non-zero annotation. With other words, the result of the evaluation of e does not contain parts of the input.

5 How to Use the Results

We adopt a notion known from strictness analysis (see for instance [Myc80]) in order to use the abstract interpretation.

Given a sort $s \in S$ we define the set of *test vectors* for s

$$T_s := \{(1, 0, \ldots, 0), (0, 1, \ldots, 0), \ldots, (0, 0, \ldots, 1)\} \subset \mathfrak{A}_s$$

as those vectors from $\mathfrak{A}_s$ with exactly one 1 entry. The set of test arguments T_F for a function $F \in DF^{s_1, \ldots, s_n \to s}$ consists of those arguments where there is exactly one 1 in a single argument position:

$$(T_{s_1} \times \{\bot_{s_2}\} \times \cdots \times \{\bot_{s_n}\}) \cup (\{\bot_{s_1}\} \times T_{s_2} \times \cdots \times \{\bot_{s_n}\}) \cup \cdots \cup (\{\bot_{s_1}\} \times \{\bot_{s_2}\} \times \cdots \times T_{s_n})$$

If $A[\![F]\!](t_{i,j}) = \bot_s$, where $t_{i,j} \in T_F$ is the test argument with a 1 at the j-th position for the i-th argument, we can be sure, that there is no member of level i of the j-th argument inherited to the result of the function F, since this would propagate the 1 to the result of the abstract function.

Recall the quicksort example from Sec. 2. The evaluation of $A[\![\text{append}]\!]$ for the four elements of T_{append} yields:

$$A[\![\text{append}]\!]((0,0),(0,1)) = (0,1) \quad A[\![\text{append}]\!]((0,0),(1,0)) = (1,0)$$
$$A[\![\text{append}]\!]((0,1),(0,0)) = (0,1) \quad A[\![\text{append}]\!]((1,0),(0,0)) = (0,0)$$

We can conclude that the constructors of the first argument are not inherited to the result. Similarly, we obtain that the functions filterle, filterge and quicksort do not inherit the constructors of their argument. With this knowledge we can insert commands for deallocation into the code for quicksort immediately after the recursive calls. At those positions every constructor created by filter* can be deallocated, since it is not be referenced from the result of quicksort.

The general scheme is to detect positions in a given program, where we have a subexpression $F(e_1, \ldots, e_n)$ on the right hand side of a function definition such that $A[\![F]\!](t_{i,j}) = \bot$. Therefore we know that all cells of level j created during evaluation of e_i is garbage after the execution of F. This information can be used for a better code generation. We can insert commands for the deallocation of all these cells after the termination of $F(e_1, \ldots, e_n)$. This could easily be done by traversing the graph representing the intermediate result e_i. However, this method is quite time consuming. We can increase performance by a simple one place cache strategy, where a data structure is not directly deallocated but reused for subsequent allocations. If there is already a structure in this cache it is traversed and its cells are deallocated. The next section will show some experimental results for this strategy.

Another possibility is to trace all allocations of cells of the appropriate level during execution of e_i. The advantage is that deallocation can be done without traversing the result. Of course, there is some overhead during the allocation and we need additional memory.

A more sophisticated method would be to allocate intermediate results in a more stack-like way on a frame which is associated to the function which is currently evaluated. All intermediate results can then be deallocated by releasing the stack frame. Of course, we need additional analyses in order to determine at compile-time how much intermediate space in terms of the input size is needed for the evaluation of a function definition. This is due to the fact that during the evaluation of a function f, which will create intermediate data, we may need space for 'inter-'intermediate data. This will be allocated in the stack frame associated with f. But this stack frame is on top of the stack frame of the function calling f, where the result of f will be located. Therefore we need to reserve enough space for the result of f before evaluating f.

6 Efficiency

One of the key targets of this work was to keep both the compile-time and run-time overhead as small as possible. The compile-time efficiency is represented by the complexity of the computation of the abstract information. On the other hand, the run-time efficiency is characterised by two parameters: the time lost by the execution of the additional commands for the deallocation and the gain in memory utilisation.

6.1 Compile-Time Efficiency

In general, we need to evaluate all abstract functions associated with a given program for all possible arguments and recompute until stability. The finiteness of all abstract domains guarantees that a stable state is reached after a finite number of iterations. Firstly, we will estimate the number of iterations for this naïve approach.

Let $n \in N$ be the number of defined functions in a program P, $m \in N$ the maximal arity of a defined function and $l \in N$ the maximal number of levels of a sort in P. This is also the length of the longest ascending chain in the abstract domain for this sort. An upper bound for the number of values computed in each iteration is $2^l nm$. The longest chain contains at most l elements, which means that each computed value can change at most l times during the computation until the maximal element is reached. Altogether, the worst case are $2^l mnl$ iterations, which is far from being practicable.

Since we are only interested in the values of the functions for the test arguments, it is convenient to compute only these. This optimisation reduces the number of iterations to l^2mn, because there are only lmn test arguments. But we may need the value of a function for a non-test argument. Luckily, we can compute such values from the values of the functions for the test arguments.

Lemma 2 (Distributivity of A and ⊔). *Let* $F \in DF^{s_1,\ldots,s_k \to s}$ *and* $a_i, a_i' \in \mathfrak{A}_i$ *for* $1 \le i \le k$. *The abstract semantics distributes with the lub-operator:*

$$A[\![F]\!](a_1 \sqcup a_1', \ldots, a_k \sqcup a_k') = A[\![F]\!](a_1, \ldots, a_k) \sqcup A[\![F]\!](a_1', \ldots, a_k')$$

Together with the next corollaries, this enables us to represent the values for a non-test argument as a 'linear' combination of test argument values.

Corollary 3 (Consistency of A). *Let* $F \in DF^{s_1,\ldots,s_k \to s}$:

$$A[\![F]\!](\bot_{s_1}, \ldots, \bot_{s_n}) = \bot_s$$

Corollary 4 (Generating set T_s). *Let* $s \in S$. *The set of test values* T_s *is a generating set for* $\mathfrak{A}_s$ *in the following sense:*

$$\mathfrak{A}_s = \bigcup_{T \subseteq T_s} \{\sqcup T\}$$

If we examine the number of function evaluations instead of iterations, the advantage is even more dramatical. While the naïve implementation requires $2^l mn$ function evaluations per iteration, the better version only needs lmn, which implies a total of $2^{2l}m^2n^2l$ evaluations versus $l^3m^2n^2$ evaluations.

Since the number of function definitions in a program is the most characteristic value, we can interpret this result as the proof for a quadratic worst case complexity of our method.

Realistic programs, however, do not reach this worst case, since the function definitions are not that enigmatic. Especially those functions, which only depend on themselves, like for instance the append function, are interesting. If we look at the corresponding transformation, we can observe, that the value of $A[\![append]\!]$ for an argument (a_1, a_2) in the $i + 1$-th iteration depends only on the value for (a_1, a_2) in the i-th iteration and not on the value for any other argument.

$$A[\![append]\!](a_1, a_2) = A[\![Nil]\!] \sqcup A[\![Cons]\!](Cons_{sel}^1(a_1), A[\![append]\!](Cons_{sel}^2(a_1), a_2))$$
$$= A[\![Nil]\!] \sqcup A[\![Cons]\!](Cons_{sel}^1(a_1), A[\![append]\!](a_1, a_2))$$

The reason for this is that $A[\![Cons_{sel}^2]\!]$ is the identity function, since this corresponds to the access to a substructure of type ListOfInt. Functions which are defined via *structural recursion* on the underlying type have this property. The impact on the complexity of the computation is enormous. Because the value for each argument is independent from the other values the computation of the fixpoint is stable after at most l steps. For a program where all definitions have this form the number of evaluations drops to l^2mn, which is essentially linear in the size of the program.

A single evaluation can be done very efficiently, since it consists of a sequence of lub-operations which can be implemented by a "bitwise or".

6.2 Run-Time Efficiency

We have done some experiments with programs which were enriched with deal-location commands. The programs are compiled by hand from our example language to C. A detailed description of this translation can be found in [Moh95b].

The first example (see Table 1) is the qsort program. The points of deallocation are those which were described above. The first line of the table shows the number of heap cells used for the input, which is the list from n down to 1. The remaining three lines show additional amount of heap used by the program without ctgc, at the end of the ctgc version, and the maximal heap usage during execution of the ctgc version, respectively. Note, that the maximal heap usage of the ctgc version is only a third

		1	10	50	100
input		2	11	51	101
w/o ctgc		7	196	3376	15451
ctgc (end)		4	22	102	202
ctgc (max)		4	58	1278	5053

Table 1.: Memory statistic for qsort

of the memory consumption of the version without optimisation. The ctgc version of the program is optimal in the sense, that all intermediate data is deallocated; only the result is still in memory at the end. If qsort would be called from a position where the input could not be accessed any more, the ctgc version would effectively be an in-place version.

In Fig. 4 we have compared the different memory consumptions during the evaluation of qsort(10). The diagram shows the number of heap cells used at each function call performed. After an initial phase, where only alternating calls to filter* and qsort are executed until an empty list is reached, the two curves start to differ. The ctgc version continuously allocates and deallocates intermediate data. The point of maximal memory consumption of the ctgc version is the end of the initial phase. Until then no deallocation is performed.

Fig. 4.: qsort(10) with and w/o ctgc

Although we do not (yet) have a real implementation we have also done some experiments on the run-time impact of our approach. In order to get realistic results, we have used the public domain garbage collecting storage allocator that is intended to be used as a plug-in replacement for C's malloc [BW88, Boe91]. Figure 5 shows the run-times (in seconds) without gc/ctgc, with gc, and with both gc and ctgc for vary-

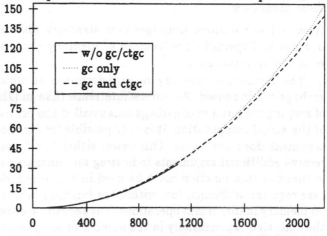

Fig. 5.: Run-times for different versions of qsort

ing list length from 100 to 2200. We have used the cached approach described in the last section.

Of course, the unaltered version has best run-times. However, the differences in run-time are only about 4% for lists up to 1100 elements. For larger lists the memory requirement of this version can not be satisfied. There we can see that the combination of ctgc and traditional gc is the best. Since additional deallocations reduce the need for normal garbage collection, the run-time is decreased (wrt. to traditional gc only) in situations where much intermediate data is created.

Our second example (Table 2) is an implementation of queens. This particular example is remarkable, because it contains only a single point where an additional deallocation can be incorporated. This point is a call to append. Surprisingly, the effect on memory usage is immense.

n	1	2	4	8
w/o ctgc	5	10	92	20446
ctgc (end)	4	4	20	2150
ctgc (max)	5	6	23	2239

Table 2.: Memory statistic for queens

These results show clearly, that ctgc is worth the (small) effort.

7 Related Work

Several methods have been proposed to reduce the run-time memory consumption by ctgc. A simple classification can be done by two characteristics:

Which kind of information is approximated by the underlying abstract interpretation(s). By definition, a heap cell is garbage if it is (part of a structure) not reachable from the main expression. Consequently, there are four actions which influence the state of a cell: **generation** by a constructor, **sharing** by non-linear function bodies, **inheriting** by returning as (part of) the result of a function, and **dereferencing** by pattern matching. Accordingly, ctgc methods use abstract interpretation in order to retrieve information on some of these properties.

How is the memory management strategy altered by the information obtained. Especially, the *location of memory reuse* and the *way of memory reuse* is of importance.

The location of memory reuse determines at which point of execution a garbage cell is reused. An *immediate* reuse (like in [JM89]) has the advantage of keeping the number of garbage cells small at the price of frequent interruptions of the actual computation. It is only possible for functions, where sharing of the argument does not occur. Otherwise, either the function must be altered to receive additional arguments indicating an "unshared situation", or be special versions of this function must be used in the appropriate situations. The first case requires additional test within the function, which increase the time spent on memory management operations. The second case avoids this but can increase the code size exponentially in the number of arguments of the function. A more *delayed* approach (like in [Hug92]) deallocates at the end of the corresponding function call. The advantage is that more deallocations are performed at the same time, which may be done more efficiently. Also, there is no need for modifications within the function, which circumvents the above problems.

The way of memory reuse can be either *deallocating*, i.e. adding the cell to the free part of the heap, or *direct reuse*. The latter can be used in situations where the deallocation is immediately followed by an allocation, like for instance the append function. The result is a "in-place" version of append. Again, there is the problem whether an argument is shared or not.

The approach taken in [Hug92] is the closest work to ours. It uses a combination of generation analysis and inheritance analysis of (nested) lists of atomic values in the setting of a higher order language. Arbitrary data structures or structures containing functional parameters are not considered. Garbage is detected if heap cells are created "locally" and are not inherited. The abstract interpretation uses domains which are also based on the notion of levels of the corresponding list structure. In contrast to our work, the levels are not handled independently, i.e. all results have the shape "all levels above and including are not inherited". As we have pointed out, this is sufficient in the setting of a purely functional language with lists, but for arbitrary data structures the reduced approach is not applicable at all.

Park and Goldberg [PG92] describe a similar approach, which they call *escape analysis*, indicating whether (parts of) the arguments can escape from the function call.

A combination of sharing analysis and inheritance analysis is used in [SY88], again for list structures only. The underlying analyses use context-free grammars.

In [JM89] a different approach is taken: list constructors which are not shared are collected as soon as they are dereferenced, i.e. as soon as they match a constructor on the left hand side of a function definition. The abstract domains are introduced as (infinite) domains I_{list}. For practical application, the height of the abstract list is restricted, the choice of a threshold is left to the user. Additionally, if it is detected that a deallocation is immediately followed by an allocation, update-in-place is performed.

Two different approaches which use backward analysis are [HJ90] and [JM90]. While the latter is essentially abstract reference counting the first infers information on whether a particular cell is going to be accessed in the future. The cell will be deallocated if this is not the case.

8 Conclusions and Future Work

In this paper we have presented a method for statically estimating positions in a functional program, where unreferenced data will occur at run-time. Our method is based on the abstract interpretation of the underlying language.

We have described how to use our method in the context of eager evaluation. Experimental results demonstrate that the storage is used in a much better way. We have quantified the gain of this optimisation by applying it to two examples: The first example, quicksort, perfectly fits this approach. Not surprisingly, the result was a program with an optimal space consumption. On the other hand, the second example, queens, only offers one point for optimisation. Nevertheless, the result was quite good. This is evidence to assume that our method works well in practice.

Apart from a optimised memory utilisation, it was a major concern to keep the overhead at compile-time and run-time as small as possible. Due to the fact that ctgc decreases the time spent for normal gc, it improves performance. Some aspects of our work, distinguishing it from other work on ctgc are:

1. We can handle arbitrary data structures instead of only lists.
2. The abstract domains can be inferred directly from the types, whereas other approaches essentially use approximations of list structures which are obtained by gratuitously restricting the height of a list.
3. We have considered correctness w.r.t. to a denotational semantics, which is more general than a proof w.r.t. a particular implementation.
4. We do not try to handle destructive updates, because this would increase the complexity of the implementation immensely, whereas the gain is only of temporary nature.
5. The 'point of optimisation' is *after* the return from a function, which has the advantage that we only need the abstract information on a particular function and not its definition. This may be of interest in the context of modules, where the definition is not available.
6. We compute the abstract function only for their test arguments, which makes the analysis fast.

More work needs to be done in the following directions:

Higher-order functions: In [Moh95a] we describe a higher-order version of our analysis. The representation of functions with functional argument or result within the abstract interpretation has two aspects: on the one hand, the interpretation must be extended to handle function as arguments and results, and on the other hand, we want to infer informations on partial applications, which are also represented in the heap in the shape of *closures*. For instance, if there is a call to filter with a partial application as functional argument, we can infer that the heap cell associated with this partial application can be deallocated after termination of filter. Alternatively, those closures can be allocated on the stack frame associated with the surrounding function. In some cases, it is even possible to allocate the closure statically, since there is always only one active *incarnation* of this closure. This can be approximated with an additional analysis of the *call structure* of the program.

Polymorphism: With easy extensions our approach is able to handle this. The major idea is that a polymorphic function can not affect the levels of data with a, so that the abstract interpretation can use the smallest set to represent those elements [BH89]. In this case we can associate the set $\{0,1\}^2$ with this polymorphic type definition and evaluate each polymorphic function once with this type. Actually, our inheritance analysis is polymorphically invariant [Abr86]. This means that given a polymorphic function, our analysis will return the same results on any two monotyped instances of that function. Therefore we can analyse a polymorphic function by analysing the simplest monotyped instance of this function.

Lazy evaluation: Since the abstract interpretation itself does not depend on the evaluation strategy, the informations obtained by the analysis are still valid for lazy languages. Since in a lazy language there is no particular point in the program associated with the termination of a function, our approach is not directly applicable. But in combination with a strictness analysis, it will be possible to determine 'eager' situations, i.e. positions in a lazy program where we have $f(g(e))$ such that f is strict. In these cases we can insert deallocation commands for those heap cells created in e, which are not inherited by g.

Additionally, it would be interesting to investigate the relationship with other ctgc approaches and other approaches to decrease memory consumption like deforestation ([Wad90, GLP93]) in more detail.

References

[Abr86]   S. Abramsky. Strictness analysis and polymorphic invariance. In G. Goos and J. Hartmanis, editors, *Workshop on Programs as Data Objects*, number 217 in LNCS, pages 1—24, 1986.

[BH89]   G. Baraki and J. Hughes. Abstract interpretation of polymorphic functions. In K. Davis and J. Hughes, editors, *Functional Programming, Glasgow*, 1989.

[Boe91]   H. Boehm. Space efficient conservative garbage collection. In *Proc. of the ACM SIGPLAN '91 Conference*, number 28 in SIGPLAN Notices, 1991.

[BW88]   H. Boehm and M. Weiser. Garbage collection in an uncooperative environment. *Software Practice & Experience*, pages 807—820, September 1988.

[GLP93]   A. Gill, J. Launchbury, and S. L. Peyton Jones. A short cut to deforestation. In *Proc. of FPCA*, 1993.

[HJ90]   G. W. Hamilton and S. B. Jones. Compile-time garbage collection by necessity analysis. In S. L. Peyton Jones, G. Hutton, and C. Kehler Holst, editors, *Functional Programming, Glasgow*, 1990.

[Hug92]   S. Hughes. Compile-time garbage collection for higher-order functional languages. *Journal of Logic and Computation*, 2(4):483–509, 1992.

[JM89]   S. B. Jones and D. Le Métayer. Compile-time garbage collection by sharing analysis. In *Proceedings of FPCA*, 1989.

[JM90]   T. P. Jensen and T.Æ. Mogensen. A backward analysis for compile-time garbage collection. In G. Goos and J. Hartmanis, editors, *Proceedings of ESOP 90*, number 432 in LNCS, pages 227—239, 1990.

[Moh95a] M. Mohnen. Efficient closure utilisation by higher-order inheritance analysis. To appear in Proceedings of SAS 95 (LNCS), 1995.

[Moh95b] M. Mohnen. Functional specification of imperative programs: An alternative point of view of functional languages. Technical Report 95-9, RWTH Aachen, 1995.

[Myc80]   A. Mycroft. Theory and practice of transforming call-by-need into call-by-value. In *Proceedings of the International Symposium on Programming*, number 83 in LNCS, pages 269–281, 1980.

[PG92]   Y. G. Park and B. Goldberg. Escape analysis on lists. In *PLDI 92*, ACM SIGPLAN, pages 116—127, 1992.

[SY88]   K. Inoue H. Seki and H. Yagi. Analysis of functional programs to detect run-time garbage cells. *TOPLAS*, 10(4):555—578, October 1988.

[Wad90]   P. Wadler. Deforestation: Transforming programs to eliminate trees. *Theoretical Computer Science*, 1(73):231—248, 1990.

Efficient Multi-level Generating Extensions
for Program Specialization

Robert Glück* and Jesper Jørgensen

DIKU, Department of Computer Science, University of Copenhagen,
Universitetsparken 1, DK-2100 Copenhagen Ø, Denmark
e-mail: {glueck, knud}@diku.dk

Abstract. Multiple program specialization can stage a computation
into several computation phases. This paper presents an effective solution for multiple program specialization by generalizing conventional
off-line partial evaluation and integrating the "cogen approach" with
a multi-level binding-time analysis. This novel "multi-cogen approach"
solves two fundamental problems of self-applicable partial evaluation:
the generation-time problem and the generator-size problem. The multi-
level program generator has been implemented for a higher-order subset
of Scheme. Experimental results show a remarkable reduction of generation time and generator size compared to previous attempts of multiple
self-application.

1 Introduction

Stages of computation arise naturally in many programs, depending on the availability of data or the frequency with which the input changes. Code for later
stages can be optimized based on values available in earlier stages.

Partial evaluation has received much attention because of its ability to stage
a program into two computation phases. A self-applicable partial evaluator, or
a compiler generator, is strong enough to convert a general algorithm, say, a
general parser with two inputs (grammar, string), into the corresponding *two-
level generating extension*, i.e. a parser generator. Our goal is more general: the
automatic construction of efficient *multi-level generating extensions* (Sect. 2).

This paper presents an effective solution to the problem of automatically
generating multi-level generating extensions by generalizing conventional off-line
partial evaluation and integrating the "cogen approach" [Hol89, BW94] with a
multi-level binding-time analysis (Sect. 3). We demonstrate that efficient *multi-
level generators* can be built that automatically produce multi-level generating
extensions from arbitrary programs. For this purpose we generalized partial evaluation techniques, such as binding-time analysis, specialization points, memoization, and code generation (Sect. 4 and 5).

This novel *"multi-cogen approach"* solves two fundamental problems of
self-applicable partial evaluation [Glü91]: the generation-time problem and the
generator-size problem. The multi-level generator gives an impressive reduction
of generation time and code size compared to multiple self-application (Sect.
6.1).

* Supported by an Erwin-Schrödinger-Fellowship of the Austrian Science Foundation
(FWF) under grant J0780 & J0964.

The multi-level generator has been implemented for a higher-order subset of Scheme. It is quite remarkable that the multi-level generator can automatically convert a two-level driving interpreter (from [GJ94]) into a new compiler generator capable of a form of supercompilation, a problem which we previously failed to solve with a state-of-the-art partial evaluator (Sect. 6.2).

Our approach shares the same advantages with the 'direct approach' [BW94], but for multiple program specialization: the generator and the generating extensions can use all features of the implementation language (no restrictions due to self-application); the generator manipulates only syntax trees (no need to implement a self-interpreter); values in generating-extensions are represented directly (no encoding overhead); and it becomes easier to demonstrate correctness for non-trivial languages (due to the simplicity of the transformation). Last but not least, generating extensions are stand-alone programs that can be distributed without the generator.

We claim that our approach scales up to other languages as evidenced by the fact that generators for two-level generating extensions have been implemented for typed and untyped languages and different language paradigms (e.g. λ-calculus [BD94], ML [BW93], ANSI C [And94]). We believe that the "multi-cogen approach" is more practical and promising than previous attempts of multiple self-application of partial evaluation, and in particular for applications where generation time and generator size are paramount, such as run-time code generation.

2 Multi-Level Generating Extensions

We now formulate the properties of multi-level generating extensions more precisely and discuss the theoretical limitations and practical problems of conventional specialization tools.

Notation. For any program text, p, written in language L we let $[\![p]\!]_L$ in denote the application of the L-program p to its input in. We use typewriter font for programs and their input and output. The notation is adapted from [JGS93].

2.1 Tools for Program Specialization

Turning a general program into a specialized program can be conceived of as turning a one-stage computation into a two-stage computation. Suppose p is a source program, in_0 is data known at stage one (*static*, 'S') and in_1 is data known at stage two (*dynamic*, 'D'). Then computation in one stage is described by

$$out = [\![p]\!]_L \ in_0 \ in_1$$

Computation in two stages using a *program specializer* spec is described by

$$p_0 = [\![spec]\!]_L \ p \ 'SD' \ in_0$$
$$out = [\![p_0]\!]_L \ in_1$$

where the binding-time classification 'SD' indicates the binding-times of p's input (we classify each argument of a source program according to its binding time). Combining these two we obtain an equational definition of spec:

$$[\![p]\!]_L \ in_0 \ in_1 = [\![[\![spec]\!]_L \ p \ 'SD' \ in_0]\!]_L \ in_1$$

The main motivation is efficiency: the specialized program p_0 runs potentially much faster than the original program p on the same input. For notational convenience we assume that spec is an L→L-specializer written in language L; for multi-language specialization see e.g. [Glü94].

Two-Level Generating Extensions. A program generator cogen, which we call *compiler generator* for historical reasons, is a program that accepts a two-stage program p and its binding-time classification as input and generates a program generator p-gen, called *generating extension* [Ers78], as output. The task of p-gen is to generate a specialized program p_0, given static data in_0 for p's first input. We call p-gen a *two-level* generating extension of p because it realizes a two-staged computation of p.

$$p\text{-gen} = [\![\text{cogen}]\!]_L \ p \ \text{`SD'}$$
$$p_0 = [\![p\text{-gen}]\!]_L \ in_0$$
$$\text{out} = [\![p_0]\!]_L \ in_1$$

Combing these three we obtain an equational definition of cogen:

$$[\![p]\!]_L \ in_0 \ in_1 = [\![[\![[\![\text{cogen}]\!]_L \ p \ \text{`SD'}]\!]_L \ in_0]\!]_L \ in_1$$

The specialization of p is now done in two steps: the compiler generator cogen produces p's generating extension p-gen which is then used to generate the specialized program p_0. The generating extension p-gen produces the specialized program p_0 potentially much faster than the program specializer spec because p-gen is a program generator specialized with respect to p.

2.2 Multiple Program Specialization

Program specialization can do more than stage a computation into two phases. Suppose p is a source program with n inputs, where in_0 is the data known at stage one, in_1 the data known at the next stage, and so on. Then the computation in one stage can be described by

$$\text{out} = [\![p]\!]_L \ in_0 \ldots in_n$$

Computation in n stages using a program specializer spec is described by

$$p_0 = [\![\text{spec}]\!]_L \ p \ \text{`SDD}\ldots\text{D'} \ in_0$$
$$p_1 = [\![\text{spec}]\!]_L \ p0 \ \text{`SD}\ldots\text{D'} \ in_1$$
$$\vdots$$
$$p_{n-1} = [\![\text{spec}]\!]_L \ p_{n-2} \ \text{`SD'} \ in_{n-1}$$
$$\text{out} = [\![p_{n-1}]\!]_L \ in_n$$

In each step the program p_{i-1} obtained from the previous stage is specialized with respect to the next input in_i. Alternatively, we can use the compiler generator cogen in each specialization step. This may improve the performance of the ith specialization in case the program p_{i-1} is specialized with respect to different inputs.

$$p\text{-gen}_i = [\![\text{cogen}]\!]_L \ p_{i-1} \ \text{`SD}\ldots\text{D'}$$
$$p_i = [\![p\text{-gen}_i]\!]_L \ in_i$$

Multiple specialization implemented by such a specialization pipeline is a form of 'on-line' specialization because the ith specialization stage can take all previous values $in_0 \ldots in_{i-1}$ into account (regardless whether the specialization at the ith stage is on- or off-line). Recall that the defining feature of off-line specialization is that it does not take static values into account, only the binding-time classification of the input.

Discussion. An 'on-line' specialization pipeline has two main advantages:

- *Precision*: the specialization at the ith specialization stage can take all static *values* into account that have been provided in the previous stages.
- *Flexibility*: at each specialization stage we are free to choose *any* binding-time classification.

But the advantages of an on-line specialization pipeline are also its disadvantages. It may be too flexible in cases where the order of the inputs $in_0 \ldots in_n$ is fixed. Consider specializing a meta-interpreter with three inputs: a language definition, a program and its data. Multiple specialization does not make much sense unless the definition is available before the program, and the program is available before its data.

The precision gained by applying the full specialization power of spec to every 'intermediate program' p_i has its price: specialization time. For example, program pieces that depend on input values that will be available only in later stages, may be re-analysed each time during the earlier stages. It is also likely that these program expressions have been duplicated during a previous specialization stage, e.g. by program point specialization.

It may be difficult to perform binding-time improvements on machine-generated program because the structure of the original program may be dissolved during specialization (unless we have fully automatic methods for binding-time improvements). For example, a program resulting from specializing an interpreter with respect to an interpreted program will usually reflect the structure of both programs. This makes the specialization pipeline less predictable since the success of a specialization stage i may depend on the static values used in the previous stages.

How to binding-time improve the source program p in the first place in order to avoid the need for binding-time improvements at a later specialization stage? For this, we have to predict the staticness of program pieces independently of static values (binding-time improvements make not much sense unless they improve the binding-time for large class of values).

Multi-Level Generating Extensions. Our approach to multiple specialization is *purely off-line*. A program generator mcogen, which we call a multi-level compiler generator (or short: multi-level generator), is a program that accepts an n-stage program p and a binding-time classification $0 \ldots n$ of its input parameters and generates a *multi-level generating-extension* p-mgen. Given the first input in_0 the multi-level generating extension produces a new specialized multi-level generating extension p-mgen$_0$ and so on, until the final value out is produced from the last input in_n (illustrated in Fig. 1). Multi-level specialization using multi-level generating extensions is described by

$$p\text{-mgen} = [\![\text{mcogen}]\!]_L\ p\ \text{`0}\ldots n\text{'}$$
$$p\text{-mgen}_0 = [\![p\text{-mgen}]\!]_L\ in_0$$
$$\vdots$$
$$p\text{-mgen}_{n-2} = [\![p\text{-mgen}_{n-3}]\!]_L\ in_{n-2}$$
$$p'_{n-1} = [\![p\text{-mgen}_{n-2}]\!]_L\ in_{n-1}$$
$$out = [\![p'_{n-1}]\!]_L\ in_n$$

We call this specialization pipeline off-line because the binding-time classification is consumed by mcogen and no classification is required at a later stage. Note that the program $p\text{-mgen}_i$ returned by stage i does not become the input of a specializer at the next stage, but can be executed directly. It is easy to see that a conventional, two-level generating extension (see Sect. 2.1) is a special case of a multi-level generating extension: it returns an 'ordinary program' and never a generating extension.[2] The generating extension $p\text{-mgen}_{n-2}$ is such a two-level generating extensions.

The generation of a multi-level generating extensions pays off if it is used with various static values. In any case, it will be faster to run a multi-level generating extension at stage i than to use a full specializer.

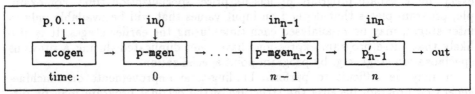

Figure 1: *Program specialization with multi-level generating extensions*

2.3 How to Generate Multi-Level Generating Extensions?

Assume that we have only the conventional specialization tools: a self-applicable specializer spec and a compiler generator cogen. Then we know of two methods that can, in principle, be used to generate a multi-level generating extension [Glü91]: *incremental self-application* and *multiple self-application*.

Consider a three-staged program, namely a meta-interpreter mint, that takes a language definition def, a program pgm and its data dat as input. Let def be written in the definition language D, let pgm be written in language P (defined by def), and let mint be written in some language L. The equational definition of mint is

$$[\![\text{mint}]\!]_L\ \text{def}\ \text{pgm}\ \text{dat} = [\![\text{def}]\!]_D\ \text{pgm}\ \text{dat} = [\![\text{pgm}]\!]_P\ \text{dat} = out$$

What we look for is the three-level generating extension mint-cogen of the metainterpreter mint to perform the computation in three stages.

$$\text{compiler} = [\![\text{mint-cogen}]\!]_L\ \text{def}$$
$$\text{target} = [\![\text{compiler}]\!]_L\ \text{pgm}$$
$$out = [\![\text{target}]\!]_L\ \text{dat}$$

---

[2] A compiler generator cogen is a three-level generating extension of a specializer spec.

The program target is an L-program which, when applied to dat, returns the final result out. The two-level generating extension compiler is a program which, when given a P-program pgm, returns an L-program target and is thus a P→L-compiler. The three-level generating extension mint-cogen is a program which, when applied to def, yields compiler and is thus a compiler generator.

Incremental Self-Application. A three-level generating extension can be constructed in two steps, using the compiler generator cogen (or using the Futamura projections together with a self-applicable specializer spec): (1) the metainterpreter mint is converted to an 'auxiliary' generating extension gen-aux with two inputs; (2) the generating extension gen-aux is converted to mint-cogen' with one input.

$$(1) \quad \text{gen-aux} = [\![\text{cogen}]\!]_L \text{ mint 'SSD'}$$
$$(2) \quad \text{mint-cogen'} = [\![\text{cogen}]\!]_L \text{ gen-aux 'SD'}$$

It can easily be verified that mint-cogen' is a three-level generating extension by composing the above equations and using the equational definition of cogen.

$$\text{out} = [\![[\![[\![\text{mint-cogen'}]\!]_L \text{ def}]\!]_L \text{ pgm}]\!]_L \text{ dat}$$

In general, we need n steps to convert an n-stage program into its n-level generating extension. In practice, incremental generation of generating extensions is more difficulties than the respecialization of residual programs (see Sect. 2.2) because cogen is not just applied to an 'ordinary' residual program but to a program generator. Binding-time improvements, should they be necessary, would have to be performed on a machine-generated program generator. Most compiler generators are not geared toward the respecialization of their generating extensions. Note that incremental generation of multi-level generating extensions is the only possibility if we are given only a conventional compiler generator cogen.

Multiple Self-Application. Given a self-applicable specializer spec there is another way to generate multi-generating extensions: by multiple self-application. This requires up to three self-applications in our example (as opposed to the Futamura projections which require only two self-applications). We now state four metasystem-transition (MST) formulas that specify the generation of a target program target'', a compiler compiler'', a compiler generator mint-cogen'' and a compiler-compiler generator cocogen''. The correctness of the formulas can be verified using the equational definition of spec. For notational convenience we distinguish between the specializers and assume that each has the corresponding arity (otherwise they are identical).

$$\text{target''} = [\![\text{spec1}]\!]_L \text{ mint 'SSD' def pgm}$$
$$\text{compiler''} = [\![\text{spec2}]\!]_L \text{ spec1 'SSSD' mint 'SSD' def}$$
$$\text{mint-cogen''} = [\![\text{spec3}]\!]_L \text{ spec2 'SSSSD' spec1 'SSSD' mint 'SSD'}$$
$$\text{cocogen''} = [\![\text{spec4}]\!]_L \text{ spec3 'SSSSDD' spec2 'SSSSD' spec1 'SSSD'}$$

It can easily be verified that the program mint-cogen'' is a three-level generating extension by composing the above equations and using the equational definitions of the corresponding specializers. The three-level generating extensions mint-cogen'' can be obtained either from the 3rd or the 4th MST-formula using cocogen″. Note that cocogen″ accepts programs written in L, while mint-cogen″ accepts definitions written in D, the language defined by mint.

$$out = [[[\text{mint-cogen}″]_L \ def]_L \ pgm]_L \ dat$$

Multiple self-application has two fundamental problems [Glü91]:

- *Generation time problem.* Assume that t is the run-time of a program p, then $t * k^n$ is the time required to run p on a tower of n self-interpreters, where k is the factor of their interpretive overhead. Since most non-trivial, self-applicable specializers incorporate a self-interpreter for evaluating static program pieces, the run-time of multiple self-application grows exponentially with the number of self-applications.
- *Generator size problem.* Each level of self-application adds one more layer of code generation to the produced generating extension, i.e. code is generated that generates code-generating code that generates code-generating code, and so on. The result is another variant of the notorious encoding problem in self-application[3]: it may lead to an exponential growth of the size of the produced generating extensions (in the number of self-applications).

To avoid the theoretical limitations and the practical problems of conventional specialization tools we are going to develop an efficient multi-level compiler generator in the remainder of this paper.

3 Construction Principles

First we give a brief review of conventional off-line partial evaluation (a comprehensive presentation can be found in [JGS93]) and then state two observations that are the starting point for our "multi-cogen approach". We use Scheme as our presentation language.

3.1 Conventional Off-Line Partial Evaluation

In *off-line partial evaluation* the transformation process is guided by a *binding-time analysis* performed prior to the specialization phase. The result of the binding-time analysis is a program in which all operations are annotated as either *static* or *dynamic*. Operations annotated as static are performed at specialization time, while operations annotated as dynamic are delayed until run-time (i.e. residual code is generated). Binding-time annotations can be represented conveniently using a *two-level syntax* [NN92], e.g. by marking all dynamic operations with an underscore _op, while leaving all static operations unchanged op.

Example 1. Assume that the first two parameters of a program computing the inner product of two vectors are known in advance: the dimension n and the first vector v. We use the function (ref i v) to access the i'th element of a vector v and leave their internal representation unspecified. The result of the binding-time analysis is shown in Fig. 2.

```
(define (iprod n v w)     ; n,v:static  w:dynamic
   (if (> n 0)
      (_+ (_* (lift (ref n v)) (_ref (lift n) w))
          (iprod (- n 1) v w))
      (lift 0)))
```

Figure 2: A two-level program of the inner product

Lifting is necessary when static values appear in a dynamic context. The operation `lift` converts static values into corresponding pieces of program code [Rom88, Mog88]. For example, the static expressions n and (ref n v) need to be lifted.

In general, specialization will fail to terminate if all function calls in a source program are unfolded unconditionally. A standard method to avoid this problem is the insertion of specialization points. A *specialization point* is a call to a named function f that is not unfolded, but treated in the following way during specialization: if the tuple $(f, \overline{arg})$ with function name f and static arguments $\overline{arg}$ has been specialized before then a call to the already specialized function is inserted; otherwise the tuple $(f, \overline{arg})$ is recorded in a *memoization table* and the function f is specialized with respect to the static arguments $\overline{arg}$. This method is known as *polyvariant program point specialization* because the same program point may be specialized with respect to different static arguments. Two well-known methods for automatically inserting specialization points are *inductive variables* [Ses88], and *dynamic conditionals* [BD91].

3.2 The Starting Point

Observation 1: Annotated Program = Generating Extension [Hol89]. An annotated program p_{ann} can be viewed as a generating extension of program p. It is only a question of the language in which one considers p_{ann} as program: static operations op can be executed, while dynamic operations _op generate program code.

Example 2. The dynamic operations in an annotated program can be defined as shown in Fig. 3. The backquote notation of Scheme is convenient for constructing list structures. For example (list '* e1 e2) ≡ `(* ,e1 ,e2)[4].

```
(define (_+ e1 e2)    `(+ ,e1 ,e2))
(define (_* e1 e2)    `(* ,e1 ,e2))
(define (_ref e1 e2) `(REF ,e1 ,e2))
(define (lift e)      `(QUOTE ,e))
```

Figure 3: Two-level code generation

A specialized version of the inner product (Fig. 4) can be obtained by evaluating the annotated program (Fig. 2) together with the definitions of the dynamic

---

[3] The other encoding problem is associated to typed languages: types in programs need to be mapped into a universal type in a partial evaluator.

[4] Upper and lower case forms of identifiers are not distinguished in Scheme. For convenience we use upper case letters when we generate code, e.g. `(REF ,e1 ,e2). The expression 'datum in Scheme is just an abbreviation for (QUOTE datum).

operators (Fig. 3). The program is specialized with respect to dimension n=3 and vector v=(7 8 9), while the dynamic variable w is bound to itself. The generation of define expressions using specialization points will be explained later.

```
(define (iprod-nv w)
  (+ (* '9 (ref '3 w))
     (+ (* '8 (ref '2 w))
        (+ (* '7 (ref '1 w)) '0))))
```

Figure 4: A specialized version of the inner product

Observation 2: Multi-Level Annotation. The two-level static/dynamic annotation of a program is, obviously, a special case of a more general multi-level annotation where binding-time values 0, 1, ..., n indicate at what time an operation can be reduced.

We conclude: if a two-level program can be viewed as generating extension, then a multi-level program can be viewed as multi-level generating extension and the corresponding multi-level binding-time analysis can be used to generate the multi-level generating extensions.

Note that annotated programs can be expressed very elegantly in Scheme, while other programming languages may require an extra translation step in order to obtain syntactically correct multi-level programs. However, this does not change the underlying principles of our approach.

4 Efficient Multi-Level Generators

We present an automatic program generator for multi-level generating extensions based on the two previous observations. More specifically, we generalize the following techniques of conventional partial evaluation: binding-time analysis, lifting, specialization points, memoization, and code generation. The multi-level generating extensions will be discussed in the next section.

4.1 The Source Language

The source language is a higher-order subset of Scheme. A program is a sequence of function definitions with expressions composed of variables x, constants c, primitive operators op, conditionals if, procedure calls, let-expressions, abstractions $\lambda x.e$ and applications $e \bullet e$[5]. Programs written in this subset have the same semantics as in Scheme. The abstract syntax is defined in Fig. 5.

4.2 Multi-Level Binding-Time Analysis

The multi-level binding-time analysis formalizes the intuition that early values may not depend on late values. It is a generalization of the two-level binding-time analysis, e.g. [JGS93]. The result of the multi-level binding-time analysis

---
[5] Without loss of generality we assume that abstractions have only one parameter.

$P \in$ Program; $D \in$ Definition; $e \in$ Expression; $c \in$ Constant;
$x \in$ Variable; $f \in$ ProcName; $op \in$ Primitive
$\quad P ::= D...D$
$\quad D ::= (\text{define } (f \; x...x) \; e)$
$\quad e ::= x \mid c \mid (\text{if } e \; e \; e) \mid (f \; e...e) \mid (op \; e...e) \mid (\text{let } ((x \; e)) \; e) \mid \lambda x.e \mid e \; @ \; e$

Figure 5: Abstract syntax of the higher-order Scheme subset

is a *multi-level program* in which all operations are annotated with *binding-time values*. *Multi-level lifting* can be used to delay values during specialization. The program obtained by erasing all annotations in a multi-level program P is called the *underlying program* and we denote it by $|P|$. We specify the multi-level binding-time analysis using a type system.

Binding-Time Values. A binding-time value $t \in 0, 1, \ldots, \mathbf{max}$ indicates at what time an operation can be reduced. The binding-time value $\mathbf{max}$ is the maximal binding-time of the specific problem considered. Each operation op in a multi-level program is annotated with its binding-time value: $\underline{op}_t$. From now on we refer to expressions with binding time $t = 0$ as *static*, and to expressions with binding time $t \geq 1$ as *dynamic*. The meta-variables r, s, t range over binding-time values. For every program P that we want to specialize we assume that the binding-time values of its input are given (in the form of a *binding-time pattern*), as well as the maximal binding-time value $\mathbf{max}$.

Lifting. The multi-level lift operator $\underline{\text{lift}}_t^s$ is defined as follows: it takes t specializations before the value of its argument becomes known and $t + s$ specializations before the value is made available to the enclosing expression ($s > 0, t \geq 0$). Thus, the binding-time value of an expression $(\underline{\text{lift}}_t^s \; e)$ is the sum of the values s and t. We require that all $\underline{\text{lift}}_t^s$ operators inserted in a multi-level program are maximized with respect to s in order to avoid redundant lift annotations. The following equivalence holds: $(\underline{\text{lift}}_t^{r+s} e) \equiv (\underline{\text{lift}}_{t+s}^r \; (\underline{\text{lift}}_t^s \; e))$.

The conventional lift operator (Sect. 3.1) is a special case of multi-level lifting where the values for s and t are constants: $\underline{\text{lift}}_0^1$.

Binding-Time Types. A type τ is well-formed binding-time type if $\vdash \tau{:}t$ is derivable from the rules in Fig. 6. The first rule states that b^t is binding-time type with binding-time value t and that this binding-time value is not greater than $\mathbf{max}$. The type b^t is assigned to *base type* objects. The rule for function types $\tau_1 \rightarrow^t \tau_2$ states that their binding-time value is the binding-time value of the arrow and that the binding-time values of the argument and result type must not be smaller than the binding-time value of the function type. Every type τ represents a binding-time value t, and we define a function from types to binding time values: $\|\tau\| = t$ **iff** $\vdash \tau{:}t$. It is easy to show that for all binding-time types τ we have $\|\tau\| \leq \mathbf{max}$. The binding-time value of an expression e in a multi-level program is equal to the binding-time value $\|\tau\|$ of its binding-time type τ.

We define an equivalence relation on binding-time types that allows us to handle programs with potential type errors (function values used as base values, base values used as functions). We defer such errors to the latest possible binding-time $\mathbf{max}$, i.e. when the final result is produced. The equality is defined by

$$\vdash b^{\mathbf{max}} \rightarrow^{\mathbf{max}} b^{\mathbf{max}} \doteq b^{\mathbf{max}}$$

$$\frac{t \leq \mathbf{max}}{\vdash b^t : t} \qquad \frac{\vdash \tau_1 : s_1 \quad \vdash \tau_2 : s_2 \quad s_i \geq t}{\vdash \tau_1 \to^t \tau_2 : t}$$

Figure 6: Binding-time types

and the usual rules and axioms for equality (symmetry, reflexivity, transitivity and compatibility with arbitrary contexts).

Well-Annotated Multi-Level Programs. The typing rules for well-annotated multi-level programs are given in Fig 7. Most of the typing rules are straightforward generalizations of the corresponding two-level rules, e.g. [JGS93]. For instance, the rule {*If*} for if-expressions annotates the if with the binding-time value t of the test-expression e_1 (the if-expression is reducible when the result of the test-expression becomes known at time t). The rule requires that the test expression has a first-order type.

The rule {*Prim*} requires that all higher-order arguments of primitive operators have binding-time **max**. This is a necessary and safe approximation since we assume nothing about the type of a primitive operator (the same restriction is used in the binding-time analysis of Similix [Bon91]). The rule {*Lift*} allows lifting only of first-order values. The rules {*Def*} and {*Prog*} type definitions and programs, respectively. Since Scheme procedures (named functions) are uncurried we write their type as $\bar{t} \to t$ where $\bar{t}$ is a possibly empty sequence of binding-time types. The binding-time value on the arrow is left out, since the decision whether to unfold (reduce) such function is not taken during binding-time analysis.

The analysis is monovariant since all calls to the same function have the same binding-time pattern (this is ensured by appropriate lift-operators).

Definition (well-annotated completion). *A multi-level program P' is a well-annotated completion of a program P, if $|P'| = P$ and if for given a binding-time value* **max** *and binding-time types (well formed with respect to* **max***) $\bar{\tau}_0$, $\bar{\tau}_1 \to \tau_1$, ... $\bar{\tau}_l \to \tau_l$ the following judgement can be derived*

$$\vdash P' : \{f_0 : \bar{\tau}_0 \to b^{\mathbf{max}}, ..., f_l : \bar{\tau}_l \to \tau_l\}$$

where f_0 is the goal function of P' and $\bar{\tau}_0 = (b^{t_0}, ..., b^{t_k})$ where $(t_0, ..., t_k)$ is the binding-time pattern for the input to f_0.

The goal of the multi-level binding-time analysis is to determine a well-annotated completion P' of a program P which is, preferably, the 'best'. A completion is the best if the binding time for every subexpression e in P is less than or equal to the binding time of e in any other well-annotated completion of P. We conjecture that, as in the two-level case, there always exists a best completion.

The correctness of two-level binding-time analysis schemes has been shown by several authors, e.g. [Wan93, Hat95]. Similar proof methods can be used in the multi-level case (beyond the scope of this paper).

$$\{Const\}\ \Gamma \vdash c:b^0 \qquad \{Var\}\ \frac{x:\tau \text{ in } \Gamma}{\Gamma \vdash x:\tau} \qquad \{Prim\}\ \frac{\Gamma \vdash e_i:b^t}{\Gamma \vdash (\underline{op}_t\, e_1...e_k):b^t}$$

$$\{If\}\ \frac{\Gamma \vdash e_1:b^t \quad \Gamma \vdash e_2:\tau \quad \Gamma \vdash e_3:\tau \quad \|\tau\| \geq t}{\Gamma \vdash (\underline{if}_t\, e_1\, e_2\, e_3):\tau} \qquad \{Lift\}\ \frac{\Gamma \vdash e:b^t \quad s>0}{\Gamma \vdash (\underline{lift}_t^s e):b^{t+s}}$$

$$\{Call\}\ \frac{\Gamma \vdash e_i:\tau_i \quad f:\tau_1...\tau_k \to \tau \text{ in } \Gamma}{\Gamma \vdash (f e_1...e_k):\tau} \qquad \{\doteq\}\ \frac{\Gamma \vdash e:\tau \quad \vdash \tau \doteq \tau'}{\Gamma \vdash e:\tau'}$$

$$\{Let\}\ \frac{\Gamma \vdash e:\tau \quad \Gamma\{x:\tau\} \vdash e':\tau' \quad \|\tau'\| \geq \|\tau\|}{\Gamma \vdash (\underline{let}_{\|\tau\|}((x\ e))\ e'):\tau'}$$

$$\{Abs\}\ \frac{\Gamma\{x:\tau_x\} \vdash e:\tau}{\Gamma \vdash \underline{\lambda}_s x.e:\tau_x \to^s \tau} \qquad \{App\}\ \frac{\Gamma \vdash e_1:\tau_2 \to^s \tau_1 \quad \Gamma \vdash e_2:\tau_2}{\Gamma \vdash e_1\underline{@}_s e_2:\tau_1}$$

$$\{Def\}\ \frac{\Gamma\{x_1:\tau_1,...,x_k:\tau_k\} \vdash e:\tau}{\Gamma \vdash (\texttt{define}\ (f\ x_1...x_k)\ e):\tau_1...\tau_k \to \tau}$$

$$\{Prog\}\ \frac{\{f_1:\overline{\tau}_1 \to \tau_1,...,f_l:\overline{\tau}_l \to \tau_l\} \vdash (\texttt{define}\ (f_i\ \overline{x}_i)\ e_i):\overline{\tau}_i \to \tau_i}{\vdash P:\{f_1:\overline{\tau}_1 \to \tau_1,...,f_l:\overline{\tau}_l \to \tau_l\}}$$

Figure 7: Typing rules for the multi-level binding-time analysis

Extensions. Other multi-level binding-time analyses can be defined in a similar way. For example, using also *recursive types* $\mu\alpha.\tau$ would allow more liberal annotations (earlier binding-time values for expressions). This can be done without changing the rules in Fig. 7. It is sufficient to extend the equality on binding-time types in such a way that all types that have the same regular type are equal (according to the translation described in [CC91]).

4.3 Insertion of Multi-Level Specialization Points

We chose dynamic conditionals and dynamic abstractions for inserting specialization points because this strategy is straightforward, surprisingly effective in practice, and easily generalized to the multi-level case. We will only explain this for conditionals.

- If a test has the binding time $t = 0$ then the test can be decided at specialization time. No specialization point is inserted.
- If a test has the binding time $t \geq 1$ then the test can not be decided at specialization time. To avoid the risk of infinite unfolding, a multi-level specialization point multi-memo is inserted and the dynamic conditional is made into a named function whose parameters are the free variables of the conditional and whose body is the entire conditional.

As in the two-level case, this strategy avoids infinite unfolding at specialization time, but does not avoid infinite specialization if static data varies unboundedly under dynamic control. Strategies for automatically guaranteeing termination without being overly conservative are a topic of current research.

Multi-Level Specialization Points. A multi-level specialization point is more general than its traditional counterpart:

1. It is sufficient to divide the argument expressions of a specialization point into two groups ($t = static$, $t = dynamic$) in the two-level case, while the argument expressions may have several different binding times $t \in \{0, 1, ...\}$ in the multi-level case.
2. The binding time of a dynamic test can only be $t = dynamic$ in the two-level case, while a dynamic test may have any binding time $t \in \{1, 2, ...\}$ in the multi-level case.

A multi-level specialization point **multi-memo** has the following arguments: the binding time t of the dynamic test, the name f of the function, and the *groups* g of f's arguments with equal binding times. Each group is tagged with the binding time t of the arguments it contains. Empty group are omitted, so the total number of groups is limited by the number of arguments of f (and does not depend on the number of specialization levels). We assume that the groups are ordered w.r.t. increasing binding time values. The use of multi-level specialization points in generating-extensions is explained in Sect. 5. See Fig. 8 for the syntax of multi-level specialization points.

4.4 Implementation

We implemented the multi-level generator for the Scheme subset. The efficiency of the implementation depends on efficiency of the multi-level binding-time analysis and the insertion of specialization points.

Based on the constraint-based binding-time analysis for higher-order languages [Hen91, BJ93] we developed an efficient multi-level binding-time analysis which has an almost-linear complexity [GJ95]. In particular, for a first-order language (without partially static structures) the multi-level binding-time analysis can be reduced to a *graph reachability problem* that runs in linear time $\mathcal{O}(n)$, where n is the number of edges in the dependency graph (bounded by the size of the source program).

The insertion of specialization points is bounded by the number of conditionals and thus by the size of the source program.

Limitations. In the present implementation of our multi-level generator we chose a few 'poor-man's' solutions (since our aim was to study the principles underlying multiple program specialization): (i) Unrestrained unfolding may result in programs that terminate more often than the original program. Termination-preserving methods for two-level partial evaluation can be found in [Bon91]. (ii) All function calls, except memoization points, are unfolded. This may lead to computation duplication. Methods for preventing this problem are well-known, e.g. inserting let-expressions [Bon91], and can easily be generalized to the multi-level case.

5   Efficient Multi-Level Generating Extensions

Representation of Multi-Level Programs. We choose the same approach as in Sect. 3.2: we do not put annotations on static expressions ($t = 0$), only on dynamic expressions ($t \geq 1$). Thus, static expressions can be evaluated directly by the underlying implementation (their semantics is the same as in Scheme).

The concrete syntax is shown in Fig. 8. We could, in principle, provide an interpreter to run multi-level programs, but this would be less efficient. Dynamic expressions require three kinds of operations:

1. *Code generation*: functions that emit program code,
2. *Lifting*: a function that turns a value into a piece of code.
3. *Memoization*: a function that maintains the memoization table and specializes functions if necessary.

```
e ::= x | c | (f e...e) | (if e e e) | (op e...e)
    | (let ((x e)) e) | (lambda (xs) e) | (e...e)
    | (_var t x) | (lift t e) | (_ 'if t e e e) | (_ 'op t e...e)
    | (_let t x e e) | (_lambda t (x...x) e) | (_app t e...e)
    | (multi-memo t f g...g)
g ::= (t e...e)
```

Figure 8: Concrete syntax of multi-level programs

Structure. A multi-level generating extension consist of two parts: the multi-level program and a library that includes the code-generating functions, lifting, and memoization[6]. No extra interpretive overhead is introduced since the definitions in the library are linked with the multi-level program at loading/compile-time. The library adds only a constant size of code to the multi-level program (in contrast to [Hol89] who used macro-extensions). Thus, the size of the multi-level generating extension depends only on the size of the multi-level program (and the static data) and is independent of the number of specialization levels.

Generic Code Generation. A generic code-generating function _ can be used that takes three arguments (Fig. 9): an operator op, a binding-time value t, and the arguments as (already pieces of code). Generic code generation works as follows: if the binding time t equals 1 then the function produces an executable expression for op; otherwise it reproduces a code-generating function for op with binding-time argument t decreased by 1. Repeatedly applying a code-generating function corresponds to a 'count-down' until the binding-time argument reaches 1 which means that the arguments for op will be static next time the program runs.

Multi-Level Lifting. We use the following translation for the lift operator:

$$\underline{\text{lift}}_0^s\, e \equiv (\text{lift } s\, e) \quad \text{and} \quad \underline{\text{lift}}_t^s\, e \equiv (\_\ \text{lift } t\, s\, e)$$

In the first case the argument e is static, but needs to be lifted s times. In the second case the argument e has the dynamic binding time t and the lift operation has to be delayed t times, using the code generating function _, before lifting the value s times. The lift operation (Fig. 9) is similar to the code generating function: it counts the binding-time value s down to 1 before releasing the value val.

---

[6] Generating extensions for languages with side-effects, such as C, require an additional management of the static store in order to restore previous computation states, e.g. when specializing the branches of a dynamic conditional (see [And94]).

The code-generation operations for abstraction and application cannot in a simple way be expressed by using the generic operation _. Application does not have a Scheme operation app, say, associated with it so we cannot just write something like (_ 'app e_1 e_2), so we have chosen to define a special code-generating function _app for application. This is shown in Fig. 9. The problem with abstractions is another: the formal parameter of an abstraction is free in the body of the abstraction and when the body is evaluated to some code the formal parameter must evaluate to itself. This is ensured by generating code for variables that, similarly to lifting, counts down to 1 before "releasing" the variable. If x is a variable bound in an expression (let-expression or abstraction) with binding time t then the following code is generated (_var t x). The code generating function _let for let-expressions is similar to _lambda.

To ensure safe unfolding and prevent variable capturing trivial let-expressions, i.e. (let ((x x)) e), are for all formal parameters in abstractions and procedures. This is similar to what Similix does.

```
(define (_ op t . as)                 (define (lift s val)
  (if (= t 1)                           (if (= s 1)
    `(,op . ,as)                          `(QUOTE ,val)
    `(_ (QUOTE ,op) ,(- t 1) . ,as))))    `(LIFT ,(- s 1) (QUOTE ,val)))))

(define (_lambda t xs e)              (define (_app t e . es)
  (if (= t 1)                           (if (= t 1)
    `(LAMBDA ,xs ,e)                      `(,e . ,es)
    `(_LAMBDA ,(- t 1) (QUOTE ,xs) e)))   `(_APP ,(- t 1) ,e . ,es)))

(define (_let t x e1 e2)             (define (_var t x)
  (if (= t 1)                           (if (= t 1)
    `(LET ((,x ,e1)) ,e2)                 x
    `(_let ,(- t 1) (QUOTE ,x) ,e1 ,e2))) `(_var ,(- t 1) (QUOTE ,x)))))
```

Figure 9: Multi-level code-generation and lifting

Example 3. Assume that the program computing the inner product has three different binding times: dimension n:0, vector v:1, and vector w:2. The multi-level program of the inner product is shown in Fig. 10.

```
(define (iprod3 n v w)
  (if (> n 0)
      (_ '+ 2 (_ '* 2
                  (_ 'lift 1 1 (_ 'ref 1 (lift 1 n) v))
                  (_ 'ref 2 (lift 2 n) w))
         (iprod3 (- n 1) v w))
      (lift 2 0)))
```

Figure 10: A three-level program of the inner product

Multi-Level Memoization. A multi-level specialization point may give rise to several function specializations over time, as opposed to conventional partial evaluation where all specialization points are specialized only once. The function has two parts:

274

1. The function maintains a memoization table (in the form of an association list called a memo-list) of previously encountered tuples (f,as0). If the tuple (f,as0) has not been specialized before, then the function f is specialized with respect to the static arguments and the tuple is recorded in the memo-list together with the name of the new function. Otherwise, a call to the already specialized function is inserted. This is similar to conventional specialization.
2. A multi-level specialization point reproduces itself after decreasing the binding-time values t and the binding-time tags by 1. It disappears if its binding-time value t becomes 1.

The function multi-memo is defined in pseudo-code in Fig. 11. The multi-level specialization function multi-memo takes the following arguments: a binding-time value t, a function name f, and lists of argument groups that are tagged with binding-time values 0, 1, 2,... where the list with static arguments has the tag 0. We assumed that there is an argument group for every binding-time value 0...t, but in the actual implementation only non-empty argument groups need to be present. The multi-level specialization is polyvariant and depth-first.

A few auxiliary functions are used in the definition: the function notSeenB4 checks whether the current specialization point exists in the memo-list; addToMemoList adds a new specialization point to the memo-list and generates a new residual function name which can be retrieved by function getResName. The function updateMemoList updates the memo-list with the specialized body of f. The function call applies a function to a list of arguments.

```
multi-memo [t f (0 as0) (1 as1)...(t ast)] =
  let pp = `(,f ,as0) in
    if notSeenB4 pp then
      addToMemoList pp;
      updateMemoList pp (call f `(,as0...,ast))
    endif;
  let fn = getResName pp in
    if t=1 then `(,fn . ,ast)
    else
      `(multi-memo ,(t-1) ,fn (0 ,as1)...(,(t-1) ,ast))
    endif
```

Figure 11: Multi-level memoization

Example 4. Consider a multi-level specialization point with a binding-time value 4 and four groups of arguments tagged with the binding-time values 0, 2, 3, and 4 respectively. The multi-level specialization of f develops in several steps.

$$(\text{multi-memo } 4 \ f \ (0 \ e_a) \ (2 \ e_b \ e_c) \ (3 \ e_d) \ (4 \ e_e))$$

$$(\text{multi-memo } 3 \ f_a \ (1 \ e_b \ e_c) \ (2 \ e_d) \ (3 \ e_e))$$

$$(\text{multi-memo } 2 \ f_a \ (0 \ e_b \ e_c) \ (1 \ e_d) \ (2 \ e_e))$$

$$(\text{multi-memo } 1 \ f_{abc} \ (0 \ e_d) \ (1 \ e_e))$$

$$(f_{abcd} \ e_e)$$

First, the initial generating extension p-mgen specializes the function f with respect to the static argument e_a and the specialization point is kept. The next generating extension p-mgen$_0$ reproduces the specialization point without specializing f_a because it receives no static arguments. The generating extension p-mgen$_1$ specializes the function f_a with respect to the static arguments e_b and e_c. Finally, the last generating extension p-mgen$_2$ specializes the function with respect to the static argument e_d and produces a direct call to f_{abcd}.

6   Results

The main advantages of the multi-level generator are the speed of the generation process and the compact size of the generated multi-level generating extensions. We demonstrate some of its advantages by comparing our approach with multiple self-application and solving a problem that we previously attacked unsuccessfully with a state-of-the-art partial evaluator.

6.1   Comparison with Multiple Self-Application

Some experiments with multiple self-application were reported in [Glü91], and we repeated these experiments with our multi-level generator in order to make a direct comparison: splitting a program for matrix transposition into 2 – 5 levels. The function transpose transposes a 5×n matrix where each of its five parameters takes one line of the original matrix. The algorithm can be transformed into a generating extension p-mgen that takes the first line of the matrix and generates a generating extension p-mgen$_0$ that takes the second line ... until the last generating extension generates a residual program that takes the last line and returns the transposed matrix as final result.

The differences in generation time (generator size) between 2 and 5 binding-time levels are drastically reduced by the multi-level compiler generator: from 1:9000 to 1:1.8 in generation-time, and from 1:100 to 1:2 in generator size. The absolute numbers for multiple self-application of a partial evaluator [Glü91] range from less than 100 ms to over 15 minutes (on a different hardware), and the size of the produced generating extensions ranges from 112 to over 12000 cons cells. The generation times using our multi-level compiler generator are given in ms and the sizes of the produced generating extensions as the number of cons cells needed to represent the program (Fig. 12).[7] The times t_{bta}, t_{ann} and t_{total} are the times for doing the multi-level binding-time analysis, the multi-level program annotation (including the insertion of specialization points), and the total run time of the compiler generator, respectively. Fig. 13 shows the size of the generating extensions p-mgen to p-mgen$_3$ when applied to a 5×3 matrix taking one line at a time.

6.2   Generating a Compiler Generator

A multi-level generator can transform multi-staged programs directly into their generating extension. In particular, we used the multi-level generator to convert a

---

[7] Times are given in cpu-seconds using Chez Scheme 3.2 and a Sparc Station 2/Sun OS 4.1 (excluding time for garbage collection, if any).

levels	2	3	4	5
t_{bta}/ms	5.5	5.4	5.1	5.1
t_{ann}/ms	8.0	11.0	15.1	19.5
t_{total}/ms	13.5	16.4	20.2	24.6
size/cells	180	236	295	357

Figure 12: Performance results for the transpose example

extension	p-mgen	p-mgen$_0$	p-mgen$_1$	p-mgen$_2$	p-mgen$_3$
size/cells	357	810	608	427	150
time/ms	28	19	12	5.6	0.1

Figure 13: Sizes and run-times of the generating extensions

two-level driving interpreter (the one defined in [GJ94]) into a compiler generator that is capable of a form of supercompilation: given a naive pattern matcher the new compiler generator returns a generating extension that, given a fixed pattern, produces specialized matchers that are equivalent to those generated by the Knuth, Morris, and Pratt algorithm (cf. [SGJ94]). We *failed* to achieve the same goal by incremental generation (Sect. 2.3) using Similix, a state-of-the-art partial evaluator for Scheme. It became impossible to respecialize the generating-extension gen-aux (see Sect. 2.3) with Similix's cogen because the generating extension was produced in continuation-passing style where static and dynamic values flow together [GJ94]. The necessary (manual) binding-time improvement of the machine-generated generating-extension turned out to be very difficult for non-trivial programs. The multi-level generator mcogen solved this problem elegantly — in *one* step!

The size of the two-level driving interpreter is 1494 cons cells, the size of the generated compiler generator is 3364 cons cells and the total generation time was 1.77 s.

7 Related Work

Two-Level Generators. The first hand-written compiler generator based on partial evaluation principles was, in all probability, the system *RedCompile* for a dialect of Lisp [BHOS76]. Romanenko [Rom88] gave transformation rules that convert annotated first-order programs into their two-level generating extensions.

Holst [Hol89] was the first to observe that the annotated version of a program is already the generating extension. What Holst called "syntactic currying" is now known as the "*cogen approach*" [BW94]. Holst and Launchbury [HL92] studied this approach for a small LML-like language in order to overcome the notorious encoding problem associated with typed languages in self-application (types in programs need to be mapped into an universal type in the partial evaluator).

Birkedal and Welinder [BW93] used the "cogen approach" for the Core Standard ML language, Bondorf and Dussart [BD94] combined this approach with cps-based specialization for the λ-calculus, and Andersen [And94] developed a generating extension generator for the ANSI C language.

Self-Application. Two methods have been used for improving the performance of self-application: *actions trees* [CD90], an off-line method that compiles the information obtained by the binding-time analysis into directives for the partial evaluator, and the *freezer* (metasystem jumps) [Tur89] in supercompilation, an on-line method that allows the evaluation of partially known expressions by the underlying implementation.

Both methods are less effective for multiple partial evaluation than the "multi-cogen approach" because (i) they do not avoid the transformation of dynamic expressions via the self-application tower (thus do not completely remove the interpretive overhead), and (ii) they do not address the generator size problem. However, we should note that a fair comparison with the freezer outside partial evaluation is difficult because of the different transformation paradigms.

Acknowledgments. Special thanks to Anders Bondorf for stimulating discussions about multiple self-application. Thanks to Dirk Dussart, Carsten K. Holst, Neil D. Jones, Torben Mogensen, Kristian Nielsen, Peter Sestoft, and Morten Welinder for comments.

References

[And94] Lars Ole Andersen. *Program Analysis and Specialization for the C Programming Language*. PhD thesis, DIKU, University of Copenhagen, May 1994. (DIKU report 94/19).

[BD91] Anders Bondorf and Olivier Danvy. Automatic autoprojection of recursive equations with global variables and abstract data types. *Science of Computer Programming*, 16:151–195, 1991.

[BD94] Anders Bondorf and Dirk Dussart. Improving cps-based partial evaluation: writing cogen by hand. In *ACM SIGPLAN Workshop on Partial Evaluation and Semantics-Based Program Manipulation*, pages 1–9, Orlando, Florida, 1994.

[BHOS76] Lennart Beckman, Anders Haraldson, Östen Oskarsson, and Erik Sandewall. A partial evaluator and its use as a programming tool. *Artificial Intelligence*, 7:319–357, 1976.

[BJ93] Anders Bondorf and Jesper Jørgensen. Efficient analyses for realistic offline partial evaluation. *Journal of Functional Programming, special issue on partial evaluation*, 11:315–346, 1993.

[Bon91] Anders Bondorf. Automatic autoprojection of higher order recursive equations. *Science of Computer Programming*, 17(1-3):3–34, December 1991. Revision of paper in ESOP'90, LNCS 432, May 1990.

[BW93] Lars Birkedal and Morten Welinder. Partial evaluation of Standard ML. DIKU Report 93/22, DIKU, Department of Computer Science, University of Copenhagen, 1993.

[BW94] Lars Birkedal and Morten Welinder. Hand-writing program generator generators. In M. Hermenegildo and J. Penjam, editors, *Programming Language Implementation and Logic Programming. Proceedings*, volume 844 of *LNCS*, pages 198–214, Madrid, Spain, 1994. Springer-Verlag.

[CC91] Felice Cardone and Mario Coppo. Type inference with recursive types: Syntax and semantics. *Information and Computation*, 92:48–80, 1991.

[CD90]   Charles Consel and Olivier Danvy. From interpreting to compiling binding
           times. In N. D. Jones, editor, *ESOP '90*, volume 432 of *LNCS*, pages 88–105,
           Copenhagen, Denmark, 1990. Springer-Verlag.

[Ers78]   Andrei P. Ershov. On the essence of compilation. In E.J. Neuhold, edi-
           tor, *Formal Description of Programming Concepts*, pages 391–420. North-
           Holland, 1978.

[GJ94]   Robert Glück and Jesper Jørgensen. Generating transformers for defor-
           estation and supercompilation. In B. Le Charlier, editor, *Static Analysis.
           Proceedings*, volume 864 of *LNCS*, pages 432–448, Namur, Belgium, 1994.
           Springer-Verlag.

[GJ95]   Robert Glück and Jesper Jørgensen. Constraint-based multi-level binding-
           time analysis of higher-order languages. Unpublished, 1995.

[Glü91]   Robert Glück. Towards multiple self-application. In *Proceedings of the Sym-
           posium on Partial Evaluation and Semantics-Based Program Manipulation*,
           pages 309–320, New Haven, Connecticut, 1991. ACM Press.

[Glü94]   Robert Glück. On the generation of specializers. *Journal of Functional
           Programming*, 4(4):499–514, 1994.

[Hat95]   John Hatcliff. Mechanically verifying the correctness of an off-line partial
           evaluator. In *PLILP'95*, LNCS. Springer-Verlag, 1995.

[Hen91]   Fritz Henglein. Efficient type inference for higher-order binding-time ana-
           lysis. In John Hughes, editor, *Conference on Functional Programming and
           Computer Architecture, Cambridge, Massachusetts. LNCS 523*, pages 448–
           472. Springer-Verlag, August 1991.

[HL92]   Carsten Kehler Holst and John Launchbury. Handwriting cogen to avoid
           problems with static typing. Working paper, 1992.

[Hol89]   Carsten Kehler Holst. Syntactic currying: yet another approach to partial
           evaluation. Technical report, DIKU, Department of Computer Science, Uni-
           versity of Copenhagen, 1989.

[JGS93]   Neil D. Jones, Carsten K. Gomard, and Peter Sestoft. *Partial Evaluation
           and Automatic Program Generation*. Prentice-Hall, 1993.

[Mog88]   Torben Æ. Mogensen. Partially static structures in a self-applicable partial
           evaluator. In Dines Bjørner, Andrei P. Ershov, and Neil D. Jones, editors,
           Partial Evaluation and Mixed Computation, pages 325–347. North-Holland,
           1988.

[NN92]   Flemming Nielson and Hanne R. Nielson. *Two-Level Functional Languages*,
           volume 34 of *Cambridge Tracts in Theoretical Computer Science*. Cam-
           bridge University Press, Cambridge, 1992.

[Rom88]   Sergei A. Romanenko. A compiler generator produced by a self-applicable
           specializer can have a surprisingly natural and understandable structure. In
           Dines Bjørner, Andrei P. Ershov, and Neil D. Jones, editors, *Partial Evalu-
           ation and Mixed Computation*, pages 445–463. North-Holland, 1988.

[Ses88]   Peter Sestoft. Automatic call unfolding in a partial evaluator. In Dines
           Bjørner, Andrei P. Ershov, and Neil D. Jones, editors, *Partial Evaluation
           and Mixed Computation*, pages 485–506. North-Holland, 1988.

[SGJ94]   Morten Heine Sørensen, Robert Glück, and Neil D. Jones. Towards unify-
           ing partial evaluation, deforestation, supercompilation, and GPC. In Donald
           Sannella, editor, *Programming Languages and Systems — ESOP '94. Pro-
           ceedings*, volume 788 of *LNCS*, pages 485–500, Edinburgh, Scotland, 1994.
           Springer-Verlag.

[Tur89]   Valentin F. Turchin. *Refal-5, Programming Guide and Reference Manual*.
           New England Publishing Co., Holyoke, Massachusetts, 1989.

[Wan93]   Mitchell Wand. Specifying the correctness of binding-time analysis. *Journal
           of Functional Programming*, 3(3):365–387, 1993.

Mechanically Verifying the Correctness of an Offline Partial Evaluator

John Hatcliff *

DIKU, Computer Science Department, Copenhagen University **

Abstract. We show that using deductive systems to specify an offline partial evaluator allows one to specify, prototype, and mechanically verify correctness *via* meta-programming — all within a single framework.

For a λ-mix-style partial evaluator, we specify binding-time constraints using a natural-deduction logic, and the associated program specializer using natural (aka "deductive") semantics. These deductive systems can be directly encoded in the Elf programming language — a logic programming language based on the LF logical framework. The specifications are then executable as logic programs. This provides a prototype implementation of the partial evaluator.

Moreover, since deductive system proofs are accessible as objects in Elf, many aspects of the partial evaluator correctness proofs (*e.g.*, the correctness of binding-time analysis) can be coded in Elf and mechanically checked.

1 Introduction

Offline partial evaluation consists of two phases: a binding-time analysis phase (where information is gathered about which parts of the source program depend on known or unknown data), and a specialization phase (where constructs depending on known data are reduced away) [3,15]. Recent work specifies the analysis phase using *type systems* [7] and the specialization phase using *operational semantics* [14,25,26]. The type system and operational semantics formalisms can be unified if one emphasizes their logical character: a type-based analysis is a logic for deducing program properties, and an operational semantics is a logic for deducing computational steps or input/output behaviour of programs. However, in program specialization systems that use these formalisms, this logical character has neither been emphasized nor exploited.

In this paper, we exploit this logical character and obtain a uniform framework for specifying, prototyping, and mechanically verifying the correctness of program specialization systems. Specifically, we consider offline partial evalua-

* This work is supported by the Danish Research Academy and by the DART project (Design, Analysis and Reasoning about Tools) of the Danish Research Councils.
** Universitetsparken 1, 2100 Copenhagen Ø, Denmark. E-mail: hatcliff@diku.dk

tion in the style of the partial evaluator λ-*mix* [7].[3] λ-mix is a good illustrative case since it is simple, and one of the few partial evaluators with a rigorous semantic foundation. It has also spawned additional work on the correctness of binding-time analysis [19,28] and specialization [16].

Our results are as follows.

- We give novel specifications of binding-time constraints and specialization as natural-deduction style logics. These specifications simplify meta-theory activities such as proving the correctness of binding-time analysis and specialization.
- We formalize the specifications using LF — a meta-language (a dependently-typed λ-calculus) for defining logics [12]. In LF, judgements (assertions) are represented as types, and deductions are represented as objects. Determining the validity of a deduction is reduced to checking if the representing object is well-typed. Since LF type-checking is decidable, purported deductions can be checked automatically for validity.
- We obtain prototypes directly from the formal specifications using Elf — a logic programming language based on LF [20]. Elf gives an operational interpretation to LF types by treating them as goals. Thus, the LF specifications of the binding-time analysis and specializer are directly executable in Elf.
- We formalize and mechanically verify much of the meta-theory of offline partial evaluation (*e.g.*, correctness of binding-time analysis and soundness of the specializer) *via* meta-programming in Elf. Correctness conditions are formalized as judgements about "lower-level" deductions describing object-language evaluation and transformation. Proofs of correctness are formalized as deductions of the correctness judgements. Elf type-checking mechanically verifies that these deductions (and hence the correctness proofs) are valid.

This methodology of specification/implementation/verification using LF and Elf has been successfully applied in other problem areas [11,17]. In particular, we build on Hannan and Pfenning's work on compiler verification in Elf [11]. They conjectured that their techniques could also be applied to partial evaluation [11, p. 416]. They also identify the verification of transformations based on flow analyses as a "challenging problem, yet to be addressed" [11, p. 415]. The present work addresses both of these points. We confirm their conjecture that LF and Elf can be used for specification/implementation/verification of partial evaluators. Moreover, we give one instance where transformations based on flow analyses can be verified — namely, the specialization of programs based on binding-time analysis.

The rest of the paper is organized as follows. Section 2 summarizes LF and Elf. Section 3 presents the object language and its encoding in Elf. Section 4 presents the specifications of the binding-time analysis and specializer. Section

---

[3] Our setting differs from that of λ-mix in two ways: (1) we are not concerned with self-applying the partial evaluator, and (2) our object language is typed while λ-mix's is untyped. However, the techniques here apply equally well to the untyped object language of λ-mix (in fact, they are simpler in the untyped case).

5 illustrates how these specifications are executable in Elf. Section 6 shows how meta-theoretic properties of the specifications can be mechanically verified. Section 7 surveys related work, and Section 8 concludes.

2 LF and the Elf Programming Language

2.1 LF — a framework for defining logics

The LF calculus has three levels: *objects*, *families*, and *kinds*. Families are classified by kinds, and objects are classified by *types*, *i.e.*, families of kind Type.

$$Kinds \quad K ::= \mathsf{Type} \mid \Pi x : A.\, K$$

$$Families \;\; A ::= a \mid \Pi x : A_1.\, A_2 \mid \lambda x : A_1.\, A_2 \mid A\, M$$

$$Objects \;\; M ::= c \mid x \mid \lambda x : A.\, M \mid M_1\, M_2$$

Family-level constants are denoted by a, and object-level constants by c. $A_1 \to A_2$ abbreviates $\Pi x : A_1.\, A_2$ when x does not appear free in A_2 (similarly for $\Pi x : A.\, K$). The typing rules for LF can be found in [12]. We take $\beta\eta$-equivalence as the notion of definitional equality in LF [12, Appendix A.3]. For all the languages we consider, we identify terms up to renaming of bound variables.

One defines a logic in LF by specifying a *signature* which declares the kinds of family-level constants a and types of object-level constants c. These constants are constructors for the logic's syntax, judgements, and deductions. Well-formedness is enforced by LF type-checking. The LF type system can represent the conditions associated with binding operators, with schematic abstraction and instantiation, and with the variable occurrence and discharge conditions associated with rules in systems of natural deduction.

2.2 Elf — an implementation of LF

The syntax of Elf is as follows (the last column lists the corresponding LF term, and optional components are enclosed in $\langle \cdot \rangle$).

kindexp ::=	**type**	Type
	$\mid \{id\langle :famexp\rangle\}\ kindexp$	$\Pi x : A.\, K$
	$\mid famexp\ \text{->}\ kindexp$	$A \to K$
famexp ::=	*id*	a
	$\mid \{id\langle :famexp_1\rangle\}\ famexp_2$	$\Pi x : A_1.\, A_2$
	$\mid [id\langle :famexp_1\rangle]\ famexp_2$	$\lambda x : A_1.\, A_2$
	$\mid famexp\ objexp$	$A\, M$
	$\mid famexp_1\ \text{->}\ famexp_2$	$A_1 \to A_2$
	$\mid famexp_2\ \text{<-}\ famexp_1$	$A_1 \to A_2$
objexp ::=	*id*	c
	$\mid [id\langle :famexp\rangle]\ objexp$	$\lambda x : A.\, M$
	$\mid objexp_1\ objexp_2$	$M_1\, M_2$

Object language:

$e \in Exp$ $\tau \in Typ$

$e ::= 0 \mid x \mid \mathbf{lam}\, x.e \mid \mathbf{app}\, e_0\, e_1$ $\tau ::= \mathbf{nat} \mid \tau_1 \to \tau_2$

Elf encoding:

```
exp : type.                        typ  : type.

z   : exp.                         nat  : typ.
lam : (exp -> exp) -> exp.         arrow : typ -> typ -> typ.
app : exp -> exp -> exp.
```

Fig. 1. The object language Λ

The terminal *id* ranges over variables, and family and object constants. Bound variables and constants in Elf can be arbitrary identifiers, but free variables in a declaration or query must begin with an upper case letter. Free variables act as logic variables and are implicitly Π-abstracted. Elf's term reconstruction phase (preprocessing) inserts these abstractions as well as appropriate arguments to these abstractions. It also fills in the omitted types in quantifications $\{x\}$ and abstractions $[x]$ and omitted types or objects indicated by an underscore _. The <- is used to improve the readability of some Elf programs. B <- A is parsed into the same representation as A -> B; -> is right associative, while <- is left associative. An Elf program is a representation of an LF signature. Although we are implicitly encoding logics in LF, we give all encodings using the syntax of Elf.

3   The Object Language

Figure 1 presents the syntax of the object language Λ (a small subset of PCF)[4] and its encoding in Elf.[5] The Elf signature for the object language syntax includes family constants **exp** and **typ**, and object constants for each term and

---

[4] In the extended version of this paper [13], we treat full PCF, *i.e.*, simply-typed λ-terms with primitive operations (*e.g.*, succ, pred), conditionals, and fixpoint constructs [8]. In the present version, space constraints force us to consider only those constructs which best illustrate principles. Including the other constructs is a straightforward extension of the work here (the encodings are very similar to those given by Michaylov and Pfenning for standard type systems and deductive semantics [17]). At no point in the presentation do we take advantage of the fact that the subset Λ is strongly normalizing and computationally incomplete.

[5] Technically, one must ensure that such encodings are *adequate*, *i.e.*, that there is a compositional bijection between the syntactic entities in the logical system and well-formed LF $\beta\eta$-normal forms under the given signature. An adequate encoding ensures that each entity is encoded uniquely and that no representations of additional entities are introduced. All the encodings we use are adequate. See Harper *et al.*[12] for a detailed discussion and proofs of adequacy for encodings similar to the ones used here.

type constructor. Binding in lam $x \cdot e$ is represented using binding in the meta-language (*i.e.*, using higher-order abstract syntax [22]). Variables in the object language are identified with variables in the meta-language, so there is no explicit representation of identifiers in the Elf signature for the object language. For example, the expression

$$\mathsf{app}\,(\mathsf{lam}\,x_1\,.\,\mathsf{lam}\,x_2\,.\,\mathsf{app}\,(\mathsf{lam}\,x_3\,.\,x_3)\,x_1)\,0$$

is encoded as

$$\mathsf{app}\,(\mathsf{lam}\,[\mathtt{x1}]\,\mathsf{lam}\,[\mathtt{x2}]\,\mathsf{app}\,(\mathsf{lam}\,[\mathtt{x3}]\,\mathtt{x3})\,\mathtt{x1})\,\mathtt{z}.$$

We omit the standard typing rules and call-by-name operational semantics for Λ^6 — the binding-time rules and operational semantics for the program specializer (given in the following section) generalize these.

4 Specifying an Offline Partial Evaluator for Λ

A partial evaluator takes a *source program* p and a subset s of p's input, and produces a *residual program* p_s which is specialized with respect to s. The correctness of the partial evaluator implies that running p_s on p's remaining input d gives the same result as running p on the complete input s and d. The data s and d are often referred to as *static* and *dynamic* data (respectively) since s is fixed at specialization time whereas one may supply various data d during runs of p_s.

The specialized program p_s is obtained from p by evaluating constructs that depend only on s, while rebuilding constructs that may depend on dynamic data. Offline partial evaluation accomplishes this in two phases: (1) a binding-time analysis phase, and (2) a specialization phase.

1. **Binding-time analysis**: Given assumptions about which program inputs are static and dynamic, binding-time analysis assigns each source program construct a *specialization directive* and a *specialization type*. This information is expressed by constructing an annotated version of the source program.

   - **Specialization directives**: A construct is assigned a directive of *eliminable* if it depends only on static data and thus can be completely evaluated during the specialization phase. A construct is assigned a directive of *residual* if it may depend on dynamic data and thus must be reconstructed in the specialization phase.

---

[6] These can be found in the extended version of this paper [13]. Michaylov and Pfenning [17] give Elf encodings of typing rules and call-by-value operational semantics for a language similar to Λ.

Annotated object language:

$$w \in Sexp \qquad\qquad m \in Sder$$
$$w ::= 0_m \mid y \mid \mathsf{lam}_m\, y\,.\, w \mid \mathsf{app}_m\, w_0\, w_1 \mid \mathsf{lift}\, w \qquad m ::= s \mid d$$

Elf encoding:

```
sexp : type.                                sder : type.

z    : sder -> sexp.                        s : sder.
lam  : sder -> (sexp -> sexp) -> sexp.       d : sder.
app  : sder -> sexp -> sexp -> sexp.
lift : sexp -> sexp.
```

Fig. 2. The annotated object language Λ_{bt}

Specialization types: *Elf encoding:*

$$\varphi_{\mathsf{nat}} \in Styp[\mathsf{nat}]$$
$$\varphi_{\mathsf{nat}} ::= \mathsf{sta} \mid \mathsf{dyn}_{\mathsf{nat}}$$

$$\varphi_{\tau_1 \to \tau_2} \in Styp[\tau_1 \to \tau_2]$$
$$\varphi_{\tau_1 \to \tau_2} ::= \varphi_{\tau_1} \to \varphi_{\tau_2} \mid \mathsf{dyn}_{\tau_1 \to \tau_2}$$

```
styp  : typ -> type.

sta   : styp nat.
dyn   : styp T.
barrow : styp T1 -> styp T2
         -> styp (arrow T1 T2).
```

Fig. 3. The specialization types for Λ_{bt}

- **Specialization types**: The specialization type assigned to a construct describes the directive structure of terms to which it may reduce during specialization. The specialization types are the carriers of information during the analysis phase.
2. **Specialization**: During the specialization phase, the specializer simply follows the directives assigned during binding-time analysis: eliminable constructs are evaluated (and thus eliminated); residual constructs are reconstructed (and thus appear in the residual program).

4.1 Binding-time analysis

A binding-time analysis associates each Λ term with a term in the annotated language Λ_{bt} of Figure 2. An annotated term is indexed by a specialization directive s or d indicating if it is eliminable (*i.e.*, it depends only on static data) or residual (*i.e.*, it may depend on dynamic data). Identifiers are not indexed since the appropriate information can be determined from the environment. A coercion construct lift is added to Λ_{bt} to residualize the result of evaluating an eliminable term. This allows static computation to occur in a residual context. A term $w \in \Lambda_{bt}$ is *completely residual* if it consists of only d-indexed constructs and identifiers. Intuitively, the specializer will output completely residual terms — all eliminable constructs will have been evaluated. The Elf encoding of Λ_{bt} terms follows that of Λ terms (except that directive indexing is captured by supplying an extra argument of type **sder** to a constructor).

Figure 3 presents a τ-indexed family of specialization types for Λ_{bt}. A specialization type φ is *dynamic* if $\varphi = $ dyn; otherwise φ is *static*. We omit type indices on specialization types when they can be inferred from the context. Specialization types are encoded in Elf *via* the type family **styp:typ->type**. For the **dyn** and **barrow** constructors, the type indices are encoded as logic variables **T**, **T1**, and **T2**, which will be instantiated by unification. This formalizes the above convention of allowing type indices to be inferred from the context.

Figure 4 presents a natural-deduction-style logic for deriving binding-time analysis constraints.[7] Parentheses in the hypotheses of the rules *bta_lam_s* and *bta_lam_d* indicate the discharging of zero or more occurrences of assumptions. We write $\Gamma \vdash bta\ e : \tau\ [w : \varphi_\tau]$ when $bta\ e : \tau\ [w : \varphi_\tau]$ is derivable under undischarged assumptions Γ. Intuitively, if $\Gamma \vdash bta\ e : \tau\ [w : \varphi_\tau]$, then given initial binding-time assumptions Γ, a binding-time analysis may map $e \in Terms[\Lambda]$ of type $\tau \in Types[\Lambda]$ to a directive annotated term $w \in Terms[\Lambda_{bt}]$ of specialization type $\varphi_\tau \in Spec\text{-}types[\tau]$. We only consider assumptions involving identifiers (*e.g.*, $bta\ x : \tau\ [y : \varphi_\tau]$) since the logic only allows discharging of assumptions of this form. For $\Gamma = \{bta\ x_1 : \tau_1\ [y_1 : \varphi_{\tau_1}], ..., bta\ x_n : \tau_n\ [y_n : \varphi_{\tau_n}]\}$, the x_i are required to be pairwise distinct (similarly for the y_i). A *static assumption* (resp. *dynamic assumption*) is an assumption $bta\ x : \tau\ [y : \varphi_\tau]$ where φ_τ is static (resp. dynamic). Γ_s denotes a set of only static assumptions; Γ_d denotes a set of only dynamic assumptions.

A simple induction over the structure of deductions for $\Gamma \vdash bta\ e : \tau\ [w : \varphi_\tau]$ shows that the relation between e and w is one-to-many, *i.e.*, there may be many valid annotations of e. One may always obtain a valid annotation of a type correct e by annotating all components as residual. Since the relation is one-to-many, it defines an *annotation forgetting function* from Λ_{bt} to Λ. Intuitively, this function removes directives and lift constructs.

The binding-time judgement is encoded in Elf as follows.

```
bta : exp -> {t : typ} sexp -> styp t -> type.
```

The dependent function type (*i.e.*, {t : typ} ...) expresses the dependency of the indexed specialization type on the type of the Λ expression.

Figure 5 presents the Elf encoding of the binding-time logic. The implicit universal quantification of the upper case variables captures the schematic nature of the rules. The rules for binding constructs (which involve hypothetical proofs in the premises) are encoded in Elf as proof constructors that take proofs of hypothetical judgements as arguments. Such proofs are represented as functions mapping proofs of assumption judgments to proofs of consequent

---

[7] We use the notation of Prawitz [24].

$$bta\_z\_s: \qquad\qquad bta\ 0 : \mathsf{nat}\ [0_s : \mathsf{sta}]$$

$$bta\_lam\_s: \qquad\qquad \frac{\begin{array}{c}(bta\ x : \tau_1\ [y : \varphi_1])\\ bta\ e : \tau_2\ [w : \varphi_2]\end{array}}{bta\ \mathsf{lam}\ x\,.\,e : \tau_1 \to \tau_2\ [\mathsf{lam}_s\ y\,.\,w : \varphi_1 \to \varphi_2]}$$

$$bta\_app\_s: \quad \frac{bta\ e_0 : \tau_1 \to \tau_2\ [w_0 : \varphi_1 \to \varphi_2] \qquad bta\ e_1 : \tau_1\ [w_1 : \varphi_1]}{bta\ \mathsf{app}\ e_0\ e_1 : \tau_2\ [\mathsf{app}_s\ w_0\ w_1 : \varphi_2]}$$

$$bta\_z\_d: \qquad\qquad bta\ 0 : \mathsf{nat}\ [0_d : \mathsf{dyn}]$$

$$bta\_lam\_d: \qquad\qquad \frac{\begin{array}{c}(bta\ x : \tau_1\ [y : \mathsf{dyn}])\\ bta\ e : \tau_2\ [w : \mathsf{dyn}]\end{array}}{bta\ \mathsf{lam}\ x\,.\,e : \tau_1 \to \tau_2\ [\mathsf{lam}_d\ y\,.\,w : \mathsf{dyn}]}$$

$$bta\_app\_d: \quad \frac{bta\ e_0 : \tau_1 \to \tau_2\ [w_0 : \mathsf{dyn}] \qquad bta\ e_1 : \tau_1\ [w_1 : \mathsf{dyn}]}{bta\ \mathsf{app}\ e_0\ e_1 : \tau_2\ [\mathsf{app}_d\ w_0\ w_1 : \mathsf{dyn}]}$$

$$bta\_lift: \qquad\qquad \frac{bta\ e : \mathsf{nat}\ [w : \mathsf{sta}]}{bta\ e : \mathsf{nat}\ [\mathsf{lift}\ w : \mathsf{dyn}]}$$

Fig. 4. The binding-time logic

```
bta_z_s : bta z nat (bz s) sta.

bta_lam_s : bta (lam E) (arrow T1 T2) (blam s W) (barrow P1 P2)
              <- {x:exp} {y:sexp} (bta x T1 y P1) -> (bta (E x) T2 (W y) P2).

bta_app_s : bta (app E0 E1) T2 (bapp s W0 W1) P2
              <- bta E0 (arrow T1 T2) W0 (barrow P1 P2)
              <- bta E1 T1 W1 P1.

bta_z_d : bta z nat (bz d) dyn.

bta_lam_d : bta (lam E) (arrow T1 T2) (blam d W) dyn
              <- {x:exp} {y:sexp} (bta x T1 y dyn) -> (bta (E x) T2 (W y) dyn).

bta_app_d : bta (app E0 E1) T2 (bapp d W0 W1) dyn
              <- bta E0 (arrow T1 T2) W0 dyn
              <- bta E1 T1 W1 dyn.

bta_lift : bta E nat lift W dyn
              <- bta E nat W sta.
```

Fig. 5. The Elf encoding of the binding-time logic

$$0_s \Downarrow_{spec} 0_s \qquad\qquad \text{spec\_z\_s : spec (bz s) (bz s).}$$

$$\text{lam}_s\, y \cdot w \Downarrow_{spec} \text{lam}_s\, y \cdot w \qquad \text{spec\_lam\_s : spec (blam s W)}$$
$$\text{(blam s W).}$$

$$\frac{w_0 \Downarrow_{spec} \text{lam}_s\, y \cdot w'_0 \qquad w'_0[y := w_1] \Downarrow_{spec} a}{\text{app}_s\, w_0\, w_1 \Downarrow_{spec} a} \quad \begin{array}{l}\text{spec\_app\_s : spec (bapp s W0 W1) A} \\ \text{<- spec W0 (blam s W0')} \\ \text{<- spec (W0' W1) A.}\end{array}$$

$$0_d \Downarrow_{spec} 0_d \qquad\qquad \text{spec\_z\_d : spec (bz d) (bz d).}$$

$$\frac{\begin{array}{c}(y \Downarrow_{spec} y) \\ w \Downarrow_{spec} a\end{array}}{\text{lam}_d\, y \cdot w \Downarrow_{spec} \text{lam}_d\, y \cdot a} \quad \begin{array}{l}\text{spec\_lam\_d : spec (blam d W) (blam d A)} \\ \text{<- \{y : sexp\}} \\ \text{(spec y y ->} \\ \text{spec (W y) (A y)).}\end{array}$$

$$\frac{w_0 \Downarrow_{spec} a_0 \qquad w_1 \Downarrow_{spec} a_1}{\text{app}_d\, w_0\, w_1 \Downarrow_{spec} \text{app}_d\, a_0\, a_1} \quad \begin{array}{l}\text{spec\_app\_d : spec (bapp d W0 W1)} \\ \text{(bapp d A0 A1)} \\ \text{<- spec W0 A0} \\ \text{<- spec W1 A1.}\end{array}$$

$$\frac{w \Downarrow_{spec} 0_s}{\text{lift } w \Downarrow_{spec} 0_d} \quad \begin{array}{l}\text{spec\_lift : spec (lift W) (bz d)} \\ \text{<- spec W (bz s).}\end{array}$$

Fig. 6. The specialization logic

judgements. The higher-order syntax representations requires that judgements involving identifiers be represented as schematic judgements (*i.e.*, identifiers are Π-quantified). This is the case in rules bta_lam_s and bta_lam_d where *e.g.*, the judgement bta (E x) T2 (W y) P2 expresses that E and W may be instantiated to representations of terms with free occurrences of x and y respectively [12, Section 3.1].

4.2 Specialization

Figure 6 presents the specialization logic for Λ_{bt} terms.

- The first four rules describe the evaluation of eliminable constructs. These rules correspond to the usual call-by-name "natural" or "deductive" operational semantics for PCF [8, Chapter 4].
- The next four rules describe the reconstruction of residual constructs after subexpressions have been specialized. The rule for lam$_d$ is noteworthy because it introduces operation on open terms (*i.e.*, bodies of lam$_d$ constructs). Following Hannan [9, p. 144], we specialize the body of a lam$_d$ under the assumption that the bound variable evaluates to itself.

– The final rule coerces an eliminable result to a residual expression.

We write $\Sigma \vdash w \Downarrow_{spec} a$ when $w \Downarrow_{spec} a$ is derivable under undischarged assumptions Σ. Intuitively, if $\Sigma \vdash w \Downarrow_{spec} a$, then the specializer maps $w \in Terms[\Lambda_{bt}]$ to answer $a \in Terms[\Lambda_{bt}]$ in context Σ. We only consider assumptions involving identifiers (*e.g.*, $y \Downarrow_{spec} y$) since the logic only allows discharging of assumptions of this form. For $\Sigma = \{y_1 \Downarrow_{spec} y_1, ..., y_n \Downarrow_{spec} y_n\}$ the y_i are required to be pairwise distinct. Σ *is compatible with* $\Gamma = \{bta\ x_1 : \tau_1\ [y_1 : \varphi_{\tau_1}], ..., bta\ x_n : \tau_n\ [y_n : \varphi_{\tau_n}]\}$ if $\Sigma = \{y_1 \Downarrow_{spec} y_1, ..., y_n \Downarrow_{spec} y_n\}$. It is easy to check that the relation induced by $\Downarrow_{spec}$ is a partial function (given the constraints on assumptions above).

Figure 6 also gives the Elf encoding of the specialization logic. The encoding techniques are similar to those used in the previous section. In the rule **spec_app_s**, we take advantage of the higher-order abstract syntax representation and use Elf application (*i.e.*, β-reduction) to implement capture-free substitution.

In a conventional deductive semantics for Λ, one has 0 and $lam\ x\ .\ e$ as canonical terms, *i.e.*, these terms are the results of evaluation. Intuitively, a specializer is part evaluator and part compiler. Therefore, the canonical terms of the specializer are 0_s and $lam_s\ y\ .\ w$ (corresponding to evaluation results), and completely residual terms or *code* (corresponding to compilation results). We will give a mechanically verified proof of this claim in Section 6. In preparation, we formalize the notion of canonical term or *answer* by defining a judgement $ans\ w : \varphi_\tau$. Intuitively, if $ans\ w : \varphi_\tau$ holds, then w is a canonical term of specialization type φ_τ. In particular, if $\vdash ans\ w : dyn$, then w is completely residual. We omit the direct statement of rules and simply give the following Elf encoding using the type $ans : sexp \rightarrow styp\ T \rightarrow type$.

```
                                       ans_lam_d: ans (blam d W) dyn
ans_z_s: ans (bz s) sta.                  <- {y} ans y dyn ->
                                              ans (W y) dyn.
ans_lam_s: ans (blam s W) (barrow P1 P2).

                                       ans_app_d: ans (bapp d W1 W2) dyn
ans_z_d: ans (bz d) dyn.                  <- ans W1 dyn
                                          <- ans W2 dyn.
```

We write $A \vdash ans\ w : \varphi_\tau$ when $ans\ w : \varphi_\tau$ is derivable under undischarged assumptions A.

A *is compatible with* $\Gamma = \{bta\ x_1 : \tau_1\ [y_1 : \varphi_{\tau_1}], ..., bta\ x_n : \tau_n\ [y_n : \varphi_{\tau_n}]\}$ if $A = \{ans\ y_1 : \varphi_{\tau_1}, ..., ans\ y_n : \varphi_{\tau_n}\}$.

4.3 Partial evaluation

We outline how the logics above define offline partial evaluation using the following object term.

$$e \stackrel{\text{def}}{=} app\ (lam\ x_1\ .\ app\ x_2\ (app\ (lam\ x_3\ .\ x_3)\ x_1))\ x_0$$

The free variables x_0, x_2 represent input parameters. The assumptions $\Gamma_s = \{bta\ x_0 : \text{nat}\ [y_0 : \text{sta}]\}$ and $\Gamma_d = \{bta\ x_2 : \text{nat} \rightarrow \text{nat}\ [y_2 : \text{dyn}]\}$ identify x_0 of type nat as *known* and x_2 of type nat $\rightarrow$ nat as *unknown*. They also associate x_0 and x_2 with annotated language identifiers y_0 and y_2. The fact that $\Gamma_s \cup \Gamma_d \vdash bta\ e : \text{nat}\ [w : \text{dyn}]$ where

$$w \overset{\text{def}}{=} \text{app}_s\ (\text{lam}_s\ y_1 \cdot \text{app}_d\ y_2\ (\text{lift}\ (\text{app}_s\ (\text{lam}_s\ y_3 \cdot y_3)\ y_1)))\ y_0$$

expresses that a binding-time analysis may associate e with w based on assumptions $\Gamma_s \cup \Gamma_d$.

To prepare for specialization, we supply known input *via* substitution.

$$\Gamma_d \vdash bta\ e[x_0 := 0] : \text{nat}\ [w[y_0 := 0_s] : \text{dyn}]$$

Now taking $\Sigma = \{y_2 \Downarrow_{spec} y_2\}$ compatible with Γ_d (expressing that the dynamic parameter evaluates to itself), the specialization logic gives

$$\Sigma \vdash w[y_0 := 0_s] \Downarrow_{spec} \text{app}_d\ y_2\ 0_d.$$

Theorem 1 (correctness of binding-time analyis) of Section 6.1 tells us there exists $e' \in \Lambda$ such that $\Gamma_d \vdash bta\ e' : \text{nat}\ [\text{app}_d\ y_2\ 0_d : \text{dyn}]$. As noted in Section 4.1, e' must be app x_2 0 — the unannotated version of $\text{app}_d\ y_2\ 0_d$. Theorem 2 (soundness of specialization) of Section 6.2 tells us that $e[x_0 := 0]$ is convertible to e' (denoted $e[x_0 := 0] =_\Lambda e'$) in the program calculus for Λ (defined in Section 6.2). Thus, for all closed inputs $d \in \Lambda$ of type nat $\rightarrow$ nat,

$$e[x_0 := 0,\ x_2 := d] =_\Lambda e'[x_2 := d].$$

This reflects the correctness criteria for partial evaluation given at the beginning of Section 4: running e on static input 0 and dynamic input d is operationally equivalent to running e' on d.

5 Prototyping an Offline Partial Evaluator for Λ

5.1 Prototyping a binding-time analysis

The Elf encoding of the binding-time logic (see Figure 5) gives a prototype of the binding-time analysis. The following Elf query returns all the possible annotations that may be assigned to the example term of Section 3 (here we make dyn the specialization type of the entire term).[8]

```
?- bta (app (lam [x1] lam [x2] app (lam [x3] x3) x1) z)
       (arrow nat nat) W dyn.
Solving...

W = bapp s
    (blam s ([y1:sexp]
```

---

[8] Some of the results of Elf evaluation are α-converted for clarity.

```
                  blam d ([y2:sexp]
                          bapp s (blam s ([y3:sexp] y3)) (lift y1))))
        (bz s).

W = bapp s
      (blam s ([y1:sexp]
              blam d ([y2:sexp]
                      lift (bapp s (blam s ([y3:sexp] y3)) y1))))
        (bz s).
```

There are actually twelve correct annotations; we show only the two above.

5.2 Prototyping a specializer

The Elf encoding of the specialization logic (see Figure 6) gives a prototype of the specializer. The following Elf query returns the result A of specializing the second annotated term above (there is only one answer since spec is a function).

```
?- S : spec (bapp s
                (blam s ([y1:sexp]
                        blam d ([y2:sexp]
                                lift (bapp s (blam s ([y3:sexp] y3))
                                                    y1))))
                (bz s)) A.
Solving...

A = blam d ([y2:sexp] bz d),
S = spec_app_s
        (spec_lam_d [y2:sexp] [S:spec y2 y2]
                    spec_lift (spec_app_s spec_z_s spec_lam_s))
            spec_lam_s.
;
no more solutions
```

In this query, we add the variable S which binds to the specialization logic deduction used to obtain A.[9]

6 Verifying an Offline Partial Evaluator for Λ

6.1 Binding-time analysis

A binding-time analysis is correct if it always produces consistent specialization directives. Directives are consistent if the specializer does not "go wrong". A

---

[9] Note that the order of arguments to the deduction constructors is the reverse of what one might expect since we use <- (instead of ->) in the encodings of the binding-time and specialization logics.

specializer may go wrong for two reasons: 1) it trusts a part of the program to be eliminable when in fact it is residual, and 2) it trusts a part of the program to be residual when in fact it is eliminable.

The specializer goes wrong for the first reason on $\mathsf{app}_s\,(\mathsf{lam}_d\,x\,.\,x)\,0_s$. In this case, it attempts evaluation using the rule *spec_app_s* (see Figure 6) but "hangs" (*i.e.*, the result of specialization is undefined) since $\mathsf{lam}_d\,x\,.\,x$ is not an eliminable abstraction. Similar problems may occur with the rule *spec_lift* (*e.g.*, if $w \Downarrow_{spec} 0_d$).

The specializer goes wrong for the second reason on $\mathsf{app}_d\,(\mathsf{lam}_s\,x\,.\,x)\,0_d$. In this case, it attempts residualization using the rule *spec_app_d* and incorrectly produces an output program that is not completely residual (since $\mathsf{lam}_s\,x\,.\,x$ is eliminable).

The property that the specializer never goes wrong (*i.e.*, the binding-time analysis always yields consistent directives) is a generalization of the type-soundness property for standard type systems. To establish type-soundness, one typically proves a *subject-reduction* result showing that typing is maintained under evaluation. To establish consistency of directives, we prove that specialization typing is maintained under specialization. Furthermore, we show that specialization results are always answers of the appropriate specialization type, *i.e.*, that the *ans* is always satisfied (as promised in Section 4.2).

This ensures that the specializer will not go wrong for the first reason. For example, in the rule *spec_app_s*, the rule *bta_app_s* guarantees that w_0 always has specialization type $\varphi_1 \rightarrow \varphi_2$ and if $w_0 \Downarrow_{spec} a_0$ then $a_0 \equiv \mathsf{lam}_s\,x\,.\,w_0'$ since only appropriate answers are produced.

This also ensures that the specializer will not go wrong for the second reason. For example, in the rule *spec_app_d*, the rule *bta_app_d* guarantees that w_0 always has specialization type dyn and if $w_0 \Downarrow_{spec} a_0$ then a_0 must be completely residual (similarly for a_1).

Theorem 1 Correctness of binding-time analysis. *If* $\Sigma \vdash w \Downarrow_{spec} a$ *and* $\Gamma_d \vdash bta\ e : \tau\,[w : \varphi_\tau]$ *and* Σ *is compatible with* Γ_d, *then there exists* $e' \in \Lambda$ *such that* $\Gamma_d \vdash bta\ e' : \tau\,[a : \varphi_\tau]$ *and* $A \vdash ans\ a : \varphi_\tau$ *where* A *is compatible with* Γ_d.

Proof. For notational convenience, we write $\mathcal{D} :: \mathcal{J}$ when $\mathcal{D}$ is a deduction of judgement $\mathcal{J}$ and state deduction rules in a linear format.[10]

For the theorem hypotheses, let $\mathcal{S} :: w \Downarrow_{spec} a$ and $\mathcal{B} :: bta\ e : \tau\,[w : \varphi_\tau]$. We show an effective method for constructing deductions $\mathcal{C} :: bta\ e' : \tau\,[a : \varphi_\tau]$ and $\mathcal{D} :: ans\ a : \varphi_\tau$ where the compatibility constraints on undischarged assumptions are satisfied. The proof proceeds by induction on the pair of deductions $\mathcal{S}$ and $\mathcal{B}$, *i.e.*, the method is primitive recursive. Although this primitive recursive method

---

[10] For example, a specialization deduction that ends with the rule *spec_app_d* is written
spec_app_d$(\mathcal{S}_0\,,\,\mathcal{S}_1) :: \mathsf{app}_d\,w_0\,w_1 \Downarrow_{spec} \mathsf{app}_d\,a_0\,a_1$ where $\mathcal{S}_i :: w_i \Downarrow_{spec} a_i\ (i = 1, 2)$. The notation is somewhat imprecise since we do not use an explicit discharge function for assumptions [12, Section 4.1]. However, the Elf encoding makes matters sufficiently clear.

cannot be represented in Elf as a function (since it is not schematic), it can be represented as a relation *via* the following judgement.

`t1 : spec W A -> bta E T W P -> bta E' T A P -> ans A P -> type.`

Each case of the constructive proof is formalized as a rule for the `t1` judgement. Below we show three illustrative cases (each increasing in complexity).

case $S = spec\_z\_s :: 0_s \Downarrow_{spec} 0_s$ and $B = bta\_z\_s :: bta\ 0 : nat\ [0_s : sta]$:

The required deductions are $bta\_z\_s :: bta\ 0 : nat\ [0_s : sta]$ and $ans\_z\_s :: ans\ 0_s : sta$. This is formalized by the following axiom.

    `t1_z_s : t1 (spec_z_s) (bta_z_s) (bta_z_s) (ans_z_s).`

case $S = spec\_app\_d(S_0, S_1) :: app_d\ w_0\ w_1 \Downarrow_{spec} app_d\ a_0\ a_1$
    $B = bta\_app\_d(B_0, B_1) :: bta\ app\ e_0\ e_1 : \tau_2\ [app_d\ w_0\ w_1 : dyn]$:

Applying the inductive hypothesis to S_0 and B_0 gives deductions $C_0 ::$ $bta\ e_0' : \tau_1 \rightarrow \tau_2\ [a_0 : dyn]$ and $D_0 :: ans\ a_0 : dyn$. Applying the inductive hypothesis to S_1 and B_1 gives deductions $C_1 :: bta\ e_1' : \tau_1 \rightarrow \tau_2\ [a_1 : dyn]$ and $D_1 :: ans\ a_1 : dyn$. The required deductions are $bta\_app\_d(C_0, C_1)$ and $ans\_app\_d(D_0, D_1)$. This is formalized by the following rule (the arguments to the proof constructors appear in reverse order (*e.g.*, `spec_app_d S1 S0` instead of `spec_app_d S0 S1`) due to the use of `<-` (instead of `->`) in the encodings of binding-time and specialization logic).

    `t1_app_d : t1 (spec_app_d S1 S0) (bta_app_d B1 B0)`
                  `(bta_app_d C1 C0) (ans_app_d D1 D0)`
                   `<- t1 S0 B0 C0 D0`
                   `<- t1 S1 B1 C1 D1.`

In operational terms, the inductive hypotheses are manifested as recursive calls to the function represented by `t1`.

case $S = spec\_app\_s(S_0, S_1) :: app_s\ w_0\ w_1 \Downarrow_{spec} a$
    $B = bta\_app\_s(B_0, B_1) :: bta\ app\ e_0\ e_1 : \tau_2\ [app_s\ w_0\ w_1 : \varphi_2]$:

Applying the inductive hypothesis to S_0 and B_0 gives deductions $C_0' ::$ $bta\ e_0' : \tau_1 \rightarrow \tau_2\ [a_0 : \varphi_1 \rightarrow \varphi_2]$ and $D_0 :: ans\ a_0 : \varphi_1 \rightarrow \varphi_2$. An examination of the rules of Figure 4 shows that we must have $C_0' = bta\_lam\_s(C_0 ::$ $bta\ e_0'' : \tau_2\ [w_0' : \varphi_2]) :: bta\ lam\ x\ .\ e'' : \tau_1 \rightarrow \tau_2\ [lam_s\ y\ .\ w_0' : \varphi_1 \rightarrow \varphi_2]$ where the undischarged assumptions of C_0 include $bta\ x : \tau_1\ [y : \varphi_1]$. A simple substitution lemma (which we omit for lack of space) allows us to replace each of these assumptions in C_0 with the deduction $B_1 :: bta\ e_1 : \tau_1\ [w_1 : \varphi_1]$ to obtain a deduction $B_2 :: bta\ e_0''[x := e_1] : \tau_2\ [w_0'[y := w_1] : \varphi_2]$. Applying the inductive hypothesis to S_1 and B_2 gives the required deductions $C :: bta\ e' : \tau_2\ [a : \varphi_2]$ and $D :: ans\ a : \varphi_2$. This is formalized as follows.

    `t1_app_s : t1 (spec_app_s S1 S0) (bta_app_s B1 B0) C D`
                  `<- t1 S0 B0 (bta_lam_s C0) D0`
                  `<- t1 S1 (C0 _ _ B1) C D.`

The "examination of the rules" (that told us deduction C_0' must have *bta_lam_s* as its last rule) is captured by matching (*i.e.*, (bta_lam_s C0)). The required substitution lemma appears for free due to the higher-order abstract syntax representation and the *Transitivity* derived rule of the LF calculus [12, Section 2.3]; it is captured by (C0 _ _ B1). The underscores correspond to the terms e_1 and w_1 above. Using the underscores and letting Elf reconstruct allows us to encode the rule more concisely.

The complete set of rules is given in Appendix A.

The following Elf query illustrates the function induced by the t1 rules. Given the specialization deduction S of Section 5.2 and a binding-time deduction for the term being specialized (*i.e.*, the second annotated term of Section 5.1), t1 constructs a binding-time deduction C for the specialization answer as well as an answer deduction D proving that the answer is completely residual.

```
?- t1 (spec_app_s
          (spec_lam_d [y2:sexp] [S:spec y2 y2]
                        spec_lift (spec_app_s spec_z_s spec_lam_s))
          spec_lam_s)
       (bta_app_s bta_z_s
                  (bta_lam_s [x1:exp] [y1:sexp]
                   [B1:bta x1 nat y1 sta]
                   bta_lam_d [x2:exp] [y2:sexp]
                    [B2:bta x2 nat y2 dyn]
                    bta_lift
                       (bta_app_s B1
                          (bta_lam_s [x3:exp] [y3:sexp]
                             [B3:bta x3 nat y3 sta] B3)))))
       C D.
Solving...

D = ans_lam_d [y2:sexp] [D2:ans y2 dyn] ans_z_d,
C = bta_lam_d [x2:exp] [y2:sexp] [B2:bta x2 nat y2 dyn] bta_z_d.
```

Elf type-checking mechanically verifies that deductions C and D are well-formed when they exist. However, verification that such deductions *always* exist (*i.e.*, that t1 is *total*) cannot be captured in Elf. This phase of verification (called *schema checking*) must be done by hand, although its automation is the subject of current research [23]. Our definition makes t1 total because we give a rule for each possible pair of *spec* and *bta* rules (giving primitive recursive structure). In addition, the deduction of answer judgements tell us that matching is used (in rules t1_app_s and t1_lift) only when it will always succeed.

6.2 Soundness of specialization

A specializer is sound if its steps reflect a meaning preserving transformation on the unannotated object program. We prove specializer soundness by appealing to a *program calculus for Λ* (*i.e.*, an equational theory for deducing operational equivalences). Gunter [8, Chapter 4] gives a calculus for PCF, and we may take an appropriate sub-theory of this as the calculus for Λ (this is essentially the traditional λβ-calculus). Let $=_Λ$ denote the convertibility relation for the Λ calculus. The following theorem captures the fact that the unannotated object program e is operationally equivalent to the unannotated specialization result e'.

Theorem 2 Soundness of specializer.

If $Σ ⊢ w ⇓_{spec} a$ and $Γ_d ⊢ bta\ e : τ\ [w : φ_τ]$ and $Σ$ is compatible with $Γ_d$, then there exists $e' ∈ Λ$ such that $Γ_d ⊢ bta\ e' : τ\ [a : φ_τ]$ and $e =_Λ e'$.

Proof. (summary) To formalize the proof, we give an encoding of the Λ-calculus into Elf based on a similar encoding given by Pfenning [21]. Next, we construct a function (*via* a relation as in Theorem 1) which constructs a deduction showing e converts to e'.

```
t2 : spec W A -> bta E T W P -> bta E' T A P -> conv E E' -> type.
```

A strategy similar to the one used in the proof of Theorem 1 gives the rules defining t2. The details are given in [13].

7 Related work

Despeyroux first emphasized using deductive systems to define transformations [5]. She specified a compiler, and source and target language semantics using deductive systems. The specifications were executed *via* encodings into Typol. Informal proofs of correctness were given as relations between deductions. However, these proofs could not be formalized in Typol because it does not support the direct manipulation of its own deductions.

Hannan and Pfenning [11] improved upon this by formalizing similar proofs of correctness in Elf. Elf (unlike *e.g.*, Typol, and λ-Prolog) supports the direct manipulation of its own deductions. Furthermore, Elf type checking mechanically verifies that deductions corresponding to correctness proofs are well-formed.

Despeyroux [5, Section 8], and Hannan and Pfenning [11, Section 7], suggested that their methods could be used to specify "mixed computation" and partial evaluation, respectively. Hannan and Miller [10] carried out Despeyroux's suggestion; they use deductive systems encoded in λ-Prolog to obtain executable specifications of mixed computation (in their work, "mixed computation" = non-deterministic on-line partial evaluation).

Our contributions include using deductive systems to specify *program specialization directed by type-based analysis*. Moreover, we adapt the techniques of Hannan and Pfenning [11], and Michaylov and Pfenning [17] to mechanically verify correctness.

In doing so, we obtain verified proofs similar in scope to the ones given by Gomard and Jones [7] for λ-mix. Their denotational meta-language is significantly more complex than our logic-based meta-language and it is unlikely that the proofs there could be formalized or mechanically verified to the extent that we have done here. However, it must be noted that one of their goals was self-application. Relying on the similarity between their meta-language and object-language, they obtain an object-language specification of the partial evaluator (giving self-applicability) by a fairly easy (though informal and unverified) translation from the meta-language specification. In our setting this is more difficult since the character of our meta-language (logical) is quite different from our object language (functional).

Building upon the work of Gomard, Jones, and Mogensen, Palsberg [19] and Wand [28] give detailed presentations of the correctness of binding-time analysis. Palsberg presents a generalization of Gomard and Jones criteria [7] for consistent binding-time annotations. Wand studies Mogensen's self-applicable partial evaluator [18] for the pure λ-calculus. His binding-time analysis is essentially the same as Gomard and Jones's as well the one presented here. Wand's goals with respect to correctness of the analysis and specializer are more ambitious than those here, because his correctness criteria is strong enough to specify the behaviour of the partial evaluator when self-applied. It would be interesting to see to what extent the meta-theory used by Palsberg and Wand could be formalized in Elf.

In recent work, Davies and Pfenning [4] give a type system for expressing staged computation based on the intuitionistic modal logic S4. They have implemented the type system and a portion of the associated correctness proofs in Elf.

Our work focuses on partial evaluation of functional programs, but similar techniques can be applied to imperative languages as well. Blazy and Facon [2] specify a simple specializer for FORTRAN using natural semantics, and derive a prototype from the specification using the Centaur programming environment. However, their correctness proofs are not formalized. Bertot and Fraer [1] show how similar correctness proofs (for a specializer for an imperative language) can be formalized and mechanically checked using Coq.

8 Conclusion

We have specified the main components of an offline partial evaluator (*i.e.*, binding-time constraints and a program specializer) using natural-deduction style logics. These specifications were formalized by encoding them in LF. Prototypes were obtained directly from the formal specifications using Elf. We formalized and mechanically verified a significant portion of the meta-theory of offline partial evaluation (*e.g.*, correctness of binding-time analysis and soundness of the specializer) *via* meta-programming in Elf. A certain degree of synergy is obtained by using the declarative formalism of deductive systems: one may specify, prototype and mechanically verify correctness *via* meta-programming — all within a single framework.

These techniques can be used to verify other forms of partial evaluation for functional programs. For example, if one takes the view that an annotated program is its own generating extension, the techniques here are very close to what one would use to verify the correctness of a *hand-written cogen* [15]. In fact, in a preliminary investigation we have prototyped a higher-order version of the *multi-level cogen* of Glück and Jørgensen [6]. It remains to be seen if these techniques scale up to *(a)* type-based analyses that include conjunctive types, polymorphism, and more general forms of subtyping, and *(b)* more robust forms of partial evaluation that include sophisticated folding strategies (one approach might be to use Sand's calculus for sound fold/unfold transformations [25]).

Acknowledgements

Thanks to Olivier Danvy for his encouragement and comments on various drafts, and to Frank Pfenning for several stimulating conversations and for providing easy access to his materials on Elf. Robert Glück, Kristian Nielsen, David Sands, and other members of the DIKU TOPPS group gave valuable feedback and support. Thanks to the PLILP referees for helpful comments.

References

1. Yves Bertot and Ranan Fraer. Reasoning with executable specifications. In TAP-SOFT'95 [27], pages 531–545.
2. Sandrine Blazy and Philippe Facon. Formal specification and prototyping of a program specializer. In TAPSOFT'95 [27], pages 666–680.
3. Charles Consel and Olivier Danvy. Tutorial notes on partial evaluation. In Susan L. Graham, editor, *Proceedings of the Twentieth Annual ACM Symposium on Principles of Programming Languages*, pages 493–501, Charleston, South Carolina, January 1993. ACM Press.
4. Rowan Davies and Frank Pfenning. A modal analysis of staged compuation. In *Proceedings of the Workshop on Types for Program Analysis*, Aarhus, Denmark, 1995.
5. Jöelle Despeyroux. Proof of translation in natural semantics. In *Proceedings of the First Annual IEEE Symposium on Logic in Computer Science*, pages 193–205, Cambridge, Massachusetts, 1986. IEEE Computer Society Press.
6. Robert Glück and Jesper Jørgensen. Efficient multi-level generating extensions. 1995. To appear in *The Proceedings of the Seventh International Symposium on Programming Languages, Implementations, Logics and Programs.* Utrecht, The Netherlands, September 20-22, 1995.
7. Carsten K. Gomard and Neil Jones. A partial evaluator for the untyped lambda-calculus. *Journal of Functional Programming*, 1(1):21–69, 1991.
8. Carl A. Gunter. *Semantics of Programming Languages: Structures and Techniques.* MIT Press, 1992.
9. John Hannan. Extended natural semantics. *Journal of Functional Programming*, 3(2):123–152, 1993.
10. John Hannan and Dale Miller. Deriving mixed evaluation from standard evaluation for a simple functional language. In J. van de Snepscheut, editor, *Mathematics of Program Construction*, number 375 in Lecture Notes in Computer Science, pages 239–255, 1989.

11. John Hannan and Frank Pfenning. Compiler verification in LF. In *Proceedings of the Seventh Symposium on Logic in Computer Science*, pages 407–418. IEEE, 1992.

12. Robert Harper, Furio Honsell, and Gordon Plotkin. A framework for defining logics. *Journal of the ACM*, 40(1):143–184, 1993. A preliminary version appeared in the proceedings of the First IEEE Symposium on Logic in Computer Science, pages 194–204, June 1987.

13. John Hatcliff. Mechanically verifying the correctness of an offline partial evaluator (extended version). DIKU Report 95/14, University of Copenhagen, Copenhagen, Denmark, 1995.

14. John Hatcliff and Olivier Danvy. A computational formalization for partial evaluation. DIKU Report 95/15, University of Copenhagen, Copenhagen, Denmark, 1995. Presented at the *Workshop on Logic, Domains, and Programming Languages*. Darmstadt, Germany. May, 1995.

15. Neil D. Jones, Carsten K. Gomard, and Peter Sestoft. *Partial Evaluation and Automatic Program Generation*. Prentice-Hall International, 1993.

16. Julia L. Lawall and Olivier Danvy. Continuation-based partial evaluation. Technical Report CS-95-178, Computer Science Department, Brandeis University, Waltham, Massachusetts, January 1995. An earlier version appeared in the proceedings of the 1994 ACM Conference on Lisp and Functional Programming.

17. Spiro Michaylov and Frank Pfenning. Natural semantics and some of its metatheory in Elf. Report MPI-I-91-211, Max-Planck-Institute for Computer Science, Saarbrücken, Germany, August 1991.

18. T. Mogensen. Self-applicable partial evaluation for the pure lambda calculus. In Charles Consel, editor, *ACM SIGPLAN Workshop on Partial Evaluation and Semantics-Based Program Manipulation*, Research Report 909, Department of Computer Science, Yale University, pages 116–121, San Francisco, California, June 1992.

19. Jens Palsberg. Correctness of binding-time analysis. *Journal of Functional Programming*, 3(3):347–363, 1993.

20. Frank Pfenning. Logic programming in the LF logical framework. In Gérard Huet and Gordon Plotkin, editors, *Logical Frameworks*, pages 149–181. Cambridge University Press, 1991.

21. Frank Pfenning. A proof of the church-rosser theorem and its representation in a logical framework. Technical Report CMU-CS-92-186, Carnegie Mellon University, Pittsburgh, Pennsylvania, September 1992. To appear in Journal of Automated Reasoning.

22. Frank Pfenning and Conal Elliott. Higher-order abstract syntax. In *Proceedings of the ACM SIGPLAN'88 Conference on Programming Languages Design and Implementation*, pages 199–208, June 1988.

23. Frank Pfenning and Ekkehard Rohwedder. Implementing the meta-theory of deductive systems. In D. Kapur, editor, *Proceedings of the 11th Eleventh International Conference on Automated Deduction*, number 607 in Lecture Notes in Artificial Intelligence, pages 537–551, Saratoga Springs, New York, 1992. Springer-Verlag.

24. Dag Prawitz. *Natural Deduction*. Almquist and Wiksell, Uppsala, 1965.

25. David Sands. Total correctness by local improvement in program transformation. In Ron Cytron, editor, *Proceedings of the Twenty-first Annual ACM Symposium on Principles of Programming Languages*, pages 221–232, San Francisco, California, January 1995. ACM Press.

26. Morten Heine Sørensen, Robert Glück, and Neil Jones. Towards unifying partial evaluation, deforestation, supercompilation, and GPC. In *Proceedings of the Fifth European Symposium on Programming*, pages 485–500, Edinburgh, U.K., April 1994.
27. *TAPSOFT '95: Theory and Practice of Software Development*, number 915 in Lecture Notes in Computer Science, Aarhus, Denmark, May 1995.
28. Mitchell Wand. Specifying the correctness of binding-time analysis. *Journal of Functional Programming*, 3(3):365–387, 1993.

A   Correctness of binding-time analysis

```
% definition of t1 judgement

t1 : spec W A -> bta E T W P -> bta E' T A P -> ans A P -> type.


t1_z_s : t1 (spec_z_s) (bta_z_s) (bta_z_s) (ans_z_s).

t1_lam_s : t1 (spec_lam_s) (bta_lam_s B)
              (bta_lam_s B) (ans_lam_s).

t1_app_s : t1 (spec_app_s S1 S0) (bta_app_s B1 B0) C D
              <- t1 S0 B0 (bta_lam_s C0) D0
              <- t1 S1 (C0 _ _ B1) C D.

t1_z_d : t1 (spec_z_d) (bta_z_d) (bta_z_d) (ans_z_d).

t1_app_d : t1 (spec_app_d S1 S0) (bta_app_d B1 B0)
              (bta_app_d C1 C0) (ans_app_d D1 D0)
                <- t1 S0 B0 C0 D0
                <- t1 S1 B1 C1 D1.

t1_lam_d : t1 (spec_lam_d S) (bta_lam_d B)
              (bta_lam_d C) (ans_lam_d D)
          <- {x} {y}
             {S': spec y y} {B' : bta x T1 y dyn} {D': ans y dyn}
                t1 S' B' B' D'
                  -> t1 (S y S') (B x y B') (C x y B') (D y D').

t1_lift : t1 (spec_lift S) (bta_lift B) (bta_z_d) (ans_z_d)
              <- t1 S B (bta_z_s) D.
```

A Semantic Model of Binding Times
for Safe Partial Evaluation

Fritz Henglein   David Sands

University of Copenhagen[*]

Abstract

In program optimisation an analysis determines some information about a portion of a program, which is then used to justify certain transformations on the code. The correctness of the optimisation can be argued *monolithically* by considering the behaviour of the optimiser and a particular analysis in conjunction. Alternatively, correctness can be established by finding an interface, a *semantic property*, between the analysis and the transformation. The semantic property provides modularity by giving a specification for a systematic construction of the analysis, and the program transformations are justified via the semantic properties.

This paper considers the problem of partial evaluation. The safety of a partial evaluator ("it does not go wrong") has previously been argued in the monolithic style by considering the behaviour of a particular binding-time analysis and program specialiser in conjunction. In this paper we pursue the alternative approach of justifying the binding-time properties semantically. While several semantic models have been proposed for binding times, we are not aware of any application of these models in proving the safety of a partial evaluator. In this paper we:

- identify problems of existing models of binding-time properties based on projections and partial equivalence relations (PERs), which imply that they are not adequate to prove the safety of simple off-line partial evaluators;

- propose a new model for binding times that avoids the potential pitfalls of projections/PERs;

- specify binding-time annotations justified by a "collecting" semantics, and clarify the connection between extensional properties (local analysis) and program annotations (global analysis) necessary to support binding-time analysis;

- prove the safety of a simple but liberal class of monovariant partial evaluators for a higher-order functional language with recursive types, based on annotations justified by the model.

1 Introduction

1.1 Transformations supported by Program Analysis

Program optimisation usually takes the following form: an analysis determines some information about a portion of a program, and the information is then used to justify certain

---

[*]DIKU, Universitetsparken 1, 2100 København Ø; {henglein,dave}@diku.dk

transformations on the code. We consider two basic methods for establishing the correctness of such a process, which we call *monolithic* and *model-based*, respectively:

Monolithic The monolithic view considers the correctness of the analysis and the transformation simultaneously. The pair of the analysis and the transformation is correct if the transformation "works."

Model-based The model-based approach associates some semantic property with the information domain of the analysis. The correctness of the analysis, and the correctness of the transformation are then considered independently, but relative to this semantic property.

The monolithic approach has attracted much interest in the last few years. Its advocates, *e.g.* Wand [Wan93], Amtoft [Amt93], and Steckler [Ste94], argue that considering the correctness of the algorithm and transformation together leads to a much simpler proof. The slogan is:

The analysis is correct because the transformation works!

It is notable that these kinds of proofs are greatly aided by the non-algorithmic specification of the analysis in terms of non-standard type systems, or constraint systems. The disadvantages of this approach are that: as the name suggests, variations in either the analysis or the transformation require that the proof must be re-established for each change or combination of analysers and transformers; there is currently no support for systematic design of correct analyses; similar analysis may be used in justifying quite different kinds of transformation, but there are no "reusable" components in the correctness proof.

In principle, the model-based approach addresses each of these deficiencies. By associating a semantic property with each piece of information from a static analysis, one obtains an intermediary between the analysis and the transformation. This, in turn, achieves a factorisation of correctness of the analysis and the transformation with respect to the semantic property. This means that independent changes to either the analysis or transformation can be justified independently. Furthermore, it enables utilisation of techniques for systematic design of correct analyses, namely *abstract interpretation* [CC79]. Finally, it facilitates reuse of analyses for different transformations which rely on a common semantic property.

The problem in practice is, to quote Wand [Wan93]:

> "While program analyses of various sorts have been studied intensively for many years, it has proven remarkably difficult to specify the correctness of an analysis in a way that actually justifies the resulting transformation."

In this paper we address this problem for a particular transformation, *off-line partial evaluation*, in the setting of higher-order functional programs. The associated analysis is called *binding-time analysis*, and the core of the correctness problem is to verify that a partial evaluator does not "go wrong" when following binding-time annotations.

While there are numerous proofs of correctness using the monolithic approach, and several candidate semantic models for binding-time properties, we know of no correctness proof for a partial evaluator based on a semantic model of binding-time properties. In this paper we:[1]

- identify problems of existing models of binding-time properties based on projections and partial equivalence relations (PERs), which imply that they are not adequate to prove the correctness of even simple off-line partial evaluators;

---

[1] Due to space limitations most proofs have been omitted. They are contained in a full report obtainable from the TOPPS (DIKU programming language group) archive at http://www.diku.dk/research-groups/topps.

- propose a new model for binding times which avoids the potential pitfalls of projections/PERs;

- clarify the connection between extensional properties (local analysis) and program annotations (global analysis) necessary to support binding-time analysis;

- prove the correctness of a simple but liberal class of partial evaluators based on the soundness of the annotations with respect to the model, and demonstrate the applicability to both off-line and on-line partial evaluation.

Model-based analysis is the hallmark of abstract interpretation. Transformations often require an abstraction not directly of the standard semantics of a language, but of its *collecting semantics*, however. Expressing the collecting semantics in denotational models has proved to be difficult. Cousot and Cousot show how powersets support this step at higher type and apply it to what they call *comportment analysis* [CC94].

2 Off-line Partial Evaluation: Related Work

An off-line partial evaluator determines which parts of a program to evaluate, and which parts to leave as residual code, by following annotations produced by a binding-time analysis.

Given a description of the parameters in a program that will be known at partial evaluation time, a binding-time analysis must determine which parts of the program are dependent solely on these known parts (and therefore also known at partial evaluation time). A binding-time analysis performed prior to the partial evaluation process can have several practical benefits (see [Jon88]), and plays an essential rôle in most approaches to the generation of efficient compilers from interpreters [BJMS88].

2.1 Approaches to Correctness

Monolithic The monolithic view for partial evaluation considers the correctness of the off-line partial evaluator and the binding-time analysis simultaneously. The annotations produced by the binding-time analysis are considered to be correct if the partial evaluator, whose actions are governed by the annotations, behaves in the intended manner, e.g. it does not "go wrong" by expecting to be able to produce a value from a program fragment dependent on an unbound variable. Examples of this approach are seen in the work of Gomard [Gom92], based on a denotationally-specified partial evaluator for a lambda calculus with constants ("λ-mix"), Wand [Wan93] and Palsberg [Pal93], based on the pure lambda calculus, Henglein and Mossin [HM94] for a typed functional language and a denotationally specified partial evaluator, Consel et al. [CJØ94] for a rewriting-based approach, and more recently [Hat95] who considers the mechanical verification of the correctness proof for a λ-mix style partial evaluator.

Model-based The model-based approach has its roots in Jones' definition of *congruence* [Jon88], which specifies correctness of binding-time analysis by focusing on semantic dependency between different parts of a program. Launchbury adapted this idea to a functional setting, using the idea of domain projections to model binding times of structured data in a first-order language [Lau88, Lau89]. This domain-based approach was subsequently adopted and extended by Mogensen [Mog89] and De Niel *et al.* [NBV91]. Hunt and Sands [HS91] showed that Launchbury's analysis could be smoothly extended to higher types using partial equivalence relations (PERs) as a model of binding times, following Hunt [Hun90]. Davis [Dav94] considers a closely related extension to higher types with general recursive types.

There is a third approach to proving correctness, closely related to the model-based view, but arguably different: an off-line partial evaluator is viewed as an abstraction of an *on-line* one. An on-line partial evaluation does not follow static binding-time annotations, but computes the necessary information on the fly. Off-line partial evaluation is then viewed as a *restriction* of the actions of the on-line version, since it makes decisions about what to partially evaluate based on annotations, rather than actual data. (As we mention later, it might also be considered an *optimisation*, since it removes the need for many tests on the nature of the data manipulated.) This approach has been considered by Consel and Khoo [CK92] and Bulyonkov [Bul93]. Consel and Khoo give an abstract denotational specification of the values encountered by an on-line specialiser for a first-order functional language, and show that a binding-time analysis correctly abstracts these values. Off-line partial evaluation is then obtained by the restriction of the actions of the on-line version. Their highly abstract and non-operational specification of an on-line specialiser resembles a collecting interpretation (a *static* semantics in the terminology of [CC79]).

3 The Problem with Projections and PERs

In this section we consider the existing proposals for modelling binding times, including partially-static structures, in functional languages. The principal technique uses domain projections [Lau88]. We will argue that this model has potential flaws from the point of view of proving the correctness of a partial evaluator. These problems carry over to the PER model [HS91], and motivate the introduction of a new model in the next section.

3.1 Uniform Congruence

In his "re-examination" paper, Jones [Jon88] defines a semantic-based condition, *congruence*, which specifies when a binding-time analysis is correct. The essence of the definition is that the parts of a program that are deemed to be static will only ever depend on the static values (and the other parts of the program which are deemed static). Launchbury adapted this idea to a functional setting, and derived what he considered to be a necessarily stronger condition, called *uniform congruence*, and expressed this condition with the help of domain projections.

In Launchbury's setting, a first-order language of recursion equations, the job of a binding-time analysis is to determine a *program division*. A division Δ is a mapping which assigns a binding time to each function symbol defined in the program. (It is therefore *monovariant* because it assigns just one binding time for each definition.)

A binding-time is identified with (modelled by) a domain projection. A projection is a continuous map $\alpha : D \to D$ on a cpo D, such that $\alpha \sqsubseteq \lambda x \in D. x$ and $\alpha \circ \alpha = \alpha$. An intuition behind the use of projections to describe binding times is that the parts of its argument that a projection discards (replaces by bottom) represent the parts about which no information is known, where "no information" is equated with "dynamic." This interpretation is used to define when a program division is safe, *i.e.* uniformly congruent.

A program division Δ is deemed to be uniformly congruent, if for each definition and call instance of the form

$$f\ x = \ldots (g\ e) \ldots$$

which occurs in the program, then

$$\Delta(g) \circ [\![\lambda v.e[v/x]]\!] \circ \Delta(f) = [\![\lambda v.e[v/x]]\!] \circ \Delta(f)$$

This means that if $\Delta(f)$ describes the static-ness of the argument to f, then $\Delta(g)$ correctly predicts the static-ness of the argument e in the call $g\ e$.

In Hunt and Sands' terminology [HS91], in which binding times $\Delta(f)$ and $\Delta(g)$ are interpreted as equivalence relations (and at higher-types, as *partial* equivalence relations), this property would be written equivalently as

$$(\lambda v.e[v/x]) : \Delta(f) \Rightarrow \Delta(g)$$

which by definition means that for all v_1, v_2 in the semantic domain associated with parameter x,

$$v_1 \Delta(f) v_2 \Rightarrow [\![\lambda v.e[v/x]]\!]v_1 \Delta(g) [\![\lambda v.e[v/x]]\!]v_2.$$

The "uniformity" in Launchbury's definition refers to the fact that no contextual assumptions are made about the possible value of x in the expression e. This reflects a simple but aggressive view of partial evaluation, which assumes that we can begin specialising the call to $(g\ e)$ without using knowledge of either the context "..." or the range of possible values of x in that context. This uniformity requirement is a strengthening of Jones' condition, and so fewer program divisions are permitted.

3.2 The Problem with Uniform Congruence

Launchbury's specialiser (for present purposes, a specialiser is just a partial evaluator which produces specialised variants of program text) specialises function calls with respect to the static parts of their arguments.

Clearly then, from the point of view of the specialiser the binding-time analysis is correct if the parts of the arguments that are deemed static can indeed be evaluated, and their evaluation either terminates with a value, or the specialiser goes into a loop in the attempt—in any case it must not get "stuck" trying to evaluate something dynamic, such as a free variable. (A consequence of the fact that Launchbury's language is statically typed is that there are no run-time errors in the standard interpreter.)

The job of the semantic specification of uniform congruence is to give an analysis-independent specification of a correct program division. This can be used to justify binding-time analyses independently of a specific partial evaluator (just as Launchbury has done). But for this to be adequate, one must be able to argue the correctness of a partial evaluator with respect to a uniformly congruent program division (something Launchbury did not do).

We argue that the semantic condition of uniform congruence is not sufficient to guarantee the correctness of a simple "mix-style" partial evaluator. That is to say, there are uniformly congruent program divisions which can cause a specialiser to "go wrong." What is more, we claim that Launchbury's *own* specialiser will go wrong on an instance of this example.

Consider the following (abstract) program:

```
letrec
  g(v, w)  =  e
  f(x, y)  =  g(if y then Ω else Ω', y)
in f(i, j)
```

where g is non-recursive, Ω and Ω' are any expressions not involving y, but which diverge (fail to terminate) for all values of x.

Now suppose we specify that i is static and j is dynamic. This property of the pair (i, j) can be represented by a projection $fst \stackrel{\text{def}}{=} \lambda\langle x, y\rangle.\langle x, \bot\rangle$. Based on this specification, the division

$$[f \mapsto fst, g \mapsto fst]$$

is uniformly congruent. To see intuitively why, first note that since Ω and Ω' do not contain y, and under the assumption that x is static any sub-expressions of Ω and Ω' are also static.

The surprise is that g's first argument can be considered static. Intuitively, this is correct because the value of its only argument (the call instance in f's definition) does not depend on the value of y — since it is always undefined ($\bot$)! The potential problem with this uniformly congruent division is readily apparent. The term if y then Ω else Ω' is deemed to be static even though y is dynamic. This means that a partial evaluator may begin to evaluate the conditional, and thereby "go wrong" by either:

1. attempting to evaluate y, or

2. by expecting that the expression if y then Ω else Ω' can be compared with other static "values", for example in a pending list of specialised function calls.

For Launchbury's simple partial evaluator it is the latter case. We can realise an instance of this scheme in Launchbury's PEL language and show that his specialiser "goes wrong" when given the above safe (= uniformly congruent) program division.

In the binding-time model we present in the next section domains contain "extra" elements that force if y then Ω else Ω' to be dynamic. More concretely, y may not only be bound to $\bot$, *true* or *false*, but also to δ, an "anonymous" dynamic value. Whereas the result of the first three bindings is $\bot$, in the latter case it is the special value δ. Thus, in the extended domain with δ, if y then Ω else Ω' does depend on y; in particular, if y is dynamic then so is the whole expression.[2]

Note that we *do not* claim that Launchbury's *analysis* will produce this program division (it will not). The point is that the safety condition which specifies a correct analysis must be adequate to prove the correctness of the transformation. We conclude that this is not the case for Launchbury's projection-based safety condition. The problem is inherited by Hunt and Sands' PER-based extension to higher-order functions — in fact we came across this problem in a higher-order setting, but a superficially quite different context: attempting to prove the correctness of a λ-mix style partial evaluator [Gom92] using the PER-model of binding times.

4 An Ideal Model of Binding Times

An appealing property of the projection/PER model of binding times is that it is purely extensional, relying as it does on the standard semantics.[3] As we showed in the previous section, using "bottom" to represent absence of information at partial-evaluation time confuses termination properties with neededness properties. Confusion arises because the property "static" does not necessarily mean "terminating."[4] So bottom is overloaded to denote static computation (though nonterminating) and static nonavailability of dynamic data, making it impossible to distinguish nonterminating computations that depend on dynamic inputs from those that don't.

Our solution is a natural one, once we accept that we are modelling properties that only have meaning at partial evaluation time. To be able to make finer distinctions than in

---

[2]Our extended interpretation of *bool* contains yet another value, a top element $\top$; $\top$, however, is only required for a *higher-order* language. Note that the problem with Launchbury's model of binding times already occurs for a first-order language. Addressing this problem in the same language context only requires the additional element δ, not $\top$.

[3]However, in order to model anything interesting about binding times the model for the language under consideration *must* be lazy. Furthermore, we claim that for this purpose the laziness must be taken to its logical conclusion — i.e. function spaces should be lifted (contrary to the model in [HS91]) as well as tuples (contrary to [Lau89][HS91][Dav94]).

[4]In principal we have no objection to an interpretation of "static" which implies "terminating"; but this is neither the interpretation used in practice in existing partial evaluators, nor the interpretation used in this paper.

the standard semantics our solution is to augment the domain constructors in the standard semantics to provide extra elements: δ and $\top$. Intuitively, δ stands for an "anonymous dynamic value", and $\top$ denotes an abortive error; that is, the result of encountering an error situation that leads to the abort of the whole program evaluation.

Partially static structures are handled by allowing data structures to contain δ-elements at component types without them being identified with δ (or $\top$) of the compound type. This model enlarges the domains, and so gives rise to a choice as to how the operators of the language should be extended. The choices reflect choices in the partial evaluation strategy — but at a much more abstract level than if we were to describe a particular partial evaluator.

In the remainder of this section we describe the language and the model of binding times and associated binding-time annotations.

4.1 A Higher-order Functional Programming Language

Our setting is a higher-order simply typed programming programming language with unit, sum, pair and recursive types.

Types The *types* are described by the closed expressions in the following grammar:

$$\tau ::= \alpha \mid \text{unit} \mid \tau' \to \tau'' \mid \tau' \times \tau'' \mid \tau' + \tau'' \mid rec\,\alpha.\,\tau'$$

Recursive types must be *formally contractive*; that is, in $rec\,\alpha.\,\tau$ the type τ must not be a type variable.

Syntax and typing rules for expressions The syntax of our language, including its "static" semantics, is given by rather standard typing rules, not presented here. They are for typing judgements $A \vdash e : \tau$ where A is a *type environment* mapping *program variables* to types, e is an *expression*, and τ is a type.

Standard denotational semantics To give a standard semantics for this language we can interpret types by Scott domains and type constructors by domain constructors appropriate for a call-by-name (lazy) language [Gun92].

Specifically, we can interpret unit by the one-element domain 1; $\to$ by the domain constructor for continuous functions; $\times$ by lifting the result of the *Cartesian product* constructor $(\cdot \times \cdot)_\perp$; $+$ by the *separated sum* constructor; and $rec\,\alpha.\,\tau$ by the inverse-limit of the domain constructor denoted by τ (as a function of α).

Expressions are denoted by domain elements. If $\vdash e : \tau$ for closed expression e then $[e]_{std} \in [\tau]_{std}$, where $[e]_{std}$ is the domain element denoted by e. This yields a denotational semantics that is *observationally adequate* for a call-by-name operational semantics relative to observing termination at all nontrivial first-order types (which excludes unit).

4.2 Extended Domains

Previous models of binding times have built upon the standard denotational semantics either by interpreting binding times as projections or PERs on the *standard* domains. As shown in Section 3 this leads to problems in that these models may be too aggressive about what they classify as static. The reason is that the standard domains do not have any "room" for intensional information that captures the essential control dependencies — neededness information — that a (simple) partial evaluator must respect. This problem is even more pronounced in a call-by-value language, where the PER and projection analyses become trivial (useless).

In this section we *extend* the standard domains by adding *extra elements* in a structural fashion: For every type τ possessing a destructor operation we add a *dynamic element* δ_τ, which, intuitively, represents *completely dynamic* values at τ. This new value lets us distinguish a dynamic value of type $\tau \times \tau'$, say a variable, from a *pair* of dynamic values. In the latter case we can perform a static (partial-evaluation time) decomposition of the pair, whereas in the former we cannot. It is the ability to distinguish between being able to perform a destructive operation (in this case π_1 or π_2) at partial evaluation time that necessitates and explains the role of the dynamic element.

Furthermore, for every type τ we add a *top element* $\top_\tau$, which represents an error at type τ. We call the resulting domains *topped domains (with δ)*, in order to distinguish them from the *standard domains* introduced earlier. The elements of the standard domain are then embedded "naturally" in the topped domains; in particular functions are extended to map the new elements δ and $\top$ to $\top$.

Structurally topped domain constructions Our binding time domains are Scott-domains with isolated δ- and $\top$-elements. In what follows we will define the topped interpretation for types and terms. We will write $[\cdot]$ to denote these mappings and use the following domain constructions on topped domains D, D'. Let $\top_D, \top'_D$ be the top elements in D and D', respectively.

- The *co-strict function domain* $D \hookrightarrow D'$ consists of the continuous functions from D to D' mapping $\top_D$ to $\top_{D'}$, plus a new top element $\top_{D \hookrightarrow D'}$. The partial ordering on non-$\top$ elements is inherited point-wise from D'.

- The *co-strict product domain* $D \otimes D'$ consists of the pairs (d, d') with $\top_D \neq d \in D$ and $\top_{D'} \neq d' \in D'$, extended with new bottom and top elements $\bot_{D \otimes D'}$ and $\top_{D \otimes D'}$, respectively. The partial ordering on pairs is inherited componentwise from D and D'.

- The *co-strict sum domain* $D \oplus D'$ consists of the elements $inl\,d$ and $inr\,d'$ for all $\top_D \neq d \in D$ and $\top_{D'} \neq d' \in D'$, extended with new bottom and top elements $\bot_{D \oplus D'}$ and $\top_{D \oplus D'}$, respectively. The partial ordering on elements $inl\,d$ is inherited from D, and on elements $inr\,d'$ from D'; elements $inl\,d$ and $inr\,d'$ are incomparable.

- The *topped domain* $D^\top$ consists of all the elements from D, plus a new top element, the partial ordering on non-top elements being the same as in D.

- The *dynamic domain* D_S^δ is D, extended with an additional element δ. S must be a set of pairwise incomparable elements from D. The partial order relation on non-δ elements is inherited from D. It is extended to δ as follows: δ is "immediately below" $\top_D$; that is, $\delta \sqsubset \top_D$, and $\top_D$ is the only element greater than δ. Furthermore, the elements of S are immediately below δ. That is, $d \sqsubset \delta$ if and only if $d \sqsubseteq s$ for some $s \in S$. This defines the partial order relation on D_S^δ. Note that, if $\top_D$ is isolated in D then D_S^δ is a domain in which both δ and $\top_D$ are isolated.

Every domain interpreting a type τ below has distinguished δ- and $\top$-elements. We shall denote these by δ_τ and $\top_\tau$, respectively.

$$
\begin{aligned}
[\text{unit}] &= \mathbf{1}^\top \\
[\tau \to \tau'] &= ([\tau] \hookrightarrow [\tau'])^\delta_{\{f\}} && \text{where } f = \lambda d \in [\tau].\,\text{if } d \sqsubseteq \delta_\tau \text{ then } \delta_{\tau'}\text{ else } \top_{\tau'} \\
[\tau \times \tau'] &= ([\tau] \otimes [\tau'])^\delta_{\{d\}} && \text{where } d = (\delta_\tau, \delta_{\tau'}) \\
[\tau + \tau'] &= ([\tau] \oplus [\tau'])^\delta_{\{d_1, d_2\}} && \text{where } d_1 = inl\,\delta_\tau \text{ and } d_2 = inr\,\delta_{\tau'} \\
[rec\,\alpha.\,\tau] &= \lim_{i \in \omega} F^i(\mathbf{1}^\top) && \text{where } F(D) = [\tau][\alpha \mapsto D]
\end{aligned}
$$

The last clause denotes the inverse-limit construction for topped domains with co-strict projection/embedding-pairs. We state without proof that this yields a domain with distinguished δ- and $\top$-elements.

Note that we add a new top element for every constructed domain. Furthermore for every type constructor with the exception of unit we add a new element δ. Recall that the possibility of distinguishing completely dynamic values from partial dynamic values in a destructive context motivated our introduction of a distinct element δ in the first place. There is no destructor for unit — and hence no need to add a distinct δ-element.

As an example, let us define $bool = \text{unit} \oplus \text{unit}$. The standard interpretation of $bool$ has the three elements $true = inl\,()$, $false = inr\,()$ and the bottom element $\perp_{bool}$. In the above extended interpretation, $[\![bool]\!]$ is the five-element domain consisting of elements $\{\perp, true, false, \delta_{bool}, \top_{bool}\}$, ordered by $\perp \sqsubseteq x \sqsubseteq \delta \sqsubseteq \top$ for $x = true$ or $x = false$.

Extended interpretation of expressions We have extended the interpretations of types, so now we must extend the interpretation of terms over these new types.

The extension of the standard interpretations of terms essentially follows the strictness properties of the basic syntactic constructs, so that any *destructor* (e.g. $\pi_1\cdot$, *case · of* ...) maps the elements δ and $\top$ of the type being "destructed" to δ and $\top$, respectively, of the resulting domain. Without giving the full details, the essence of the extension is characterised by the following semantic equations for $\top$. They are completely analogous for δ.

$$
\left.
\begin{array}{r}
\left[\!\!\left[
\begin{array}{l}
case\ e\ of \\
\quad inl\ x \Rightarrow e' \\
\quad inr\ y \Rightarrow e''
\end{array}
\right]\!\!\right]\rho \;=\; \top \\[3ex]
[\![e\ e']\!]\rho \;=\; \top \\
[\![\pi_1 e]\!]\rho \;=\; \top \\
[\![\pi_2 e]\!]\rho \;=\; \top
\end{array}
\right\} \quad \text{if } [\![e]\!]\rho = \top
$$

4.3 Binding times as ideals

A *binding time* at type τ is modelled by a nonempty, nonfull *ideal (closed set, inclusive set)* in the Scott-topology of $[\![\tau]\!]$; that is, it is a subset of $[\![\tau]\!]$ which is:

1. neither empty nor full: it contains $\perp_\tau$, but *not* $\top_\tau$;

2. downwards closed: $x \sqsubseteq y \in I \Longrightarrow x \in I$; and

3. closed under ω-chains: if $\{x_i\}$ is an ascending ω-chain with $x_i \in I$ for all $i \in \omega$ then $\bigsqcup_{i \in \omega} x_i \in I$.

For each type τ a set I is said to be a *(semantic) binding time* at type τ if I is a nonempty, nonfull ideal over the domain $[\![\tau]\!]$. We will say that an element d (of some domain E) has binding time I (an ideal over E) whenever $d \in I$. Let I and I' be arbitrary binding times at types τ and τ', respectively. The binding times are closed under the following operations:

$$
\begin{aligned}
I \to I' &\;\stackrel{\text{def}}{=}\; \{f \in [\![\tau \to \tau']\!] \mid f \neq \top, f \neq \delta \text{ and } (\forall d \in I)\, f(d) \in I'\} \\
I \times I' &\;\stackrel{\text{def}}{=}\; \{(d,d') \in [\![\tau \times \tau']\!] \mid d \in I, d' \in I'\} \cup \{\perp\} \\
I + I' &\;\stackrel{\text{def}}{=}\; \{inl\,d \in [\![\tau + \tau']\!] \mid d \in I\} \\
&\qquad \cup \{inr\,d' \in [\![\tau + \tau']\!] \mid d' \in I'\} \cup \{\perp\}
\end{aligned}
$$

Furthermore, the ideals of $[\![\tau[rec\,\alpha.\,\tau/\alpha]]\!]$ and $[\![rec\,\alpha.\,\tau]\!]$ are in a one-to-one correspondence.

Being Scott-closed sets, binding times at the same type are closed under finite unions $I_1 \cup \ldots \cup I_n$ and (finite or infinite) intersections $\bigcap_{k \in K} I_k$.

4.4 Binding-time statements

We say closed expression e has binding time I and write $\models e : I$ if $[e] \in I$. For open expressions, let B be a mapping from program variables to binding times. We write $B \models e : I$ if for all environments ρ and variables x in the domain of B such that $\rho(x) \in B(x)$ we have $[e]\rho \in I$.

"Dynamic" and "Static" At each non-trivial type τ we define an ideal Δ_τ which represents the property "completely dynamic" at τ, and Σ_τ, which represents "surface" static. We define Δ_τ to be the *downwards closure* $\downarrow\delta_\tau$ of δ_τ; that is, the least ideal containing δ_τ: $\downarrow\delta_\tau = \{d \in [\tau] : d \sqsubseteq \delta_\tau\}$. Note in particular that Δ_τ contains δ_τ, but not $\top_\tau$. The binding time Σ_τ at τ denotes $\Delta_\tau - \{\delta_\tau\}$; that is, all of Δ_τ except for its maximal element δ_τ. (Note that δ is isolated in every domain, so this *is* an ideal.) Intuitively, the elements in Σ_τ are those that are "surface" static, in the sense that one can apply the corresponding destructor at partial evaluation time without getting an error. We usually write Δ and Σ without subscripts whenever the type is derivable from the context.

Taking a domain such as *bool* $\times$ *bool*, we can represent the property that the pair is statically known (intuitively, available to the partial evaluator to destruct) by the ideal $\Sigma_{bool \times bool}$, which is equal to $\Delta_{bool} \times \Delta_{bool}$. Figure 1 sketches part of the Hasse diagram for *bool* $\times$ *bool*, and indicates some example binding times.

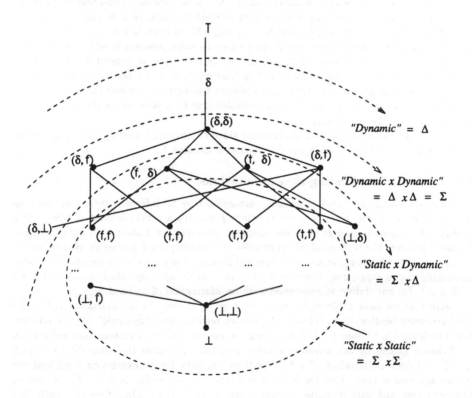

Figure 1: Example binding times in *bool* $\times$ *bool*

5 Internalising Binding-time Properties: Semantics-based Annotations

In partial evaluation and other transformations it is important to know not only *what* *extensional* property a program has, but also *how* it is established; in particular, what properties have to hold internally, for the individual *parts* of the program, since it is this *intensional* information that is usually exploited in optimizing transformations.

In partial evaluation it is rather useless in itself to find out that an expression has binding time "dynamic." Indeed this is usually stipulated from the outset. What is desired is a *proof* whose structure captures what the binding-time properties of individual parts of the expression are and how they can be combined to yield the binding time of the overall expression.

What is required then is an *internalisation* of what it means for an expression e to have a certain binding time. That is, the subexpressions of e must have certain binding times if the whole expression e is to have its final, desired binding time.

Ideally one would hope for a *complete* internalisation, in essence a *sound and complete* logic for inferring binding times. This may be undesirable (on top of being difficult to accomplish at all) since it expects the ensuing (automated) transformation process to exploit or at least to "understand" all proofs in the logic.

In this section we present an internalisation for *monovariantly well-annotated* expressions. Intuitively, monovariancy requires that the binding time of a bound variable be a fixed binding time (ideal), which must be "big enough" to contain the values the variable may ever be bound to. Similarly, the binding time of a subexpression must be a single ideal big enough to contain the value of the subexpression in every environment it is evaluated.

This is in contrast to *polyvariant binding-time analysis* where bound variables can be associated with several binding times (for different contexts) and where the binding times of (sub)expressions in the scope of bound variables may be *dependent* on (functions of) the binding times of the actual values.

We restrict our attention to monovariancy since monovariant binding-time analyses are currently better and easier understood in partial evaluation.[5]

5.1 Monovariant binding-time annotations

Monovariantly well-annotated expressions are defined by an inference system on *binding-time judgements* of the form $B \vartriangleright e : I$, where I denotes a *semantic binding time*, e is a well-typed expression, and B is an environment associating variables with binding times. Figure 2 presents the *nonlogical* inference rules for inferring binding-time properties for the constructs of the language in a syntax-directed fashion. Figure 3 gives a *logical* rule that is applicable to *any* expression. It lets us *weaken* an ideal to any larger ideal. Finally, Figure 4 adds a rule for *annotating* an expression with an *abstraction* of an ideal.

Note that we have no *formal language* of binding times: the metavariables I, I' range over *semantic* binding times (ideals), Δ denotes binding time "dynamic" (at the relevant type) as defined in Section 4.4, and $\times$, $+$ and $\rightarrow$ are the ideal constructors from Section 4.3.

What is gained by this inference system, and what role does the annotation rule play in it? Consider a derivation of a binding-time property for an expression e *without* use of the annotation rule. This results in a binding-time judgement $A \vartriangleright e : I$. Discarding the derivation and only retaining its conclusion we know the binding time of e itself, but nothing about the binding times of its subexpressions! The practical problem for a specialiser

---

[5]We believe the monovariant inference system given here can be extended to a polyvariant logic that is semantically complete, but leave this to future work.

$$B\{x \mapsto I\} \triangleright x : I \qquad \frac{B\{x \mapsto I\} \triangleright e : I'}{B \triangleright \lambda x.\, e : I \to I'} \qquad \frac{B \triangleright e : I \to I' \quad B \triangleright e' : I}{B \triangleright e\, e' : I'}$$

$$\frac{B \triangleright e : I \quad B \triangleright e' : I'}{B \triangleright (e, e') : I \times I'} \qquad \frac{B \triangleright e : I_1 \times I_2}{B \triangleright \pi_i e : I_i} \qquad \frac{B \triangleright e : \Delta}{B \triangleright \pi_i e : \Delta} \quad (i = 1,2)$$

$$\frac{B \triangleright e : I}{B \triangleright inl\, e : I + \emptyset} \qquad \frac{B \triangleright e : I}{B \triangleright inr\, e : \emptyset + I}$$

$$\frac{\begin{array}{c} B \triangleright e : I + \emptyset \\ B\{x \mapsto I\} \triangleright e' : I'' \\ B\{x' \mapsto \Delta\} \triangleright e'' : \Delta \end{array}}{B \triangleright \left(\begin{array}{l} case\ e\ of \\ \quad inl\, x \Rightarrow e' \parallel \\ \quad inr\, x' \Rightarrow e'' \end{array} \right) : I''} \qquad \frac{\begin{array}{c} B \triangleright e : \emptyset + I' \\ B\{x \mapsto \Delta\} \triangleright e' : \Delta \\ B\{x' \mapsto I'\} \triangleright e'' : I'' \end{array}}{B \triangleright \left(\begin{array}{l} case\ e\ of \\ \quad inl\, x \Rightarrow e' \parallel \\ \quad inr\, x' \Rightarrow e'' \end{array} \right) : I''}$$

$$\frac{\begin{array}{c} B \triangleright e : I + I' \\ B\{x \mapsto I\} \triangleright e' : I'' \\ B\{x' \mapsto I'\} \triangleright e'' : I'' \end{array}}{B \triangleright \left(\begin{array}{l} case\ e\ of \\ \quad inl\, x \Rightarrow c' \parallel \\ \quad inr\, x' \Rightarrow e'' \end{array} \right) : I''} \qquad \frac{\begin{array}{c} B \triangleright e : \Delta \\ B\{x \mapsto \Delta\} \triangleright e' : \Delta \\ B\{x' \mapsto \Delta\} \triangleright e'' : \Delta \end{array}}{B \triangleright \left(\begin{array}{l} case\ e\ of \\ \quad inl\, x \Rightarrow e' \parallel \\ \quad inr\, x' \Rightarrow e'' \end{array} \right) : \Delta}$$

$$\frac{B\{f \mapsto I\} \triangleright e : I}{B \triangleright fix\, f.\, e : I}$$

Figure 2: Monovariantly annotated expressions, nonlogical rules

$$\frac{B \triangleright e : I}{B \triangleright e : I'} \quad (I \subseteq I')$$

Figure 3: Logical rules

$$\frac{B \triangleright e : I}{B \triangleright e^b : I} \quad (b = \alpha(I))$$

Figure 4: Annotation rule

is that it needs to process program *parts* on the basis of *their* binding-time properties. Knowing only the binding-time property of the whole expression is *too little* information. One way of remedying this is to retain the whole derivation since it contains binding-time judgements for the subexpressions — this is, in essence, what the monolithic approach does: the result of a binding-time analysis is the (whole) derivation, and a specialiser does not operate on the program expression itself, but on a representation of a derivation for it. As we have remarked earlier, this ties the specialiser closely to a particular formalisation of a binding-time analysis. Furthermore, it may retain *too much* information, information that is not really relevant to partial evaluation or other transformations. Since there is a potentially large "semantic distance" between the original expression and the product of the analysis (the whole derivation), it furthermore complicates establishing the correctness of the specialiser since the specialiser is defined on derivations, whereas the standard semantics is defined on the original expression.

By using the annotation rule we can discard the derivation of a binding-time property for an expression and yet retain *relevant semantic information* about its subexpressions. This information is extensional in the sense that annotations only abstract the derived binding time for a subexpression, not a particular way the binding time can be established. We can think of the annotations as a "semantic trace" of how an expression has a certain binding-time property. The *abstraction function* α is a partial function mapping binding times to annotations. It expresses how much semantic information is retained from a binding time. The standard annotations we shall consider in the following section are $A = \{S, D\}$ where S stands for "static" and D for "dynamic."

It is easy to check that $B \rhd e : I$ implies $B \models e : I$; that is, our monovariant internalisation is *sound*. For every (open) expression e and assumptions B mapping the free variables of e to binding-times contained in Δ there is at least one binding time I such that $B \rhd e : I$ is derivable using the rules of Figures 2 and 3. The internalisation is not *complete*, however: $B \models e : I$ does not generally imply $B \rhd e : I$.

5.2 Preservation of Binding-Time Properties under Reduction

In this section we show that well-annotated expressions are closed under *reduction* in *any* context; that is, they have the *Subject Reduction Property*. As shown in the following section this is the crucial connection that lets us argue the safety of partial evaluation for expressions that are — via a translation into our annotation scheme — semantically well-annotated.

An *annotated expression* is a (well-typed) expression where subexpressions may carry arbitrary annotations from the range of α.

Definition 5.1 (Reduction rules) *The one-step reduction,* $\rightarrow$, *for annotated expressions is given by the following rules,*

$$(\lambda x.\, e_1)\, e \rightarrow e_1\{e/x\} \qquad \pi_1(e, e') \rightarrow e \qquad \pi_2(e, e') \rightarrow e'$$

$$\begin{pmatrix} case\ inl\ (e)\ of \\ inl\ x \Rightarrow e_1\ \| \\ inr\ y \Rightarrow e_2 \end{pmatrix} \rightarrow e_1\{e/x\} \qquad \begin{pmatrix} case\ inr\ (e)\ of \\ inl\ x \Rightarrow e_1\ \| \\ inr\ y \Rightarrow e_2 \end{pmatrix} \rightarrow e_2\{e/y\}$$

$$fix\ f.\, e_1 \rightarrow e_1\{fix\ f.\, e_1/f\} \qquad e^b \rightarrow e$$

The expressions to the left are called redexes *and those on the right the corresponding* reducts. *We close* $\rightarrow$ *under arbitrary contexts; that is,* $e \rightarrow e'$ *if* $e = C[r]$ *for some (single-hole) context* $C[]$ *and redex* r *with reduct* r', *where* $e' = C[r']$. *We write* $e \rightarrow^+ e'$ *if* $e \rightarrow \ldots \rightarrow e'$ *in one or more reduction steps;* $e \rightarrow^* e'$ *if* $e = e'$ *or* $e \rightarrow^+ e'$.

The main theorem of this section is that well-annotated terms are closed under reduction; that is, if $e \to^* e'$ and e is well-annotated then e' is also well-annotated. This is the critical connection between our semantic model of binding times and what one can do with this information.

Due to the annotations that (may) occur inside expressions the subject reduction property also has to hold *locally*; that is, intuitively, we must be able to establish that, if $e \to e'$ then $B \rhd e' : I$ can be obtained by "local" transformation of *any* proof of $B \rhd e : I$. For this to hold in the presence of the weakening rule of Figure 3, it is necessary and sufficient to establish that containments between constructed ideals hold only if suitable containment relations hold for the component ideals; that is,

$$I \times I' \subseteq J \times J' \;\Rightarrow\; I \subseteq J \text{ and } I' \subseteq J'$$
$$I + I' \subseteq J + J' \;\Rightarrow\; I \subseteq J \text{ and } I' \subseteq J'$$
$$I \to I' \subseteq J \to J' \;\Rightarrow\; J \subseteq I \text{ and } I' \subseteq J'$$

The technically crucial point is that the first two implications hold for *arbitrary* nonempty ideals, whereas the last implication does not: if J' is full (the whole domain) then $I \to I' \subseteq J \to J'$ for any choice of I, I', J!

This problem is the — sole — reason why our extended domains do not only contain an element δ representing dynamic values, but also an element $\top$, and why binding times, by definition, are nonfull ideals. The following lemma shows that the problematic implication *does* hold if J' is *not* full.

Lemma 5.2 *Let D be a domain with a top element $\top$. Let I, I', J, J' be nonempty ideals such that J' is a proper subset of D.*
Then $I \to I' \subseteq J \to J'$ if and only if $J \subseteq I$ and $I' \subseteq J'$.

Thus all three implications above hold for binding times, and we can establish our main theorem:

Theorem 5.3 (Subject Reduction Theorem) *If $B \rhd e : I$ and $e \to^* e'$ then $B \rhd e' : I$.*

6 Standard and Partial Evaluation

In this section we give a definition of off-line partial evaluation for our language, built by removing restrictions on reductions possible in the standard evaluation rules, and prove its safety from our semantic definition of a sound binding-time annotation.

6.1 Standard Call-by-name Evaluation

The operational semantics of the language is specified by reduction rules, which represent the basic computation steps, plus a description of the syntactic contexts in which the rules can be applied.

The standard operational semantics is built using the reduction rules of Definition 5.1, applying them to closed unannotated expressions. We need to specify the syntactic contexts in which we allow the reduction rules to be applied. We do not specify a deterministic reduction order, since it is sufficient (and more general) simply to constrain the reduction rules so that we: (i) do not reduce under a λ-abstraction, (ii) do not reduce under a fix-expression, (iii) do not reduce in the branches of a case expression, and (iv) do not reduce under a constructor (pairing, or sum injections). Let $\to_n$ be the resulting relation (call-by-name reduction). We state the following properties of $\to_n$ without proof: for all closed expressions $e : \tau$

- If $e \to_n e'$ then $e' : \tau$ and $[e] = [e']$

- If there is an infinite reduction sequence starting from e then all reduction sequences are infinite.

- For all e of first-order type, but excluding unit, we have that $[e] = \perp$ if and only if there is an infinite reduction sequence starting from e.

Thus denotational semantics is sound with respect to a definition of observational equivalence which observes termination at any first-order type except unit.

6.2 Partial Evaluation

We define a class of partial evaluators by describing the possible reductions that a partial evaluator *may* perform. A particular partial evaluator could then be built by choosing some reduction strategy from these reductions.

We view partial evaluation as standard evaluation extended rather straightforwardly to handle "symbolic" values; that is, open expressions. "Dynamic" inputs are then just modelled by free variables.

Binding-time analysis is used to guide either the actions of a specialiser (so-called off-line partial evaluation) or to optimise its actions (on-line partial evaluation). In both cases it is important to guarantee that the specialiser can "trust" the information provided by the binding-time analysis in order to avoid costly checks of data at partial evaluation time (such as whether data are static or dynamic values). In this section we show how semantically consistent binding-time annotations ensure that a specialiser cannot "go wrong" as long as its actions (reduction steps) *respect* the (semantic) binding-time annotations.

Partial evaluation applies the same reduction rules as standard evaluation, but with fewer constraints. Firstly, we allow evaluation to be more eager, so evaluation of the components of pairs, or of the argument of the sum-injections, is now possible. Secondly, we draw upon the binding-time annotations to allow reductions to reach even deeper into an expression — namely, under lambda abstractions or in the branches of case-expressions. Due to this, and the presence of dynamic inputs (free variables), it must be guaranteed that a specialiser that executes a "static" reduction can be assured that it does not encounter dynamic data where it expects to find static data. In off-line partial evaluation only static reductions are performed. In on-line partial evaluation dynamic reductions may also be performed (see the next section), but require a check as the nature of the data (static or dynamic).

PE-annotations   The definition of a monovariant binding-time annotation provided in the previous section is particularly simple because it only describes annotations on sub-expressions. The price of the simplicity of this definition is that there may not be an exact correspondence between the kind of annotations which are followed by a given partial evaluator, and the notion of a monovariant binding-time annotation. In particular, the partial evaluator we will describe operates on expressions with annotations on the binding occurrences of variables—and these are not proper sub-expressions. To avoid confusion we will call the annotations expected by our partial evaluator *PE-annotations*. When we come to prove the safety of the partial evaluator we will show how the PE-annotations can be interpreted as semantic annotations.

The PE-annotated expressions handled by our partial evaluator have possible annotations in two places: on all destructors (projections, case-expressions, applications), indicating a binding-time property of the expression in the "destructed" position, and on some binding occurrences of variables (lambda abstractions, fix-expressions and case expressions). Furthermore, the partial evaluator knows about only two annotations: dynamic (D) and static (S). Destructors annotated by S can be reduced (the partial evaluator expects that the destructed expression will be static), but dynamic destructors will not be reduced (since it cannot be trusted that the expression in the hole will be static).

Reduction now occurs in a more liberal class of contexts. In particular we can reduce under lambdas, or inside the branches of a case expression whenever the variable is annotated with dynamic.

The definition is divided into *static reductions* $\rightarrow_{pe}$, and *partial evaluation contexts P*. Let variables b, b_1, b_2 range over annotations $\{S, D\}$, and let $[D]$ denote either annotation D or "no annotation". The static reductions $\rightarrow_{pe}$ are given in Fig. 5 and the partial evaluation contexts P are given in Fig. 6.

$$\lambda x^{[D]}.\, e_1 \,@^S e \rightarrow_{pe} e_1\{e/x\} \qquad \pi_1^S(e, e') \rightarrow_{pe} e \qquad \pi_2^S(e, e') \rightarrow_{pe} e'$$

$$\begin{pmatrix} case^S\, inl\,(e)\ of \\ inl\, x^{[b_1]} \Rightarrow e_1 \,\| \\ inr\, y^{[b_2]} \Rightarrow e_2 \end{pmatrix} \rightarrow_{pe} e_1\{e/x\} \qquad \begin{pmatrix} case^S\, inr\,(e)\ of \\ inl\, x^{[b_1]} \Rightarrow e_1 \,\| \\ inr\, y^{[b_2]} \Rightarrow e_2 \end{pmatrix} \rightarrow_{pe} e_2\{e/y\}$$

$$fix\, x^S.\, e_1 \rightarrow_{pe} e_1\{fix\, x^S.\, e_1/x\}$$

Figure 5: Static Reduction Rules

$$P \quad ::= \quad [] \quad | \quad P\,@^b e' \quad | \quad e\,@^b P \quad | \quad \pi_1^b P \quad | \quad \pi_2^b P$$

$$| \quad \begin{array}{l} case^b\, P\ of \\ inl\, x^{[D]} \Rightarrow e_1 \,\| \\ inr\, y^{[D]} \Rightarrow e_2 \end{array} \quad | \quad \begin{array}{l} case^b\, e\ of \\ inl\, x^D \Rightarrow P \,\| \\ inr\, y^{[D]} \Rightarrow e_2 \end{array} \quad | \quad \begin{array}{l} case^b\, e\ of \\ inl\, x^{[D]} \Rightarrow e_1 \,\| \\ inr\, y^D \Rightarrow P \end{array}$$

$$| \quad \lambda x^D.\, P \quad | \quad fix\, x^D.\, P \quad | \quad (P, e') \quad | \quad (e, P) \quad | \quad inl\, P \quad | \quad inr\, P$$

Figure 6: Partial Evaluation Contexts

Definition 6.1 *One step partial evaluation relation $\mapsto_{pe}$ is defined by closing the static reductions under partial evaluation contexts. In other words, for all annotated expressions e, e', $e \mapsto_{pe} e'$ iff $e \equiv P[e_1]$, for some P, e_1 such that $e_1 \rightarrow_{pe} e_2$ and $P[e_2] \equiv e'$.*

Restricted to pure lambda-terms, the reductions permitted by our definition include those of Palsberg's definition of a *top-down partial evaluator* [Pal93].

6.3 Safety of the Partial evaluator

The definition of safety focuses on the statically annotated destructors. It is convenient to give a formal definition of these expressions:

Definition 6.2 (Destructors) *Define the destructors D to be the following single-holed contexts:*

$$D \quad ::= \quad []\, e \quad | \quad \pi_1[] \quad | \quad \pi_2[] \quad | \quad \begin{array}{l} case\, []\ of \\ inl\, x \Rightarrow e_1 \,\| \\ inr\, y \Rightarrow e_2 \end{array}$$

All destructors occurring in a PE-annotated expression must carry an annotation $(S$ or $D)$. We call these expressions the PE-destructors. Let D^S and D^D respectively denote the static and dynamically annotated PE destructors. For example, if D is the destructor $\pi_1[]$, then $D^D[x]$ is the expression $\pi_1^D x$. A static destructor tells the partial evaluator that it can expect that the expression in the hole can be evaluated to a constructor of the right type (or we loop in the attempt). What this means in an implementation (*e.g.* a partial evaluator like Similix [Bon91]) is that when partial evaluation of the expression in the destructor position has finished, it will be trusted that the result will be of the right kind to be "destructed". The partial evaluator "goes wrong" and reaches a possible *error state* if this is not the case. Before we give a syntactic characterisation of these error states, we note the following properties, where we assume that the standard semantics of an annotated expression is defined to be that of the corresponding unannotated version.

Proposition 6.3 *If* $A \vdash e : \tau$ *and* $e \to_{pe} e'$ *then* $A \vdash e' : \tau$, *and for all environments* ρ *matching type environment* A, $[e]\rho = [e']\rho$

So partial evaluation preserves the type and denotation of an expression. Note that if we wish to consider the underlying language to be call-by-value, then this partial evaluator increases termination properties (so $[e]_{val} \sqsubseteq_{val} [e']_{val}$) in the manner of lambda-mix [GJ91]. However, this is not the aspect of safety that concerns binding-time analysis.

The fact that we always have a well-typed program leads to the conclusion that the error states are those for which a variable, or a dynamic destructor, appears in the hole of a static destructor (and that this occurs in some partial evaluation context). The following proposition helps characterise the error states:

Proposition 6.4 *If* $D[e]$ *is a well-typed expression, and* $D^S[e]$ *is not a partial evaluation redex, then either* $e \equiv x$ *or* $e \equiv D'[e']$ *for some destructor* D'.

Definition 6.5 (Error states) *A PE-annotated expression e is in an* error state *if either*

1. $e \equiv P[D_1^S[x]]$ *for some* P, D_1, x, *or*

2. $e \equiv P[D_1^S[D_2^D[e']]]$ *for some* P, D_1, D_2, e'.

Our goal is to show that applying the reduction rules of the partial evaluator on a semantically well-annotated program can never lead to an error state. To do this we must give a definition of "a well-annotated expression" by interpreting PE-annotated expressions as semantic annotations.

Definition 6.6 *We define a mapping,* $\hat{\cdot}$, *from PE-annotated expressions to (ordinary) annotated expressions by induction on the syntax:*

$$\hat{x} = x \qquad \widehat{(e_1, e_2)} = (\hat{e_1}, \hat{e_2}) \qquad \widehat{inl\, e} = inl\, \hat{e} \qquad \widehat{inr\, e} = inr\, \hat{e}$$

$$\widehat{\lambda x^{[D]}.\, e} = \lambda x.\, (\hat{e}\{x^{[D]}/x\}) \qquad \widehat{e_1\, @^b e_2} = (\hat{e_1})^b\, \hat{e_2} \qquad \widehat{\pi_1^b e} = \pi_1(\hat{e}^b) \qquad \widehat{\pi_2^b e} = \pi_2(\hat{e}^b)$$

$$\widehat{\left(\begin{array}{l} case^b\ c\ of \\ \quad inl\, x_1^{[D]} \Rightarrow c_1 \,\| \\ \quad inr\, x_2^{[D]} \Rightarrow c_2 \end{array}\right)} = \begin{array}{l} case\ (\hat{e}^b)\ of \\ \quad inl\, x_1 \Rightarrow (\hat{c_1})\{x_1^{[D]}/x_1\} \,\| \\ \quad inr\, x_2 \Rightarrow (\hat{e_2})\{x_2^{[D]}/x_2\} \end{array}$$

So, for example, if e is the PE-annotated expression $\lambda x^D.\, (x, x)$, then $\hat{e} = \lambda x.\, (x^D, x^D)$.

Next we must give an interpretation of the annotations $\{S, D\}$ as abstractions of ideals. In what follows we assume the following definition for the abstraction map α:

$$\alpha(I_\tau) = \begin{cases} D, & \text{if } I = \Delta_\tau \\ S, & \text{if } I \subseteq \Sigma_\tau \\ \text{undefined}, & \text{otherwise} \end{cases}$$

Now we can define when a PE-annotated expression is semantically well-annotated.

Definition 6.7 *A PE-annotated open expression* e, *such that* $A \vdash e : \tau$ *for some type environment* A, *is* well-annotated *if there exists an* I *such that* $B_0 \rhd \hat{e} : I$, *where* $B_0 = \{x \mapsto \Delta_{A(x)} \mid x \text{ in the domain of } A\}$.

The structure of the proof of safety of the partial evaluator is as follows: first we show that anything appearing in a partial evaluation context is well-annotated, providing that the whole expression is well-annotated. We use this to argue that the error states are not well-annotated, and hence that the partial evaluator never starts out in an error state. The proof is completed by using the subject reduction property, which states that well-annotatedness is preserved by partial evaluation steps.

Theorem 6.8 *If* e *is well-annotated and* $e \mapsto_{pe}^* e'$ *then* e' *is not in an error state.*

7 On-line Partial Evaluation

Off-line partial evaluation is *defined* to be partial evaluation which uses a binding-time analysis. Conversely, the term *on-line* is used for partial evaluators which do not. This means that before attempting to do a reduction, an on-line partial evaluator must always check to see if the object being destructed is of the appropriate kind. It is generally able to perform more reductions than an off-line evaluator, but is potentially less efficient (and less simple in structure) because of the extra checking necessary.

In this section we show that the safety condition—that we never reach an error state—also holds for a form of on-line partial evaluation. This result is significant because it shows that an on-line partial evaluator could be optimised by using a binding-time analysis, since it removes partial-evaluation-time checks on the argument to a static destructor.

Definition 7.1 *Define an on-line reduction relation* $\to_{on}$ *on PE-annotated terms by extending the partial evaluation reductions* $\mapsto_{pe}$ *to include the following rewrite:*

$$\text{If } \mathbb{D}^S[e] \to_{pe} e' \text{ then } \mathbb{P}[\mathbb{D}^D[e]] \to_{on} \mathbb{P}[e']$$

Note then that we still do not permit reduction under non-dynamically annotated binding operators (not a severe restriction, since because they are static they are likely to be eliminated by reduction anyway). But now if a dynamic-annotated destructor is a redex, then it can be reduced.

Theorem 7.2 *If* e *is well annotated and* $e \to_{on}^* e'$ *then* e' *is not in an error state.*

8 Correctness of binding-time analyses

We have focused on proving safety (correctness) of partial evaluation from a semantic model. A reasonable question is whether it is possible to design analyses and show that they are sound *w.r.t.* this semantic model. In this section we briefly claim that this is so, by outlining how existing monovariant binding-time analyses can be justified in our model and monovariantly internalised.

8.1 λ-mix

Gomard and Jones describe a simple, but illustrative off-line partial evaluator for the kernel of an untyped higher-order programming language [GJ91, Gom92]. We shall only consider the (pure) lambda calculus subset of the language. It is untyped, but can be understood to be typed by giving every expression the type $rec\,\alpha.\ \alpha \to \alpha$. Our extended domain interpretation maps this type to the "smallest" domain D_∞ such that $D_\infty \cong (D_\infty \hookrightarrow D_\infty)_{\{f\}}^\delta$

$$A\{x \mapsto I\} \vdash x : I$$

$$\frac{A\{x \mapsto I\} \vdash e : I'}{A \vdash (\lambda x.\, e)^S : I \to I'} \qquad \frac{A\{x \mapsto \Delta\} \vdash e : \Delta}{A \vdash (\lambda x.\, e)^D : \Delta}$$

$$\frac{A \vdash e : I \to I' \quad A \vdash e' : I}{A \vdash (e^S)\, e' : I'} \qquad \frac{A \vdash e : \Delta \quad A \vdash e' : \Delta}{A \vdash (e^D)\, e' : \Delta}$$

Figure 7: Binding-time analysis a la Gomard and Jones

via top-strict continuous isomorphism $\Psi : (D_\infty \hookrightarrow D_\infty)^\delta_{\{f\}} \to D_\infty$, where $f = \lambda d \in D_\infty.\, if\ d \sqsubseteq \delta_{D_\infty}\ then\ \delta_{D_\infty}\ else \top_{D_\infty}$.

As before, the semantic ideal Δ modelling "dynamic" (D) is $\downarrow\!\delta_{D_\infty}$, whereas "(surface) static" (S) is $\Sigma = \Delta - \{\delta_{D_\infty}\} = \Psi(\Delta \to \Delta)$.

The binding-time analysis of Gomard and Jones can be described by an inference system consisting of rules that are derivable in our monovariant internalisation. We give the rules using the ordinary annotated expressions. These correspond, via a translation as in Definition 6.6, to PE-annotated terms where not only destructors, but also constructors are annotated by either S or D.

This shows that the binding-time analysis of Gomard and Jones is *sound* with respect to our model of binding times and our monovariant internalisation. An immediate consequence of the Subject Reduction Theorem is that no partial evaluator, in particular λ-mix, whose actions can be modelled by the reductions of Section 5 can reach an error state.

8.2 Other analyses

Mogensen Mogensen extends the binding times in Gomard and Jones' analysis with recursively specified binding times [Mog92]. His off-line partial evaluator has been shown correct relative to the analysis by Wand [Wan93]. Mogensen's analysis can also be justified by the rules of Figure 7. The safety of his partial evaluator follows from our Subject Reduction Theorem.

Palsberg/Schwartzbach The binding-time annotations of Palsberg and Schwartzbach [Pal93] are the same as for Gomard/Jones and Mogensen. Their binding-time analysis, however, cannot be shown correct relative to our monovariant internalisation since our inference system lacks the ability of propagating *disjunctive* properties. Adding rules for *unions* of ideals, in the style of Jensen [Jen92], to our internalisation, seems to provide a — still monovariant — internalisation of binding-time properties that subsumes their analysis. This extension promises interesting applications to constructor specialisation and closure analysis.

Launchbury and Hunt/Sands The binding-time models of Launchbury [Lau89] and Hunt and Sands [HS91] can be expressed in our model in the sense that their (syntactic descriptions of) binding times can be interpreted as ideals in our extended domains, giving valid binding-time statements for expressions. The analysis of Hunt and Sands is polyvariant, however, and thus cannot be expressed in our monovariant internalisation. We conjecture that Launchbury's analysis is expressible in our monovariant internalisation.

9 Conclusions and Further Work

In this paper we have considered a model-based approach to the safety of off-line partial evaluation. We have motivated a new model for structured binding times in higher-order functional languages, illustrating problems with the previous models using projections and PERs. The model is based on an extension of the standard domains with "extra" elements δ (anonymous dynamic value) and $\top$ (error value) at each type, reflecting the finer distinctions that we are able to make between programs at partial evaluation time than we are able to make during normal execution. We tackle the problem of program annotation by showing that semantic properties can be expressed in a structural syntax-directed style. This is essentially a collecting interpretation [CC79], but avoids the cumbersome details of an explicit "sticky" semantics mapping properties to program points (cf. [Nie85])

The model is able to represent partially static data structures in the manner of [Mog88] and [Lau88], as well as properties of higher-order functions. Furthermore, we are not dependent on any assumption of lazy data structures and non-strict evaluation in the underlying language.

We have shown that the model is adequate to prove safety for a class of partial evaluators for this language. This class of partial evaluators is similar in spirit to Palsberg's definition of top-down partial evaluators for the pure lambda-calculus. We believe this is the first proof of its kind—based on a semantic specification of a safe binding time annotation, rather than on a particular analysis. We have also shown that sound binding-time annotations are preserved by dynamic reductions, a fact which has implications for the optimisation of on-line partial evaluators. Finally, we have argued that existing analyses can be shown to be sound with respect to the model given here.

9.1 Limitations and Further Work

There are some fundamental limitations in the definition of a sound annotation which are necessary in order to prove the correctness of simple-minded partial evaluators. One such limitation is the "uniformity" assumption [Lau89], which is implicit in the structural nature of our conditions for a safe annotation[6]. This restriction is fundamental in the sense that it would actually be *unsafe* to perform, for example, constant propagation or relational analyses between variables unless the partial evaluator where to employ exactly the same flow analysis "on-line" (as in Turchin's *driving* [Tur86]). This also means that we cannot account for partial evaluators which perform arbitrary algebraic manipulations (*e.g.* code propagation across dynamic conditionals [Bon92]), unless they can be factored out in a pre-processing stage (*e.g.* [CD91]).

Polyvariant binding-time analysis is intimately connected to (finitary or infinitary) conjunctive properties of functions. This can be modelled by taking intersections of ideals. Finitary conjunctive properties of functions capture the polyvariant binding-time analyses of Gengler and Rytz [GR92] and Consel [Con93]. Infinitary conjunctive properties can be expressed as binding-time functions [HS91, CJØ94] or polymorphic types [HM94]. We plan on extending the monovariant internalisation of this paper to a sound and complete polyvariant internalisation using infinitary conjunction. This, we hope, will enable us to justify both polyvariant analyses as well as the safety of partial evaluators driven by polyvariant analyses.

---

[6] This does *not* imply that the analysis must give a uniform treatment to the components of recursive types (in the sense of e.g. [EM91]). The definition of a safe annotation permits, for example, properties such as "the first ten elements of the list are static."

Acknowledgements

Thanks to the referees for numerous useful suggestions for improvements. This work was partly funded by the Danish Science Council (SNF), project "DART".

References

[Amt93]   T. Amtoft. Minimal thunkification. In *Proceedings of the 3rd International Symposium on Static Analysis*, number 724 in LNCS. Springer-Verlag, 1993.

[BEJ88]   D. Bjørner, Ershov, and N. D. Jones, editors. *Partial Evaluation and Mixed Computation. Proceedings of the IFIP TC2 Workshop, Gammel Avernæs, Denmark, October 1987*. North-Holland, 1988. 625 pages.

[BJMS88] Anders Bondorf, Neil D. Jones, Torben Mogensen, and Peter Sestoft. Binding time analysis and the taming of self-application. Draft, 18 pages, DIKU, University of Copenhagen, August 1988.

[Bon91]   Anders Bondorf. Automatic autoprojection of higher order recursive equations. *Science of Computer Programming*, 17(1-3):3-34, December 1991. Selected papers of ESOP '90, the 3rd European Symposium on Programming.

[Bon92]   Anders Bondorf. Improving binding times without explicit cps-conversion. In *1992 ACM Conference on Lisp and Functional Programming. San Francisco, California*, pages 1-10, June 1992.

[Bul93]   Mikhail Bulyonkov. Extracting polyvariant binding time analysis from polyvariant specializer. In *Proc. ACM SIGPLAN Symp. on Partial Evaluation and Semantics-Based Program Manipulation (PEPM), Copenhagen, Denmark*, pages 59-65. ACM, ACM Press, June 1993.

[CC79]   P. Cousot and R. Cousot. Systematic design of program analysis frameworks. In *ACM 5th Symposium on Principles of Programming Languages*, 1979.

[CC94]   P. Cousot and R. Cousot. Higher-order abstract interpretation (and application to comportment analysis generalizing strictness, termination, projection and PER analysis of functional languages). In *Proc. 1994 Int'l Conf. on Computer Languages, Toulouse, France*, pages 95-112. IEEE Computer Society Press, May 1994.

[CD91]   C. Consel and O. Danvy. For a better support of static data flow. In J. Hughes, editor, *Proc. 5th ACM Conf. on Functional Programming Languages and Computer Architecture (FPCA), Cambridge, Massachusetts*, number 523 in Lecture Notes in Computer Science, pages 496-519. Springer-Verlag, Aug. 1991.

[CJØ94]   Charles Consel, Pierre Jouvelot, and Peter Ørbæk. Separate polyvariant binding time reconstruction. CRI Report A/261, Ecole des Mines, Oct. 1994.

[CK92]   Charles Consel and Siau Cheng Khoo. On-line & off-line partial evaluation: Semantic specifications and correctness proofs. Technical Report YALEU/DCS/RR-912, Yale University Department of Computer Science, June 1992.

[Con93]   Charles Consel. Polyvariant binding-time analysis for applicative languages. In *Proc. Symp. on Partial Evaluation and Semantics-Based Program Manipulation (PEPM), Copenhagen, Denmark*, pages 66-77, June 1993.

[Dav94]   K. Davis. Pers from projections for binding-time analysis. In *Proceedings of the ACM Workshop on Partial Evaluation and Semantics-Based Program Manipulation*, 1994. (Proceedings available as Tech Report 94/9, University of Melbourne).

[EM91]   C. Ernoult and A. Mycroft. Uniform ideals and strictness analysis. In *Proc. 18th Int'l Coll. on Automata, Languages and Programming (ICALP), Madrid, Spain*, number 510 in Lecture Notes in Computer Science. Springer-Verlag, 1991.

[GJ91]   C. Gomard and N. Jones. A partial evaluator for the untyped lambda calculus. *J. Functional Programming*, 1(1):21-69, 1991.

[Gom92] C. Gomard. A self-applicable partial evaluator for the lambda calculus: Correctness and pragmatics. *ACM Transactions on Programming Languages and Systems*, 14(2):147–172, 1992.

[GR92] Marc Gengler and Bernhard Rytz. A polyvariant binding time analysis handling partially known values. In *Proc. Workshop on Static Analysis (WSA), Bordeaux, France*, pages 322–330, Sept. 1992.

[Gun92] Carl Gunter. *Semantics of Programming Languages — Structures and Techniques*. Foundations of Computing. MIT Press, 1992.

[Hat95] J. Hatcliff. A mechanised proof of correctness of off-line partial evaluation. In *These proceedings*, 1995.

[HM94] Fritz Henglein and Christian Mossin. Polymorphic binding-time analysis. In Donald Sannella, editor, *Proceedings of European Symposium on Programming*, volume 788 of *Lecture Notes in Computer Science*, pages 287–301. Springer-Verlag, April 1994.

[HS91] S. Hunt and D. Sands. Binding Time Analysis: A New PERspective. In *Proceedings of the ACM Symposium on Partial Evaluation and Semantics-Based Program Manipulation (PEPM'91)*, pages 154–164, September 1991. ACM SIGPLAN Notices 26(9).

[Hun90] S. Hunt. PERs generalise projections for strictness analysis. In *Proceedings of the Third Glasgow Functional Programming Workshop*, Ullapool, 1990. Springer Workshops Series.

[Jen92] T. Jensen. *Abstract interpretation in logical form*. PhD thesis, Department of Computing, Imperial College, November 1992. (Available as DIKU tec. report 93/11).

[Jon88] N.D. Jones. Automatic program specialization: A re-examination from basic principles. In *[BEJ88]*, 1988.

[Lau88] J. Launchbury. Projections for specialisation. In *[BEJ88]*, 1988.

[Lau89] J. Launchbury. *Projection Factorisations in Partial Evaluation*. PhD thesis, Department of Computing, University of Glasgow, 1989.

[Mog88] T. Mogensen. Partially static structures in a self-applicable partial evaluator. In *[BEJ88]*, 1988.

[Mog89] T. Mogensen. Binding time analysis for polymorphically typed higher order languages. In J. Diaz and F. Orejas, editors, *TAPSOFT '89 (LNCS 352)*, pages 298–312. Springer-Verlag, 1989.

[Mog92] Torben Æ. Mogensen. Self-applicable partial evaluation for pure lambda calculus. In Charles Consel, editor, *ACM SIGPLAN Workshop on Partial Evaluation and Semantics-based Program Manipulation*, pages 116–121. ACM, Yale University, 1992.

[NBV91] A. De Niel, E. Bevers, and K. De Vlamnick. Program bifurcation for polymorphically typed functional languages. In *Proceedings of the ACM Symposium on Partial Evaluation and Semantics-Based Program Manipulation (PEPM'91)*, September 1991. ACM SIGPLAN Notices 26(9).

[Nie85] F. Nielson. Program transformation in a denotational setting. *ACM Transactions on Programming Languages and Systems*, 7(3):359–379, 1985.

[Pal93] J. Palsberg. Correctness of binding time analysis. *Journal of Functional Programming*, 3(3):347–363, 1993.

[Ste94] P. Steckler. *Correct Higher-Order Program Transformations*. PhD thesis, College of Computer Science, Northeastern University, Boston, 1994. Tech Report NU-CCS-94-15.

[Tur86] V. F. Turchin. The concept of a supercompiler. *ACM Transactions on Programming Languages and Systems*, 8:292–325, July 1986.

[Wan93] Mitchell Wand. Specifying the correctness of binding-time analysis. *Journal of Functional Programming*, 3(3):365–387, July 1993. preliminary version appeared in *Conf. Rec. 20th ACM Symp. on Principles of Prog. Lang.* (1993), 137–143.

Gadgets: Lazy Functional Components for Graphical User Interfaces

Rob Noble and Colin Runciman
email: {rjn,colin}@minster.york.ac.uk

Dept. of Computer Science, University of York, UK

Abstract. We describe a process extension to a lazy functional programming system, intended for applications with graphical user interfaces (GUIs). In the extended language, dynamically-created processes communicate by asynchronous message passing. We illustrate the use of the language, including as an extended example a simple board game in which squares are implemented as concurrent processes. We also describe a window manager, itself implemented in the extended functional language.

Keywords: functional language, processes, concurrency, window manager, Gofer.

1 Introduction and Motivation

Most of the time, elements of a graphical user interface (GUI) operate independently. For example, a menu doesn't interact with the rest of the program until the user selects an option. The user can highlight options, open up further menus or move the menu around the screen, all without doing anything that should concern any other element of the program. Popular languages such as C do not readily lend themselves to the job of programming interfaces like these. Programs end up contorted by event-polling loops or a multitude of callback functions.

The Fudgets [CH93] system has shown that lazy functional languages have advantages over imperative languages when it comes to writing programs with GUIs. Fudgets interact with each other by message passing. A Fudget may have a representation on the screen, or may just work in the background (an *abstract* Fudget). Fudgets are linked or composed by *combinators* that also specify their relative layout. Each Fudget may have a private state associated with it. A library of useful Fudgets ranges from simple integer displayers to a complete text editor. It is simple to piece together these *building blocks* into a program with a graphical user-interface. Aside from the interface, the rest of the program may also take advantage of conceptual concurrency by being implemented across a number of abstract Fudgets.

However, each Fudget has only a single input and a single output through which to pass messages. Multiple channels of communication must be multiplexed explicitly. A Fudget has a pair of channels for I/O (eg. the screen representation) that are multiplexed automatically by the combinators. However, the general connections of a program are more varied, and any change in the communication will need the multiplexing to be re-worked. A further effect of this restriction is that the layout combinators can only place together Fudgets that have a channel of communication between them.

1.1 Improving on Fudgets

We have extended a lazy functional language to support concurrent execution of component processes. Like Fudgets, these processes communicate through typed channels, but we have removed the restriction to single channels. In our *Gadget Gofer* system programs with a GUI may be written. Not only are application programs written in Gadget Gofer, but the window manager is too. We have developed a basic library of useful interface Gadgets (buttons, windows, etc.) and functions for glueing Gadgets together.

As a brief example of what can be done, Fig. 1 shows the program and screen image of an up-down bargraph display. It is simplified slightly; Sect. 4.3 gives a complete version. Both button and bargraph are defined in a Gadget library; these are *not* calls to an existing C library.

```
updown :: Gadget
updown =
    wire $ \u ->
    wire $ \d ->
    let b1 = button (\x->x+1) (op u) upArrow
        b2 = button (\x->x-1) (op d) downArrow
        bg = bargraph [ip u,ip d] in
    wrap (b1 <|> bg <|> b2)
```

Fig. 1. An up-down bargraph display.

In Sect. 2 we describe our extended language. In Sect. 4 we see how we have used the language to implement programs with graphical user interfaces. Section 5 describes the screen manager. A longer example program is given in Sect. 6. Some issues of the implementation are discussed in Sect. 7. Section 8 briefly reviews related work. Section 9 concludes and suggests future work.

2 Components

Programs are networks of processes that communicate by message passing. Processes have input and output *pins* that are connected together with *wires* that convey the messages. This makes a diagram of the processes in a program look like a circuit of electronic components, hence we call processes with pins *components*. A component may have any number of input and output pins. Message passing is asynchronous. Wires act like buffers, holding a queue of messages. Messages can be any first class value and are passed unevaluated. So functions, wires or even processes themselves may be passed as messages.

Predefined components that implement I/O devices provide a means to communicate with the user. For example, the *mouse* component has a single output pin that delivers *mouse-click* messages.

2.1 Component Definitions

The exact definition of primitive `Wire` and `Process` types is not important here. It is enough to know that a process is of type `Process s` where `s` is the type of a state value kept by the process. A wire of type `Wire a` will pass messages of type `a` only. Any attempt to send a message of the wrong type along a wire will result in a static type error. The pins of a component can be seen in its type. For example a component that has one input pin for messages of type `Int` and one output pin for messages of type `Char` has type `In Int -> Out Char -> Component`.

2.2 The Continuation Passing Style

One of the strengths of lazy functional languages is that the programmer can ordinarily abstract away from order of evaluation. In defining the operation of a component however, we need to state the order in which messages are sent and received. To do this we use the *continuation passing style* (CPS, described in [Kar81, HS88]). To avoid deeply nested brackets in continuations a right associative operator `$` for function application is useful: we write `f $ g $ h` instead of `f (g h)`.

2.3 Message Transmission

The function `tx` transmits a message from a component:

```
tx :: Out m -> m -> Process s -> Process s
```

Example: A simple component with one output pin sends the message "Hello World!" before ending using the primitive `end :: Process s`.

```
component :: Out String -> Component
component o = tx o "Hello World!" $ end
```

The style may look imperative and sequential. But there is no loss of referential transparency and it is only the parts in which the order of I/O is important that use continuation-based sequencing.

2.4 Message Reception

Because message passing is asynchronous and a component may have more than one input pin, we cannot tell where the next message will arrive from. A non-deterministic choice must be made when accepting the next message received. How can we introduce non-deterministic choice, whilst maintaining the referential transparency that is fundamental to a lazy functional language? We introduce a primitive that is like a restricted form of the ALT construct of Occam or select in PICT [PRT93]. The primitive is from, and is used *only* with another primtive, rx:

```
from :: In m -> (m -> Process s) -> Guarded s
rx :: [Guarded s] -> Process s
```

The function rx takes a list of *guarded continuations* of the form:

```
from i $ c
```

where i is an input pin and c :: m -> Process s is the continuation that will be used if the next message is of type m and from pin i. Making guarded continuations an abstract type ensures that from cannot be used *outside* of an rx construct, and that only guarded continuations can be used *inside*. The rx primitive waits for a message to arrive on a pin, then picks the first guarded continuation that matches that pin. Referential transparency is maintained because the choice of pin to receive from is made outside the functional program, by the rx primitive.

Example: Figure 2 shows a component that implements a memory cell. A message received on one input pin causes the value stored to be emitted from the single output pin. A message on the other input pin replaces the stored value.

```
memory :: m -> In m -> In t -> Out m -> Component
memory s i t o =
    claim i $
    claim t $
    m s where
    m s =
        rx [
            from i $ \s' -> m s',
            from t $ \_ -> tx o s $ m s
        ]
```

Fig. 2. A memory cell component.

The claim primitive is used by components to declare that they will receive messages from a particular wire. Components are sometimes passed wires that they do not receive messages on, because they pass the wire on to another component. In this case, the component does not claim the wire. At the implementation level, a `claim` enables a simpler and more efficient scheduling algorithm to be used.

2.5 Composing Components

Components are often defined as compositions of other components. For this we use two more primitive functions:

```
wire  :: (Wire a -> Process s) -> Process s
spawn :: Process a -> Process s -> Process s
```

where wire generates a new wire, passing it to its continuation argument, and spawn creates a new process out of the given process function. Generating a new wire or process requires a new value of the *world state* to be calculated, to which only processes have access. Therefore a composition of processes must itself be created by a process. Often this composing process exists only long enough to wire together and "switch on" the constituents of a composition. The assembled components remain separate as far as process scheduling is concerned, but appear from the outside to be a single component.

Example: Two memory components can be composed to form a memory storing two values, both released by the same trigger. Figure 3 shows the definition and wiring diagram. The principal tools used here are the primitives wire (to obtain a new wire) and spawn (to create a new process). A wire has two ends. Applying the function ip (or op) to a wire yields the end attached to an input (or output) *pin*.

```
two_memory :: m -> n -> In m -> In n ->
       In t -> Out m -> Out n -> Component
two_memory s1 s2 i1 i2 t o1 o2 =
    wire $ \a ->
    wire $ \b ->
    spawn (tee t (op a) (op b)) $
    spawn (memory s1 i1 (ip a) o1) $
    spawn (memory s2 i2 (ip b) o2) $
    end
```

Fig. 3. A composition of three components.

2.6 Duplex Wires

Wires communicate messages in one direction only, but often two-way communication is required. It is convenient to define a type `Duplex a b` of paired wires that communicate messages of type a in one direction and of type b in the other.

```
type InOut a b = (In a, Out b)
type Duplex a b = (InOut a b, InOut b a)
```

The `duplex` function supplies a new duplex wire and is defined in terms of `wire`:

```
duplex :: (Duplex a b -> Process s) -> Process s
duplex c =
    wire $ \x ->
    wire $ \y ->
    c ((ip x,op y),(ip y,op x))
```

2.7   Some Useful Higher-Order Functions

We call a function of type `Process s -> Process s` an *action* because it describes one action in the operation of a process. When combined with a continuation, an action forms a complete process. Some actions pass on a value and have the type:

```
(r -> Process s) -> Process s
```

They must be applied to a continuation that expects a value of type `r`.

The higher-order functions `sequence` and `accumulate` are two of many useful *glue* functions:

```
sequence :: [Process s -> Process s] -> Process s -> Process s
accumulate :: [(r->Process s) -> Process s] ->
                 ([r] -> Process s) -> Process s
```

Given a list of actions that *don't* pass on a value, and a continuation c, `sequence` performs the actions in the list one after another, then continues with c. Given a list of actions that *do* pass on a value, and a continuation c, `accumulate` performs the actions in the list one after another, storing each value passed on, then continues with c applied to the list of values collected.

2.8   Components With Lists of Pins

What if we want to define a component with several pins of the same type, but we won't know how many until the component is used in a program? A component can take a *list* of pins as argument. When receiving a message from one of a list of pins, the `froms` function is a useful alternative to `from`. Instead of a single input pin, `froms` expects a list of input pins each paired with some information that will be useful when responding to a message. The continuation given to `froms` expects a message (as with `from`) *plus* another value — the information that was paired with the corresponding input pin. For example, the `tag` component takes a list of input pins and a single output pin. It copies messages from any of the input pins to the output pin, pairing the message with an index number that indicates which pin the message arrived on.

```
tag :: [In m] -> Out (Int,m) -> Component
tag is o = sequence (map claim is) $ t
    where
    t = rx [froms (zip is [1..]) $ \m i -> tx o (i,m) $ t]
```

3 The Program Environment

All but the simplest of components are given a connection to a component called the *operating system* (OS) when they are created. Connections to resources such as a display screen are then obtained via the OS. A component with a connection to the OS is called a *System* component.

4 Gadgets

A Gadget is a component with an image on the screen. The image displays output from the Gadget. Mouse-clicks over the image generate input messages to the Gadget. Gadgets are hierarchical (the area occupied by a Gadget can contain further, smaller Gadgets) and may overlap on the screen (Gadget images are flat but may be placed at different depths). The screen hardware provides only a two-dimensional picture, gives no support for sharing the screen between multiple Gadgets, and cannot route user-input to the relevant Gadget. Instead, a *screen manager* component (SM) provides these facilites. It has one duplex connection to the **screen** and one duplex connection to each Gadget. When a Gadget sends a message to the SM, it can be thought of as communicating directly with its image on the screen.

Gadgets too can be defined directly or as a composition. In both cases, the wire connections to the SM are made implicitly and communication with the SM occurs without explicitly referring to these wires (using **txSM** and **fromSM** functions). This is possible because the SM wires are hidden within the state value held by the Gadget.

4.1 Gadget Compositions

As well as describing the connections between Gadgets, a Gadget composition must also define the relative layout of the Gadgets on the screen. This is normally achieved using *layout combinators* such as `<->` and `<|>`:

```
<->, <|> :: Gadget -> Gadget -> Gadget
```

that place Gadgets side-by-side (centred vertically) and one above the other (centred horizontally), respectively. Other combinators provide, for example, horizontal placement with Gadgets aligned at the top. Figure 4 shows a composition of a button and a bargraph. The composed Gadget operates like a bargraph, but has a button beside it that resets the bargraph to the lowest level. This is like a component composition, but instead of spawning the Gadgets to be combined, they are composed with the layout combinator `<->` and wrapped in a box large enough to hold the two.

```
resetable_bargraph :: [Int -> Int] -> Gadget
resetable_bargraph is =
    wire $ \r ->
    let b = button (op r)
                (const 0) blank
        g = bargraph (ip r:is) in
    wrap (b <-> g)
```

Fig. 4. A Resettable Bargraph.

4.2   Gadget Attributes

Gadgets have *attributes* such as width, height and colour. With many attributes to set, passing them all as arguments could lead to clutter and confusion. Even with only the attributes of width, height and colour a button application might be:

```
button (op w) Click 50 60 c
```

where "Click" is the message the button sends when clicked, "50" is the width, "60" is the height and "c" a colour. With only three attributes the button has five parameters, and it must be remembered which parameter controls each one.

Instead, suppose we define an algebraic type that specifies all a Gadget's parameters, and assume that the Gadget has a *default value* of these *attributes*. We can pass the gadget a function to alter the default value. Take the button for example:

```
button :: (ButtonAttributes -> ButtonAttributes) ->
    Out m -> m -> Gadget
```

This is the *default Gadget*. It must be applied to some attribute-changing function before it becomes a Gadget. To leave the attributes at the default, we apply the default Gadget to the function id. Alternatively, *attribute modifying functions* may be composed and used to alter the attributes. For example, a button of default width and colour, but with a height of 100 and a *momentary* action (ie. the button pops out again when the mouse button is released):

```
button (height 100.buttonMomentary) (op w) Click
```

Some aspects of a Gadget's operation are peculiar to the type of Gadget and have no sensible default value (eg. for a button, the output wire and message), so these values are separate parameters to the Gadget. Some Gadget attributes are specific to a particular Gadget (eg. buttonMomentary) but can have a default value. Others are common to many Gadgets (eg. width and height), and we would like to use the same name regardless of the particular type of Gadget. Here the *type classes* of Haskell prove useful [HHPJW94]. Attributes are made a type class, of which an attribute type such as `ButtonAttributes` is an instance. Each type of Gadget defines its own overloaded versions of common modifiers such as `width` to alter its particular attribute datatype. Modifiers for attributes specific to a particular Gadget are not overloaded.

4.3 A Gadget Program

We are now in a position to look back at the updown bargraph display first introduced in Sect. 1.1 and see how it works. Figure 5 gives the complete version.

```
updown :: Gadget
updown =
    wire $ \u ->
    wire $ \d ->
    let b1 = button (picture upArrow) (op u) (\x->x+1)
        b2 = button (picture downArrow) (op d) (\x->x-1)
        bg = bargraph id [ip u,ip d] in
    wrap id (b1 <|> bg <|> b2)
```

Fig. 5. The Wiring of `updown`.

Two wires, u and d, carry messages from each button to the bargraph display. The messages sent each time a button is clicked are *functions* of type `Int -> Int` that alter the current level of the bargraph. Altering the level with functions instead of simply stating a new level is more flexible and reduces the number of components required. It would be a simple matter to add a further button, for example, to reset the level to zero (as in Sect. 4.1).

The `<|>` combinator takes two Gadgets and places them one above the other. Layout components continue to exist throughout the life of the program, in case the Gadgets need to be re-positioned — eg. when a Gadget changes size. A layout component has no screen area of its own, but relies on a *container* supplied by `wrap` to hold the neighbouring Gadgets. By default, `wrap` supplies a blank box to contain its Gadget parameter, but the default can be changed with the `picture` attribute modifier, as used by the buttons.

5 The Screen Manager

Rather than provide an interface to an imperative window manager such as The X Windows System, we chose to program a screen manager in Gadget Gofer. Our motivation for doing this was to move the boundary between the functional and imperative worlds a step closer towards the hardware, increasing the applicability of functional languages, and to avoid the design constraints of another programming paradigm (eg. event-polling loops or callbacks). Besides, others have already produced interfaces between lazy functional languages and imperative window managers (eg. [CH93, AvGP92, RS93, Sin92]).

Our aim is to hide the work involved in sharing the screen between layers of images (eg. redrawing images that become uncovered) from the Gadget programmer. A Gadget's connection to the SM may be viewed as a connection to the image itself. A new drawing sent to the image (really to the SM) changes the way it is displayed. Mouse-clicks or key-presses arrive (as if) from the image. We use the term *image* rather than *window* because these areas are used for a variety of interface items such as icon buttons, windows and window containers. The SM manages images in a hierarchy — a Gadget's image is created within the image of its parent. The Gadget end of the duplex wire connecting Gadget to SM is hidden within the Gadget state automatically when the Gadget is created.

Rather than sending actual drawing commands to the screen area, a Gadget sends a *drawing function* that the SM applies to *size* and *placement* values to generate a list of drawing commands. The benefit is that the SM can automatically redraw screen areas that become exposed without the need for further communication with relevant Gadgets.

```
type DrawFun = Size -> Placement -> [ScreenCmnd]
type Size = (Int,Int)
type Placement = (Coord,Coord)
type Coord = (Int,Int)
```

Examples of ScreenCmnds are DrawLine (Coord,Coord), DrawSetColour Colour, and DrawText Coord String. The Size argument of a DrawFun indicates the size of the Gadget's screen area. The Placement argument indicates the position of a rectangle within the screen area that needs redrawing. When the SM uses a DrawFun, it only draws part of the area if another area is overlapping. The DrawFun can use the placement information to speed redrawing by filtering out commands that draw outside of the required region. If a DrawFun returns commands that draw outside of the required region they are clipped away.

Sending a new DrawFun results in the whole screen area of a Gadget being redrawn. In cases where only a small change is being made (eg. one extra line added) the Gadget may send a DrawFun that draws only what is being added.

The SM redraws the screen areas front-most first[1]. A list of rectangles that need redrawing, rs, is maintained. Initially rs contains one rectangle the size of the whole screen. As each window is redrawn, the rectangle it occupies is subtracted from rs. The intersection of the position of a screen area with rs gives the placement argument for its DrawFun. The intersection may consist of more than one rectangle so the DrawFun may be called several times. But only the SM is involved, not the Gadget that supplied the DrawFun.

---

[1] The method is based on an unpublished algorithm due to Roger Took

5.1 Gadget Layout Management

Gadgets can specify their position relative to a corner of their parent's screen area, but they need not know about their on-screen neighbours. Decisions about their position are delegated to *layout components* (already described in Sect. 4.1). A parent Gadget specifies the layout of its children by wiring them to a layout manager using layout combinators. Gadgets with their respective layout managers are connected in an ancestral tree. At the top of the tree is the OS, as it owns the *root* window and is responsible for creating any program Gadgets the user wants to run.

When a program starts, layout managers collect a description of each image they are responsible for, package them together, and pass them up the tree. At the root of the tree this information is passed to the SM, where it forms the initial image data structure held by the SM. A Gadget may change the size of its image, sending the new size up the layout tree: this may affect the size of ancestors too (for example, if a Gadget is wrapped in a box that stretches to fit its contents), and several Gadgets may need to be repositioned. These changes are all handled transparently by the layout managers.

5.2 Gadget Connections

The SM, OS and layout wires of a Gadget are provided by its parent when a Gadget is created. They are kept in a hidden state and used by library functions such as `setSize` that sends new-size messages to the SM and the Gadget's layout manager.

In addition, a Gadget has connections specific to the job it performs. Typically, these pins are wired by the next level up the Gadget hierarchy. For example, in the updown bargraph display of Sect. 4.3, the wires u and d are connected by the parent, `updown`. Abstraction allows the programmer to forget about most of the wiring, concentrating on the main application wires only.

6 A Longer Example: Grid Explode!

6.1 The Game

The game of Grid Explode is played on a rectangular grid of cells. Each player has a supply of stones of a unique colour. Players take turns placing a stone into any cell that does not already contain another player's stones. The *neighbours* of a cell are the cells above, below, to the left and to the right of the cell. The *capacity* of a cell is the number of neighbours it has. If the total number of stones in a cell remains below the capacity after a player places a stone into it, then the turn is over and it is the next player's turn. When a cell reaches its capacity, it *explodes* invading each of its neighbours with one stone. A stone invading a cell turns any occupants to its colour, and may cause further explosions. The turn is over when all cells are below their capacity. A player wins if his or her move results in perpetual explosions or the explosions cease leaving every cell containing at least one stone of their colour.

6.2 The Gadget Version

The Gadget version assumes two players only. The picture on the right-hand side of Fig. 6 shows an example screen image. Each cell is a gadget, with a duplex connection for each of its four sides. This is the type of a *Grid Gadget* with four duplex wires, one for each side:

InOut m m -> InOut m m -> InOut m m -> InOut m m -> Gadget

A higher-order function, grid, wires grid Gadgets to their neighbours and places them in a grid on the screen:

```
grid :: [InOut m m] -> [InOut m m] -> [InOut m m] -> [InOut m m] ->
    [[InOut m m -> InOut m m -> InOut m m -> InOut m m -> Gadget]]
    -> Gadget
```

grid takes a list of duplex connections for each side of the grid and a list of rows of grid Gadgets. In this application, we do not use the edge connections in the grid, so give a list of (nci,nco) values for each edge, where nci and nco are special values of type In a and Out a, respectively, indicating that a pin is a *not-connected input* or *output*. In the main explode game function, we set the capacity of cells and wire them to a referee component, then pass all the cells to grid. The cells are wired as shown on the left-hand side of Fig. 6.

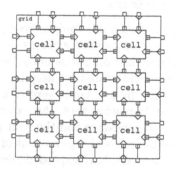 

Fig. 6. The Programmer and Player Views of the Grid

To achieve a similar wiring in the Fudgets system would require a significant amount of multiplexing. Our initial design of the Explode game did not foresee the need for a referee component, so the links were added at a later stage. A new duplex pin was added to the cell and wired to the referee by a partial application of each cell function, before passing the resulting grid Gadgets to grid. To modify the wiring in this way with Fudgets would require the Fudget equivalent of grid to be completely re-written.

Stones are transmitted in messages to adjacent cells when an explosion occurs. Each cell knows when it has been clicked, how many stones it contains (if any) and what colour they are, but knows nothing about the other cells in the grid. There are three problems to solve in this distributed implementation: (1) How does a cell determine what a click represents (ie. which player)? (2) How is termination detected? (3) How is the next player prevented from moving until the explosions from the previous move have completed?

To solve these problems each cell has a further duplex connection to a *referee* that negotiates the players turns and decides when a player has won. Cells send the referee messages of type Notify:

```
data Notify = ClickOver (Maybe Player)
            | Bonk (Maybe Player)
            | Boom Int (Maybe Player)
```

A ClickOver message indicates the colour of any stones already within the cell; a "Boom n c" message indicates an explosion, where n is the number of neighbouring cells exploded into and c is the colour of any stones previously in the cell; a cell that receives a stone, but does *not* explode, sends a "Bonk c" message, where c is the colour of any stones previously in the cell. The cell (Fig. 7) remembers its contents with a state value of type Maybe (Player,Int) where Maybe a = Yes a | None indicating that either the contents is empty with "None", or that there are n stones belonging to player p with "Yes (p,n)".

The referee (Fig. 8) sends cells Rulings:

```
data Ruling = Invasion Player | ClearSmoke
```

The referee remembers whose turn it is and when a cell sends a message saying it has been clicked and currently contains colour c, the referee checks that the move is valid, and if so returns an Invasion ruling, but otherwise makes no response. The referee is able to tell that explosions have ceased by keeping a tally. On receipt of a "Boom n c", it increases the tally by n-1; On receipt of a "Bonk c", it decreases the tally by one. When the tally reaches zero, the explosions have stopped, exploded cells are sent a ClearSmoke message and the next player may take a turn. To detect the end of the game the referee keeps a count of how many cells contain stones of each colour and which cells have exploded this move.

7 Implementation Outline

7.1 Communication Between Components

We have already seen that components are programmed using CPS. As well as enforcing an order on the operations of a component, the continuation functions also maintain a *state* for the process. Each continuation function, or *Action*, is able to modify this state — it is a state-transformer (ST). The state value consists of two or three parts:

1. the *world* of the component — a record of communications between this component and others. The world state is hidden in the definition of a component and not directly accessible. Special primitive ST's describe message

```
cell :: Int -> InOut Ruling Notify ->
          Invade -> Invade -> Invade -> Invade -> Gadget
cell capacity (fromReferee,toReferee) l t r b =
   wire $ \d ->
   wire $ \c ->
   giveImage (button (pictureIn (ip d)) (op c) ()) $
   let draw = op d
       click = ip c
       cell' o =
           let neighbours = [l,t,r,b]
               invasion p o =
                   let (op,n') = case o of
                           None -> (None,1)
                           Yes (p,n) -> (Yes p,n+1) in
                   tx draw (stones p n') $
                   if n' < capacity then
                       tx toReferee (Bonk op) $
                       cell' (Yes (p,n'))
                   else
                       tx draw boom $
                       tx toReferee (Boom capacity op) $
                       sequence [tx n p | (_,n) <- neighbours] $
                       cell' None in
           rx [
               from click $ \_ ->
                   let o' = case o of
                        None -> None
                        Yes (p,_) -> Yes p in
                   tx toReferee (ClickOver o') $
                   cell' o,
               from fromReferee $ \m -> case m of
                   Invasion p -> invasion p o
                   ClearSmoke -> tx draw blank $ cell' o,
               froms neighbours $ \p _ ->
                   invasion p o
           ]
   in
   claim fromReferee $
   claim click $
   sequence (map (claim.fst) [l,t,r,b]) $
   cell' None
```

Fig. 7. The Operation of cell Defined.

```
referee :: [InOut Notify Ruling] -> Out Turn -> Component
referee cs turn =
  sequence (map (claim.fst) cs) $
  tx turn (Turn Black) $ p Black 0 everyone [] ih
  where
  p::Player->Int->[CellID]->[CellID]->[(Player,Int)]->Component
  p c t b s h =
    rx [
      froms cs $ \r toCell ->
        let newcount h o n = map (adj (+1) n.adj (\x->x-1) o) h
            adj _ None h = h
            adj f (Yes np) (p,n) = (p,if p==np then f n else n)
    t' = t - 1 in
        case r of
          ClickOver v ->
            if (v == None || v == Yes c) && t == 0 then
              tx turn NoTurn $
              tx toCell (Invasion c) $
              p c 1 b s h
            else p c t b s h
          Bonk pc -> let h' = newcount h pc (Yes c)
                         s' = s \\ [toCell] in
            if t' == 0 then
              if (any (==nc) (map snd h')) then
                tx turn (Win c) $
                end
              else
                let c' = opponent c in
                tx turn (Turn c') $
                sequence [tx e ClearSmoke | e <- s'] $
                p c' 0 everyone [] h'
            else p c t' b s' h'
          Boom n pc -> let b' = b \\ [toCell] in
            if b' == [] then tx turn (Win c) $ end
            else p c (t'+n) b' (toCell:s) (newcount h pc None)
    ]
  everyone = map snd cs
  ih = zip players (repeat 0)
  nc = length cs
```

Fig. 8. The Operation of referee Defined.

transmission or reception by returning an altered world state. When a process is *executed* the changes in world state result in the I/O described being performed.

If we can ensure that only a single world state is in existence at any one time, then the primitive I/O functions can safely update the world value *in-place*. A similar optimisation is discussed in [PJW93].

2. the part used to record *default wire connections* in certain types of component (for example, the OS, SM and layout connections in a Gadget). These are hidden from component definitions but may be read or altered by the component. The type of this part of the state is the parameter to the `Process` type.

3. an optional part maintained at the component definition level and directly used by the component. For example, in the memory component of Sect. 2.4 an explicit state is used to hold the current memory contents.

7.2 Concurrency

Processes operate concurrently: evaluation is time-sliced between them. Each process is a separate evaluation with a private stack. Processes may share portions of the heap, and if two processes share an expression it is evaluated at most once. To ensure the integrity of the heap, a context-switch can only occur at certain points in the execution of a process, such as after a message has been sent.

7.3 Scheduling

When first created, a process is *initialising*. A process that has tried to receive when there are no messages waiting for it is *suspended*. The rest are *running*. Initialising processes are kept in a FIFO queue, and move to the running queue when they are about to receive or emit a message. Running processes are kept in a queue, ordered according to the number of messages waiting for them. Suspended processes are kept in a suspended list (in no particular order). Suspended processes that receive a message are moved to the running queue.

Whenever there is a change of context, the *scheduler* picks a process to be evaluated. Processes do not have a priority, but are scheduled according to the number of messages waiting for each process. The scheduler picks the top process from the initialising queue. If there are no processes initialising, then it picks the top process from the running queue. If there are no running processes, then no process is evaluated until a process becomes available on the running queue — for example, as a result of some user-input. After a running process has been evaluated to a point where a change of context can occur, it moves down the running queue to just below the lowest process with the same number of messages waiting. This has the effect that processes with the same number of messages waiting are chosen in a round-robin fashion.

This strategy was chosen to minimise the build-up of messages in wires. However, a situation could arise where a group of components starve those with fewer messages by communicating heavily with each other. In practice we have not seen this happen, but the risk could be reduced by including in the scheduling decision the length of time since a process last had a time-slot.

7.4 I/O Device Primitive Components

I/O devices such as the screen and mouse are implemented as primitive components. For each output pin of an I/O component (ie. input to the program), the wire-number[2] is noted so that when some input becomes available it is inserted into the relevant wire, ready to be received by the component at the other end. For the input pins of an I/O component (ie. output from the program), a C function is attached to the wire that will be called whenever a message arrives at the component. The message becomes an argument to the C function. Whenever the scheduler is called to make a change of context, it first calls a routine that polls to deliver any input (eg. mouse-click) to the relevant wire.

8 Related Work

There are many examples of work relating to message-passing concurrency in functional languages. One of the earliest is [Kar81]. One of the most recent is [JH93]. This system has a common ancestor with the system described in [WR95]. A few are concerned with the use of concurrency in programming applications with GUIs:

- Fudgets [CH93] behave in many respects like concurrent processes, but are actually implemented in a sequential language by exploiting the fact that Fudgets are restricted to one input and one output for messages, and must observe the *Fudget Law*. This makes composing and re-using Fudgets complicated. Fudgets are implemented in the lazy functional language Haskell. For a more detailed review, see [NR94].

- eXene [GR93] is an interface to The X Windows System, making full use of light-weight concurrent threads in the language (based on the strict functional language SML). It supports synchronous operations as first-class values, so new synchronisation abstractions can be built. The system is well developed and supports most of the functionality of The X Windows System.

- Erlang [AWV93] is a programming language for communicating processes, with an interface to The X Window System, used for programming real-time and control systems. The language borrows ideas from functional and concurrent languages, though it does not feature lazy evaluation or type-checking and is not a pure language.

---

[2] Internally, a wire is identified by an integer

- Pict [PRT93] is based on the π-calculus, so concurrency is central to the language. An interface to the The X Windows System is under development.

- Concurrent Clean [AP93] is a lazy functional language in which *unique types* are used to allow side-effecting I/O functions to operate safely. An interface to the The X Windows System is provided, using a system similar to callbacks and relying heavily on a global state for communication between elements of the program. More details can be found in [NR94].

- The Haggis system ([FJ95]) is based on a concurrent version of Haskell ([Jon95]) and features interface parts that are composable, extensible and concurrent. Communication between parts is achieved through synchronised access to shared memory.

Gadget Gofer is closest to the Fudgets system. It builds on the strengths of Fudgets by removing the restriction on channels per process: this simplifies the task of composing processes. It also untangles the interface layout from the communication structure of the application. The communication channels in Gadget Gofer are first class values, increasing the expressive power of the language.

9 Conclusions and Future Work

We have implemented an extension to a lazy functional language that gives us concurrent processes that communicate by passing messages via an unrestricted number of pins. This enables us to define independent components that are simple to compose, and program structures that are easy to change. Screen layout is independent of application communication structures, unlike the Fudgets system where layout and communication are closely tied. A screen manager built from the same components allows us to make good use of laziness and abstraction in defining interfaces. A small library of components and Gadgets, along with several example applications (such as the Explode game) have demonstrated the success of the system.

With the recent rise in popularity of the use of monads to encapsulate I/O (and more generally state-transformers) in lazy functional languages, why did we choose to use CPS? The reason is partly due to history — our system is based on some earlier work incorporating processes into a functional language that used CPS — and partly because the nature of continuation-based programs enables us to simplify the process implementation. However, we would like to benefit from the use of *lazy state threads* [LPJ94]. It is not yet clear whether this would require us to change from CPS to the monadic style of combinators and I/O, or whether we could implement lazy state threads under CPS.

There are shortcomings of the current implementation. For example, the creator of a component cannot destroy it at some later stage; each process can only **end** its operation of its own accord. A component cannot select a subset of its input pins to receive the next message from, but must be ready to receive a message from any input at all times. We have found situations where it would be useful to wait for a message from a particular pin or subset of pins before receiving any messages from the rest. A revised **rx** construct could permit guards for only a subset of the input pins.

Generating user interfaces for applications is simplified a great deal by the use of higher-order functions provided in a library (like **sequence**) and written for particular applications (like **grid**), but the job of wiring together compositions still seems a little *low-level*. We plan to write an application (using Gadgets of course) to allow a programmer to piece together an interface of ready-built components graphically, connecting pins with wires by clicking on them, and placing Gadgets by dragging them across the screen. This should speed up the development of Gadget programs; it will also serve as a further application study.

The creation of wires and components in a composition is *dynamic*, but this is not immediately apparent because it is the first and only thing a composing process does. A more interesting case is where a component creates another component in response to receiving a message. We hope to explore this possibility more fully in due course.

10 Acknowledgements

We thank Malcolm Wallace, Roger Took, Sigbjørn Finne, the anonymous PLILP referees and others from the universities of York, Bristol, Glasgow and Kent, whose comments on this system have helped to shape its development.

References

[AP93]   P. Achten and R. Plasmeijer. The Beauty and the Beast, 93.

[AvGP92]   P. M. Achten, J. H. G. van Gronigen, and M. J. Plasmeijer. High level specification of I/O in functional languages. In *Fifth Annual Glasgow Workshop of Functional Programming, Ayr 6th-8th July 1992.*, July 1992.

[AWV93]   J. Armstrong, M. Williams, and R. Virding. *Concurrent Programming in Erlang.* Prentice-Hall, 1993.

[CH93]   M. Carlsson and T. Hallgren. Fudgets - a graphical user interface in a lazy functional language. In *Functional Programming and Computer Architectures*, pages 321-330. ACM Press, June 1993.

[FJ95]   Sigbjørn Finne and Simon Peyton Jones. Composing Haggis. In *Proceedings of the Fifth Eurographics Workshop on Programming Paradigms in Computer Graphics*, Maastrict, Netherlands, September 1995.

[GR93]   Emden R. Gansner and John H. Reppy. A Multithreaded Higher-order User Interface Toolkit. *User Interface Software: Software Trends*, 1:61-80, 1993.

[HHPJW94] Cordelia V. Hall, Kevin Hammond, Simon L. Peyton Jones, and Philip Wadler. Type classes in haskell. In *European Symposium on Programming*, volume 788 of *LNCS*. Springer-Verlag, April 94.

[HS88] P. Hudak and R. S. Sundaresh. On the expressiveness of purely functional I/O systems. Technical report, Yale University Research Report YALEU/DCS/RR-665, Dept. of Computer Science, December 1988.

[JH93] Mark Jones and Paul Hudak. Implicit and explicit parallel programming in haskell. Technical Report YALEU/DCS/RR-982, Department of Computer Science, Yale University, Aug 93.

[Jon95] Simon Peyton Jones. Concurrent haskell. In *Haskell Workshop*, La Jolla, June 1995.

[Kar81] K. Karlsson. Nebula: A functional operating system. Technical report, Laboratory for Programming Methodology, Chalmers University of Technology and University of Goteb, 1981.

[LPJ94] J. Launchbury and S. Peyton Jones. Lazy functional state threads. In *Programming Languages Design and Implementation*, Orlando, 1994. ACM Press.

[NR94] Rob Noble and Colin Runciman. Functional languages and graphical user interfaces — a review and a case study. Technical Report YCS-94-223, Department of Computer Science, University of York, 1994.

[PJW93] Simon L. Peyton Jones and Philip Wadler. Imperative functional programming. In *Principles of Programming Languages*, Jan 93.

[PRT93] Benjamin C. Pierce, Didier Rémy, and David N. Turner. A typed higher-order programming language based on the pi-calculus. In *Workshop on Type Theory and its Application to Computer Systems, Kyoto University*, July 1993.

[RS93] A. Reid and S. Singh. Implementing Fudgets with standard widget sets. In *Glasgow functional programming workshop*, pages 222–235. Springer-Verlag, 93.

[Sin92] Duncan C. Sinclair. Graphical user interfaces for Haskell. In J. Launchbury and Patrick M. Samson, editors, *Glasgow functional programming workshop*. Springer-Verlag, 1992.

[WR95] Malcolm Wallace and Colin Runciman. Extending a functional programming system for embedded applications. *Software Practice & Experience*, 25(1):73–96, January 1995.

Lightweight GUIs for Functional Programming

Ton Vullinghs, Daniel Tuijnman, Wolfram Schulte

Fakultät für Informatik, Universität Ulm, D-89069 Ulm, Germany
email: {ton,daniel,wolfram}@informatik.uni-ulm.de

Abstract. Graphical user interfaces (GUIs) are hard to combine with functional programming. Using a suitable combination of monads, we are able to tame the imperative aspects of graphical I/O in a straightforward and elegant way. We present a concept to integrate lightweight GUIs into the functional framework, together with a library of basic functions and layout combinators to manipulate the GUI. An implementation of this library, using a set of high-level graphical I/O routines, is outlined. Examples demonstrate the simple way in which applications can be written.

1 Introduction

Everybody wants to use graphical user interfaces. And everybody wants to use functional programming languages. Unfortunately, these concepts are hard to combine: I/O, and graphical I/O in particular, is imperative in nature, and thus contradictory to the functional paradigm. Our goal is to reconcile these two concepts, by developing a framework in which a functional programmer smoothly can provide his program with a graphical user interface.

There are various ways to integrate I/O in general with a functional language [NR94, PJW93]. We have chosen for monadic I/O. This choice is firstly motivated by the advantage that monadic I/O can be combined with threading a global state through programs. Secondly, monadic programming expresses the sequentialization of I/O actions better than other styles do. This makes monads the natural choice for our system.

For the implementation of graphical I/O we have deliberately chosen not to reinvent the wheel, but to use an existing high quality tool, Tcl/Tk [Ous94], instead. This tool visualizes the user interface and handles events from the console and it communicates with the functional program in the form of strings. Thus we have reduced, on the side of the functional program, graphical I/O to textual I/O — for which I/O primitives are provided in, e.g., Glasgow Haskell [HPJW92] and Gofer [Jon93b].

We define a library of functions for manipulating the GUI. The programmer can write the GUI for his application in a structured and well-defined way, without having knowledge of the tool used. Also, the communication that takes place between the tool and the functional program is completely hidden for the user.

Advantages of our approach:

- *Easy to use.* Thanks to the convenient abstraction level of the library functions the main problems one has to consider upon creation of the GUI are aesthetical ones.
- *Unity of framework.* The application programmer can now write the user interface and the application proper in the same language.
- *Modularity.* The GUI state and the application state are independent. This makes it possible to write modular, reusable building blocks to create larger applications.
- *Generally applicable.* As long as monadic I/O is supported, our framework can be implemented in any functional language, lazy or eager.
- *Simple to implement.* Except for a few lines, we have written all the code in Gofer. Moreover, it is easy to extend the libraries.

Organization of this paper. Section 2 describes the way monads are used to reach the desired abstraction level. Section 3 gives a glance at the structure and interface of the functional GUI library. Then we present example applications in Sect. 4. Relevant implementation aspects are discussed in Sect. 5. A comparison with related work is given in Sect. 6, and finally, conclusions are drawn in Sect. 7.

Notation. The programs in this paper are written in Gofer [Jon93b], which is a subset of Haskell [HPJW92]. Characteristics of the syntax are: lambda abstraction \x-> , the use of function names as infix operators by placing them between backquotes, and the use of placeholders _ for values that are discarded.

2 Monads

Since the seminal work of Moggi [Mog89], monads have become increasingly popular. After Wadler's paper [Wad90] on their use in functional programming, numerous applications have been published, e.g., [Wad92, JD93, KL94, LPJ94, Wad94].

In general, a monad serves to record *side-effects* of functions. A good example is the *I/O monad* as it is implemented in the Glasgow Haskell compiler [PJW93] — and recently laid down in the new definition of the Haskell language — and in the Gofer language [Jon93b]. Another one is the *State Reader monad* that enables the user to thread the reading of a state through his program. Likewise, there are monads to handle exceptions, non-determinism, continuations, etc.; but only the I/O and State Reader monads are relevant for the sequel.

2.1 Concepts and Laws

A monad consists of three parts: a parameterized datatype M and two polymorphic functions

```
result :: a -> M a
bind   :: M a -> (a -> M b) -> M b
```

The datatype M a encapsulates a value of type a into some action, e.g., an I/O operation. Using bind, one performs the action, retrieves the proper value and passes that to a function into the monad. The result function encapsulates a value into a void action.

When programming with monads, one writes functions with typing a -> M b and strings them together with bind. So, bind replaces function application which is reflected by the *monad laws*: bind must be associative, and result is its left and right identity.

In this paper, we use the do notation as described by [Jon93a]. An expression do {e1; x <- e2; e3} replaces the equivalent e1 `bind` _ -> e2 `bind` \x -> e3, i.e., the subexpression x <- e2 binds x to the result of e2, whereas the result of subexpression e1 is discarded.

2.2 Monadic I/O and Mutable Variables

Clearly, we need to perform I/O. To this end, we use the predefined IO monad, which comes with the primitive functions

```
putchar :: Char -> IO ()
getch   :: IO Char
```

for writing and reading of characters.

Secondly, we need some means of storing a state. Therefore, we use the *mutable variables* [LPJ94] as defined in Glasgow Haskell and Gofer, which also act on the IO monad.[1] A mutable variable is manipulated with the functions

```
newVar   :: a -> IO (Var a)
readVar  :: Var a -> IO a
writeVar :: Var a -> a -> IO ()
```

for creating one with an initial value, for reading its value, resp. for writing a new value to it. Of course, when multiple functions want to access a state stored in a mutable variable, the variable must be passed to these functions as an explicit argument.

As an example, consider a program fragment for a simple adder. The state consists of an integer, viz. the running total. The function that reads a number and calculates the new running total then looks like:

```
add :: Var Int -> IO ()
add v = do { tot <- readVar v ; s <- getLine
           ; newtot <- result (tot + numval s)
           ; putLine (newtot) ; writeVar v newtot
           }
```

First, the value of the running total is fetched from the variable v and the next line is read. Then the new running total is calculated; it is written on the console, and stored in the same variable v.

---

[1] Actually, they are defined more generally, but for our purposes this simplified version suffices.

2.3 Combining Monads: Monadic I/O and State Reader

Next, we integrate the GUI into the monadic framework. As the GUI will be changed dynamically, part of it must be stored — in a mutable variable. This variable must be accessible anywhere in the program, hence it must be threaded through all functions that are provided to the application. This problem is not specific for GUIs: any library that gives access to some global state, e.g. a library of database functions, must cope with this.

Therefore, we need the state reader monad which consults some fixed environment. As our environment is a variable, it is fixed, though its contents may change. Hence, to consult it, we do not need a state transformer, but a state reader will suffice. As we now have two monads, IO and State Reader, we need to combine them. This can be done in the following way [BW85, JD93]:

```
data RIO s a = RIO (s -> IO a)

bindRIO :: RIO s a -> (a -> RIO s b) -> RIO s b
bindRIO (RIO x) f = RIO (\s -> x s `bindIO` \a ->
                                let RIO g = f a in g s)

embIO2RIO :: IO a -> RIO s a
embIO2RIO a = RIO (\s -> a)
```

The definition of **bind** for **RIO** takes care that the same state **s** is used by **x** and again by **f**, and that the corresponding I/O operations are performed sequentially. The described functions that act on the **IO** monad, such as **putchar** and **newVar** can be embedded into the **RIO** monad with the function **embIO2RIO**. The embedded counterparts of **newVar**, **readVar** and **writeVar** will appear in the sequel as **newState**, etc. The programmer needs these functions to manipulate his own (application) state which also must be stored in a **Var**.

Thus, the GUI state remains completely hidden for the programmer; we only provide him with the datatypes

```
type GIO a  = RIO (Var GUI) a
type Action = GIO ()
```

and the corresponding monad functions on **GIO**. The type **Action** is useful as most actions on the GUI only have a side effect and no proper result. The programmer only has to worry about defining his own state(s), global to the whole application or local to a window, as illustrated in Sect. 4.

3 A Functional Library for Graphical I/O

This section presents a relevant subset of our Gofer library for graphical user interfaces. A GUI consists of entities like windows and buttons, on which events are defined. These events are caused by the user or some other external process (e.g. a clock) and result in some action. The underlying application that is controlled in this way is called event driven.

Writing a library for graphical I/O is a complex and intensive job to do. In our approach we therefore decided to use an existing library of graphical I/O routines and to integrate this library within our concept. Because of its elegance and power we based our implementation on Tcl/Tk [Ous94] (see Sect. 5). Our library offers a high level interface to this tool by providing a powerful set of widget constructors and layout combinators.

3.1 Windows and Widgets

The basic building blocks of a graphical user interface are windows and widgets. A window is a container for widgets. A widget is a graphical entity with a particular appearance and behaviour. We can distinguish several kinds of widgets like buttons, labels, and entries.

Windows. A user interface may contain one or more windows. Possible actions on windows are closing and opening.

```
openWindow  :: [Config] -> Widget -> Action
closeWindow :: Ident -> Action

type Ident  = String
```

These functions yield actions. An action essentially produces side effects, i.e., changes to the GUI (cf. Sect. 2.3).

Windows are identified by a unique name. This name can be specified in the configuration list (see below). When closing a window, this name must be provided.

Widgets. In our view, a widget is either atomic or composite. For composite widgets, see Sect. 3.2. Widgets form an abstract datatype. For each atomic widget type, our library offers a constructor function:

```
buttonW, labelW, entryW, textW :: [Config] -> Widget
menuW :: [Config] -> [MenuItem] -> Widget
```

In this paper, we use buttons, labels (static text of one line), entry fields (edit text of one line), text fields (edit text of several lines) and pull-down menus.

Specifying the external outline of individual widgets and windows is done by giving appropriate values for the configuration options: the name identifying the window or widget, the textual contents, the command to be performed upon pressing a button, etc.:

```
data Config = Name Ident | Text String | Command Action | ...
```

If the widget is identified with a name, the configuration of the widget can be read or modified, e.g., to read or write its textual contents, the functions

```
getText :: Ident -> GIO String
setText :: Ident -> String -> Action
```

are given.

3.2   Layout Combinators

Widgets can be composed vertically and horizontally using layout combinators and functions. Our basic combinators are

    (<<), (^^) :: Widget -> Widget -> Widget

These combinators are associative and have the following meaning:

- v << w places widget w to the right of widget v;
- v ^^ w places widget w below widget v.

The resulting new widget is called the father of v and w.

With every widget we can associate an inherited and an occupied area. The inherited area is the area a widget gets from its father. The occupied area is actually used for displaying information, and is always a centered subarea of the inherited one.

Initially, the occupied and inherited area equal the minimum dimensions needed by the widget to display its information. After combination with some other widget, the occupied area of the father is minimal again. His concatenated sons are placed in the left uppermost corner of his occupied area. If widget v is bigger than widget w, the inherited area of v will equal its occupied area, and the inherited area of w will equal the rest of the occupied area of the father.

The fill functions make a widget occupy its inherited area either horizontally (fillX) or vertically (fillY). The expand function makes a widget claim from its father all occupied area that is not inherited by one of his (other) sons.

    fillX, fillY, fillXY, expand :: Widget -> Widget

In Fig. 1 we see three possible layout situations after application of (variants of) the combinators and fill functions. In the first picture, A and B are composed horizontally. Together they are combined vertically with C. In the second one, we let A << B and C occupy their inherited area in a horizontal direction. As a result of this, the father of A and B grows over the full length of C. In the third one, we let A and B grow vertically.

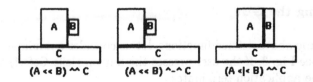

A << B) ^^ C   (A << B) ^-^ C   (A <|< B) ^^ C

Fig. 1. *Layout combinators and fill functions*

In Fig. 2 we show the result of expanding widgets. In the first picture, we let A claim and take the area of its father. Likewise, in the second one, this area is

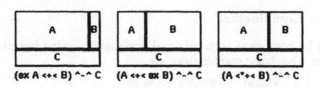

Fig. 2. *Layout combinators and expand functions*

claimed and taken by B. Finally, in the last picture, the area is claimed, taken and divided by both A and B.

In these examples, **ex** abbreviates **expand**, and the additional combinators are defined by:

```
(^-^)  = both (^^) fillX       (<+<)  = both (<<) fillXY
(<|<)  = both (<<) fillY        (<*+<) = both (<<) (expand . fillXY)

both comb f a b = f a `comb` f b
```

Using the above sketched layout mechanism and combinators we can lay out user interfaces in an almost arbitrary way. Additionally, we can develop more powerful layout functions:

```
hfill :: widget
hfill = (expand.fillX) (labelW [])

matrix :: Int -> [Widget] -> Widget
matrix n = foldl1 (^^) . map (foldl1 (<<)) . splitSegs n
```

The function `hfill` stretches an 'empty' widget as much as it can, pushing everything else aside. The function `matrix` takes a number of columns and a list of widgets and composes them in row major order.

The above gives us a basic library with which the user can already write his window-oriented applications. It may be extended with more widget types or configuration options, etc. But the essential structure remains the same.

3.3 Running the GUI

To illustrate the above defined functions, let us consider a minimal "hello world" application, see Fig. 3. If the user presses the button Hello Gofer!, the window is closed and the application terminates.

In Gofer, a program must be called `main` and have type `IO ()`. In our concept, `main` must call the library function `startProg` with an initial window setup. It sets up the user interface and starts the event loop. The first window that is opened is called the main window. If the main window is closed, termination follows.

```
main :: IO ()
main = startProg w where
  w = openWindow [Name "h", Title "Hello"]
                 (buttonW [Text "Hello Gofer!",
                  Command (closeWindow "h")])
```

Fig. 3. *The hello world application*

4 Applications

In order to demonstrate the usefulness of the presented approach, we work out three more illustrating examples. We describe a calculator, an editor and a combination of these two.

4.1 The Calculator

We take a look at a simple desk-calculator (see Fig. 4). For this calculator we define a state (cf. Sect. 2.3) consisting of the actual value of the display and an accumulator function.

```
type CalcState = Var (Int, Int -> Int)
```

The function `calculator` initializes the state, opens the window and resets the display.

```
main       = startProg calculator
calculator = do { st <- newState (0, id)
                ; openWindow [Title "Calculator"] (calc st)
                ; setText "display" "0"
                }
```

The user interface is built using an entry widget and a matrix of buttons for the keypad. Whenever the user presses a digit, it is displayed and the value component of the state is updated. When an operator is pressed, the display is reset and the accumulator function is modified. After pressing the '=' button, the calculator evaluates the accumulator function.

The code listed below implements the calculator completely.

```
calc :: CalcState -> Widget
calc st = disp ^-^ keys  where
  disp = entryW [Name "display", Relief Sunken, Width 12]
  keys = matrix 4 (map wid [ '1', '2', '3', '+',
                             '4', '5', '6', '-',
                             '7', '8', '9', '*',
                             'C', '0', '=', '/'
                           ])
  wid c = buttonW [Text [c], Command (lift (cmd c)), Width 3]
```

```
lift f = do { (disp, accu)  <- readState st
            ; (disp',accu') <- (result . f) (disp, accu)
            ; setText "display" (show disp')
            ; writeState st (disp',accu')
            }

cmd 'C' (d,a) = (0, id)
cmd '=' (d,a) = (a d, const (a d))
cmd  c  (d,a) | isDigit c = (10*d + ord c - ord '0', a)
              | otherwise = (0, ((char2op c).a) d)

char2op '+' = (+)      char2op '-' = (-)
char2op '*' = (*)      char2op '/' = (div)
```

The function wid maps each displayed character to a widget with a corresponding action. This action is defined by lift, which embeds the associated command into the GIO monad.

4.2 The Editor

The editor example illustrates the use of layout combinators, menus and text fields (see Fig. 4).

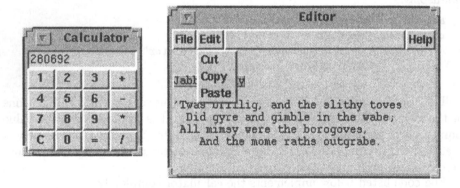

Fig. 4. *The calculator and the editor*

The editor state is a cut-copy-paste buffer, implemented by a sequence of strings. All other details concerning the state of the text widget, e.g., the actual position of the insertion cursor, are managed by the GUI.

```
type Buffer = Var [String]

main    = startProg editor
```

```
editor = do { buf <- newState []
            ; openWindow [Title "Editor", Name "ed"] (edit buf)
            }
```

The menubar of the editor has three pull-down menus. The first two are ragged left, the last one is ragged right. By applying a configuration-map (<@), we raise the relief of the menubar. Below the menubar the textfield is placed.

```
edit :: Buffer -> Widget
edit buf =
  (fileM << editM << hfill << helpM) <@ ((Relief Raised):)
  ^-^
  textW [Name "text", Width 40, Height 10]
  where
    fileM = menuW [Text "File"]
                  [cmdI "New"  (warning "Really clear?" clear),
                   cmdI "Quit" (warning "Really quit?"
                                        closeWindow "ed")]
    editM = menuW [Text "Edit"]
                  [cmdI "Cut"   (cutE buf),
                   cmdI "Copy"  (copyE buf),
                   cmdI "Paste" (pasteE buf)]
    helpM = menuW [Text "Help"] []

cmdI :: String -> Action -> MenuItem
cmdI t c = commandM [Text t, Command c]
```

If Cut, Copy or Paste is chosen from the Edit menu, the buffer is read or over-written. The implementation of the corresponding functions is straightforward and therefore not listed here.

When New or Quit is chosen, a modal warning dialog opens. If the user presses No, the window is closed again; otherwise the confirmed action is performed.

```
warning :: String -> Action -> Action
warning s c =
  openWindow [Modal, Name "w", Title "Warning"] (l1 ^-^ b1 <*+< b2)
  where l1 = labelW  [Text s, Relief Raised, Width 20]
        b1 = buttonW [Text "No ", Command (closeWindow "w")]
        b2 = buttonW [Text "Yes", Command (do {closeWindow "w";c})]
```

4.3 Combining the Examples

To demonstrate the modularity and reuseability of GUI components, we combine the editor and the calculator. We let the calculator and the editor share the editor's buffer. The user now is able to copy the result of a calculation into the text editor and vice versa. To handle this extension we have to make some minor changes to the previously presented programs.

We add the menu Options to the editor, containing an item to invoke the calculator with the buffer.

```
edit´ buf =
  (fileM << editM << optM << hfill << helpM) <@ ((Relief Raised):)
  ^-^
  textW [Name "text", Width 40, Height 10]
  where
    optM = menuW [Text "Options"]
                 [cmdI "Calculator ..." (calculator´ buf)]
  ...
```

The definition of the calculator is extended with the, now global, buffer. Furthermore, the calculator needs its own copy and paste commands. Therefore, we simply add a menubar to the calculator.

```
calculator´ buf =
  do { st <- newState (0, id)
     ; openWindow [Name "cc", Title "Calculator"] (calc´ st buf)
     ; setText "display" "0"
     }

calc´ st buf = (fileM << editM << hfill) ^-^ (calc st) where
  fileM = menuW [Text "File"]
                [cmdI "Quit" (closeWindow "cc")]
  editM = menuW [Text "Edit"]
                [cmdI "Copy"  (copyC st buf),
                 cmdI "Paste" (pasteC st buf)]
```

The Copy command writes the value of the display into the global buffer, the Paste command replaces the contents of the buffer by the displayed value. Their implementation is not shown here.

Note that we were able to reuse the original definition of `calc`. This nicely demonstrates how we can define modular building blocks for attractive user interfaces.

5   Implementation

This section discusses the runtime system, in particular the communication between the GUI library routines from Sect. 3 and the Tcl/Tk toolkit.

5.1   Tcl and Tk

The combination of Tcl and Tk [Ous94] provides a simple and comfortable programming system for developing small applications and graphical user interfaces.

Tcl. Tcl (tool command language) is an interpretive script language for controlling and extending applications. The syntax and semantics are close to C and the Unix shell. Tcl is an embeddable language, i.e., the language is in fact a library, designed to be linked together with other applications.

Tk. Tk is a toolkit for the X Window System [SG86] based on Tcl. It offers a set of widget commands for the creation of Motif-like user interfaces. All of the functionality of Tk-based applications is available through Tcl, i.e., evaluation of X events in Tk is done by invoking Tcl commands. The underlying event loops, call-back and display routines are all hidden away from the programmer.

5.2 Communication

It is surprising how little Tcl scripting is necessary to link a functional program to Tcl/Tk. In fact, the most simple, but already quite effective method is to run the program and Tcl/Tk as separate processes which are linked by a bidirectional pipe. Tcl/Tk sends events to the program (via stdin), which for its part returns its actions (on stdout) to Tcl/Tk.

Separating the GUI manager and the application in two processes poses two problems: which communication protocol should be used, and, where will information be stored?

We decide to store a minimal amount of GUI information in the functional program, viz. a call-back list. This list associates widget identifiers with actions. It suffices to communicate events by sending these identifiers from the GUI manager to the application. The identifier describes on which widget an event has happened. The event loop of the functional program then evaluates the corresponding command stored in the call-back list.

As Tcl/Tk is an interpretive language and the Tk operations are on a very high abstraction level, we represent actions that change the GUI as executable Tcl/Tk scripts.

However, not all user manipulation of the GUI triggers an event, e.g., typing text in a text field. So, information in the functional program is not always up to date. If we need the actual information from the GUI, a request is sent to the GUI manager: a Tcl/Tk script which, when evaluated, responds with the actual value. In any case, this strategy leads to synchronous communication.

Example: A trace of the editor. For the functional program, see Sect. 4.2.

First, the application must have the user interface set up. It generates the following script:

```
▷    wm title . "Editor"
▷    pack [frame .@1 ] -in .
▷    pack [frame .@2 ] -in .@1 -fill x
...
▷    pack [menubutton .@5 -relief raised \
▷              -menu .@5.m -text "Edit"] -in .@2 -side left
▷    menu .@5.m
▷    .@5.m add command -label "Cut" -command {communicate "C .@5m0"}
▷    .@5.m add command -label "Copy" -command {communicate "C .@5m1"}
...
▷    pack [text .text  -width 40 -height 10 ] -in .@1 -fill x
▷    EOM
```

Tcl/Tk interprets these commands and displays window and widgets. The EOM
terminates the action.

Second, suppose that the user has entered some text and marked the word
"Jabberwocky" — this does not generate an event. When the user chooses the
menu item Copy, the GUI communicates the event

◁   C .05m1

In order to fill its buffer, the application must get the actual value of the marked
text from the GUI. The application therefore sends the action

▷   write_event [.text .get sel.first sel.last]

which, when evaluated, responds with the actual marked part, namely

◁   Jabberwocky

The application reads this string and updates the buffer. Hereafter, it sends

▷   EOM

to terminate the (inter)action.

5.3   The GUI

The GUI starts with evaluating the init procedure. It uses the Tcl command
open to create the Gofer application $prog as its subprocess. Once it is created,
interaction is possible via the pipe channel, using the commands puts and gets.

```
proc init {prog} {
  global channel
  set channel [open "|$prog" r+]; read_actions
}
```

Calling the read-eval loop read_actions is the last command of the init procedure. This procedure reads and evaluates actions, i.e., Tcl/Tk scripts, until the
message EOM is read. Obviously, the set-up script of the GUI must be the first
message from the application.

```
proc read_actions {} {
  global channel
  set act [gets $channel]
  while {"$act" != "EOM"} {eval "$act"; set act [gets $channel]}
}
```

It then gives control to the Tcl/Tk event handler.

Whenever the user presses the button, the Tcl/Tk procedure communicate
is invoked, which writes events into the pipe and starts the read-eval loop of
Tcl/Tk again, to interpret the replied actions of the application.

```
proc communicate {msg} {
  write_event $msg; read_actions
}
```

```
proc write_event {msg} {
  global channel
  puts $channel $msg; flush $channel
}
```

This script is all you have to write in Tcl.

6 Related Work

The integration of referential transparent I/O in functional languages has been studied for a long time. We consider here only research to integrate GUIs in functional languages in a referential transparent way.

Among the first who integrated a GUI in a functional language was S. Singh [Sin92]. This approach already used a back to back arrangement, i.e., the application, written in Miranda, and the GUI, written in C, communicate over pipes. Both sides of the pipe supply primitive interpreters. However, (nearly) no abstraction is used: the graphical library XView is too low level – thus requiring a lot of 'non-portable' implementation work – and the GUI is implemented as an ordinary state variable. Thus, it is not asserted that the GUI representation in Miranda is used single threaded, whereas our monadic framework ensures this.

An alternative solution to this last problem was proposed by the Clean group [AvGP93], in the form of unique types. A unique type is guaranteed to be used only single threaded. Thus, the danger of copying GUIs is avoided. They developed a library with primitives similar to those of the Macintosh Toolbox, and implemented this GUI as a part of their runtime system.

A rather different approach was presented by M. Carlsson and T. Hallgren [CH93]. Instead of using the traditional widgets, they proposed fudgets as a functional equivalent. Fudgets may be thought as processes having two streams between them: one for high-level messages, connected to the application, and one for low-level X-messages. A button, for instance, is a process that, when pressed (low-level message), emits a click (high-level message). Fudget GUIs, based on streams, are programmed in a continuation style. Although this approach is rather nice, it has some shortcomings too. Problems mainly concern the inherent combination of stream based coupling and visualization of widgets (cf. [RS93]).

A work that also picks up Tcl/Tk as its GUI basics was already presented by D.C. Sinclair [Sin93]. He proposes to combine Haskell and Tcl/Tk using pipes, too. However, he does not hide Tcl/Tk behind the curtain — as we do — but uses a command-line oriented protocol between both processes.

Basically the same approach as Sinclair's was pursued in our own previous work [SV94]. However, we did not have to modify the Tcl shell as he did. Furthermore, we showed how structured functional code is achievable using this approach. The disadvantage, however, remains that we had to program the user interface itself in pure Tcl/Tk.

Lastly, Caml Light[PR94] was also linked to Tcl/Tk, but in a purely imperative way.

7 Conclusion

In this paper, we presented a referential transparent functional library to develop modular applications with a lightweight graphical user interface. We showed how abstraction mechanisms like monads and layout combinators can be used to realize such a library. The here presented solution enabled us to make a very smooth and high-level interface to the application programmer — who can now fully concentrate on the core of his system instead of fuzzy details of window programming.

The resulting system, composed of the Gofer GUI library and the Tcl/Tk script is small, easily extendable, reasonably fast and, due to the robustness of Tcl/Tk, stable.

Furthermore, our technique does not require any changes in runtime systems, as many other approaches do. Moreover, the needed I/O primitives are limited to the bare basics: reading and writing characters. Thus, implementation is not limited to Gofer, but can be adapted to any functional language, as long as it provides textual (monadic) I/O.

Alternatively, to improve efficiency, one could choose to embed Tcl/Tk into the Gofer runtime system. Since Tcl/Tk is particularly suited for such, this requires only a few local changes in the Gofer runtime system. For reasons of simplicity and clarity, we decided to present the transparent solution via pipes.

Acknowledgements. We would like to thank Erik Meijer for encouraging us to pursue this research; Klaus Achatz, Max Geerling, Magnus Carlsson and Thomas Hallgren for their helpful comments.

References

[AvGP93] P.M. Achten, J.H.G. van Groningen, and M.J. Plasmeijer. High-level specification of I/O in functional languages. In *Glasgow Workshop on Functional Programming 1992.* Springer Verlag, 1993.

[BW85] M. Barr and C. Wells. *Toposes, Triples and Theories.* Springer-Verlag, 1985.

[CH93] M. Carlsson and Th. Hallgren. Fudgets – a graphical user interface in a lazy functional language. In *Conference on Functional Programming and Computer Architecture.* ACM Press, 1993.

[HPJW92] P. Hudak, S.L. Peyton Jones, and Ph. Wadler (eds.). Report on the programming language Haskell, Version 1.2. *ACM SIGPLAN Notices,* 27(5), 1992.

[JD93] M.P. Jones and L. Duponcheel. Composing monads. Research Report RR-1004, Yale University, December 1993.

[Jon93a] M. Jones. *Release notes for Gofer 2.28,* 1993. Included as part of the standard Gofer distribution.

[Jon93b] M.P. Jones. *An introduction to Gofer (draft),* 1993.

[KL94] D.J. King and J. Launchbury. Lazy depth-first search and linear graph algorithms in haskell. Technical report, University of Glasgow, 1994.

[LPJ94] J. Launchbury and S.L. Peyton Jones. Lazy functional state threads. Technical report, University of Glasgow, November 1994.

[Mog89] E. Moggi. Computational lambda-calculus and monads. In *Symposium on Logic in Computer Science,* pages 14–23, Washington DC, 1989. IEEE.

[NR94] R. Noble and C. Runciman. Functional languages and graphical user interfaces — a review and a case study. Available by ftp at `ftp.york.ac.uk`, February 1994.

[Ous94]   J.K. Ousterhout. *Tcl and the Tk toolkit*. Addison Wesley, 1994.

[PJW93]   S.L. Peyton Jones and Ph. Wadler. Imperative functional programming. In *Proc. 20th ACM Symposium on Principles of Programming Languages*, Charlotte, North Carolina, January 1993.

[PR94]   F. Pessaux and F. Rouaix. The CamlTk interface — release beta2. Available by ftp at ftp.inria.fr, February 1994.

[RS93]   A. Reid and S. Singh. Implementing fudgets with standard widget sets. In *Glasgow Workshop on Functional Programming 1993*, 1993.

[SG86]   R.W. Scheiffler and J. Getty. The X window system. *ACM Transactions on Graphics*, 5(2), 1986.

[Sin92]   S. Singh. Using XView/X11 from Miranda. In *Glasgow Workshop on Functional Programming 1991*, Workshops in Computing. Springer Verlag, 1992.

[Sin93]   D.C. Sinclair. Graphical user interfaces for Haskell. In *Glasgow Workshop on Functional Programming*. Springer Verlag, 1993.

[SV94]   W. Schulte and T. Vullinghs. Linking reactive software to the X-Window system. Technical Report 94-14, Universität Ulm, December 1994.

[Wad90]   Ph. Wadler. Comprehending monads. In *Proc. 1990 ACM Conference on Lisp and Functional Programming*, 1990.

[Wad92]   Ph. Wadler. The essence of functionnal programming. In *ACM Principles of Programming Languages*, 1992.

[Wad94]   Ph. Wadler. Monads and composable continuations. *Lisp and Symbolic Computation*, pages 39–56, January 1994.

Local Linear Logic for Locality Consciousness in Multiset Transformation

Hugh McEvoy and Pieter H. Hartel

Department of Computer Systems, University of Amsterdam,
Kruislaan 403, 1098 SJ Amsterdam, The Netherlands
{hugh,pieter}@fwi.uva.nl

Abstract. We use Girard's linear logic (LL) to produce a semantics for Gamma, a multiset transformation language. The semantics improves on the existing structured operational semantics (SOS) of the language by highlighting Gamma's inefficiencies, which were hidden by the SOS. We propose a new logic called local linear logic (Local LL), which adds locality-consciousness to the resource-consciousness of linear logic. As a case study, we use this logic to propose a new semantics for Gamma. The new semantics suggests an annotation of Gamma which increases its efficiency without compromising its programming style. We show how the new semantics also gives us a better understanding of parallel Gamma and its implementation, and offers insight into the nature of chemical-reaction based computational models in general.

1 Introduction

Languages based upon the chemical reaction model combine terse expression of parallel programs with terrible efficiency problems. Gamma [9] is such a language. The tendency has been to view the efficiency problem as either a matter for implementors to overcome, or a reason to abandon such languages altogether. Abandoning such languages altogether would be a shame, as they possess some pleasing mathematical properties [22]. They are also showing themselves useful as abstract languages for describing physical and biological phenomena [28]. The persistence of Linda [17, 16] in the programming community also bears witness to the usefulness of the paradigm. We believe that instead of abandoning these languages, we should attempt to reap the benefits of these languages without reaping the whirlwind.

A traditional SOS of these languages doesn't help clarify the issue, because it captures possible execution traces while ignoring the cost of termination detection (linked to the cost of finding data). For example, extant SOS of Gamma [21, 15] use quantifiers to express the presence in the multiset of data with particular properties, which obscures that it is the cost of *finding* these data which cripples the implementations. This is unfortunate, for an SOS is otherwise an ideal vehicle for studying languages without the need to produce an implementation with all its associated clouding of the essential characteristics of the language for which it was written. We show that Girard's Linear Logic [18] can be used to provide a semantics for Gamma which highlights its parallelism, non-determinism and inefficiencies. We propose a locality-conscious variant of LL which enables us to reason about both the cost of reproducing data *and* the cost of relocating data, and show the benefits of this approach for understanding the costs of maintaining an ordered data structure and the partitioning of data on a parallel platform. This logic provides the theoretical underpinning for adding an annotation to Gamma to improve its efficiency without altering the basic, compelling, style of the language and of the programs written therein.

This paper is structured as follows. We first give a brief introduction to the language Gamma and to linear logic. The next two sections describe how we translate Gamma functions into linear logic formulae, and how the execution of these programs in Gamma correspond to steps in linear logic proof trees. Following that, we describe the shortcomings of the approach and suggest a solution, in the form of a refinement both of linear logic and of the Gamma model. We discuss the benefits yielded by these refinements. Finally, we draw our conclusions, survey related work, and discuss possibilities for future research.

2 Gamma

Gamma is a non-deterministic language for the transformation of multisets proposed by Le Métayer and Banâtre [9]. It combines a pleasing declarative programming style with the possibility of massively parallel implementations. Its abstract syntax is shown in Fig. 1. Gamma is designed as a general-purpose language, and has been used to describe a number of programs [10, 27]. The computational model is based upon a chemical reaction: an initial multiset is transformed repeatedly as functions are applied to its elements. The transformations continue, with the products of old reactions being re-used in new reactions, until no more reactions are possible. This occurs either when the multiset contains too few elements to provide enough arguments to any function, or when no permutation of the elements satisfies the boolean conditional of any function. The arguments of a function can be thought of as the reagents and the results as the products of a chemical reaction.

Multiple functions in a single program can be combined using two connectives. The first is written 'o' and is a right-associative composition of two functions. f1 o f2 performs f2 as many times as possible and then f1 as many times as possible. The second connective is written '+' and indicates the parallel interleaving of two functions. For example, f1 + f2 will repeatedly apply either f1 or f2 to the set until neither can be applied. In the case that both functions can be applied, a choice between them is made non-deterministically.

An SOS-style semantics for Gamma, as given in [21], is shown in Fig. 2. A pair (R, A) is a function (with its name elided). R is the reaction condition, which is a boolean conditional corresponding to the 'if ...' clause in the abstract syntax. A is the action, which is a function returning a multiset of expressions to be added to the multiset. Its syntax corresponds to '[exp [: type]]$^+$' in the abstract syntax. M is the multiset.

$$\text{system} \quad = \quad \text{function} \, \{ \, [\text{exp}]^* \, \}$$

$$\text{function} \quad = \quad \text{fn} \, (\, [\text{tup} \, [:: \text{type}]]^+ \,) \Rightarrow [\text{exp} \, [: \text{type}]]^+ \, \text{if exp}$$
$$| \quad \text{function} + \text{function}$$
$$| \quad \text{function} \circ \text{function}$$

$$\text{tup} \quad = \quad \text{id}$$
$$| \quad \langle \, \text{tup}, \text{tup} \, \rangle$$

Fig. 1. The essentials of the abstract syntax of Gamma. Expressions ('exp') are built out of constants, variables ('id') and operators. We do not allow function names ('fn') in expressions.

$$((R, A), M) \to ((R, A), (M - \{x_1 \ldots x_n\} \quad (1) \qquad ((R, A), M) \to M \qquad (2)$$
$$\cup \ A(x_1 \ldots x_n))) \qquad \qquad \text{if} \ \neg \exists x_1 \ldots x_n \in M.R(x_1 \ldots x_n)$$
$$\text{if} \ \exists x_1 \ldots x_n \in M.R(x_1 \ldots x_n)$$

$$\frac{(P_2, M) \to M}{(P_1 \circ P_2, M) \to (P_1, M)} \quad (3) \qquad \frac{(P_2, M) \to (P_2', M')}{(P_1 \circ P_2, M) \to (P_1 \circ P_2', M')} \quad (4)$$

$$\frac{(P_2, M) \to (P_2', M')}{(P_1 + P_2, M) \to (P_1 + P_2', M')} \quad (5) \qquad \frac{(P_1, M) \to (P_1', M')}{(P_1 + P_2, M) \to (P_1' + P_2, M')} \quad (6)$$

$$\frac{(P_1, M) \to M \quad (P_2, M) \to M}{(P_1 + P_2, M) \to M} \quad (7)$$

Fig. 2. An SOS-style semantics for Gamma

The Gamma model is related to the models of UNITY [12], Action Systems [6, 7], Linda and the Chemical Abstract Machine of Berry and Boudol [11]. All four related formalisms use the metaphor of a chemical reaction as their computational model, although only the latter two use multisets (tuple spaces) as their only data types.

The chemical reaction metaphor allows the expression of algorithms which contain a high degree of data parallelism. For example, the Gamma implementations described in [13, 32, 20] all exhibit potential data parallelism up to the limits imposed by the hardware (the nature of the data may restrict this in practice). Furthermore, in Gamma, multiple functions can be applied simultaneously to arbitrary disjoint subsets of the multiset. This latter property might incline one to believe that parallel implementations of Gamma could achieve good scalability. Unfortunately, this has transpired not to be the case. We discuss why this is not the case in due course.

2.1 An Example Gamma Program

An example Gamma program 'addup' is given in Fig. 3. It finds the sum of a multiset of numbers, by repeatedly replacing each pair of numbers with their sum. It continues until there are not enough elements in the set to provide it with its arguments: this single element is the sum of all the elements in the original multiset. We can 'execute' this program using the SOS rules of Fig. 2. A possible execution trace is shown in Fig. 3.

$$\text{addup } (x,y) \Rightarrow x + y \text{ if True } \{\ldots\}$$

$$\cfrac{\cfrac{\cfrac{\cfrac{((\text{True}, x + y), \{1, 2, 3, 4\})}{((\text{True}, x + y), \{3, 3, 4\})} \; \textbf{(1)}}{((\text{True}, x + y), \{6, 4\})} \; \textbf{(1)}}{((\text{True}, x + y), \{10\})} \; \textbf{(1)}}{\{10\}} \; \textbf{(2)}$$

Fig. 3. The program 'addup' (on the left) and an example execution according to the SOS (on the right). The boldface numbers on the right-hand side indicate which SOS rule was used for each step. Many other execution trees are possible, as the SOS is non-deterministic.

The SOS shows the result of executing the program, without having to implement the language. However, the SOS fails to highlight Gamma's inefficiency, for the presence of quantifiers in the SOS hides the cost of finding data with the required properties for a function to be applied. We show how linear logic can be used to generate a semantics which highlights this information, as well as a number of other interesting properties which were not immediately obvious from the SOS given above. The SOS is also an interleaving semantics rather than a parallel semantics. It may well be possible to address the first of these criticisms by refining the SOS, but the second criticism cannot be so addressed.

3 Linear Logic

Linear logic was proposed as an alternative to conventional intuitionistic and classical logics, wherein the structural rules of weakening and contraction were removed and replaced with more restricted variants. The result is a logic of resources: the effect of using a formula in a derivation is that the formula is destroyed. If it is desired that a particular formula be copied or discarded, it must be labelled with an 'exponential' symbol to indicate that this is the case. Otherwise, no copying or discarding of formulae is allowed. If the mantra for classical logic is 'truth is free', then that for linear logic is 'computation is not free'. The current work is not intended to be an introduction to linear logic. Excellent introductions are to be found in many of the papers cited above, and in [31, 4].

The contribution of linear logic, from the point of view of computer science, is that it allows reasoning about the number of storage operations required and formulae which must be discarded in order that a particular result be generated. In functional languages in particular, this allows reasoning about the costs of particular evaluation strategies and has inspired work in type systems [24, 8, 33, 1].

We use a somewhat nonstandard set of rules for our presentation of classical linear logic. We use a combination of Girard's original notation for the multiplicative fragment of the logic, Yetter's κ rule [35], from his Cyclic LL, and variants of the polynomially-bounded exponentials from Girard, Scedrov and Scott's Bounded LL [19]. We also extend the latter notion by allowing a special case of an exponential bounded by ω. This is because we wish to be able to produce infinitely deep proofs for non-terminating computations (either ones which cycle endlessly or those which continue to generate new results forever). We require all our sequences of wff (an abbreviation of 'Well-Formed Formulae') to be finite, so we have no use for Gentzen's Ω-rule.

We use completely non-standard exchange laws. The reason for this is that we wish to allow our one-dimensional sequence of wff to mimic the multiset, which contains elements of different types. From the point of view of efficiency, it is beneficial to partition the multiset into a number of subsets; one for each type. We therefore allow elements of different types to 'pass over' each other as if all elements of different types were invisible to each other. Although this was already possible with the original LL exchange law, we wish to distinguish this use of exchange from other uses. The reasons for this will become clear when we discuss the inefficiencies of the Gamma model. However, it is easy to see that the union of these exchange laws apply to precisely the same set of formulae as that to which the normal exchange rule applied.

The sequent calculus for the multiplicative fragment of our version of Classical Linear Logic is shown in Fig. 5. The reader is referred to Fig. 4 for the readings of the symbols. Some useful equivalences in LL are given in Fig. 6 (following [1]). The τ in our exchange laws stands for 'type-different', indicating that it applies to elements of different types. The ς stands for 'stirring', indicating that it applies to elements of the same type and corresponds to the magic-stirring mechanism (explained in section 6). The κ exchange laws apply only to formulae prefixed by a κ, and allow us to make further distinctions in our uses of exchange laws.

4   Casting Gamma into LL

Gamma possesses a notion of resource-consciousness. That is, data are destroyed when a function is successfully applied to them. To have multiple copies of a

⊗	= multiplicative conjunction	⅋	= multiplicative disjunction
⊸	= multiplicative linear implication	1	= multiplicative truth
⊥	= multiplicative falsehood	!, ?	= exponentials
∀, ∃	= quantifiers	$a, b, \ldots$	= terms
$A, B, \ldots$	= types	$x, y, \ldots$	= variables
∅	= empty set of wff	A	= Axiom rule
E	= Exchange rule	C	= Contraction rule
CD	= Contraction-Dereliction rule	W	= Weakening rule

Fig. 4. The readings of the symbols of the multiplicative fragment of LL. The notation follows Girard. The last five entries are names of rules. The rules themselves are shown in Fig. 5.

datum, it must be explicitly copied. In this sense, Gamma is linear. Furthermore, Gamma possesses several properties which make it simple to translate into LL, and which we use to minimise the size of the formulae generated:

- It is first order.
- It is eager.
- It has no recursive data types: only tuples may be constructed.

These properties serve to make the translation extremely terse: so terse that the statements of linear logic produced in the translation are often no longer than the Gamma functions from which they were translated. This enables us to reason about real Gamma programs, even by hand, and gives a better appreciation of the simplicity and compelling nature of the chemical-reaction model, and of some of its disadvantages. However, a possible difficulty looms:

- Gamma is non-deterministic.

Fortunately our logic-based semantics is inherently non-deterministic, and allows us to capture the non-determinism in the language without difficulty. Indeed, it would be more difficult to capture determinism than to capture non-determinism: we would have to ensure that only one logical rule be applicable to the sequence of wff at any one time.

4.1 The Translation of Gamma Programs into LL

We can translate Gamma programs into formulae of linear logic, following the rules shown in Fig. 7. Note that we choose to translate functions of the form 'if A then B' into the unconventional $A^\perp ⅋ B$ instead of the usual $A \multimap B$. These two forms are equivalent, as shown in Fig. 6. We choose to adopt the unconventional rendering for reasons which will become apparent later on.

$$\overline{a{:}A \vdash a{:}A}\;\textbf{A} \qquad\qquad \frac{\Gamma_1, \Gamma_2 \vdash \Delta}{\Gamma_1, 1, \Gamma_2 \vdash \Delta}\;\textbf{1L} \qquad\qquad \overline{\emptyset \vdash 1}\;\textbf{1R}$$

$$\frac{\Gamma_1, t{:}A, b[t/x]{:}B, \Gamma_2 \vdash \Delta}{\Gamma_1, t{:}A, \forall(x{:}A).b{:}B, \Gamma_2 \vdash \Delta}\;\forall\textbf{L} \qquad \frac{\Gamma_1, \Gamma_2 \vdash a{:}A, \Delta}{\Gamma_1, a^{\perp}{:}A^{\perp}, \Gamma_2 \vdash \Delta}\;\perp\textbf{L} \qquad \frac{\Gamma_1, a{:}A, \Gamma_2 \vdash \Delta}{\Gamma_1, \Gamma_2 \vdash a^{\perp}{:}A^{\perp}, \Delta}\;\perp\textbf{R}$$

$$\frac{\Gamma_1, !_n a{:}A, !_m a{:}A, \Gamma_2 \vdash \Delta}{\Gamma_1, !_{n+m} a{:}A, \Gamma_2 \vdash \Delta}\;\textbf{C} \qquad \frac{\Gamma_1, a{:}A, !_n a{:}A, \Gamma_2 \vdash \Delta}{\Gamma_1, !_{1+n<\omega} a{:}A, \Gamma_2 \vdash \Delta}\;\textbf{CD} \qquad \frac{\Gamma_1, \Gamma_2 \vdash \Delta}{\Gamma_1, !_n a{:}A, \Gamma_2 \vdash \Delta}\;\textbf{W}$$
$$\text{if } n > 1 \text{ and } m > 1 \qquad\qquad\qquad \text{if } n > 0$$

$$\frac{\Gamma_1, !_z a{:}A, !_\omega a{:}A, \Gamma_2 \vdash \Delta}{\Gamma_1, !_\omega a{:}A, \Gamma_2 \vdash \Delta}\;\textbf{C}_\omega \qquad \frac{\Gamma_1, a{:}A, !_\omega a{:}A, \Gamma_2 \vdash \Delta}{\Gamma_1, !_\omega a{:}A, \Gamma_2 \vdash \Delta}\;\textbf{CD}_\omega \qquad \frac{\Gamma_1, \Gamma_2 \vdash \Delta}{\Gamma_1, !_\omega a{:}A, \Gamma_2 \vdash \Delta}\;\textbf{W}_\omega$$
$$\text{if } z > 1$$

$$\frac{\Gamma_1, a{:}A, b{:}B, \Gamma_2 \vdash \Delta}{\Gamma_1, a \otimes b{:}A \otimes B, \Gamma_2 \vdash \Delta}\;\otimes\textbf{L} \qquad \frac{\Gamma_1 \vdash \Delta_1, a{:}A, \Delta_2 \quad \Gamma_2 \vdash \Delta_3, b{:}B, \Delta_4}{\Gamma_1, \Gamma_2 \vdash \Delta_3, \Delta_1, a \otimes b{:}A \otimes B, \Delta_4, \Delta_2}\;\otimes\textbf{R}$$
$$\text{if } \Delta_2 = \Delta_3 = \emptyset \text{ or } \Delta_2 = \Gamma_1 = \emptyset \text{ or } \Delta_3 = \Gamma_2 = \emptyset$$

$$\frac{\Gamma \vdash \Delta_1, a{:}A, b{:}B, \Delta_2}{\Gamma \vdash \Delta_1, a \,\bar{\otimes}\, b{:}A \,\bar{\otimes}\, B, \Delta_2}\;\bar{\otimes}\textbf{R} \qquad \frac{\Gamma_1, a{:}A, \Gamma_2 \vdash \Delta_1 \quad \Gamma_3, b{:}B, \Gamma_4 \vdash \Delta_2}{\Gamma_3, \Gamma_1, a \,\bar{\otimes}\, b{:}A \,\bar{\otimes}\, B, \Gamma_4, \Gamma_2 \vdash \Delta_1, \Delta_2}\;\bar{\otimes}\textbf{L}$$
$$\text{if } \Gamma_2 = \Gamma_3 = \emptyset \text{ or } \Gamma_2 = \Delta_1 = \emptyset \text{ or } \Gamma_3 = \Delta_2 = \emptyset$$

$$\frac{\Gamma_1, a{:}A, \Gamma_2 \vdash \Delta}{\Gamma_1, \kappa a{:}A, \Gamma_2 \vdash \Delta}\;\kappa \qquad \frac{\Gamma_1, b{:}B, a{:}A, \Gamma_2 \vdash \Delta}{\Gamma_1, a{:}A, b{:}B, \Gamma_2 \vdash \Delta}\;\textbf{E}_\tau \qquad \frac{\Gamma_1, b{:}A, a{:}A, \Gamma_2 \vdash \Delta}{\Gamma_1, a{:}A, b{:}A, \Gamma_2 \vdash \Delta}\;\textbf{E}_\varsigma$$
$$\text{if } A \neq B$$

$$\frac{\Gamma_1, \kappa b{:}B, a{:}A, \Gamma_2 \vdash \Delta}{\Gamma_1, a{:}A, \kappa b{:}B, \Gamma_2 \vdash \Delta}\;\textbf{E}_\kappa\,(\text{left}) \qquad \frac{\Gamma_1, b{:}B, \kappa a{:}A, \Gamma_2 \vdash \Delta}{\Gamma_1, \kappa a{:}A, b{:}B, \Gamma_2 \vdash \Delta}\;\textbf{E}_\kappa\,(\text{right})$$

Fig. 5. The multiplicative fragment of the sequent calculus for classical linear logic, showing both types and formulae. **C** is slightly modified from the traditional presentation of contraction, and **CD** is a combination of the traditional contraction and dereliction laws. In the structural rules, n and m are polynomials. z may be a polynomial or ω.

The reader's attention is called to the translation of functions which throw away their arguments. An example would be the function 'pointless', defined thus:

$$\text{pointless } (x, y) \Rightarrow \{\} \text{ if True};$$

As it is not possible to use uncontrolled applications of the Weakening rule in linear logic, it would appear that such functions are untranslatable. However, Fig. 8 shows that this view is mistaken, and that translation of such functions is possible. Part of the reason for this is that our translation rules in Fig. 7 render functions in a form equivalent to $!(A \multimap B)$ rather than the more conventional $(!A) \multimap B$.

$$1^\perp \stackrel{\text{def}}{=} \perp \qquad \perp^\perp \stackrel{\text{def}}{=} 1 \qquad (\forall x.a)^\perp \stackrel{\text{def}}{=} \exists x.(a^\perp) \qquad (!a)^\perp \stackrel{\text{def}}{=} ?(a^\perp)$$

$$\perp \,\eightpar\, a = a \qquad 1 \,\eightpar\, a = 1 \qquad \perp \otimes a = \perp \qquad 1 \otimes a = a$$

$$a^{\perp\perp} = a \qquad a \multimap b \stackrel{\text{def}}{=} a^\perp \,\eightpar\, b \qquad (a \otimes b)^\perp = a^\perp \,\eightpar\, b^\perp$$

Fig. 6. Some important equivalences in Linear Logic. The last four entries in the table indicate the distinguished elements of each multiplicative connective. $\otimes$ and $\eightpar$ are commutative.

$$
\begin{aligned}
\mathcal{T}[\![f_1 + f_2]\!] \quad &= \mathcal{T}[\![f_1]\!], \mathcal{T}[\![f_2]\!] & (1)\\
\mathcal{T}[\![f \{e^+\}]\!] \quad &= \mathcal{T}[\![e^+]\!], \mathcal{T}[\![f]\!] & (2)\\
\mathcal{T}[\![fn(x_1 \ldots x_n) \to e^+ \text{ if } e']\!] &= & (3)\\
&\quad !_\infty \kappa \forall_{i=1}^n \mathcal{T}[\![x_{n-i+1}]\!].(\mathcal{T}[\![e']\!])^\perp \eightpar ((\bigotimes_{j=1}^n \mathcal{T}[\![x_j]\!])^\perp \eightpar \mathcal{T}[\![e^+]\!])\\
\mathcal{T}[\![e_1, e_2]\!] \quad &= \mathcal{T}[\![e_1]\!] \otimes \mathcal{T}[\![e_2]\!] & (4)\\
\mathcal{T}[\![\langle e_1, e_2 \rangle]\!] \quad &= \mathcal{T}[\![e_1]\!] \eightpar \mathcal{T}[\![e_2]\!] & (5)\\
\mathcal{T}[\![True]\!] \quad &= 1 & (6)\\
\mathcal{T}[\![False]\!] \quad &= \perp & (7)\\
\mathcal{T}[\![e_1 \wedge e_2]\!] \quad &= \mathcal{T}[\![e_1]\!] \otimes \mathcal{T}[\![e_2]\!] & (8)\\
\mathcal{T}[\![e_1 \vee e_2]\!] \quad &= \mathcal{T}[\![e_1]\!] \eightpar \mathcal{T}[\![e_2]\!] & (9)\\
\mathcal{T}[\![\neg e]\!] \quad &= \mathcal{T}[\![e]\!]^\perp & (10)\\
\mathcal{T}[\![x]\!] \quad &= x & (11)\\
\mathcal{T}[\![e_1 + e_2]\!] \quad &= e_1 + e_2 & (12)\\
&\qquad \vdots \\
\mathcal{T}[\![e_1 \bmod e_2]\!] \quad &= e_1 \bmod e_2 & (13)\\
\mathcal{T}[\![\{\}]\!] \quad &= \perp & (14)
\end{aligned}
$$

Fig. 7. The translation rules from Gamma programs to LL. Angle brackets indicate tupling. Note that there is no translation of "o" in this table; an explanation of this is given in the text. One- and many-shot functions can be translated used a trivial variant of (3), not shown.

Bounded modalities allow us to give an upper bound to the number of times a function can be applied. In other words, we can identify the minimum number of times a function must be applied to give a particular answer. The mechanism can be used to allow one-shot (or n-shot) functions, if so desired, and functions which may be applied an ω times. A one-shot function, by definition, may only be applied *at most* once; it is then discarded. An n-shot function is applied *at most* n times before being discarded. Programming experience has shown that these are a useful addition to the language, but we do not use them in the examples presented in this paper. They are also useful for calculating complexity measures, as Girard et al. have shown [19].

$$\mathcal{T}[\![\text{pointless } (x, y) \Rightarrow \{\} \text{ if } \textit{True}]\!]$$
$$= \{(3)\}$$
$$!_\omega \kappa \forall \mathcal{T}[\![y]\!], \mathcal{T}[\![x]\!].(\mathcal{T}[\![\textit{True}]\!])^\perp \mathfrak{B} ((\mathcal{T}[\![x]\!] \otimes \mathcal{T}[\![y]\!])^\perp \mathfrak{B} \mathcal{T}[\![\{\}]\!])$$
$$= \{(11), (11), (6), (11), (11), (14)\}$$
$$!_\omega \kappa \forall y, x.1^\perp \mathfrak{B} ((x \otimes y)^\perp \mathfrak{B} \perp)$$
$$= \{\text{equivalence rules}\}$$
$$!_\omega \kappa \forall y, x.\perp \mathfrak{B} ((x \otimes y)^\perp \mathfrak{B} \perp)$$
$$= \{\text{equivalence rules, commutativity}\}$$
$$!_\omega \kappa \forall y, x.(x \otimes y)^\perp$$

Fig. 8. The translation of 'pointless'.

We have not translated the "∘" operator. Looking at the SOS for Gamma in Fig. 2, we see that a dot requires that the first function terminates before the second is applied. It is interesting to note that Hankin and Le Métayer have both intimated, during private conversations, that they do not feel that the sequential composition of Gamma fits well with the rest of the language. It is certainly the case that sequential composition requires global knowledge of the state of the computation. This contrasts with the spirit of the model, which allows functions to have only local knowledge of the multiset. This is the very reason why the operator is difficult to encode in LL: it needs to know something about the sequence of wff of which it is a member, viz. whether any of the other functions in the sequence can successfully be applied. We hope to provide a translation for "∘" in the future. However, even without dot composition, we can translate a large number of programs. In fact, Gamma is perhaps unique in possessing a sequential composition operator: other related formalisms such as UNITY and Action Systems have no such operator. Certainly it is possible to code a class IV cellular automata ([34]) in Gamma without sequential composition, thus showing that Gamma without sequential composition is capable of universal computation.

5 Executing Gamma Programs in LL

The eagerness of Gamma allows us to write an extremely simple set of expression reduction laws, as shown in Fig. 9. The eagerness implies that expression reduction occurs immediately after the variables in an expression are bound. We shall see this in our later derivations.

We can now translate 'addup' into an LL wff.

$$\mathcal{T}[\![addup]\!] = !_\omega \kappa \forall y{:}Int, x{:}Int.((x \otimes y)^\perp \mathfrak{B} ((x + y){:}Int))$$

An example 'execution' of the program, with an initial multiset, is given in Fig. 10. Notice that function application is realised as an application of **CD**

$$\llbracket constant \rrbracket = \ constant \qquad \llbracket e_1 > e_2 \rrbracket = 1 \ \text{if} \ e_1 > e_2$$
$$\llbracket e_1 + e_2 \rrbracket \ = \llbracket e_1 \rrbracket + \llbracket e_2 \rrbracket \qquad \llbracket e_1 > e_2 \rrbracket = \bot \ \text{if} \ e_1 \not> e_2$$
$$\vdots \qquad\qquad\qquad\qquad \vdots$$

Fig. 9. The reduction laws for expressions.

or CD_ω followed by (in a bottom-up reading) an application of the expression reduction laws (if applicable), neutral element laws and $\wp L$. There may be instances of the contraction rules applied to different formulae between these, however. The neutral element laws were given in Fig. 6. The four phases of function application: variable binding, condition evaluation, expression evaluation and storage reclamation, are thus clearly distinguished. We can therefore analyse the effects on each of these caused by changes in the order of the wff in the sequence, or by changes in the order of application of rules of LL.

Comparing the LL proof in Fig. 10 with the SOS execution in Fig. 3, we see that the results of executing the program are the same, as we would expect. In fact, we can show that the set of possible results is always the same for both the LL semantics and the SOS, by an easy argument. The main difference is that LL proof trees highlight the possibilities for parallel function application which are not visible in the SOS semantics, because the former is a true concurrency semantics whereas the latter is an interleaving semantics.

Examination of the proof trees generated for this 'program' with different numbers of data, gives us a number of insights. We first give some intuitions about how these trees should be read.

- Contraction corresponds to copying a function for parallel execution. Several functions could be created in this way, all of which move to their data *before* the application of the ∀ rule. Several ∀ rules can then be applied together. This implements parallel composition of functions on disjoint subsets of the multiset, as can be seen at the top of the proof tree.
- Applications of Weakening correspond to discarding of unwanted functions, as can be seen towards the bottom of the tree.
- Bounded exponentials can be used to determine a lower bound on the number of function applications needed to ensure termination of a computation, as can be seen at the bottom of the proof tree.
- We can identify trees with partially-applied functions at their leaves with trees in which such functions were discarded *before* they were partially applied. For Gamma does not return partial functions as values of computations, and our semantics must reflect the fact that functions will only be applied when all of their arguments are present. There is no danger in our identification, except that it allows functions to be wasted in an attempted application when they lack sufficient data. This feature highlights the necessity, in an implementation, of checking the availability of a function's arguments *before* it is applied.

$$
\cfrac{
  \cfrac{
    \cfrac{\cfrac{}{1\vdash 1}\,\mathbf{A} \quad \cfrac{}{2\vdash 2}\,\mathbf{A}}{1,2\vdash 1\otimes 2}\,\otimes\mathbf{R}
  }{1,2,(1\otimes 2)^{\perp}\vdash\emptyset}\,{\perp}\mathbf{L}
  \qquad
  \cfrac{
    \cfrac{
      \cfrac{
        \cfrac{\cfrac{}{3\vdash 3}\,\mathbf{A}\quad\cfrac{}{7\vdash 7}\,\mathbf{A}}{3,7\vdash 3\otimes 7}\,\otimes\mathbf{R}
      }{3,7,(3\otimes 7)^{\perp}\vdash\emptyset}\,{\perp}\mathbf{L}\quad\cfrac{}{10\vdash 10}\,\mathbf{A}
    }{3,7,((3\otimes 7)^{\perp}\mbox{\small$\wp$}10)\vdash 10}\,\mbox{\small$\wp$}\mathbf{L}
  }{3,7,\forall y,x.((x\otimes y)^{\perp}\mbox{\small$\wp$}(x+y))\vdash 10}\,\forall\mathbf{L},\forall\mathbf{L},\mathbf{Arith.}
}{\cdots}
$$

$$\dfrac{3,4,(3\otimes 4)^{\perp},3,7,\forall y,x.((x\otimes y)^{\perp}\mbox{\small$\wp$}(x+y))\vdash 10}{\,}\;\mbox{\small$\wp$}\mathbf{L}$$

$$\dfrac{1,2,((1\otimes 2)^{\perp}\mbox{\small$\wp$}3),7,\forall y,x.((x\otimes y)^{\perp}\mbox{\small$\wp$}(x+y))\vdash 10}{\,}\;\mbox{\small$\wp$}\mathbf{L}$$

$$\dfrac{1,2,(1\otimes 2)^{\perp}\mbox{\small$\wp$}(1+2),7,\forall y,x.((x\otimes y)^{\perp}\mbox{\small$\wp$}(x+y))\vdash 10}{\,}\;\mathbf{Arith.}$$

$$\dfrac{1,2,\kappa\forall y,x.((x\otimes y)^{\perp}\mbox{\small$\wp$}(x+y)),7,\kappa\forall y,x.((x\otimes y)^{\perp}\mbox{\small$\wp$}(x+y))\vdash 10}{\,}\;\kappa * 2,\forall\mathbf{L},\forall\mathbf{L}$$

$$\dfrac{1,2,7,\kappa\forall y,x.(\cdots\mbox{\small$\wp$}\cdots),\kappa\forall y,x.(\cdots\mbox{\small$\wp$}\cdots),\kappa\forall y,x.(\cdots\mbox{\small$\wp$}\cdots),!_{n-2}\forall\cdots\vdash 10}{\,}\;\mathbf{W}$$

$$\dfrac{1,2,7,\kappa\forall y,x.((x\otimes y)^{\perp}\mbox{\small$\wp$}(x+y)),\kappa\forall y,x.((x\otimes y)^{\perp}\mbox{\small$\wp$}(x+y)),!_{n-2}\forall\cdots\vdash 10}{1,2,7,!_{n}\kappa\forall y,x.((x\otimes y)^{\perp}\mbox{\small$\wp$}(x+y))\vdash 10}\;\mathbf{E}_\tau,\mathbf{E}_\tau \;\; \mathbf{CD},\mathbf{CD}$$

Fig. 10. A proof showing the execution of 'addup'. '**Arithmetic**' corresponds to the use of expression-rewrite rules and neutral element laws.

For quantitative results, we note the equivalences shown in Fig. 11. For 'addup', these equivalences allow us to derive the results shown in Fig. 12. All of these may be gained by easy arguments.

Proof sketch for the case of '# functions applied': Define 'initial sequence' as the root of the derivation tree. Our functions are the only elements in our initial sequence which contain quantifiers. No rule introduces quantifiers in a premise except the structural rules, so no quantifiers can be present in any sequence unless they are present in wff unchanged during the derivation, or are present in wff generated through applications of **CD**, **C**, **CD**$_\omega$, **C**$_\omega$. Therefore, all of the quantifiers originate in functions in the initial sequence. Quantified variables can only be instantiated to numbers in this proof, because of the type constraints on the $\forall L$ rule. Therefore, the number of quantified variables which can be instantiated in the proof is equal to the number of numbers in the proof (assuming that we have enough copies of the function). For every pair of numbers added together, a new number (their sum) is created. Therefore, if our initial sequence of wff contains n numbers, we generate a binary tree of numbers with $2n - 1$ nodes. However, at the root the function cannot be successfully applied because it does not have enough arguments. Therefore, there are $2n - 2$ successful function applications. $\square$

The number of elements permuted reflects the attempt to gain the maximal

Runtime information	LL proof tree property
# functions successfully applied	$\#\mathbf{CD} + \#\mathbf{CD}_\omega$
# tuples reclaimed	$\#^\perp\mathbf{L}$
# elements permuted	$\#\mathbf{Exchange}_\varsigma$
max. parallelism	max. $\#\forall$ on a line $\div$ fn. arity

Fig. 11. Some correspondences between features of LL proof trees for the addup program and features of an execution of the corresponding program. 'Elements' in the above table indicates one-tuples, not wff. By 'permuted' , we do *not* mean the flowing over each other of elements with different types, but only applications of exchange on pairs of adjacent elements with the same type.

Runtime information	Number of occurrences	
	This run	Per n numbers
# functions successfully applied	2	$2n - 2$
# tuples reclaimed	2	$2n - 2$
max. parallelism	1	n div 2

Fig. 12. Results from the LL derivation of 'addup' with an initial set of three elements.

parallelism of the program. Were the reductions done sequentially, the number of elements to which the exchange laws were applied would have been zero.

5.1 Storage Reclamation of Used Data

We translated our functions in an unconventional way, using $A^\perp \wp B$ instead of the more usual, and equivalent, $A \multimap B$. Examining the derivation tree of 'addup' shows why this particular translation was chosen. When a function is applied, it produces 'anti-information' (negated-wff), whose job it is to track down and annihilate one of the arguments of the function. In LL, anti-information allows us to apply the **A** rule, thus removing from the multiset the wff which is the negated counterpart of the anti-wff. We therefore see the operational interpretation of the negated formulae in the translation of a function: they correspond to the notion of storage reclamation of unwanted data. Notice that this reading implies that the storage reclamation need not take place when the wff first reacts, but is only required to take place at some point in the future *before* the wff of which it is the anti-wff is re-used in another reaction (for if that happened, then linearity would be violated). Similarly, it does not matter if the actual wff is destroyed by the anti-wff, only that a wff with the same type and value is destroyed by the anti-wff. This opens the possibility for very intelligent storage reclamation, which should be further explored. Using $A \multimap B$ for functions removes this freedom.

6 Magic Stirring: Gamma's Efficiency Haemorrhage

Detecting termination of a Gamma computation requires checking that every datum performs all the reactions of which it is capable. Such behaviour requires an $O\left(\frac{n!}{(n-m)!}\right)$ search of all possible permutations of the multiset, where n is the cardinality of the multiset and m is the largest arity of any function. Clearly, this affects both sequential and parallel implementations adversely. The implementations in [13, 32] 'solve' the problem by partitioning the multiset into individual values, which move around the network and interact when they meet at a node (and certain conditions are satisfied). Obviously, the communication costs involved in transporting so many tiny packets of data (single multiset elements) are prohibitively high.

Consider the following example; an asynchronous one-dimensional cellular automaton. At each function application, the value at a point on a one-dimensional surface is updated, depending on the values of its neighbours. The Gamma program is shown in Fig. 13. The translation into LL of part of this program, is shown in Fig. 14.

$f + g \{\dots\}$

$f(\langle x', y'\rangle, \langle x, y\rangle, \langle x'', y''\rangle)$
$\quad \Rightarrow \langle x, (y' \vee y \vee y'')\rangle, \langle x', y'\rangle, \langle x'', y''\rangle$
$\quad\quad \text{if } (x \bmod 2 = 0) \wedge (x' = x - 1) \wedge (x'' = x + 1)$

$g(\langle x', y'\rangle, \langle x, y\rangle, \langle x'', y''\rangle)$
$\quad \Rightarrow \langle x, y\rangle, \langle x', y'\rangle, \langle x'', y''\rangle$
$\quad\quad \text{if } (x \bmod 2 \neq 0) \vee (x' \neq x - 1) \vee (x'' \neq x + 1)$

Fig. 13. 1-D Cellular automaton in Gamma

We can derive a proof tree for the LL translation of the program, as shown in Fig. 15. We use the notation $\mathbf{E}_\varsigma^{O(n!)}$ to indicate the application of $O(n!)$ permutations to order the data, where n is the cardinality of the largest set whose type appears in the argument list of the function. We see that such reorderings of the data occur every time a function has been successfully applied. This is clearly

$!_\omega \kappa \forall (x'' \mathbin{\rotatebox[origin=c]{180}{\&}} y''), (x \mathbin{\rotatebox[origin=c]{180}{\&}} y), (x' \mathbin{\rotatebox[origin=c]{180}{\&}} y').$
$\qquad (((x \bmod 2 = 0) \otimes (x' = x - 1) \otimes (x'' = x + 1))^{\perp} \mathbin{\rotatebox[origin=c]{180}{\&}}$
$\qquad\qquad (((x \mathbin{\rotatebox[origin=c]{180}{\&}} y) \otimes (x' \mathbin{\rotatebox[origin=c]{180}{\&}} y') \otimes (x'' \mathbin{\rotatebox[origin=c]{180}{\&}} y''))^{\perp}$
$\qquad\qquad \mathbin{\rotatebox[origin=c]{180}{\&}} ((x \mathbin{\rotatebox[origin=c]{180}{\&}} (y' \mathbin{\rotatebox[origin=c]{180}{\&}} y \mathbin{\rotatebox[origin=c]{180}{\&}} y'')) \otimes (x' \mathbin{\rotatebox[origin=c]{180}{\&}} y') \otimes (x'' \mathbin{\rotatebox[origin=c]{180}{\&}} y''))))$

Fig. 14. LL translation of function f from the 1-D Cellular automaton program into linear logic. Function g is similar, and has been elided. It is not used in the later proofs.

the major source of inefficiencies in Gamma. We shall address this problem in the next section.

Fig. 15. Fragment of execution of the automaton program in Linear Logic. **Arith.** is an abbreviation for **Arithmetic.** $\mathbf{E}_\varsigma^{O(n!)}$ is an abbreviation for $O(n!)$ occurrences of $\mathbf{E}_\varsigma$.

Recent work [20] has shown that Gamma can be implemented more efficiently on a shared memory machine, with the generation of permutations replaced by a PROLOG-like search strategy. It is not yet clear whether or not their approach will carry over well onto a distributed memory machine, but their reliance on a homogeneous data set would seem to indicate otherwise: backtracking search across a distributed data set suffers from the same efficiency shortcomings as permutation generation.

6.1 Locality-Conscious Multisets

We believe that the inherent inefficiencies in Gamma cannot be overcome by clever implementations. We believe that progress will only be made by moving the problem into the semantic domain and solving it there. Therefore, we outline a more radical alternative to those discussed above. We claim that the Gamma model conflates two issues, which we believe should be separated. Solutions (multisets) in Gamma are constantly permuted to ensure that all reagents come into contact with one another. Yet there are really *two* notions here. The first is the notion of locality: that reagents cannot react unless they are in contact with one another. The second is the notion of permutation generation: all elements come into contact with one another sooner or later (if they don't react with something else first). These two notions are confused, both in the quantifier-laden SOS of Gamma, and in the permutation-driven implementations. Both neglect to distinguish data which *have* locality and data which *change* locality, and account for the inefficiency of the language.

Therefore, we distinguish the locality of reagents from their motion. We make locality *primitive* to the model, and allow re-orderings of the data only when they do not violate the locality constraints. We shall discuss the details of this in the next section, and explain how arbitrary permutations of the data can be generated, if they are required. This return to, and refinement of, the original chemical-reaction metaphor, is the insight behind local linear logic and Local Gamma.

7 Local LL and Local Gamma

LL (and particularly Bounded LL) gave us a more fine-grained notion of resource-consciousness. Local linear logic extends this notion with locality-consciousness, so we offer the mantra 'movement is not free', to complete a trio. Local LL is linear logic with the normal exchange rule replaced with those shown in Fig. 16. The logic gives us a notion of locality by giving indices to elements in the sequence of wff and ensuring that the sequence remains ordered relative to some metric. Extending LL with a notion of order allows us to reason about structured data and partitioning of data on parallel machines. When used as a semantics for Gamma, we therefore have an improvement over the SOS of Gamma given earlier in that we have specific information at the semantic level concerning the costs of relocating data. We will use this power to show how the efficiency of Gamma can be improved. An illustration of the difference between the old and the new exchange laws is shown in Fig. 17. The laws of Local LL always move an incorrectly-ordered element *towards* the correct place in the sequence. The best that the traditional exchange rule could manage was to generate permutations of the solution until the correct ordering appeared. The traditional exchange rule of LL corresponds to the permutation generation of multiset languages. In both cases, it can be difficult to perform it intelligently when it is automated. Notice that the new exchange laws apply to a strict subset of the sequences to which the exchange law of Classical LL applied.

Yetter [35] and Abrusci [2, 3] effectively proposed a topology of the sequence of wff. We generalise the topology of the sequence to a multi-dimensional, open, surface. That is, formulae have (multi-dimensional) neighbourhoods within which they can move freely, but may only move outside of these neighbourhoods in certain circumstances. Thus, at a local level—when we are only concerned with one place—the behaviour of wff is the same as in LL. In a wider context, however, movement cannot be performed freely and, in some cases, not at all (see section 7 for a discussion).

$$\frac{\Gamma_1, b_j{:}A, a_i{:}A, \Gamma_2 \vdash \Delta}{\Gamma_1, a_i{:}A, b_j{:}A, \Gamma_2 \vdash \Delta} \; \mathbf{E_\epsilon} \qquad \frac{\Gamma_1, b_j{:}A, a_i{:}A, \Gamma_2 \vdash \Delta}{\Gamma_1, a_i{:}A, b_j{:}A, \Gamma_2 \vdash \Delta} \; \mathbf{E_\lambda} \qquad \frac{\Gamma_1, b_j{:}B, a_i{:}A, \Gamma_2 \vdash \Delta}{\Gamma_1, a_i{:}A, b_j{:}B, \Gamma_2 \vdash \Delta} \; \mathbf{E_\tau}$$
$$\text{if } i > j \qquad\qquad\qquad \text{if } i = j \qquad\qquad\qquad \text{if } A \neq B$$

$$\frac{\Gamma_1, \kappa b_j{:}B, a_i{:}A, \Gamma_2 \vdash \Delta}{\Gamma_1, a_i{:}A, \kappa b_j{:}B, \Gamma_2 \vdash \Delta} \; \mathbf{E_\kappa} \text{(left)} \qquad \frac{\Gamma_1, b_j{:}B, \kappa a_i{:}A, \Gamma_2 \vdash \Delta}{\Gamma_1, \kappa a_i{:}A, b_j{:}B, \Gamma_2 \vdash \Delta} \; \mathbf{E_\kappa} \text{(right)}$$

Fig. 16. The exchange rules in Local LL.

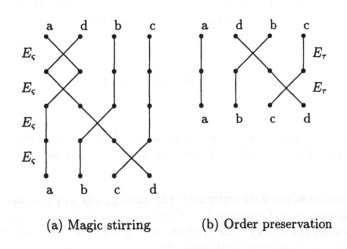

(a) Magic stirring   (b) Order preservation

Fig. 17. The difference between the exchange rules of LL and of Local LL.

Formulae of LL can be translated into formulae of Local LL with the same behaviour in two ways: either by prefixing them with a κ, or by giving every wff in a sequence the same index. However, by giving all formulae indices, then we restrict the possible derivations which can be made from the formulae. This is achieved because the rules given in Fig. 5 require the formulae to be in certain spatial configurations in order that they apply. In Local LL, then, we can generally only obtain a subset of the proofs available in LL. We aim to show that this subset is interesting.

Local Gamma is a variant of Gamma which replaces the unordered multiset with an ordered multiset. The syntax and translation rules given for Gamma in Fig. 1 are therefore unchanged. However, we add an annotation which indicates whether a particular expression should be regarded as an index on some ordering. Any ordering can be expressed which defines a metric space. Adding this annotation requires more translation rules, of the following form:

$$\mathcal{T}[\![\langle \widehat{x_1 \ldots x_n}, y \rangle]\!] = \mathcal{T}[\![y]\!]_{(x_1 \ldots x_n)}$$

In other words, the annotated vector $x_1 \ldots x_n$ is used as the index of the translated formula.

As in Local LL, elements can be permuted freely if they are at the same location. Elements at different locations can only move in such a way that they preserve the ordering of the multiset. Thus, new elements created at runtime can only move *towards* the position in the multiset dictated by their index. An example Local Gamma program is given in Fig. 18. Notice that we do not have to indicate the values of indices, only their presence. We believe that adding this annotation to the language will not result in programming errors, for the programmer should know which elements of which tuples are to be regarded as indices. In fact, forcing the programmer to be explicit about this should be regarded as good Local Gamma programming style.

$$order(x) \Rightarrow \langle \hat{x}, x \rangle$$

Fig. 18. An example local Gamma program. The $^\wedge$ is the annotation indicating an element of a tuple which should be regarded as an index.

Programs whose data ordering is captured in this way become more efficient, in the general case. Of course, it is always possible to write a program (such as a permutation generator) whose behaviour is unaffected. However, in the author's experience such pathological programs are seldom written. An example of a program whose efficiency improves is the one-dimensional cellular automaton, whose time complexity reduces drastically. The Local Gamma version of the program is shown in Fig. 19, its translation into Local LL is shown in Fig. 20, and its 'execution' in Fig. 21.

The derivation tree for the cellular automaton in Local Linear Logic yields a quantity of interesting information. Firstly, we give the intuitive readings of the trees.

$f + g \{\dots\}$

$f(\langle \hat{x}', y' \rangle, \langle \hat{x}, y \rangle, \langle \hat{x}'', y'' \rangle)$
$\quad \Rightarrow \langle \hat{x}, (y' \vee y \vee y'') \rangle, \langle \hat{x}', y' \rangle, \langle \hat{x}'', y'' \rangle$
$\qquad$ if $(x \bmod 2 = 0) \wedge (x' = x - 1) \wedge (x'' = x + 1)$

$g(\langle \hat{x}', y' \rangle, \langle \hat{x}, y \rangle, \langle \hat{x}'', y'' \rangle) \Rightarrow \langle \hat{x}, y \rangle, \langle \hat{x}', y' \rangle, \langle \hat{x}'', y'' \rangle$
$\qquad$ if $(x \bmod 2 \neq 0) \vee (x' \neq x - 1) \vee (x'' \neq x + 1)$

Fig. 19. 1-D Cellular automaton in Local Gamma.

$!_\omega \kappa \forall y''_{x''}, y_x, y'_{x'} . (((x \bmod 2 = 0) \otimes (x' = x - 1) \otimes (x'' = x + 1))^\perp \mathbin{\mathscr{B}}$
$\qquad ((y_x \otimes y'_{x'} \otimes y''_{x''})^\perp \mathbin{\mathscr{B}} ((y' \mathbin{\mathscr{B}} y \mathbin{\mathscr{B}} y'')_x \otimes y'_{x'} \otimes y''_{x''})))$

Fig. 20. LL translation of function f from the 1-D Cellular automaton program into Local LL.

- Instances of Exchange$_\epsilon$ correspond to the preservation of ordering through moving new elements to their correct position in the solution.
- We may wish to examine the effect of partitioning the data over a distributed-memory machine. Ordering the set using indices gives the most natural way of partitioning data: all the elements whose indices are between certain bounds are in the same partition. For example, in a cellular automaton we know that with the Margolus or Von Neumann neighbourhoods ([30]) we need the values of only those elements within one unit's distance of the cell in question. By examining possible partitioning strategies, we can investigate their effects in terms of the lengths of the borders and the relative sizes of the partitions.
- In a parallel implementation of the language, we have a measure of the number of elements which will be communicated from one partition of the set to another, given a particular program. For example, if the number of the node is given as an index to a formula, then the number of formula for which that value changes during a single generation of the automata is exactly equal to the number of elements which have to be communicated from one place to another.

Comparing the LL version of the automaton with the Local LL version highlights the reduction in the number of permutations which have been generated by introducing the locality awareness: the exponential searches of the data space are replaced by $O(n)$ steps for each new element. An actual implementation of the language would probably use a range-tree ([23]) or similar to reduce this to

$$
\cfrac{
\cfrac{1_1,0_2,0_3 \vdash 1_1 \otimes 0_2 \otimes 0_3}{1_1,0_2,0_3,(1_1 \otimes 0_2 \otimes 0_3)^\perp \vdash \emptyset}\,{}^\perp\mathbf{L}
\qquad
\cfrac{
\cfrac{
\cfrac{\cdots,1_1,1_2,0_3,!_\omega\kappa\forall\cdots}{\cdots,1_2,1_1,0_3,!_\omega\kappa\forall\cdots\vdash\Delta}\,\mathbf{E}_\epsilon^{O(n)}
}{\cdots,(1_2\otimes 1_1\otimes 0_3),!_\omega\kappa\forall\cdots\vdash\Delta}\,\otimes\mathbf{L}
}{}\,\mathscr{B}\mathbf{L}
}{
1_1,0_2,0_3,(1_1\otimes 0_2\otimes 0_3)^\perp\mathscr{B}(1_2\otimes 1_1\otimes 0_3),!_\omega\kappa\forall\cdots\vdash\Delta
}\;\mathbf{Arith.}
$$

$$
\cfrac{
\cfrac{
\cfrac{
\cfrac{\cdots 1_1,0_2,0_3,((1\bmod 2{=}0)\otimes(1{=}2{-}1)\otimes(3{=}2{+}1))^\perp\mathscr{B}(\cdots\mathscr{B}\cdots),!_\omega\kappa\forall\cdots\vdash\Delta}{\cdots,1_1,0_2,0_3,\forall y''_{x''},y_x,y'_{x'}.(\cdots\mathscr{B}\cdots),!_\omega\kappa\forall\cdots\vdash\Delta}\;\forall\mathbf{L}*3
}{\cdots,1_1,0_2,0_3,!_\omega\kappa\forall y''_{x''},y_x,y'_{x'}.(\cdots\mathscr{B}\cdots)\vdash\Delta}\;\mathbf{CD},\kappa
}{\cdots,0_2,\cdots,1_1,\cdots,0_3,!_\omega\kappa\forall y''_{x''},y_x,y'_{x'}.(\cdots\mathscr{B}\cdots)\vdash\Delta}\;\mathbf{E}_\epsilon^{O(n)}
}{}
$$

Fig. 21. Fragment of execution of the automaton program in Local Linear Logic. Note that $\mathbf{E}_\epsilon$ has replaced $\mathbf{E}_\varsigma$.

$O(\log n)$. The quantitative results for one iteration of the automaton are shown in Fig. 22, and are obtained by easy arguments.

Runtime information	Number of occurrences per function application	
	LL	Local LL
# tuples reclaimed	3	3
# elements reordered	$n!$	$3n$

Fig. 22. Results comparing the automaton program in LL and Local LL, for a single iteration of an automaton with three elements. n is the number of tuples of type Num $\mathscr{B}$ Num in the multiset. Notice how much more efficient the Local LL version is.

8 Related Work

In LL, Girard recognised that resources are not infinite—but he did not take into account that in many circumstances (for example on a distributed-memory machine), not all resources are equally easy to access. So while LL went a long way towards a logic of computation, it did not address issues of data access. To adequately reflect the behaviour of a real system, we must factor in the cost of accessing or relocating remote data. This is what Local LL does. So as well as charging the customer for the product, we must charge her for the delivery too.

Yetter [35] proposed a variant of LL which he called Cyclic LL. In Cyclic LL, wff may not be consumed except in the order in which they are presented. Cyclic permutations of the wff *are* allowed, however. Yetter's work amounts to introducing a weak notion of locality to the sequence of wff: every wff is in a different location on a one-dimensional torus. In order to embed LL in Cyclic LL, Yetter allowed formulae prefixed with a κ to move around freely in the sequence. We have adopted his idea.

Abrusci described a completely non-commutative linear logic [2, 3]: that is, formulae are not allowed to move around at all. This suggestion is tantamount to locating each formula at a different place on a one-dimensional, open, surface.

Guo [14] has presented a mechanism for translating terms of the lambda calculus into formulae of linear logic. Guo's system is more powerful than that here presented, for it allows the embedding of the entire lambda calculus and the description of graph reduction, laziness etc. However, his system has the disadvantage, due to its power, that in the general case, translation of complete functional programs yields unwieldy formulae. The links between these formulae and the lambda terms from which they were translated, are not always clear. Nevertheless, Guo's work has been a major inspiration for the current work.

Abramsky has produced a linear variant of the Chemical Abstract Machine [1], whose operation he describes in detail. However, our translation offers us the advantage that pieces of Gamma code can be translated into short logical sequences.

Andreoli [5] has examined a certain class of LL proof trees—focusing proofs—from the point of view of efficient proof search for a logic programming language. In the current paper, we have not concerned ourselves with focusing proofs, although an examination of these in the light of Local LL might be interesting. At this stage, we are not interested in the question of automatically executing Gamma programs in LL, and we assume the usual intelligence on the part of the prover. Providing an executable semantics is obviously an interesting goal, but we leave it for future research.

Meseguer [29] has presented a unifying framework for a number of models of concurrency, based upon category-theoretic considerations. The essential difference between our approach and his is that we are interested in an operational semantics of a rewriting language which is sufficiently low-level to give insights into the implementation of the language on a parallel architecture.

Yves Lafont [25] has proposed Interaction Nets as a generalisation of Girard's Proof Nets [18] and as a new sort of programming language in their own right. There may well be close connections between our LL semantics of Gamma and (a class of) Interaction Nets, but we have not yet explored this avenue.

Lincoln [26] has examined the complexity of LL fragments. An examination of the complexity of fragments of Local LL would also be an interesting area for further study.

9 Future Work

There are a number of directions for future work. The first is to provide a full proof- and model- theoretic examination of Local LL and an investigation of its relationship to other extant flavours of LL. The second direction is to examine in greater detail the ways in which Local LL proof trees mimic the general behaviour of an single program, multiple data (SPMD) system, by examining a number of programs and parallelisation strategies, in an attempt to extract quantitative information. Thirdly, an attempt should be undertaken to translate sequential composition. Fourthly, the definition of Local LL and the translations of Gamma programs to Local LL expressions could be used used to build an abstract machine for Local Gamma. The abstract machine instructions can be built from sequences of Local LL rule applications. From there an implementation could be produced, both for parallel and sequential machines. These implementations would provide interesting results concerning the efficiency and scalability of implementations based upon the 'locality-sensitive chemical reactions' model.

We are already working on a parallel implementation of Local Gamma. Initial results are promising but inconclusive, as only small programs can so far be compiled. More work needs to be done.

10 Conclusions

The continued existence of arrays as data structures attests to the presence in most algorithms of a notion of data locality. However, multiset transformation languages such as Gamma lack this notion. This accounts for their terrible efficiency problems. Unfortunately, Gamma's traditional SOS hid this difficulty, leading to a belief that the efficiency problem should be solved through clever implementations. We used Girard's resource-conscious linear logic (LL) to provide an alternative semantics for Gamma, which captured Gamma's non-determinism and potential parallelism through a straightforward translation scheme from Gamma programs into terms of LL. This semantics is more fine-grained than the traditional SOS of Gamma in that it highlighted the language's inefficiencies. We extended linear logic, making it locality-conscious. This logic we called Local LL. It does for data locality what LL does for resource allocation and discarding. We used the new logic to suggest an alternative Gamma in which the multiset is ordered according to some metric. We showed that the Gamma programming style is essentially unchanged, but that the inefficiencies are removed. Our semantics also gives us clues for efficient data partitioning for a parallel implementation, and yields information on the maximum parallelism we can attain from a Gamma program.

11 Acknowledgments

We would like to thank Anne Troelstra, Yi-Ke Guo, Marcel Beemster and Jon Mountjoy for proof reading this paper and providing a number of comments.

378

Thanks also go to Walter Hoffman, Eelco Visser, Wim Vree, Peter Sloot, Jaap Kaandorp and Jeroen Voogt for several helpful discussions. We would also like to thank the anonymous PLILP referees for their helpful suggestions.

References

1. S. Abramsky. Computational interpretations of linear logic. *Theoretical computer science*, 111(3):3–57, Oct 1993.
2. V. M. Abrusci. Non-commutative intuitionistic linear logic. *Zeitschrift fur Mathematische Logik und Grundlagen der Mathematik*, 36(1):297–318, 1990.
3. V. M. Abrusci. Phase semantics and sequent calculus for pure non-commutative classical linear propositional logic. *J. Symbolic Logic*, 56(4):1403–1451, Dec 91.
4. V. Alexiev. Applications of linear logic to computation. Technical Report TR93-18, University of Alberta, Saskatchewan, Canada, 1993.
5. J.-M. Andreoli. Logic programming with focusing proofs in linear logic. *J. Logic and Computation*, 2(3):297–347, 1992.
6. R. J. R. Back and R. Kurki-Suonio. Decentralisation of process nets with centralised control. In *2nd SIGACT-SIGOPS Symp. on Principles of Distr. Computing (PODC)*, pages 131–142, Montreal, Canada, 1983. Springer Verlag, Berlin.
7. R. J. R. Back and R. Kurki-Suonio. Distributed co-operation with action systems. *ACM transactions on programming languages and systems*, 10(4):513–554, 1988.
8. H. G. Baker Jr. Lively linear Lisp – "look ma, no garbage!". *ACM SIGPLAN notices*, 27(8):89–98, Aug 1992.
9. J.-P. Banâtre, A. Coutant, and D. le Métayer. Parallel machines for multiset transformations and their programming style. *Informationstechnik, Oldenbourg Verlag*, 30(2):99–109, 1988.
10. J.-P. Banâtre and D. le Métayer. Programming by multiset transformation. Research report PI 522, IRISA, Rennes, France, Mar 1990.
11. G. Berry and G. Boudol. The chemical abstract machine. In *17th Principles of programming languages*, pages 81–94, San Fransisco, California, Jan 1990. ACM, New York.
12. K. M. Chandy and J. Misra. *Parallel program design: A foundation*. Addison Wesley, Reading, Massachusetts, 1988.
13. C. Creveuil and G. Moguerou. Dérivation systématique d'un algorithme de segmentation d'images - un exemple d'application du formalisme Gamma. Research report 1049, INRIA Rocquencourt, France, Jun 1989.
14. J. Darlington and Y. Guo. Reduction as deduction. In J. R. W. Glauert, editor, *6th Implementation of Functional Languages*, pages 10.1–10.10. School of Information Systems, Univ. of East Anglia, Norwich, UK, Sep 1994.
15. L. Errington, C. Hankin, and T. Jensen. A congruence for gamma programs. In P. Cousot, M. Falaschi, G. Filè, and A. Rauzy, editors, *3rd Static Analysis (WSA)*, *LNCS 724*, pages 242–253, Padova, Italy, 1993. Springer Verlag, Berlin.
16. D. Gelernter. Generative communication in linda. *ACM transactions on programming languages and systems*, 7(1):80–112, 1985.
17. D. Gelernter and A. J. Bernstein. Distributed communication via global buffer. In *1st SIGACT-SIGOPS Symp. on Principles of Distr. Computing (PODC)*, pages 10–18, Ottawa, Canada, 1982. ACM, New York.
18. J.-Y. Girard. Linear logic. *Theoretical computer science*, 50(1):1–102, 1987.

19. J.-Y. Girard, A. Scedrov, and P. J. Scott. Bounded linear logic: a modular approach to polynomial-time computability. *Theoretical computer science*, 97:1–66, 1992.

20. K. Gladitz and H. Kuchen. Shared memory implementation of the Gamma-operation. In J. R. W. Glauert, editor, *6th Implementation of Functional Languages*, pages 26.1–26.13. School of Information Systems, Univ. of East Anglia, Norwich, UK, Sep 1994.

21. C. L. Hankin, D. le Métayer, and D. Sands. A calculus of Gamma programs. Research Report 92/22, Dept. of Computing, Imperial College London, Jul 1992.

22. C. L. Hankin, D. le Métayer, and D. Sands. Transformation of Gamma programs. In M. Billaud, P. Castéran, M-M. Corsini, K. Musumbu, and A. Rauzy, editors, *Static Analysis (WSA 92)*, pages 12–19, Bordeaux, France, Sep 1992. BIGRE, 81-82.

23. P. H. Hartel, M. H. M. Smid, L. Torenvliet, and W. G. Vree. A parallel functional implementation of range queries. In P. G. M. Apers, D. Bosman, and J. van Leeuwen, editors, *Computing science in The Netherlands*, pages 173–189, Utrecht, The Netherlands, Nov 1989. CWI, Amsterdam.

24. Y. Lafont. The linear abstract machine. *Theoretical Computer Science*, 59:157–180, 1988.

25. Y. Lafont. Interaction nets. In *17th Principles of programming languages*, pages 95–108, San Fransisco, California, Jan 1990. ACM, New York.

26. P. Lincoln and T. Winkler. Constant-only multiplicative linear logic is NP-complete. *Theoretical computer science*, 135(1):155–169, 1994.

27. H. McEvoy. Gamma, chromatic typing and vegetation. Technical report in preperation, Dept. of Comp. Sys, Univ. of Amsterdam; Presented at the ESPRIT 9102 coordination meeting, Geneva, Dec 1994.

28. H. McEvoy and J. Kaandorp. *Multisets and their transformers as models for environmentally-sensitive growth*. Dept. Comp. Sys., Univ. of Amsterdam, 1995.

29. J. Meseguer. Conditional rewriting logic as a unified model of concurrency. *Theoretical computer science*, 96(1):73–155, 1992.

30. T. Toffoli and N. Margolus. *Cellular automata machines*. MIT press, Cambridge, Massachusetts, 1991.

31. A. S. Troelstra. Lectures on linear logic. Lecture notes 29, Centre for the Study of Language and Information, Leyland Stanford Junior University, California, 1992.

32. M. Vieillot. Premiers pas de Gamma avec une PAM. Rapport de stage, IFSIC, IRISA, Univ. de Rennes, France, 1992.

33. P. L. Wadler. Linear types can change the world! In M. Broy and C. B. Jones, editors, *Programming concepts and methods*, pages 561–581, Sea of Gallilee, Israel, Apr 1990. North Holland, Amsterdam.

34. S. Wolfram. *Cellular automata and complexity*. Addison Wesley, Reading, Massachusetts, 1994.

35. D. Yetter. Quantales and (noncommutative) linear logic. *J. Symbolic logic*, 55(1):41–64, 1990.

The Notion of Floundering
for SLDNF-Resolution Revisited

Jean-Hugues Réty

INRIA - Rocquencourt, B.P. 105, 78153 Le Chesnay cedex, France.
E-mail: **jean-hugues.rety@inria.fr**

Abstract. SLDNF-resolution is the standard operational semantics for normal programs. Its safeness condition induces a notion of floundering which may be considered as a programming error. But this notion of floundering seems too weak: a program and a goal may flounder despite the existence of an SLDNF-tree and, in such a case, one computation may flounder while another one may not. We propose a novel stronger notion of floundering – called *true floundering* – which depends only on the existence of an (extended) SLDNF-tree and not on a particular strategy. This notion formalises the idea that the existence of a floundered goal is irrelevant if with another strategy, it would have been possible to continue the computation without floundering. We also introduce a new operational semantics called *extended SLDNF-resolution*. Extended SLDNF-resolution may be efficiently implemented. Moreover, with a result from Drabent ([Dra94]), extended SLDNF-resolution is complete wrt Kunen semantics for non true floundering programs and goals, which corresponds to a large class of normal programs with a declarative characterization.

1 Introduction

SLDNF-resolution is an extension of SLD-resolution which incorporates the negation as failure rule proposed by Clark ([Cla79]). Its soundness needs the application of a safeness condition (a negative literal in a goal cannot be selected until it is not ground) which involves a notion of floundering.

Current implementations consider floundering as an error. So one problem is to avoid floundering, i.e. to build programs that do not flounder. Research has mainly taken two approaches: the first one is to define (syntactic) classes of floundering-free programs and, the second one is to design analysis tools for testing floundering-freeness. Let us cite for instance [BM90] and [MSD90]. We think that the problems encountered mainly come from the notion of floundering itself which is not satisfactory.

Floundering is a computational notion and a real problem is that it depends on the selection rule and on the clause choice rule (see the examples in Sect. 4). One computation for a program and a goal may flounder while another one may not. This is obviously very unexpected: it makes difficult to statically state whether a program may flounder or not. Moreover we will show that in some cases, when a floundering occurs, it is possible to continue the computation (with a refined notion of computation) and be complete wrt Kunen semantics ([Kun87]). Indeed a program and a goal may flounder although an SLDNF-tree exists. The notion of floundering seems to be too weak. So it is worth visiting it again.

The literature usually defines floundering in the following way: Let P be a normal program and G a normal goal, The evaluation of $P \bigcup \{G\}$ flounders if a goal which contains only non-ground negative literals is reached. This is strongly operational and actually, is related to computation. There is a real lack of a declarative notion of floundering (i.e. abstracted from computation).

According to the standard definition of SLDNF-resolution ([Llo87]), for some programs and goals, no SLDNF-tree or SLDNF-derivation exists. There are three reasons: Firstly, SLDNF-trees being built in a "bottom-up" way with ranks which are natural numbers, infinite sequences of negative calls cannot be taken into account. Secondly, for a program like $\{p \leftarrow p\}$ and a goal $\leftarrow \neg p$ (p being ground), there exists neither an SLDNF-refutation nor a finitely failed SLDNF-tree for $\leftarrow p$. The third reason is floundering: no declarative notion of floundering is given and, e.g. for the goal $\leftarrow \neg p(X)$ and some program, no SLDNF-tree exists. These problems are mentioned in [AB91] and [AP91] and some re-definitions of SLDNF-resolution have been proposed ([AD92], [MT92], [MN90]) but these papers are mainly interested in solving the first two problems.

This paper proposes a stronger notion of floundering called *true floundering* which corresponds to the idea that the existence of a floundered goal is irrelevant if, with another strategy, it would have been possible to continue the computation without floundering. We prove that true-floundering is independent of the selection rule. We prove a more general result of which a particular consequence is that, for SLDNF-resolution, successes are independent of the selection rule. Moreover, true floundering can be easily and efficiently implemented.

Another approach to the problem of floundering is in [Dra94]. Drabent remarks that the notion of floundering is too general. He gives a new notion of floundering - called serious floundering. We compare true floundering and serious floundering in Sect. 5. A consequence of the main result of [Dra94] is that, for an implementation of SLDNF-resolution with true floundering, a computation which terminates and does not truly flounder is complete wrt Kunen semantics. We consider that the absence of true floundering is a desirable property of normal programs.

Section 2 defines true floundering and extends SLDNF-resolution. Section 3 shows the independence from the selection rule of extended SLDNF-resolution. Section 4 is devoted to implementation issues: we study the computational problems that arise from floundering and we discuss the implementation of true floundering.

2   True Floundering

We define a notion of floundering that we call *true floundering*. We choose to base our work on the definition of SLDNF-resolution of [Llo87]. We do not recall the definitions of SLDNF-refutation and finitely failed SLDNF-tree ([Llo87]).

We first define the preliminary notion of *fully floundered SLDNF-tree*. Intuitively, a fully floundered SLDNF-tree has no success leaf node and at least one floundering leaf node.

Definition 1. Let P be a normal program and G a normal goal. A *fully floundered SLDNF-tree of rank 0* for $P \bigcup \{G\}$ is a tree satisfying the following:

1. Each node of the tree is a non-empty normal goal.
2. The root node is G.
3. Only positive literals are selected at nodes in the tree.
4. Let $\leftarrow L_1, ..., L_m, ..., L_p$ be a non-leaf node in the tree and suppose that L_m is an atom and it is selected. Then for each program clause (variant) $A \leftarrow M_1, ..., M_q$ such that L_m and A are unifiable with mgu θ, this node has a child $\leftarrow (L_1, ..., L_{m-1}, M_1, ..., M_q, L_{m+1}, ..., L_p)\theta$.
5. Let $\leftarrow L_1, ..., L_p$ be a leaf node in the tree, then either
   (a) each L_i is a non-ground negative literal, or
   (b) an atom L_m is selected and there is no program clause (variant) in P whose head unifies with L_m.
6. There is at least one leaf node satisfying 5.(a).

Definition 2. Let P be a normal program and G a normal goal. A *fully floundered SLDNF-tree of rank k+1* for $P \bigcup \{G\}$ is a tree satisfying the following:

1. Each node of the tree is a non-empty normal goal.
2. The root node is G.
3. Let $\leftarrow L_1, ..., L_m, ..., L_p$ be a non-leaf node in the tree and suppose L_m is selected. Then either
   (a) L_m is an atom and, for each program clause (variant) $A \leftarrow M_1, ..., M_q$ such that L_m and A are unifiable with mgu θ, this node has a child $\leftarrow (L_1, ..., L_{m-1}, M_1, ..., M_q, L_{m+1}, ..., L_p)\theta$, or
   (b) L_m is a ground negative literal $\neg A$ and there is finitely failed SLDNF-tree of rank k for $P \bigcup \{\leftarrow A_m\}$, in which case the only child is $\leftarrow L_1, ..., L_{m-1}, L_{m+1}, ..., L_p$.
4. Let $\leftarrow L_1, ..., L_p$ be a leaf node in the tree, then either
   (a) each L_i is a non-ground negative literal, or a ground negative literal $L_i = \neg A_i$ satisfying: there is a fully floundered SLDNF-tree of rank k for $P \bigcup \{\leftarrow A_i\}$, or
   (b) an atom L_m is selected and there is no program clause (variant) in P whose head unifies with L_m, or
   (c) a ground negative literal $L_m = \neg A_m$ is selected and there is an SLDNF-refutation of rank k of $P \bigcup \{\leftarrow A_m\}$.
5. There is at least one leaf node satisfying 4.(a).

Definition 3. Let P be a normal program and G a normal goal. A *fully floun-dered SLDNF-tree* for $P \bigcup \{G\}$ is a fully floundered SLDNF-tree of rank k for $P \bigcup \{G\}$, for some k.

The notion of SLDNF-tree is extended in order to deal with ground negative literals for which there exists a fully floundered SLDNF-tree. This is simply achieved by adding the case 4.(c) to the definition of SLDNF-tree ([Llo87]). The notion of SLDNF-derivation may be extended in the same way.

Definition 4. Let P be a normal program and G a normal goal. An *extended SLDNF-tree* of $P \bigcup \{G\}$ is a tree satisfying the following:

1. Each node of the tree is a (possibly empty) normal goal.
2. The root node is G.
3. Let $\leftarrow L_1, ..., L_m, ..., L_p$, $(p \geq 1)$ be a non-leaf node in the tree and suppose that L_m is selected. Then either
   (a) L_m is an atom and, for each program clause (variant) $A \leftarrow M_1, ..., M_q$ such that L_m and A are unifiable with mgu θ, the node has a child $\leftarrow (L_1, ..., L_{m-1}, M_1, ..., M_q, L_{m+1}, ..., L_p)\theta$, or
   (b) L_m is a ground negative literal $\neg A_m$ and there is finitely failed SLDNF-tree for $P \bigcup \{\leftarrow A_m\}$, in which case the only child is $\leftarrow L_1, ..., L_{m-1}, L_{m+1}, ..., L_p$.
4. Let $\leftarrow L_1, ..., L_m, ..., L_p$, $(p \geq 1)$ be a leaf node in the tree. Then either
   (a) L_m is an atom, L_m is selected and there is no program clause (variant) in P whose head unifies with L_m, or
   (b) L_m is a ground negative literal $\neg A_m$, L_m is selected and there is an SLDNF-refutation of $P \bigcup \{\leftarrow A_m\}$, or
   (c) each L_i is either a non-ground negative literal or a ground negative literal $L_i = \neg A_i$ satisfying: there is a fully floundered SLDNF-tree for $P \bigcup \{\leftarrow A_i\}$.
5. Nodes which are the empty goal have no children.

A leaf node which satisfy 4.(c) is said to be *true floundering*. A leaf node which satisfy 5. (resp. 4.(a) or 4.(b)) is said to be *success* (resp. *failure*).

Definition 5. An extended SLDNF-tree is said to be *truly floundered* if it has (at least) one true floundering leaf node.

3 Independence from the Selection Rule

Independence from the selection rule is an expected property for a non-deterministic algorithm. This independence problem, despite its relevance, is not solved in general for SLDNF-resolution.

Since we are interested in an extended definition of SLNDF-resolution including a notion of floundering, a further difficulty arises from this floundering. It may be related to the difficulties encountered in SLD-resolution with delay ([Nai93],[BFR93]) or concurrent (constraint) logic programming ([CF+93]).

These papers study the independence from the selection rule (from the scheduling for [CF+93]) and have to deal with a notion of deadlock which is close to floundering. However the problem here is complicated by the presence of subsidiary derivations.

The *computed answer* of a non-failure leaf node of an extended SLDNF-tree is the composition of the substitutions used in the branch.

Proposition 6. *Let P be a normal program and G a normal goal. Let T and T' be two extended SLDNF-trees for $P \bigcup \{G\}$.*
If there exists a non-failure leaf node B (possibly empty) of T with computed answer σ, $\langle C_1, ..., C_n \rangle$ being the sequence of the variants of program clauses used in the branch. Then there exists a non-failure leaf node B' of T' with computed answer σ', $\langle C'_1, ..., C'_n \rangle$ being the sequence of the variants of program clauses used in the branch, satisfying:
σ' is a variant of σ and,
B' is a variant of B and,
$\forall i \in \{1, ..., n\}$ there exists a variant $\tilde{C}_i$ of C_i such that the sequence $\langle C'_1, ..., C'_n \rangle$ is a permutation of the sequence $\langle \tilde{C}_1, ..., \tilde{C}_n \rangle$.

The proof is based on the grammatical view of logic programming ([DM93]) which provides a general framework to prove static properties of logic programs. It is an extention of the proof of the main result of [BFR93] which states the independence from the selection rule for SLD-resolution with delay.

We give a very precise formulation of this independence in term of skeletons of partial proof trees. In order to deal with negation we extend the notion of skeleton to incorporate negative leaf nodes. We define a particular set of skeletons (a skeleton is an object which "sums up" the clause choices of a derivation) which is intrinsic to the program. Then the result is proven by the existence of a bijection between the set of non-failure leaf nodes of an extended SLDNF-tree and this set of skeletons. We don't give the proof here. The interested reader may refer to the appendix.

We now give the definition of true floundering for a program and a goal:

Definition 7. *Let P be a normal program and G a normal goal. $P \bigcup \{G\}$ truly flounders if there is a truly floundered extended SLDNF-tree for $P \bigcup \{G\}$.*

Remark. One may argue that (for a program and a goal) true floundering is irrelevant if there is a full success leaf node (its computed answer does not instantiate the goal). Definition 7 may be modified in the following way: $P \bigcup \{G\}$ *truly flounders* if there is a truly floundered extended SLDNF-tree with no full success leaf node for $P \bigcup \{G\}$.

Remark. Another improvement may be made by taking into account non-defined predicates and slightly weakening the safety condition. Indeed if a predicate p is non-defined in a program then a call such as $\neg p(X)$ (where X is not ground) may be considered successful.

4 Implementation Issues

There is a important problem of incompleteness between SLDNF-resolution and implementations. The reasons are infinite computations (this may be related to the case of SLD resolution but it is here complicated by the problem of infinite sequences of negative calls) and floundering.

In this section we give a refined notion of computation - called *extended computation* - which calculates an extended SLDNF-tree when it terminates.

We first define what a computation is and discuss the problems that arise from floundering. Then we define extended computation.

In [Llo87] a computation is defined as an attempt to build an SLDNF-derivation. For our purposes, we may say that a computation for SLDNF-resolution behaves like a computation for SLD-resolution as long as an atom is selected. When a ground negative literal $\neg A$ is selected in a goal H, a computation for A is started. If it eventually fails (a finitely failed SLDNF-tree being built) then H has one child with the literal $\neg A$ removed. If it eventually succeeds (an SLDNF-refutation for A being built) then H has failed.
A computation flounders if a goal with only non-ground negative literals is reached.

Floundering being considered as an error, actual implementations of SLDNF-resolution stop a computation as soon as a floundering occurs. The following examples show that it would be worth continuing the computation.

Example 1. Let's consider the normal program $P1$:
$s(X) \leftarrow \neg p(a), q.$
$p(a) \leftarrow \neg r(X).$

and the goal $G1$:
$\leftarrow s(X).$

Figure 1 shows two computations for $P1 \bigcup \{G1\}$ under two different selection rules. The selection of $\neg p(a)$ (computation C_1) in the goal $\neg p(a), q$ leads to a floundering in the subsidiary computation. The selection of q (computation C_2) leads to a failure and computation C_2 does not flounder.

There exists an SLDNF-tree for $P1 \bigcup \{G1\}$ but computation C_1 does not calculate it. One may remark that a correct answer could have been obtained by going back to the node $\neg p(a), q$, selecting q and then continuing the computation. Such a computation would then be correct and complete despite the existence of a floundering.

Example 2. Let's consider the normal program $P2$:
$p(a) \leftarrow \neg r(X).$
$p(a).$

and the goal $G2$:
$\leftarrow \neg p(a).$

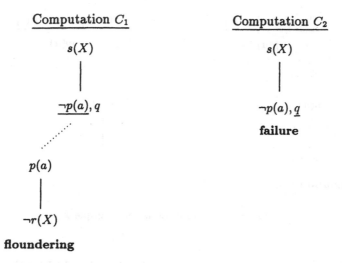

Fig. 1. Some computations for example 1

Figure 2 shows two computations for $P2\bigcup\{G2\}$. The choice of the first clause in the definition of p leads to a floundering (computation C_3) while the choice of the second clause leads to a failure and computation C_4 does not flounder.

There exists an SLDNF-tree for $P2\bigcup\{G2\}$ but computation C_3 does not calculate it. A correct answer could have been obtained by backtracking and continuing the (subsidiary) computation for $\neg p(a)$. Then, an SLDNF-refutation being built for $p(a)$, one could have conclude to a failure for $P2\bigcup\{G2\}$ despite the existence of a floundering. Such a computation would be correct and complete.

These examples show that in some cases it is possible to continue a computation (with possibly a different selection rule, with possibly a different clause choice rule) and obtain all the answers despite a floundering.

We propose a computation to apply the following rules:

1. Only select in a goal either an atom or a ground negative literal *non-marked as floundered*.
2. When a goal with only non-ground negative literals or ground negative literals marked as floundered is reached, backtrack to the previous choice point and continue the computation.
3. When a computation for a negative call $\neg A$ in a goal H builds a (finite) tree without success but with at least one leaf node with only non-ground negative literals or ground negative literals marked as floundered (rules 1.

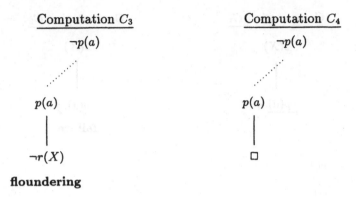

Fig. 2. Some computations for example 2

and 2. being applied) (such a tree is a fully floundered SLDNF-tree), mark $\neg A$ as "floundered" in H and select another literal in H (if that is possible).

We call *extended computation* a computation applying these rules.

Definition 8. A *truly floundered goal* of an extended computation is a goal satisfying:
- it is not in the scope of a negative call (i.e. it belongs to the main derivation).
- each literal in the goal is a non ground negative literal or a ground negative literal marked as floundered.

Extended computation closely corresponds to extended SLDNF-resolution: an extended computation either does not terminate or builds an extended SLDNF-tree. Moreover a terminated extended computation which does not contain truly floundered goals (although it may contain floundered goals) builds an SLDNF-tree (in the sense of [Llo87]). Thus floundered goals which are not truly floundered are irrelevant. In our view floundering should no longer be viewed as an error. Avoiding it may be seen as an optimisation question. The problem is now to build true floundering free programs.

We have shown in Sect. 3 that true floundering is independent of the selection rule. The definition of SLDNF-resolution of [Llo87] (and so our extended one), unlike later re-definitions, has a one-step view of subsidiary derivations. So the selection rule concerns only the main derivation. However Prop. 6 can be applied to terminated extended computations.

Let's call *true floundering answer* (of an extended computation), a pair formed by the goal itself and the computed answer at a truly floundered goal. Let's call *success answer*, the computed answer at a success goal. The following is a corollary of Prop. 6.

Corollary 9. *Let P be a normal program and G a normal goal.*
Two terminated extended computations of $P \bigcup \{G\}$ calculate the same (up to a renaming) successes and true floundering answers.

As long as no floundering occurs, extended computation is as efficient as computation because in this case an extended computation is a computation. Moreover, for some programs and goals, computations may flounder under some selection rules while extended computations terminate and are complete.

5 Related works

Drabent ([Dra94]) gives a notion of floundering called *serious floundering*. He remarks that fairness in SLDNF-resolution should be extended. For a goal ← ¬A, L̄ (L̄ being a sequence of literals) selecting ¬A may make it impossible to select any literal of L̄ (if A loops). He defines a notion of SLDNF1 pre-failed tree (using a constructive negation formulation) in which, in such a case, derivations for ← A and ← L̄ both may be built. Serious floundering is defined with SLDNF1 pre-failed trees. Unlike true floundering, serious floundering is defined wrt a selection rule.

In this section we give some comparison results between serious floundering and true floundering. Fairness in the following propositions is related to SLDNF1 pre-failed trees ([Dra94]).

We need the notion of safe selection rule (e.g. [Llo87]) but for technical convenience, we use a slightly modified version:

Definition 10. A selection rule is *safe* if it satisfies:
if a non ground negative literal is selected in a goal then the goal only contains non-ground negative literals.

Proposition 11. *Let P be a normal program and G a normal goal. Let R be a safe and fair selection rule. If $P \bigcup \{G\}$ seriously flounders under R then $P \bigcup \{G\}$ truly flounders.*

The interested reader may refer to the appendix for the proof.

A consequence of this proposition is that a non-truly floundered terminated extended computation (its selection rule R is safe and fair) does not seriously flounder under R. Thus, with the main result of [Dra94] we have (intuitively):

Let P be a normal program and G a normal goal. A non-truly floundered terminated extended computation for $P \bigcup \{G\}$ is correct and complete wrt Kunen semantics.

The reciprocal of Prop. 11 holds (i.e. if $P \bigcup \{G\}$ does not seriously flounder under R then $P \bigcup \{G\}$ does not truly flounder). Thus true floundering and serious floundering are in a sense equivalent. However, true floundering is defined in a simpler way (serious floundering is defined wrt a selection rule).

Martelli and Tricomi [MT92] give a new definition of SLDNF-resolution. They consider extended goals that enclose bracketed subgoals coming from negative calls. They define floundering leaf nodes. This notion of floundering is close to ours but some derivations are considered as floundered while we think they should be failed. True froundering could be obtained with some slight changes. However we think that stating independence of the selection rule would be more difficult in this framework than in ours.

[AB94] (which reproduces Apt and Doets's definition of SLDNF-resolution) gives a declarative definition of floundering. But this one is very weak: it involves a program and a goal floundering if there exists a selection rule for which a computation may flounder. Actually, this notion is directly abstracted from the standard notion of floundering.

Many other papers deal with negation and control but, as far as we know, none attempts to redefine floundering.

From a practical point of view, we need to find conditions under which a given program and a goal may truly flounder or not. In our approach, to state that a program (and a goal) does not truly flounder is simply achieved by building a non truly floundered extended SLDNF-tree. In addition, we have also given an efficient (sketch of) implementation which closely corresponds to extended SLDNF-resolution.

6 Conclusion

Avoiding floundering is a major problem in the context of SLDNF-resolution. However, the notion of floundering does not seem to be a suitable one. Firstly, it is too weak. Secondly, it is a computational notion. There is a lack of a new definition abstracted from computation.

We have proposed a stronger notion of floundering - *true floundering* - with a declarative understanding and an efficient computational behaviour. It seems that true floundering gives a good understanding of what floundering is in SLDNF-resolution.

Some recent works ([AD92], [MT92], [MN90]) explore the problem of giving a complete (in the sense that for a program and a goal there always exists an SLDNF-tree) and suitable re-definition of SLDNF-resolution but do not completely treat floundering. True floundering may be incorporated into these works.

The safeness condition may be weakened ([Llo87], [MN90]). True floundering with a weak safeness condition may be defined within the definition of SLDNF-resolution of [MN90].

An implementation of SLDNF-resolution with true floundering is as efficient as a standard implementation as long as no floundering occurs. In some cases, extended computations (i.e. computations in an implementation of SLDNF-resolution with true floundering) may not truly flounder while some standard computations flounder. Moreover an extended computation which terminates and does not truly flounder is correct and complete wrt Kunen semantics. Thus extended computation is more complete than standard computation.

The class of non true floundering programs is quite large because every program that does not flounder (under some selection rule) does not truly flounder. Moreover it has a declarative characterization. We can expect to build true floundering freeness analysis tools that are simpler and more efficient than floundering freeness analysis tools.

Acknowledgements

The author would like to thank Wlodzimierz Drabent for his comments on a draft version of this paper. Thanks to Pierre Deransart, Gérard Ferrand and Jan Maluszynski for some fruitful discussions. Thanks also to some anonymous referees of a previous version of this paper.

References

[AB94]   K. R. Apt, R. Bol. Logic Programming and Negation: A Survey. To appear in Journal of Logic Programming.

[AB91]   K. R. Apt, M. Bezem. Acyclic programs. *New generation computing*, 29(3) pages 335-363, 1991.

[AD92]   K. R. Apt, K. Doets. A new definition of SLDNF-Resolution. *Report CS-R9242, CWI*, 1992.

[AP91]   K. R. Apt, D. Pedreschi. Proving termination of general Prolog programs. *In T.Ito and A. Meyer editors, Proceedings of the International Conference on Theoretical Aspects of Computer Software*, Lecture notes in Computer Science 526, pages 265-289, Berlin, 1991. Springer-Verlag.

[BFR93] M. Bergère, G. Ferrand, J-H. Réty. Independence of the Computation R for SLD-Resolution with Delay. *Research Report LIFO 93-9*, Orléans, 1993.

[BM90]   R. Barbuti, M. Martelli. Recognizing non-floundering logic programs and goals. International Journal on Foundations of Computer Science, 1(2) pages 151-163, 1990.

[CF+93] M. Codish, M. Falaschi, K. Mariott, W. Winsborough. Efficient Analysis of Concurrent Constraint Logic Programs. *proceedings of ICALP'93*, pages 633-644, 1993.

[Cla79] K. L. Clark. Predicate logic as a computational formalism. *Research Report DOC 79/59*, Imperial college, Dept. of computing, London, 1979.

[DM93]   P. Deransart, J. Maluszinski. A Grammatical View of Logic Programming. *MIT Press*, 1993.

[Doe94] K. Doets. From Logic to Logic Programming. *MIT Press*, 1994.

[Dra94] W. Drabent. Completeness of SLDNF-Resolution for Non-Floundering Queries. Draft, 1994.

[Kun87] K. Kunen. Negation in Logic Programming. *Journal of Logic Programming*, 4, pp. 289-308, 1987.

[Llo87] J. W. Lloyd. Foundations of Logic Programming. *Springer Verlag, 2nd ed.*, 1987.

[MN90]   J. Maluszynski, U. Nilsonn. Logic programming and Prolog. *Wiley*, New-York, 1990.

[MSD90] K. Marriott, H. Sondergaard, P. Dart. A characterization of non-floundering logic programs, in Debray and Hermenegildo ed., Proc. NACLP, pages 661-680, Austin 1990, *MIT Press 1990*.

[MT92] M. Martelli, C. Tricomi. A new SLDNF-tree. *Information Processing Letters* 43 (1992) pages 57-62.

[Nai93] L. Naish. Coroutining and the Construction of terminating logic programs. *Australian Computer Science Communications*, 15 (1), pp. 181-190, 1993.

Appendix

A   Proof of Proposition 6

The proof is based on the grammatical view of logic programming[1]. In order to deal with negation we extend skeletons to incorporate negative leaf nodes. We define a subset of the set of skeletons - n-complete skeletons - which is intrinsic to the program (and the goal). We reformulate extended SLDNF-resolution in terms of skeletons. Then the result is proven by the existence of a bijection between the set of n-complete skeletons and the set of non-failure leaf nodes of an extended SLDNF-tree.

A.1   Skeletons with Negation.

From now on a *normal program* P and a *goal* are assumed to be fixed.

It is useful to formalize the *normal program* by a finite family $\{Cr\}_{r \in N}$ where each C_r is a *normal clause* $a_0 \leftarrow a_1, ..., a_n$ (a_i *literals*). The symbol r is the *name* of the clause C_r and n is the *arity* of r.

In the same way the *goal* is formalized by a clause $\leftarrow a_1, \cdots, a_n$ (a_i *literals*) and a symbol g ($g \notin N$) which is the name of the goal, n is the arity of g.

We consider two other symbols, '?' and '¬', with arity 0.

Definition 12. A *(partial) skeleton* (with negation) is a finite oriented tree, labelled by $N \bigcup \{g, '?', '¬'\}$, such that:

- the degree of a node (number of children) is the arity of its label,
- the root is the only occurrence of g.

A node of a skeleton is *undefined* if its label is '?'. The other nodes are *defined*. A node is *negative* if its label is '¬'. A node is *proper* if it is defined but not negative.

The set of undefined (resp. defined, proper, negative) nodes of the skeleton s is denoted by $undef(s)$ (resp. $def(s)$, $prop(s)$, $neg(s)$).

The skeleton s is said to be *total* if $undef(s) = \emptyset$.

Remark. A skeleton is completely defined by its defined nodes.

Definition 13. A *renaming* for the skeleton s is a family $(C'_u)_{u \in prop(s)}$ where:

1. if u is not the root and u is labelled by r, C'_u is a *variant* of the clause C_r,
2. if u is the root, C'_u is exactly the *goal* clause $\leftarrow a_1, \cdots, a_n$,
3. if $u \neq v$, C'_u and C'_v have no common variable.

Definition 14. Let s be a skeleton equipped with a renaming $(C'_u)_{u \in prop(s)}$.

For each node u except the root, we define a *literal* a_u and, if moreover $u \in prop(s)$, we also define an *equation* e_u as follows:

Let v be the parent node of u. u is the ith child of v ($1 \leq i \leq arity\ of\ the\ label\ of\ v$). The literal a_u is the ith member of the body of the clause C'_v.

Moreover if $u \in prop(s)$ (so C'_u is also defined) the equation e_u is $a_u = head\ of\ C'_u$.

---

[1] P. Deransart, J. Maluszinski. A Grammatical View of Logic Programming. MIT press. 1993.

$(e_u)_{u \in prop(s), \ u \ not \ the \ root}$ is said to be *the system of equations* associated with s and the renaming.

An undefined leaf whose associated literal is a positive(resp. negative) literal is said to be a positive(resp. negative) undefined leaf.

Let s be a skeleton equipped with a renaming and S the associated system of equations. The solvability of S does not depend on the particular renaming (another renaming gives only another variant of the system of equations).

Definition 15. A skeleton is said to be *solvable* if (given any renaming) its system of equations is solvable.

Definition 16. Let s be a solvable skeleton, let σ be the mgu of its system of equations. An *undefined leaf* u of the skeleton s is said to be *suspended* if its associated literal is a negative literal $\neg A$ which satisfies:
σA is not ground, or
σA is ground and there is a truely floundered SLDNF-tree for $P \bigcup \{\leftarrow \sigma A\}$.

We denote by $susp(s)$ the set of the suspended leaves of s.

Definition 17. A *skeleton* s is said to be *n-solvable* if, it is solvable (its system of equations admits an mgu σ) and each literal associated with a negative leaf of s is a negative literal $\neg A_i$ which satisfies:
σA_i is ground, and
there is a finitely failed SLDNF-tree for $P \bigcup \{\leftarrow \sigma A_i\}$.

Note that a n-solvable skeleton is not necessarily total (undefined leaves are possible).

Definition 18. A skeleton is *n-complete* if it is n-solvable and all its undefined leaves are suspended.

Definition 19. Let s, s' be two skeletons. s is an *approximation* of s', denoted by $s \leq s'$, if:

- $def(s) \subseteq def(s')$,
- for each *defined* node u of s, the labels of u in s and in s' are the same.

Note that $\leq$ is a *partial order*.

The following notations will be useful.

- If r is a clause name whose arity is n, we denote by $tree(r)$ the tree in which the root is labelled by r and the other nodes are n leaves labelled by '?'.
- We denote by $tree('\neg')$ the tree in which the root is labelled by '$\neg$' and is the only node.
- If $u \in undef(s)$ and $l \in N \bigcup \{'\neg'\}$, we denote by $s[u \leftarrow tree(l)]$ the result of the grafting of $tree(l)$ at the node u in s.
  Note that s is an approximation of $s[u \leftarrow tree(l)]$.

The notion of n-complete skeleton is defined without any reference to a top-down computation. The following technical lemma will be a key result to link, in the next section, this notion to extended SLDNF-Resolution.

Lemma 20. *Let s' be a n-complete skeleton and s a n-solvable approximation of s'.*

1. *If $u \in (def(s') \bigcap undef(s)) - susp(s)$ then $s[u \leftarrow tree(label\ of\ u\ in\ s')]$ is again a n-solvable approximation of s'.*
2. *If $s' \neq s$ then $(def(s') \bigcap undef(s)) - susp(s) \neq \emptyset$*

Sketch of proof: A renaming for s' is fixed. We denote by S' the system of equations associated with s' and this renaming, and by S the restriction of S' to $prop(s)$.

1. $prop(s[u \leftarrow tree(label\ of\ u\ in\ s')]) = prop(s) \bigcup \{u\}$.
   If *label of u in s'* is $\neg$ then let $\neg A$ be associated literal with u (it is negative because s' is n-solvable).
   Let σ be an mgu of S, $\sigma(\neg A)$ is ground because u is not suspended.
   moreover, let μ be an mgu of S', $\sigma A = \mu A$.
   So, s' beeing n-solvable, there exists a finite failure tree for σA.
   Else, Let σ be an mgu of S, σe_u has a solution (else s' could not be n-solvable).
2. Each positive undefined leaf of s belongs to $def(s')$ and is trivially not suspended.
   If there is no positive undefined leaves in s then $s = s'$.
   Let $u \in def(s') \bigcap undef(s)$ (there exists because $s \neq s'$).
   Its associated literal is a negative literal $\neg A$ and , let σ be an mgu of S and μ an mgu of S', $\sigma A = \mu A$. Thus s' being n-solvable, u is not suspended.□

Note that, as a consequence of this lemma, an approximation s of a n-complete skeleton s' with $s \neq s'$ cannot be n-complete.

Such an approximation is not necessarily n-solvable, but the approximation $tree(g)$ of s' is trivially n-solvable because there is no equation and no negative leaves.

A.2 Extended SLDNF-resolution.

We will now give a definition of extended SLDNF-resolution in terms of skeletons (with negation).

In our framework the choice of a literal becomes merely the choice of an undefined leaf of a skeleton.

Definition 21. A *selection function* is a function $\mathcal{F}$ defined for each *skeleton* s which is *n-solvable* but *not n-complete* and such that $\mathcal{F}(s) \in undef(s) - susp(s)$.

A selection function $\mathcal{F}$ being fixed, we define the notion of *derivation* (wrt $\mathcal{F}$) on the n-solvable skeletons :

If s and s' are two n-solvable skeletons, s' *derives* from s if, either, there is a clause name r such that $s' = s[\mathcal{F}(s) \leftarrow tree(r)]$, or $s' = s[\mathcal{F}(s) \leftarrow tree('\neg')]$.

A *derivation* for the goal g is a (finite or infinite) sequence $s_0, s_1, \cdots$ such that $s_0 = tree(g)$ and s_{i+1} derives from s_i for each i.

A finite derivation $(s_0, \cdots, s_n)$ is said to be *from s_0 to s_n*.

A *terminated* derivation is a finite derivation which cannot be extended. The *terminal* skeleton of a terminated derivation is:

either a n-complete skeleton (total or partial),

either a n-solvable skeleton s which has at least one positive undefined leaf, $\mathcal{F}(s)$ for which, for each clause name r, the skeleton $s[\mathcal{F}(s) \leftarrow tree(r)]$ is not n-solvable.

either a n-solvable skeleton s which has at least one negative undefined leaf $\mathcal{F}(s)$, with associated literal $\neg A$, which is not suspended, and there is an SLDNF refutation for σA (σ being an mgu of the system of equations associated with s).

A *successful* (resp. *truely floundered*) derivation is a terminated derivation to a n-complete skeleton which is total (resp. partial and not total).

A *failed* derivation is a terminated derivation which is neither *successful* nor *truly floundered*.

Extended SLDNF-derivation is expressed by this notion of derivation. The *selection rule* (choice of the selected literal) is obviously specified by the selection function $\mathcal{F}$.

Lemma 22. *The mapping* $(s_0, \cdots, s_n) \mapsto s_n$, *where* $(s_0, \cdots, s_n)$ *is a derivation wrt* $\mathcal{F}$, *($\mathcal{F}$ fixed), is* one-to-one.

Proof. At first, note that if $(s_0, \cdots, s_n)$ is a derivation then each s_i is an approximation of s_n. Note also that if $(s_0, \cdots, s_n, s_{n+1})$ is also a derivation then $s_n \neq s_{n+1}$.

Now, consider two derivations $s_0, s_1, \cdots$ and $s'_0, s'_1 \cdots$ from $s = s_0 = s'_0$ such that $s_i = s'_i$ for $i < k$ but $s_k \neq s'_k$. There is a node of s_k and s'_k (the node $\mathcal{F}(s_{k-1}) = \mathcal{F}(s'_{k-1})$) whose labels are different in s_k and s'_k. So s_k and s'_k cannot be approximation of the same skeleton s_n.$\Box$

Our major result is that this mapping is one-to-one and *onto* between *non-failure derivations* and *n-complete skeletons*. It remains to prove:

Theorem 23. *For each selection function* $\mathcal{F}$, *for each n-complete skeleton* s, *there is a derivation wrt* $\mathcal{F}$ *from* g *to* s.

Proof. Justified by Lemma 20, we can define the following derivation where each s_k is a skeleton which is an approximation of s:

$s_0 = tree(g)$,

if $s_k \neq s$ then $s_{k+1} = s_k[\mathcal{F}(s_k) \leftarrow tree(label\ of\ \mathcal{F}(s_k)\ in\ s)]$

There is a step k for which $s_k = s$. So, we have defined a derivation from $tree(g)$.$\Box$

Let us translate this result in terms of extended SLDNF-tree.

From a formal viewpoint the set of the derivations for g is already organized as a *non-oriented* tree as follows:

the *nodes* are the finite derivations,

the *label* of the node $(s_0, \cdots, s_n)$ is the skeleton s_n,

the *parent* of the node $(s_0, \cdots, s_n)$ is the node $(s_0, \cdots, s_{n-1})$ if $n \geq 1$,

the *root* is $(tree(g)) = (s_0)$.

The *derivations* which are terminated or infinite can be identified with the *branches* of the tree. So there are four kinds of branches: *infinite*, *success* when s_n is n-complete and total, *true floundering* when s_n is n-complete and not total, and *failure*.

This non-oriented tree is the *search tree* associated with the goal g. This search tree expresses the notion of extended SLDNF-tree.

So from a theoretical viewpoint this search tree is an abstract tool for expressing and proving properties where variables, renaming, variants, ... are only bothersome details which are not relevant.

The following corollary is the translation of Theorem 23 in terms of search tree: the independence wrt the selection rule is translated into an independence wrt the search tree.

Corollary 24. *Let S be the* search tree *defined by a* selection function $\mathcal{F}$. *Let $\phi_{\mathcal{F}}$ be the mapping $l \mapsto s$, where l is a non-failure leaf of S and s is the* n-complete skeleton *which is the label of l in S. $\phi_{\mathcal{F}}$ is one-to-one and onto.*

Let S_1 and S_2 be the two search trees *defined by two* selection functions $\mathcal{F}_1$ *and* $\mathcal{F}_2$. *The mapping $(\phi_{\mathcal{F}_2})^{-1} \circ \phi_{\mathcal{F}_1}$ is one-to-one and onto between the non-failure leaves of S_1 and the non-failure leaves of S_2. A non-failure leaf of S_1 and a non-failure leaf of S_2 which are associated by this mapping have the same label (n-complete skeleton) in S_1 and in S_2.*

In terms of *extended SLDNF-trees*, we can say that two extended SLDNF-trees have, in some sense, (up to a renaming of the variables), the *same success leaves*, (hence the same computed answers, instantiation of the goal), and the *same true floundering leaves*. This independence result comes from the correspondence between these leaves and the n-complete skeletons.

As a consequence of this independence, if some extended SLDNF-tree has no true floundering leaf then no extended SLDNF-tree has a true floundering leaf. In other words, true floundering does not depend on the selection rule.

B   Proof (sketch) of Proposition 11

By induction on the depth of serious floundering.

Let B be a normal goal.

- If B depth 0 seriously flounders under R then there exists a non-failure branch of an SLDNF1 pre-failed tree for $P \bigcup \{B\}$ via R such that it is finite and the last goal contains only non-ground negative literals (R is safe).

This branch is a truly floundered extended SLDNF-derivation so $P \bigcup \{B\}$ truly flounders.

If B is ground, there is no SLDNF-refutation (def. of serious floundering). So in this case, there is a fully floundered SLDNF-tree of rank 0.

- If B depth k seriously flounders under R then there exists a non-failure branch of an SLDNF pre-failed tree for $P \bigcup \{B\}$ via R such that it is finite and the last goal contains only non-ground negative literals or ground negative literals $\neg A_i$ such that A_i depth k-1 seriously flounders (R is safe and fair).

By induction hypothesis, for each A_i, there exists a fully floundered SLDNF-tree (of rank k-1) for $\leftarrow A_i$. This branch is a truly floundered extended SLDNF-derivation. So $P \bigcup \{G\}$ truly flounders.

If B is ground, there is no SLDNF-refutation. So, in this case, there is a fully floundered SLDNF-tree of rank k.$\square$

Minimal Set Unification *

Puri Arenas-Sánchez[1] and Agostino Dovier[2]

[1] Dpto. Informática y Automática, Fac. C.C. Matemáticas, (U.C.M.).
Avnda Complutense s/n, 28040 MADRID. puri@dia.ucm.es
[2] Dip. di Informatica, Univ. di Pisa. Corso Italia 40, 56100 PISA (I). dovier@di.unipi.it

Abstract. A unification algorithm is said to be minimal for a unification problem if it generates exactly a complete set of minimal unifiers, without instances, without repetitions. Aim of this paper is to describe a new set unification algorithm minimal for a significant collection of sample problems that can be used as benchmarks for testing any set unification algorithm. To this end, a deep combinatorial study for such problems has been realized. Moreover, an existing naïve set unification algorithm has been also tested in order to show its bad behavior for most of the sample problems.
Keywords: Logic Programming with Sets, CLP, Unification.

1 Introduction

The drawing up of many papers concerning *Constraint Logic Programming with Sets* (see e.g. [5, 6, 12]) has pointed out that the complexity of the (NP complete– see e.g. [9]) *set unification problem* is the real bottleneck of any attempt to extend Logic Programming with set entities.

The loss of the *uniqueness of the most general unifier* property forces any set unification algorithm to compute a *complete* set of unifiers (i.e., as it will be explained in Sect. 2, a set of unifiers which covers all possible solutions for the problem at hand) for any satisfiable input. Since the cardinality of a complete set of unifiers grows exponentially w.r.t. the size of the unification problem, it seems unconvenient to return all the unifiers at once. Rather, the non-determinism lying inside any Logic Programming interpreter suggests the use of a non-deterministic unification algorithm which returns exactly one unifier for each non-deterministic branch: in this way any non-deterministic computation can run without undergoing the unpleasant effects of the non uniqueness of the most general unifier.

In [2] it has been shown that if a representation for sets comprising also a constant symbol for the universal set and the set-minus operator is adopted, the *unique most general unifier theorem* can be recovered. However, this approach is more properly described as boolean unification rather than set unification.

---
* P. Arenas-Sánchez is partially supported by the Spanish National Project TIC92–0793–C02–01 "PDR" and the Esprit BRA Working Group Nr. 6028 "CCL". The work is partially supported by C.N.R. grant 94.00472.CT12, "Logic Programming with Sets".

Nested sets are not allowed, and the answer to a unification problem contains a large amount of information, becoming scarcely readable.

In [12] any set unification problem is delayed until it is transformed into a simple ground 'test'. This improves efficiency, however, if the two terms do not become ground, obscure answers such as $\{X_1, f(X_1, X_3)\} = \{Y_1, f(\{Y_3\}, X_1), X_2\}$ are returned.

There are different ways to represent a finite set. Among them, the *union of singletons* representation which depicts $\{s_1, \ldots, s_m\}$ as $\{s_1\} \cup \cdots \cup \{s_m\}$, and the *list* representation which uses the term $\{s_1 \,|\, \{s_2 \,|\, \{\cdots \{s_m \,|\, \{\}\} \cdots\}\}\}$ for denoting the same set. The former representation (which is associated with an *ACI* equational theory–cf. [13]) is more expressive than the latter (which is associated to the equational theory described in Sect. 2). For instance, the problem $X_1 \cup \cdots \cup X_m = \{a_1\} \cup \cdots \cup \{a_n\}$, where X_i's are pairwise distinct variables and a_j's are pairwise distinct constant symbols, admits $(2^m - 1)^n$ independent solutions. Since the semantics of $\{t \,|\, s\}$ is $s \cup \{t\}$, if $m > 1$ such problem cannot be expressed by a list representation.

Since the minimum cardinality of a complete set of unifiers expressible with the list representation is itself conspicuous (cf. Sect. 3), we prefer to deal with such representation, avoiding further problems that the "union of singletons" approach would open. The same choice has been performed in [5, 12].

In this paper we present a *new Set Unification Algorithm* SUA which is minimal (it computes a complete set of minimal unifiers) for a significant collection of sample problems. In order to prove the minimality of SUA, a deep combinatorial study has been necessary, in order to compute the minimal number of solutions for our sample unification problems. In our opinion, the selected problems can be used as benchmarks for testing any set unification algorithm. Two reasons justify our choice: their simplicity (which reflects into a simplification of the analysis) and the fact that they maximize the number of solutions for unification problems of given size (the presence of distinct variables as elements of the sets to be unified guarantees the maximum number of solutions).

The aim of getting minimal algorithms for set unification has already been treated in literature. For instance in [14] three set unification algorithms are proposed. The most efficient seems to be the third one, however, comparing its results with our minimality study, it is possible to conclude that such algorithm is not minimal for problems (6), (7), and (8) described in Sect. 3. In [3], the presented naïve set unification algorithm (based on [8]) has a minimal behavior only for the first of such benchmarks, as it will be proved in Sect. 4.

The paper is organized as follows: in Sect. 2 we comment briefly on some preliminary concepts needed in the rest of the paper. Sect. 3 presents eight sample unification problems with their corresponding minimality studies and the recursive functions computing the number of solutions. Numerical values for such functions are reported in appendix. The behavior of a naïve set unification algorithm –based on [8]– on such benchmarks is analyzed in Sect. 4. In Sect. 5 a *new set unification algorithm*, named SUA, is presented, proving its termination and minimality for all sample problems suggested. Some conclusions are finally drawn up in Sect. 6.

2 Preliminaries

We will make use of standard *CLP* (see e.g. [7]) and *unification theory* (see e.g. [13]) notations. Given a signature Σ (a set of functional symbols together with their arities) and a denumerable set of variables $\mathcal{V}$, $\tau(\Sigma \cup \mathcal{V})$ and $\tau(\Sigma)$ will denote the sets of terms and ground terms, respectively.

We require the signature Σ to contain (at least) the constant symbol $\{\}$, representing the *empty set*, and the binary symbol $\{\cdot\,|\,\cdot\}$, used as *set constructor symbol*; the intuitive semantics of $\{t\,|\,s\}$ is $\{t\} \cup s$. Similarly to lists in PROLOG, the term $\{a\,|\,\{b\,|\,\{c\,|\,\{\}\}\}\}$ will be denoted simply as $\{a, b, c\}$.

The following two equational axioms

$$(A_b)\ \{X, X\,|\,Z\} = \{X\,|\,Z\} \qquad (C_R)\ \{X, Y\,|\,Z\} = \{Y, X\,|\,Z\}$$

(A_b stands for *absorption* and C_R stands for *commutativity on the right*–see [13]) uniquely identify a finest congruence $=_\tau$ on $\tau(\Sigma)$. Following [13], two terms s and t are said to be *$\mathcal{T}$-unifiable* iff there is a substitution σ such that $s\sigma =_\tau t\sigma$; such a σ is called a *$\mathcal{T}$-unifier*. We write, for any set of variables $W \subseteq \mathcal{V}$, $\sigma =_\mathcal{T}^W \rho$ iff $\forall x \in W\ x\sigma =_\tau x\rho$. In the same way, σ is *more general than* ρ in $\mathcal{T}$ over W ($\sigma \leq_\mathcal{T}^W \rho$) iff $\exists \eta\ \eta \circ \sigma =_\mathcal{T}^W \rho$. The corresponding equivalence relation on substitutions is denoted by $\equiv_\mathcal{T}^W$; i.e., $\sigma \equiv_\mathcal{T}^W \rho$ iff $\sigma \leq_\mathcal{T}^W \rho$ and $\rho \leq_\mathcal{T}^W \sigma$. The set of all $\mathcal{T}$-unifiers of two terms s and t is denoted by $U_\mathcal{T}(s, t)$. A *complete set of $\mathcal{T}$-unifiers* for s,t is any subset $CSU_\mathcal{T}(s, t) \subseteq U_\mathcal{T}(s, t)$ such that $(\forall \sigma \in U_\mathcal{T}(s, t))(\exists \theta \in CSU_\mathcal{T}(s, t))(\theta \leq_\mathcal{T}^W \sigma)$, where $W = Var(s) \cup Var(t)$. A *complete set of minimal $\mathcal{T}$-unifiers* for s and t is any complete set of $\mathcal{T}$-unifiers $\mu CSU_\mathcal{T}(s, t)$ satisfying the minimality condition: $(\forall \sigma \tau \in \mu CSU_\mathcal{T}(s, t))(\sigma \neq \tau \Rightarrow \sigma \not\leq_\mathcal{T}^W \tau)$, where $W = Var(s) \cup Var(t)$. Note that since $CSU_\mathcal{T}(s, t)$ exists and it is finite, for any $s, t \in \tau(\Sigma \cup \mathcal{V})$, then $\mu CSU_\mathcal{T}(s, t)$ exists and it is unique up to $\equiv_\mathcal{T}^\mathcal{V}$ [11]. Similar definitions can be given for unification problems consisting of some finite system of equations $\mathcal{E}$, instead of a single equation $s = t$. When the context is clear, we will omit the prefix $\mathcal{T}$ before the word unifier. Sometimes we will speak of the set of most general unifiers when referring to $\mu CSU_\mathcal{T}(\mathcal{E})$.

For any satisfiable equation system $\mathcal{E}$ involving terms from $\tau(\Sigma \cup \mathcal{V})$, a unification algorithm should be able to compute through non-determinism *each element* of a complete set of unifiers for $\mathcal{E}$. Notice that it is not required that it computes exactly $\mu CSU_\mathcal{T}(\mathcal{E})$. However, as we will see in detail in Sect. 3, the presented theory $\mathcal{T}$ is such that, even for simple unification problems $s = t$, $\mu CSU_\mathcal{T}(s, t)$ has a huge size. This means that a valid criterion to compare two (set) unification algorithms is the analysis of the length of the *list* of solutions computed by them (the use of the word 'list' here is needed to reflect the fact that if a unification algorithm computes exactly $\mu CSU_\mathcal{T}(s, t)$, but some unifier is returned more than once, it cannot be considered minimal).

If the input system $\mathcal{E}$ is $\mathcal{T}$-unsatisfiable, any unification algorithm should conclude its computation reporting a *failure* result.

3 Sample Problems

In this section we single out eight set unification problems that we propose as 'benchmarks' to test the minimality of any set unification algorithm.

Given a unification problem between two set terms (for instance $\{\{X, f(g(Y)) \mid R\}, f(X)\} = \{A, \{B, f(C)\} \mid R\}$) it is possible, in principle, to determine the number of most general and independent unifiers for it (i.e. the cardinality of $\mu CSU_T(s,t)$). Nevertheless, it is impossible to test the *minimality* (namely the capability of returning exactly $\mid \mu CSU_T(s,t) \mid$ solutions) for all possible problems $s = t$ with s, t in $\tau(\Sigma \cup \mathcal{V})$. A criterion for selecting some sample problems must be chosen:

- problems with *nested* set terms (i.e. sets containing sets, such as the above example) give rise to confusion in the analysis. It is more important to concentrate the efforts in pointing out the new unification problems between elements of the two sets that must be generated;
- with the signature Σ, *closed* set terms (i.e. of the form $\{t_1 \mid \{t_2 \mid \cdots \{t_n \mid \{ \} \} \cdots \}\}$) and *open* set terms (i.e. of the form $\{t_1 \mid \{t_2 \mid \cdots \{t_n \mid R\} \cdots \}\}$, R variable) can be described. Any unification problem between two set terms can be in one of the following forms:
  - closed with closed (problems (1)–(4));
  - open with closed (or vice versa) (problems (5) and (6));
  - open with open (problems (7) and (8)).

In each of such cases, if the elements of the set terms are distinct variables, the number of possible solutions is maximized (for instance $\{X_1, X_2, X_3\} = \{Y_1, Y_2, Y_3\}$ admits 15 solutions, meanwhile $\{X_1, X_2, X_3\} = \{a_1, a_2, a_3\}$, $\{X_1, X_2, X_3\} = \{Y_1, X_2, X_3\}$ and $\{X_1, X_2, X_3\} = \{Y_1, Y_2, X_3\}$ admit 6, 3, and 6 solutions, respectively).

As particular cases, it is interesting to analize the cleverness of an algorithm in solving the problem in which the two set terms share elements (problems (3) and (4)). Moreover, also a shrewd treatment of the *matching* problem (namely when one of the two sets is ground) is important (problems (1) and (5)).

For the unification between open set terms we consider two cases, corresponding to identical and different 'rest' variables, respectively. The announced sample problems are the following:

$$
\begin{aligned}
&(1) && \{X_1, \ldots, X_m\} = \{a_1, \ldots, a_n\}, \\
&(2) && \{X_1, \ldots, X_m\} = \{Y_1, \ldots, Y_n\}, \\
&(3) && \{X_1, \ldots, X_m, a_1, \ldots, a_k\} = \{Y_1, \ldots, Y_n, a_1, \ldots, a_k\}, \\
&(4) && \{X_1, \ldots, X_m, Z_1, \ldots, Z_k\} = \{Y_1, \ldots, Y_n, Z_1, \ldots, Z_k\}, \\
&(5) && \{X_1, \ldots, X_m \mid Z\} = \{a_1, \ldots, a_n\}, \\
&(6) && \{X_1, \ldots, X_m \mid Z\} = \{Y_1, \ldots, Y_n\}, \\
&(7) && \{X_1, \ldots, X_m \mid Z\} = \{Y_1, \ldots, Y_n \mid Z\}, \\
&(8) && \{X_1, \ldots, X_m \mid W\} = \{Y_1, \ldots, Y_n \mid Z\},
\end{aligned}
$$

where $X_1, \ldots, X_m, Y_1, \ldots, Y_n, Z_1, \ldots, Z_k, W, Z$ are pairwise distinct variables, and $a_1, \ldots, a_{\max\{n,k\}}$ are pairwise distinct constant symbols.

In the rest of the section we will describe the functions which compute the cardinality of μCSU_T for problems (1)-(8). Since such functions are described recursively, tables reporting some values for them are presented in the appendix.

3.1 The Problem (1)

The number of solutions for problem (1) is exactly the number of surjective applications from a set of m elements onto a set of n elements (such number will be denoted as $Surj(m, n)$). A compact description of $Surj(m, n)$ can be given using *Stirling numbers* of the second type (see, for instance [10]): $\left\{\begin{matrix} m \\ k \end{matrix}\right\}$ is the number of ways to partition a set of m elements into k nonempty disjoint subsets. Any surjective function $g : \{X_1, \ldots, X_m\} \longrightarrow \{a_1, \ldots, a_n\}$ can be obtained by mapping (with a bijection) an n-partition of $\{X_1, \ldots, X_m\}$ onto the set $\{a_1, \ldots, a_n\}$. Thus, $Surj(m, n) = n! \left\{\begin{matrix} m \\ n \end{matrix}\right\}.$[3]

3.2 The Problem (2)

We will describe a function $\Phi : \mathbb{N}^2 \longrightarrow \mathbb{N}$ computing the number of most general and independent solutions for the problem (2) (notice that problem (2) admits solution for any $m, n \geq 0$). If $m = 0$ and $n = 0$ then $\Phi(m, n) = 1$; if $m > 0$ and $n = 0$, or $m = 0$ and $n > 0$, then $\Phi(m, n) = 0$.

Assume $m, n > 0$; if $m = 1$ or $n = 1$ then $\Phi(m, n) = 1$; otherwise (i.e. when $m, n > 1$) fix an element in the first set, say X_m. Two cases must be analyzed:

(i) X_m is joined to all elements of a subset T (even empty) of $\{X_1, \ldots, X_{m-1}\}$ and mapped onto one element Y_i, for some $i = 1, \ldots, n$ (n ways). This mapping is then added to any solution for the sub-problem $\{X_1, \ldots, X_{m-1}\} \setminus T = \{Y_1, \ldots, Y_{i-1}, Y_{i+1}, \ldots, Y_n\}$;

(ii) X_m is mapped to all elements of a subset S of $\{Y_1, \ldots, Y_n\}$ s.t. $|S| \geq 2$ (if $|S| = 1$ then it would be one of the cases analyzed in the previous item). This mapping is then added to any solution for the sub-problem $\{X_1, \ldots, X_{m-1}\} = \{Y_1, \ldots, Y_n\} \setminus S$.

For $m, n > 0$ the function Φ is recursively defined as follows:

$$\begin{cases} \Phi(m, 1) = 1 & m \geq 1 \\ \Phi(1, n) = 1 & n > 1 \\ \Phi(m, n) = n \sum_{i=0}^{m-2} \binom{m-1}{i} \Phi(m-1-i, n-1) + & m, n > 1 \\ \qquad \sum_{j=2}^{n-1} \binom{n}{j} \Phi(m-1, n-j). \end{cases}$$

---

[3] Notice that $Surj(m, n) = n^m - \sum_{i=0}^{n-1} \binom{n}{i} \left\{\begin{matrix} m \\ i \end{matrix}\right\} i!$, that is $O(n^m)$, while the simple ACI problem discussed in the introduction has complexity $O(2^{m \cdot n})$.

The concepts of *k-fork*, *h-cone* and *configuration* defined below will be useful for the study of problems (6) and (8). Moreover, such concepts allow an alternative analysis of Φ that we omit due to space limits. The interested reader can refer to [1]. Also, in [14], an interesting approach for computing the minimal number of solutions for problem (2) using Taylor's series, can be found. In particular, it is implicitly proved that $\Phi(m,n) = (\Delta_x^m \Delta_y^n e^{(x(e^y-1)+y(e^x-1)-xy)})_{(0,0)}$, where $\Delta_v^k f(\cdots, v, \cdots)$ means to derive f k times w.r.t. the variable v.

From items (i) and (ii) it follows that any most general unifier θ for problem (2) connects X-variables with Y-variables in one of the following ways:

(a) $X_i\theta = Y_j\theta$ and for all $h = 1, \ldots, m$, $h \neq i$ and for all $k = 1, \ldots, n$, $k \neq j$, $X_h\theta$ and $Y_k\theta$ are all distinct from $X_i\theta$. In such case, we say that $X_i = Y_j$ is a *simple binding* in the solution θ;

(b) there are $j_1, \ldots, j_q$, $q \geq 2$, s.t. $X_i\theta = Y_{j_1}\theta = \cdots = Y_{j_q}\theta$ and for all $h = 1, \ldots, m$, $h \neq i$, and for all $k = 1, \ldots, n$, $k \neq j_1, \ldots, k \neq j_q$, $X_h\theta$ and $Y_k\theta$ are all distinct from $X_i\theta$; in this case we say there is a *q-fork* in the solution θ;

(c) there are $i_1, \ldots, i_p$, $p \geq 2$, s.t. $X_{i_1}\theta = \cdots = X_{i_p}\theta = Y_j\theta$ and for all $h = 1, \ldots, m$, $h \neq i_1, \ldots, h \neq i_p$, and for all $k = 1, \ldots, n$, $k \neq j$, $X_h\theta$ and $Y_k\theta$ are all distinct from $X_{i_1}\theta$; in this case we say there is a *p-cone* in the solution θ.

Define now the concept of *configuration*. Given $n > 0$, a n-tuple $c \equiv [i_1, \ldots, i_n]$ is said to be a *configuration* for the set $\{Y_1, \ldots, Y_n\}$ if the non-negative integers $i_1, \ldots, i_n$ are such that $\sum_{j=1}^n i_j \cdot j = n$. Note that each solution θ for problem (2) has an *associated configuration* $c \equiv [i_1, \ldots, i_n]$, where i_1 is the number of *simple bindings* (item (a)) occurring in θ, i_p $(2 \leq p < n)$ is the number of *p-forks* contained in θ (item (b)), and i_n is 0 except for $m = 1$. For instance, when $m = 3$ and $n = 4$, the solution $\theta = \{Y_1 = X_1, Y_2 = X_1, X_2 = Y_3, X_3 = Y_4\}$ has as associated configuration $c \equiv [2, 1, 0, 0]$.

Let k $(1 \leq k \leq n)$ be $\sum_{j=1}^n i_j$; the configuration c is a witness of any partition of $\{Y_1, \ldots, Y_n\}$ into k nonempty disjoint subsets such that there are exactly i_1 singleton subsets, i_2 doubleton subsets, and so on. C_n is the set of all the configurations for a fixed n. For instance, when $n = 6$, C_n is composed of the following eleven configurations:

i_1	i_2	i_3	i_4	i_5	i_6
6	0	0	0	0	0
4	1	0	0	0	0
3	0	1	0	0	0

i_1	i_2	i_3	i_4	i_5	i_6
2	0	0	1	0	0
1	0	0	0	1	0
0	0	0	0	0	1

i_1	i_2	i_3	i_4	i_5	i_6
2	2	0	0	0	0
0	3	0	0	0	0
0	0	2	0	0	0

i_1	i_2	i_3	i_4	i_5	i_6
1	1	1	0	0	0
0	1	0	1	0	0

where each row represents a possible configuration.

3.3 The Problem (3)

Any solution θ of problem (3) will map every element of $\{X_1, \ldots, X_m, a_1, \ldots, a_k\}$ onto an element of $\{Y_1, \ldots, Y_n, a_1, \ldots, a_k\}$. Of course, for $i = 1, \ldots, k$, a_i is implicitly mapped onto itself.

Assume that $\theta = \{X_{i_1} = a_\ell, \ldots, X_{i_\alpha} = a_\ell, Y_{i_1} = a_\ell, \ldots, Y_{i_\beta} = a_\ell\} \cup \theta'$, where $\alpha, \beta > 1$, $\ell \in \{1, \ldots, k\}$, is a solution for problem (3). Then θ will be an instance of the substitution $\theta'' \cup \theta'$, for any θ'' solution of the problem of type (2) $\{X_{i_1}, \ldots, X_{i_\alpha}\} = \{Y_{i_1}, \ldots, Y_{i_\beta}\}$.

This means, in particular, that we do not have to count solutions in which both $X_i = a_\ell$ and $Y_j = a_\ell$ occur, for any $1 \leq i \leq m$, $1 \leq j \leq n$, and $1 \leq \ell \leq k$.

Any most general solution θ for problem (3) can be obtained as follows:

- choose two disjoint subsets S_0 and S_1 of $\{a_1, \ldots, a_k\}$;
- choose a subset T_0 of $\{X_1, \ldots, X_m\}$ and a subset T_1 of $\{Y_1, \ldots, Y_n\}$;
- θ_i is a solution for the problem of type (1) $T_i = S_i$, for $i = 0, 1$;
- θ_2 is a solution for the problem of type (2) $\{X_1, \ldots, X_m\} \backslash T_0 = \{Y_1, \ldots, Y_n\} \backslash T_1$;
- θ is $\theta_0 \cup \theta_1 \cup \theta_2$.

Hence, the number of most general and independent solutions is:

$$\Psi(m, n, k) = \sum_{i=0}^{k} \binom{k}{i} \sum_{j=0}^{k-i} \binom{k-i}{j} \sum_{a=0}^{m} \binom{m}{a} \sum_{b=0}^{n} \binom{n}{b} \left(Surj(a, i) \cdot Surj(b, j) \cdot \Phi(m-a, n-b) \right).$$

3.4 The Problem (4)

From any solution θ of problem (3) a corresponding solution for problem (4) can be obtained by replacing any occurrence of a_i with Z_i, for $i = 1, \ldots, k$.

Note that for problem (4), solutions containing p-forks, $p \leq k$, coming from the unification problems $T_0 = S_0$ and $T_1 = S_1$ could be considered. However any solution $\theta = \{Z_{i_1} = X_i, \ldots, Z_{i_p} = X_i\} \cup \theta'$, $2 \leq p \leq k$ of the form described above is an instance of any of the solutions $\theta_{i_1} = \{Z_{i_1} = X_i\} \cup \theta', \ldots, \theta_{i_p} = \{Z_{i_p} = X_i\} \cup \theta'$.

Thus problem (4) has <u>exactly</u> the same number of most general and independent solutions as problem (3).

3.5 The Problem (5)

Any solution for the problem (5) can be computed as follows: let S be a nonempty subset of $\{a_1, \ldots, a_n\}$ and let θ_1 be a solution for the problem of type (1) $\{X_1, \ldots, X_m\} = S$; then any substitution θ extending θ_1 with the mapping $Z = (\{a_1, \ldots, a_n\} \backslash S) \cup T$, for any $T \subseteq S$, is a solution of (5). It is easy to see that all the solutions obtained in this way are independent and, furthermore, the collection of them is a complete set of unifiers. The total number of solutions is therefore $\sum_{i=1}^{n} \binom{n}{i} \left(Surj(m, i) \cdot 2^i \right)$.

3.6 The Problem (6)

Let S be a nonempty subset of $\{Y_1, \ldots, Y_n\}$ and let θ_1 be a solution for the problem of type (2) $\{X_1, \ldots, X_m\} = S$; then any substitution θ extending θ_1 with the mapping $Z = (\{Y_1, \ldots, Y_n\} \backslash S) \cup T$ is a solution of (6), for any $T \subseteq S$. However, in this case, solutions are not all pairwise independent, as shown by the following example: consider the problem $\{X_1 \mid Z\} = \{Y_1, Y_2\}$. Let $S = \{Y_1, Y_2\}$, then $\{X_1\} = \{Y_1, Y_2\}$ has the unique solution $Y_1 = X_1, Y_2 = X_1$. Such a solution can be extended with $Z = \{Y_2\}$. Let now $S = \{Y_1\}$, then the problem $\{X_1\} = \{Y_1\}$ has the unique solution $X_1 = Y_1$. Such a solution can be extended with $Z = \{Y_2\}$, a more general solution than the first presented.

A deeper analysis of the solutions for the problem $\{X_1, \ldots, X_m\} = S$ must be performed in order to identify whether a solution is minimal or not.

Given a solution θ of $\{X_1, \ldots, X_m\} = S$, we want to extend it with a substitution for Z. Such substitution should be of the form $Z = (\{Y_1, \ldots, Y_n\} \setminus S) \cup T$, with $T \subseteq S$. Assume an element Y_i of S belongs to T. Two cases are possible:

1. θ does not connect Y_i with any Y_j, for $j = 1, \ldots, n$, $i \neq j$. In other words

   $X_{i_1}\theta = \cdots = X_{i_k}\theta = Y_i\theta$, for some $i_1, \ldots, i_k \in \{1, \ldots, m\}$, $k \geq 1$, and
   $Y_i\theta \neq Y_j\theta$, for $j = 1, \ldots, n$, $i \neq j$.

2. θ connects Y_i with some Y_j's, for $j = 1, \ldots, n$, $i \neq j$. This means (see problem (2)) that there is a $(k+1)$-*fork* in the solution θ:

   $X_h\theta = Y_i\theta = Y_{i_1}\theta = \cdots = Y_{i_k}\theta$,
   for exactly one $h \in \{1, \ldots, m\}$ and for some $i_1, \ldots, i_k \in \{1, \ldots, n\}$.

In the former case we can insert Y_i in T. In fact, if we consider the sub-problem obtained by removing it from S, there are no possibilities to map $X_{i_1}, \ldots, X_{i_k}$ in a way that subsumes such solution.

In the latter case the situation is radically different. Consider the sub-problem obtained by removing Y_i from S, we can get a solution θ' equal to θ except that Y_i is not considered (i.e. $X_h\theta' = Y_{i_1}\theta' = \cdots = Y_{i_k}\theta'$). This means that if we insert Y_i into T, we would generate an instance of a solution already computed. An analogous reasoning can be performed for any of $Y_{i_1}, \ldots, Y_{i_k}$ and for all k-*forks* in θ.

Thus, the number of such solutions can be computed using the concept of *configuration* defined in Sect. 3.2. Given a solution θ for the unification problem $\{X_1, \ldots, X_m\} = S$, for any $S \subseteq \{Y_1, \ldots, Y_n\}$ such that $|S| = j$, consider its associated configuration $c_\theta = [i_1, \ldots, i_j]$. There are $2^{j-forks(\theta)}$ possible values for T, where $forks(\theta) =_{Def} 2 \cdot i_2 + \cdots + j \cdot i_j$. Hence the recursive function we are looking for is:

$$\sum_{j=1}^{n} \binom{n}{j} \sum_{\substack{\theta \text{ is a solution of} \\ \{X_1, \ldots, X_m\} = \{Y_1, \ldots, Y_j\}}} 2^{j-forks(\theta)}.$$

3.7 The Problem (7)

Let T be a subset of $\{X_1, \ldots, X_m\}$ and S be a subset of $\{Y_1, \ldots, Y_n\}$. Problem (7) can be reduced to the family of problems $T = S, Z \supseteq (\{X_1, \ldots, X_m\} \setminus T) \cup (\{Y_1, \ldots, Y_n\} \setminus S)$. If θ is a solution of $T = S$, then $\theta \cup \{Z = (\{X_1, \ldots, X_m\} \setminus T) \cup (\{Y_1, \ldots, Y_n\} \setminus S) \cup N\}$, where N is a new variable, whose intended meaning is 'any set', is a solution for problem (7). Furthermore, they are all pairwise independent. The number of most general and independent solutions is

$$\sum_{i=0}^{m} \binom{m}{i} \sum_{j=0}^{n} \binom{n}{j} \Phi(i,j).$$

3.8 The Problem (8)

Problem (8) can be reduced to the family of problems $S_0 = S_1, Z = (\{X_1, \ldots, X_m\} \setminus S_0) \cup N \cup T_0, W = (\{Y_1, \ldots, Y_n\} \setminus S_1) \cup N \cup T_1$, where:

- $S_0 \subseteq \{X_1, \ldots, X_m\}$ and $S_1 \subseteq \{Y_1, \ldots, Y_n\}$, and
- T_i is a subset of S_i, for $i = 0, 1$, and
- N is a new variable (whose intended meaning is 'any set').

Similarly to problem (6), we need to bound the range of the T_i's in order to avoid the generation of instances of other generated solutions.

As shown in Sect. 3.2, any solution θ for the problem $S_0 = S_1$ (instance of problem (2)) can contain *simple bindings*, *p-forks* ($2 \leq p \leq |S_1|$), and *h-cones* ($2 \leq h \leq |S_0|$). For any *simple binding* $X_i = Y_j \in \theta$, X_i and Y_j can be inserted in T_0 and T_1, respectively, but not simultaneously. For instance, for the problem $\{X|W\} = \{Y|Z\}$, $\theta = \{X = Y, W = \{Y|N\}, Z = \{X|N\}\}$ is an instance of $\theta_1 = \{X = Y, W = \{Y|N\}, Z = N\}$.

Following a identical reasoning to the one performed in the analysis of problem (6), *Y-variables* occurring in any *k-fork* of θ can not be inserted in T_1. Similarly, *X-variables* occurring in any *k-cone* must not belong to T_0.

Given a solution θ of $S_0 = S_1$, we define as $vert_0(\theta)$ the sum of number of *k-forks* for $k = 2, \ldots, |S_1|$, and $vert_1(\theta)$ the sum of the number of *h-forks* for $h = 2, \ldots, |S_0|$. $cones(\theta)$ is defined to be $\sum_{h=0}^{|S_0|} h \cdot (\# \text{ of } h\text{-cones})$, and $forks(\theta)$ is the same function defined in the solution of problem (6).

Of course $|S_0| - cones(\theta) - vert_0(\theta) = |S_1| - forks(\theta) - vert_1(\theta)$; such number (say p) is the number of elements connected with a bijection (*simple bindings*). As it has already been explained, such elements can be inserted in T_0 and T_1 not simultaneously: there are $|\{\langle A, B \rangle : A, B \subseteq \{1, \ldots, p\}, A \cap B = \emptyset\}| = 3^p$ possibilities to extend θ. Thus, we are ready to count all the solutions:

$$\sum_{a=0}^{m} \binom{m}{a} \sum_{b=0}^{n} \binom{n}{b} \sum_{\substack{\theta \text{ is a solution of} \\ \{X_1, \ldots, X_a\} = \{Y_1, \ldots, Y_b\}}} 2^{vert_0(\theta)} 2^{vert_1(\theta)} 3^{a - vert_0(\theta) - cones(\theta)}.$$

4 Solutions Computed by a Naïve Algorithm

The following PROLOG code is the core of the general set unification algorithm presented in [8]. Such algorithm does not terminate for problem (7) (same rest variables). The algorithm presented in [3] extends it covering this case, too. Functional symbols $\{\}$ and $\{\cdot \mid \cdot\}$ are represented by [] and $[\cdot \mid \cdot]$, respectively.

```
naive(A, B) :-                    naive([T | Trest], [S | Srest]) :-        (i)
    (var(A); var(B)), !, A = B.        naive(T, S), naive(Trest, Srest).
naive(A, B) :-                    naive([T | Trest], [S | Srest]) :-        (ii)
    A =.. [F | Alist],                 naive(T, S), naive([T | Trest], Srest).
    B =.. [F | Blist],           naive([T | Trest], [S | Srest]) :-        (iii)
    F ≠ '.', !,                        naive(T, S), naive(Trest, [S | Srest]).
    naive_all(Alist, Blist).     naive([T | Trest], [S | Srest]) :-        (iv)
                                       naive([T | New], Srest),
                                       naive(Trest, [S | New]).
```

Predicate naive_all is recursively defined on lists in the natural way. naive algorithm has a minimal behavior for problem (1) only. This is stated in the following Theorems whose proofs are omitted due to space limits[4].

Theorem 1. *The* PROLOG *execution of the goal* :– naive($[X_1, \ldots, X_m], [a_1, \ldots, a_n]$) *returns exactly* $Surj(m, n)$ *solutions, namely it is minimal w.r.t. problem (1).*

Theorem 2. *The number of solutions computed by the* PROLOG *execution of the goal* :– naive($[X_1, \ldots, X_m], [Y_1, \ldots, Y_n]$) *is computed by the following function*

$$
\begin{cases}
f_2(m, 1) = 1 & m \geq 1 \\
f_2(1, n) = 1 & n > 1 \\
f_2(m, n) = f_2(m - 1, n - 1) + f_2(m, n - 1) + & m, n > 1 \\
\qquad n f_2(m - 1, n) + \sum_{i=0}^{n-2} \binom{n}{i} f_2(m - 1, i + 1).
\end{cases}
$$

The naive program treats problems (2) and (4) exactly in the same way. For instance, when $m = 2$, $n = 2$, and $k = 3$, 95401 solutions are computed instead of the 56 needed. That shows its bad behavior in some cases.

Theorem 3. *The function f_8, computing the number of solutions generated by the* PROLOG *execution of the goal* :– naive($[X_1, \ldots, X_m \mid W], [Y_1, \ldots, Y_n \mid Z]$), *can be defined as follows:*

$$
\begin{cases}
f_8(1, 1) = 4 & \\
f_8(m, 1) = 3 \cdot 2^m - 2 & m > 1 \\
f_8(1, n) = 3 \cdot 2^n - 2 & n > 1 \\
f_8(m, n) = f_8(m - 1, n - 1) + f_8(m, n - 1) + & m, n > 1 \\
\qquad (n + 1) f_8(m - 1, n) + \sum_{i=0}^{n-2} \left(\binom{n}{i} + \binom{n-1}{i} \right) f_8(m - 1, i + 1).
\end{cases}
$$

As already sketched, presented naive program is not sufficient to deal with problem (7) (same rest variables). However, referring to the PROLOG implementation of the complete algorithm presented in [3] (assume it is named naive*), it is easy to prove that

Theorem 4. *The function f_7 returning the number of solutions computed by the* PROLOG *execution of the goal* :– naive*($[X_1, \ldots, X_m \mid Z], [Y_1, \ldots, Y_n \mid Z]$) *is recursively defined as follows:*

$$
\begin{cases}
f_7(0, n) = 1 & n \geq 0 \\
f_7(m, 0) = 1 & m > 0 \\
f_7(m, n) = n(f_7(m - 1, n - 1) + f_7(m, n - 1)) + & m > 0, n > 0 \\
\qquad (n + 1) f_7(m - 1, n).
\end{cases}
$$

---

[4] As in Sect. 3, X_i, Y_j, W, Z denote pairwise distinct variables, meanwhile $a_1, \ldots, a_n$ denote distinct constants.

5 The Algorithm SUA

In order to make the description of the algorithm as clear as possible, some local notation will be defined.

The *set-operations* '$|\cdot|$' (cardinality), '$\subseteq$' (inclusion), '$\subset$' (strict inclusion), '$\cup$', (union) '$\cap$', (intersection) and '$-$' (set difference) will be used on terms denoting sets. The meaning of the set operators is purely syntactical; for instance $|\{X_1, X_2\}| = 2$: we do not need to distinguish the two cases $X_1 = X_2 \wedge |\{X_1, X_2\}| = 1$ and $X_1 \neq X_2 \wedge |\{X_1, X_2\}| = 2$. We also assume a meta-function Can which computes to any term t a canonical representation $Can(t)$ of $t's$ equivalence class w.r.t. $=_T$.

A set of equations $\mathcal{E}$ is said to be in *solved form* if it has the form $\{X_1 = t_1, \ldots, X_n = t_n\}$ where X_i's are distinct variables not occurring in t_j, for all $i, j \in \{1, \ldots, n\}$. A solved form system $\{X_1 = t_1, \ldots, X_n = t_n\}$ can be viewed as the substitution $\{X_1/t_1, \ldots, X_n/t_n\}$. An equation $X = t$ is said to be *solved* in $\mathcal{E}$ if X does occur neither in t nor in $\mathcal{E} \setminus \{X = t\}$.

Actions 1–8 of the unification algorithm SUA are identical to those used in the unification algorithm presented in [3] (for further details the reader can consult that paper). SUA takes as input a system of equations $\mathcal{E}$ between terms and returns either *fail* –$\mathcal{E}$ is not unifiable– or, non-deterministically, a substitution θ. The set of all such θ's constitutes a complete set of T-unifiers. In the algorithm, X, W, W', Z and Z' denote generic variables, $t, t_1, t_2, \ldots, s_1, s_2, \ldots$ denote generic terms, N denotes a new variable introduced by SUA. k and k' will denote non-variable terms whose main functional symbol is distinct from $\{\cdot | \cdot\}$[5]. $Var(\ell)$ represents the set of variables occurring in the term ℓ, while $\theta|_S$ constraints the domain of θ to the variables contained in S. n.d. is a shorthand for non-deterministically. The algorithm temporarily generates equations marked by '$*$'; they are called *active* equations and they are immediately removed by action 0. $\mathcal{E}^*$ denotes the set of active equations.

function SUA($\mathcal{E}$);
If $\mathcal{E}$ is in solved form then return $\mathcal{E}$
elseif $\mathcal{E}^*$ is not empty
then choose n.d. one *active* equation $s \overset{*}{=} t$ in $\mathcal{E}^*$; $\mathcal{E}' := \mathcal{E} - \{s \overset{*}{=} t\}$;
0. i. $s \equiv X$: if X occurs in t then *fail* else return SUA($\mathcal{E}'[X/t] \cup \{X = t\}$);
    ii. $s \equiv f(s_1, \ldots, s_m)$ and $t \equiv f'(t_1, \ldots, t_n)$: if $f \not\equiv f'$ then *fail* else
        (i.e. $f \equiv f'$ and $m = n$): return SUA($\{s_1 = t_1, \ldots, s_n = t_n\} \cup \mathcal{E}'$);
else choose n.d. one equation e (not in solved form) in $\mathcal{E}$; $\mathcal{E}' := \mathcal{E} - \{e\}$; case e of:
1. $X = X$: return SUA($\mathcal{E}'$);
2. $t = X$ and t is not a variable: return SUA($\{X = t\} \cup \mathcal{E}'$);
3. $X = t$, X is a variable occurring in $\mathcal{E}'$ but not in t:
   return SUA($\mathcal{E}'[X/t] \cup \{X = t\}$);
4. $X = \{t_1, \ldots, t_m | X\}$: return SUA($\mathcal{E}' \cup \{X \overset{*}{=} \{t_1, \ldots, t_m | N\}\}$);

---

[5] Such entities are named *kernels* in [3].

5. $X = \{t_1, \ldots, t_m \,|\, t\}$, where t is a variable or $t \equiv f(t_1, \ldots, t_n)$, $f \not\equiv \{\cdot \,|\, \cdot\}$, and X occurs in t_1, or $\ldots$, or in t_m, or t is not a variable and X occurs in t: *fail*;

6. $X = t$, and $t \equiv f(t_1, \ldots, t_n)$, $f \not\equiv \{\cdot \,|\, \cdot\}$ and X is a variable occurring in t: *fail*;

7. $f(s_1, \ldots, s_n) = g(t_1, \ldots t_m)$, where $f, g \in \Sigma$, $f \not\equiv g$: *fail*;

8. $f(s_1, \ldots, s_n) = f(t_1, \ldots t_n)$, $f \in \Sigma$, $f \not\equiv \{\cdot \,|\, \cdot\}$:
    return SUA($\mathcal{E}' \cup \{s_1 = t_1, \ldots, s_n = t_n\}$);

9. $\{t_1, \ldots, t_m \,|\, k\} = \{s_1, \ldots, s_n \,|\, k'\}$:
    return SUA(unify_set($\{t_1, \ldots, t_m\}, \{s_1, \ldots, s_n\}$) $\cup \{k \stackrel{*}{=} k'\} \cup \mathcal{E}'$);

10. $\{s_1, \ldots, s_n \,|\, k\} = \{t_1, \ldots, t_m \,|\, Z\}$:
    return SUA($\{\{t_1, \ldots, t_m \,|\, Z\} = \{s_1, \ldots, s_n \,|\, k\}\} \cup \mathcal{E}'$);

11. $\{t_1, \ldots, t_m \,|\, Z\} = \{s_1, \ldots, s_n \,|\, k\}$, choose n.d. $\{\} \neq S \subseteq Can(\{s_1, \ldots, s_n\})$: return SUA($\{Z \stackrel{*}{=} Can(\{s_1, \ldots, s_n\}) - S \cup Z' \cup k)\} \cup$
    limit_1($\{t_1, \ldots, t_m\}, S, Z', \mathcal{E}'$)$|_{Var([s_1, \ldots, s_n, t_1, \ldots, t_m, Z])}$);

12. $\{t_1, \ldots, t_m \,|\, Z\} = \{s_1, \ldots, s_n \,|\, Z\}$: select n.d. one of the following actions:
    i. choose n.d. $T \subseteq \{t_1, \ldots, t_m\}$ and $S \subseteq \{s_1, \ldots, s_n\}$ s.t. $T \cup S \neq \{t_1, \ldots, t_m, s_1, \ldots, s_n\}$:
        return SUA(unify_set(T, S) $\cup \{Z \stackrel{*}{=} Can((\{t_1, \ldots, t_m\} - T) \cup (\{s_1, \ldots, s_n\} - S) \cup N)\} \cup \mathcal{E}'$);
    ii. return SUA(unify_set($\{t_1, \ldots, t_m\}, \{s_1, \ldots, s_n\}$) $\cup \mathcal{E}'$);

13. $\{t_1, \ldots, t_m \,|\, W\} = \{s_1, \ldots, s_n \,|\, Z\}$ where Z and W are different variables; choose n.d. $T \subseteq Can(\{t_1, \ldots, t_m\})$, $S \subseteq Can(\{s_1, \ldots, s_n\})$:
    return SUA($\{W \stackrel{*}{=} Can(\{s_1, \ldots, s_n\} - S \cup W')$, $Z \stackrel{*}{=} Can(\{t_1, \ldots, t_m\} - T \cup Z')\} \cup$ limit_2($T, S, Z', W', \mathcal{E}'$)$|_{Var([s_1, \ldots, s_n, t_1, \ldots, t_m, Z, W])}$).

Active equations $s \stackrel{*}{=} t$ should be handled before the others in order to guarantee termination (see details in the proof of Theorem 6). They are temporarily introduced by actions 4, 9, 11, 12, and 13.

The function unify_set takes as input two terms S_1 and S_2 representing nonempty sets; it selects non-deterministically which equalities between elements of S_1 and S_2 should accompany the system $\mathcal{E}$ in a recursive call to SUA.

function unify_set(S_1, S_2);
Let $\{t_1, \ldots, t_m\} = Can(S_1); \{s_1, \ldots, s_n\} = Can(S_2)$;
1. If $\{t_1, \ldots, t_m\}$ and $\{s_1, \ldots, s_n\}$ are syntactically equal then return SUA($\mathcal{E}$)
2. elseif $m = 1$ and $n > 1$ then return $\{s_i = t_1 : 1 \leq i \leq n\}$;
3. elseif $m \geq 1$ and $n = 1$ then return $\{t_i = s_1 : 1 \leq i \leq m\}$;
4. else $Common\_part := \{t_1, \ldots, t_m\} \cap \{s_1, \ldots, s_n\}$;
        $Disagr_1 := \{t_1, \ldots, t_m\} - Common\_part$;
        $Disagr_2 := \{s_1, \ldots, s_n\} - Common\_part$;
   (a) if $Common\_part = \{\}$ then fix an $i \in \{1, \ldots, m\}$: select n.d. one of the following actions:

        i.   return $\stackrel{1}{=} (t_i, \{t_1, \ldots, t_m\}, \{s_1, \ldots, s_n\})$;
        ii. return $\stackrel{2}{=} (t_i, \{t_1, \ldots, t_m\}, \{s_1, \ldots, s_n\})$;
        iii. return $\stackrel{3}{=} (t_i, \{t_1, \ldots, t_m\}, \{s_1, \ldots, s_n\})$;

(b) if $Common\_part \neq \{\}$ then choose n.d. $S_0, S_1 \subseteq Common\_part$, $S_0 \cap S_1 = \{\}$; choose n.d. $T_0 \subseteq Disagr_1$ and $T_1 \subseteq Disagr_2$ such that $|T_0| \geq |S_0|$ and $|T_1| \geq |S_1|$: return unify$\_$set($Disagr_1 - T_0, Disagr_2 - T_1$) $\cup$ unify$\_$set2(T_1, S_1) $\cup$ unify$\_$set2(T_0, S_0).

Some comments on unify$\_$set are needed in order to relate it to the unification problems (1), (2), (3) and (4). Action 4.(a) is motivated by problems of type (1) and (2). The basic idea is that when an answer is computed by SUA without using the function $\stackrel{2}{=}$, then such answer may be considered as a surjective function from the leftmost set onto the rightmost one. Since $\stackrel{2}{=}$ can not be applied to problem (1) (otherwise SUA would produce *fail*), then the algorithm computes surjective functions as solutions for such problem. However, for problem (2), $\stackrel{2}{=}$ must be considered in order to capture *k-forks*.

The intended meaning of unify$\_$set2 is to consider non deterministically either $\stackrel{1}{=}$ or $\stackrel{3}{=}$, but never $\stackrel{2}{=}$, since, as we commented in Sect. 3.2, to consider *k-forks* produces redundant solutions for problem (4)[6]. The formal definition of unify$\_$set2 is the following:

function unify$\_$set2($\{t_1, \ldots, t_m\}, \{s_1, \ldots, s_n\}$);
1. if $m = 1$ and $n > 1$ then return $\{s_i = t_1 : i = 1, \ldots, n\}$;
2. if $m \geq 1$ and $n = 1$ then return $\{t_i = s_1 : i = 1, \ldots, m\}$;
3. if $m, n > 1$ then fix a value $i \in \{1, \ldots, m\}$; select n.d. one of the following actions:
    i. return $\stackrel{1}{=} (t_i, Can(\{t_1, \ldots, t_m\}), Can(\{s_1, \ldots, s_n\}))$;
    ii. return $\stackrel{3}{=} (t_i, Can(\{t_1, \ldots, t_m\}), Can(\{s_1, \ldots, s_n\}))$.

$\stackrel{1}{=}$ generates *simple bindings* by matching one element t_i of the first set with one element s_j of the second one. Afterwards it combines the two sets deprived of the selected elements. $\stackrel{1}{=}$ has the following structure:

function $\stackrel{1}{=}(t_i, \{t_1, \ldots, t_m\}, \{s_1, \ldots, s_n\})$;
choose n.d. one $j \in \{1, \ldots, n\}$:
return $\{t_i = s_j\} \cup$ unify$\_$set($\{t_1, \ldots, t_{i-1}, t_{i+1}, \ldots, t_m\}, \{s_1, \ldots, s_{j-1}, s_{j+1}, \ldots, s_n\}$).

$\stackrel{2}{=}$ captures the concept of *k-fork* and is defined as follows:

function $\stackrel{2}{=}(t_i, \{t_1, \ldots, t_m\}, \{s_1, \ldots, s_n\})$;
choose n.d. $S \subset \{s_1, \ldots, s_n\}$ such that $|S| \geq 2$:
return $\{s = t_i : $ for all $s \in S\} \cup$ unify$\_$set($\{t_1, \ldots, t_{i-1}, t_{i+1}, \ldots, t_m\}, \{s_1, \ldots, s_n\} - S\}$).

$\stackrel{3}{=}$ captures the concept of *k-cone* and its definition is:
function $\stackrel{3}{=}(t_i, \{t_1, \ldots, t_m\}, \{s_1, \ldots, s_n\})$;
choose n.d. $\{\} \neq T \subset \{t_1, \ldots, t_{i-1}, t_{i+1}, \ldots, t_m\}$:
return $\{t = t_i : $ for all $t \in T\} \cup \stackrel{1}{=}(t_i, \{t_1, \ldots, t_m\} - T, \{s_1, \ldots, s_n\})$.

A problem of type (6) can only be solved by using action 11 of SUA. The function limit$_$1 must simultaneously solve the unification problem $\{X_1, \ldots, X_m\} = S$, for some $S \subseteq \{Y_1, \ldots, Y_n\}$ (for such reason, the definitions of $\stackrel{1}{=}, \stackrel{2}{=}$, and $\stackrel{3}{=}$ are embedded in its definition) and bind Z to any subset of S not containing

---

[6] For problem (3), the use of function $\stackrel{2}{=}$ is not possible.

variables occurring in any *h-fork* corresponding to the solution being computed. The definition of limit_1 is the following:

function limit_1$(S_1, S_2, Z, \mathcal{E})$;

Let $\{t_1, \ldots, t_m\} = Can(S_1)$; $\{s_1, \ldots, s_n\} = Can(S_2)$;

1. If $\{t_1, \ldots, t_m\}$ and $\{s_1, \ldots, s_m\}$ are syntactically equal
   then return SUA$(\{Z = \{\ \}\} \cup \mathcal{E})$;

2. elseif $m = 1$ and $n > 1$ then return SUA$(\{s_i = t_1 : 1 \leq i \leq n\} \cup \{Z = \{\ \}\} \cup \mathcal{E})$

3. elseif $m = 1$ and $n = 1$ or $m > 1$ and $n = 1$ then choose n.d. $T \subseteq \{s_1\}$ and
   return SUA$(\{t_i = s_1 : 1 \leq i \leq m\} \cup \{Z = T\} \cup \mathcal{E})$

4. else fix $i \in \{1, \ldots, m\}$ and choose n.d. one of the following actions:

   i. choose n.d. $j \in \{1, \ldots, n\}$ and $T \subseteq \{s_j\}$:
   return SUA$(\{t_i = s_j\} \cup \{Z = T \cup Z'\} \cup$
   limit_1$(\{t_1, \ldots, t_{i-1}, t_{i+1}, \ldots, t_m\}, \{s_1, \ldots, s_{j-1}, s_{j+1}, \ldots, s_n\}, Z', \mathcal{E}))$;

   ii. choose n.d. $S \subset \{s_1, \ldots, s_n\}$ s.t. $|S| \geq 2$: return SUA$(\{s = t_i : \text{for all } s \in S\}$
   $\cup$ limit_1$(\{t_1, \ldots, t_{i-1}, t_{i+1}, \ldots, t_m\}, \{s_1, \ldots, s_n\} - S, Z, \mathcal{E}))$;

   iii. choose n.d. $T' \subset \{t_1, \ldots, t_{i-1}, t_{i+1}, \ldots, t_m\}$, $j \in \{1, \ldots, n\}$ and $T \subseteq \{s_j\}$:
   return SUA$(\{t = s_j : \text{for all } t \in T'\} \cup \{t_i = s_j\} \cup \{Z = T \cup Z'\} \cup$
   limit_1$(\{t_1, \ldots, t_{i-1}, t_{i+1}, \ldots, t_m\} - T', \{s_1, \ldots, s_{j-1}, s_{j+1}, \ldots, s_n\}, Z', \mathcal{E}))$.

The definition of limit_2 is similar to the one of limit_1. In particular, for problems of type (8), the values of W and Z are constrained in order to avoid the introduction of variables occurring in some *h-fork* or *h-cone* respectively, of the solution θ being computed. On the other hand, limit_2 must also control that those variables bounded in θ by a *simple binding* are not introduced in Z and W simultaneously.

function limit_2$(S_1, S_2, Z, W, \mathcal{E})$;

Let $\{t_1, \ldots, t_m\} = Can(S_1)$; $\{s_1, \ldots, s_n\} = Can(S_2)$;

1. If $\{t_1, \ldots, t_m\}$ and $\{s_1, \ldots, s_n\}$ are syntactically equal then
   return SUA$(\{Z = \{\ \}\} \cup \{W = \{\ \}\} \cup \mathcal{E})$;

2. elseif $m = 1$ and $n > 1$ then choose n.d. $T \subseteq \{t_1\}$:
   return SUA$(\{s_i = t_1 : 1 \leq i \leq n\} \cup \{Z = T\} \cup \{W = \{\ \}\} \cup \mathcal{E})$;

3. elseif $n = 1$ and $m > 1$ then choose n.d. $S \subseteq \{s_1\}$:
   return SUA$(\{t_i = s_1 : 1 \leq i \leq m\} \cup \{Z = \{\ \}\} \cup \{W = S\} \cup \mathcal{E})$;

4. elseif $m = 1$ and $n = 1$ then choose n.d. $T \subseteq \{t_1\}$, $S \subseteq \{s_1\}$ s.t. $T \cup S \neq \{t_1, s_1\}$:
   return SUA$(\{t_1 = s_1\} \cup \{Z = T\} \cup \{W = S\} \cup \mathcal{E})$;

5. else fix $i \in \{1, \ldots, m\}$ and choose n.d. one of the following actions:

   i. choose n.d. $j \in \{1, \ldots, n\}$ and $S \subseteq \{s_j\}$, $T \subseteq \{t_j\}$ such that $T \cup S \neq \{s_j, t_i\}$:
   return SUA$(\{t_i = s_j\} \cup \{W = S \cup W'\} \cup \{Z = T \cup Z'\} \cup$
   limit_2$(\{t_1, \ldots, t_{i-1}, t_{i+1}, \ldots, t_m\}, \{s_1, \ldots, s_{j-1}, s_{j+1}, \ldots, s_n\}, Z', W'\mathcal{E}))$;

   ii. choose n.d. $S \subset \{s_1, \ldots, s_n\}$ s.t. $|S| \geq 2$, $T \subseteq \{t_i\}$:
   return SUA$(\{s = t_i : \text{for all } s \in S\} \cup \{Z = T \cup Z'\} \cup$
   limit_2$(\{t_1, \ldots, t_{i-1}, t_{i+1}, \ldots, t_m\}, \{s_1, \ldots, s_n\} - S, Z', W, \mathcal{E}))$;

   iii. choose n.d. $T \subset \{t_1, \ldots, t_{i-1}, t_{i+1}, \ldots, t_m\}$, $j \in \{1, \ldots, n\}$ and $S \subseteq \{s_j\}$:
   return SUA$(\{t = s_j : \text{for all } t \in T\} \cup \{t_i = s_j\} \cup \{W = S \cup W'\} \cup$
   limit_2$(\{t_1, \ldots, t_{i-1}, t_{i+1}, \ldots, t_m\} - T, \{s_1, \ldots, s_{j-1}, s_{j+1}, \ldots, s_n\}, Z, W', \mathcal{E}))$.

5.1 Termination, Correctness, and Minimality of SUA

The following definition is helpful for the termination proof. Let $\mathcal{E}$ be a system of equations, and let p be the total number of occurrences of function symbols in it. Then a function $lev : Var(\mathcal{E}) \longrightarrow \mathbb{N}$ extended to non-variable terms as follows

$$lev(f(t_0, \ldots, t_n)) = \max\{1 + lev(t_0), \ldots, 1 + lev(t_n)\} \text{ when } f \in \Sigma, f \not\equiv \{\cdot \mid \cdot\}$$
$$lev(\{t \mid s\}) = \max\{1 + lev(t), lev(s)\}$$

and fulfilling condition *(i)* $lev(\ell), lev(r) \leq p$, for any equation $\ell = r$ in $\mathcal{E}$ always exists. If we require the further condition *(ii)* $lev(\ell) = lev(r)$ for any equation $\ell = r$ in $\mathcal{E}$ then such lev may not exist (e.g. when $\mathcal{E} = \{X = f(X)\}$).

Lemma 5. *Let $\mathcal{E}$ be a satisfiable equation system; then a function lev whose extension to terms fulfills both conditions (i) and (ii) always exists.*
Let $\{X = t\} \cup \mathcal{E}$ be an equation system and let p be the number of occurrences of functional symbols in it. Assume X does not occur in t and assume the function lev fulfills condition (i). Moreover, assume that lev is such that $lev(X) = lev(t)$. Then lev fulfills condition (i) also for the system $\mathcal{E}[X/t] \cup \{X = t\}$.

Algorithm SUA calls the functions unify_set, limit_1, and limit_2. To prove the termination of them is straightforward. In the proof of the following Theorem we will make use also of their semantics.

Theorem 6 (Termination). *For any input system $\mathcal{E}$, SUA always terminates, no matter what non-deterministic sequence of choices is made.*

Proof Assume that there is an infinite sequence of non-deterministic choices s.t. SUA$(\mathcal{E})$ does not terminate. Let $E^{(0)}, E^{(1)}, E^{(2)}, \ldots$ be the values of $\mathcal{E}$ at the $0^{th}, 1^{st}, 2^{nd}, \ldots$ iteration, respectively, and let p be the number of occurrences of functional symbols in $E^{(0)}$. A function $lev : Var(\bigcup_{j \geq 0} E^{(j)}) \longrightarrow \mathbb{N}$ s.t.

- it fulfills condition *(i)* for all equation sets $E^{(j)}$, and
- any time a substitution $[X/t]$ has been applied (actions 0 and 3), then $lev(X) = lev(t)$,

must necessarily exist. If it did not exist, then a failure situation caused by occur check would rise, causing the computation be finite.

Picking such lev, we define a measure of complexity $\mathcal{L}_\mathcal{E}$ for the equation set $\mathcal{E}$: $\mathcal{L}_\mathcal{E} =_{\text{Def}} [\#(2p), \#(2p-1), \#(2p-2), \ldots, \#(1), \#(0)]$ where $\#(j)$ returns the number of equations *not in solved form* $\ell = r$ in $\mathcal{E}$ s.t. $lev(\ell) + lev(r) = j$. The ordering between two lists of this form is the well-founded lexicographical ordering.

It is easy to see that actions 0, 1, 3, and 8 cause $\mathcal{L}_\mathcal{E}$ to strictly decrease. Actions 4, 9, 11, 12, and 13 do not increase $\mathcal{L}_\mathcal{E}$; however they introduce active (i.e. marked by '*') equations which will be eliminated immediately by action 0. Actions 2 and 10 leave $\mathcal{L}_\mathcal{E}$ unchanged; nevertheless they at most can double the number of actions, hence we may forget them. Since the lexicographical ordering on constant-length lists of non-negative integers is a well-ordering, this is sufficient to prove termination. □

The above Theorem proves that SUA search tree does not contain infinite branches: this guarantees that SUA computes a finite number of unifiers.

The following Theorem is a direct consequence of the corresponding one presented in [3] and therefore we omit the proof here.

Theorem 7 (Correctness and Completeness). *The function* SUA *is correct with respect to the equational theory presented in Sect. 2. Moreover, it is correct and complete with respect to the well-founded theory of hybrid sets presented in [3, 5].*

It remains to show that the unification algorithm SUA is minimal w.r.t. all chosen sample problems. We will deal only with set terms ended by the constant symbol { }. This simplifies the analysis of actions 9 and 11 of the algorithm (k and k' are both { }). The very long and technical proof is omitted due to space limits.

Theorem 8 (Minimality). SUA *is minimal w.r.t. all the eight sample problems described in Sect. 3, i.e.,* SUA *enumerates without repetitions a complete set of minimal unifiers, to any given unification problem belonging to one of the eight given kinds.*

6 Conclusions

In this paper we have presented a new set unification algorithm SUA, minimal for a significant set of sample problems, that is computing exactly the complete set of minimal unifiers for them. Such problems have been deeply analized from a combinatorial point of view, and functions computing their minimal number of solutions have been also defined. Due to the recursive nature of such functions, it has been possible to develop tables reporting some values for them. These tables show that searching minimal set unification algorithms is an interesting line of research: since the minimal number of solutions is big by itself, it is very important to avoid repeated solutions and solutions being instances of other ones. Although it has been only proved that SUA is minimal for the eight problems presented in Sect. 3, due to their relevance –they can be used for testing any set unification algorithm– it is reasonable to expect a good behavior of SUA for any instance of the set unification problem. Finally, we deem that SUA can adventageously replace any other known set unification algorithms in implementations of set-based logic programming languages (for instance in the implementation of {log} presented in [4]).

Acknowledgements

We are very grateful to Giorgio Levi for giving us the possibility of working together in Pisa. We would like to thank Eugenio Omodeo, Alberto Policriti, Enrico Pontelli, Mario Rodríguez-Artalejo, Gianfranco Rossi, and Alvaro Ruiz-Andino for their wise advises. A special 'thank you' to Ana Gil-Luezas for her unconditional help.

References

1. ARENAS-SÁNCHEZ, P., AND DOVIER, A. Minimal Set Unification. Technical Report: TR–6/95, Università di Pisa, dip. di Informatica, april 1995.

2. BÜTTNER, W., AND SIMONIS, H. Embedding Boolean Expressions into Logic Programming. *Journal of Symbolic Computation 4* (1987), 191–205.

3. DOVIER, A., OMODEO, E., PONTELLI, E., AND ROSSI, G. Embedding Finite Sets in a Logic Programming Language. In *3rd International Workshop on Extension of Logic Programming* (1993), E. Lamma and P. Mello, Eds., vol. 660 of *Lecture Notes in Artificial Intelligence*, Springer-Verlag, Berlin, pp. 150–167.

4. DOVIER, A., AND PONTELLI., E. A WAM based Implementation of a Logic Language with Sets. In *Proc. Fifth Int'l Symp. on Programming Language Implementation and Logic Programming* (1993), M. Bruynooghe and J. Penjam, Eds., vol. 714 of *Lecture Notes in Computer Science*, Springer-Verlag, Berlin, pp. 275–290.

5. DOVIER, A., AND ROSSI, G. Embedding Extensional Finite Sets in CLP. In *Proc. of International Logic Programming Symposium, ILPS'93* (1993), D. Miller, Ed., The MIT Press, Cambridge, Mass., pp. 540–556.

6. GERVET, C. Conjunto: Constraint Logic Programming with Finite Set Domains. In *Proc. of International Logic Programming Symposium, ILPS'94* (1994), M. Bruynooghe, Ed., The MIT Press, Cambridge, Mass., pp. 339–358.

7. JAFFAR, J., AND MAHER, M. J. Constraint Logic Programming: A Survey. *The Journal of Logic Programming 19-20* (1994), 503–581.

8. JAYARAMAN, B. Implementation of Subset-Equational Programs. *Journal of Logic Programming 12*, 4 (1992), 299–324.

9. KAPUR, D., NARENDRAN, P. NP-completeness of the set unification and matching problems. In *Proc. 8th CADE, 1986*, vol. 230 of *Lecture Notes in Computer Science*, Springer-Verlag, Berlin, pp. 489–495.

10. KNUTH, D. E. *The Art of Computer Programming*, vol. 1–Fundamental Algorithms. Addison-Wesley, 1968.

11. HUET, G. Résolution d'équations dans des langages d'ordre 1,2,...,ω. Thèse d'État, Univ. de Paris, VII, 1976.

12. LEGEARD, B., AND LEGROS, E. Short overview of the CLPS system. In *Proc. Third Int'l Symp. on Programming Language Implementation and Logic Programming* (august 1991), J. Maluszynsky and M. Wirsing, Eds., vol. 528 of *Lecture Notes in Computer Science*, Springer-Verlag, Berlin, pp. 431–433.

13. SIEKMANN, J. H. Unification theory. In *Unification*, C. Kirchner, Ed. Academic Press, 1990.

14. STOLZENBURG, F. An Algorithm for General Set Unification and its Complexity. In *Proc. of Workshop on Logic Programming with Sets, in conjunction with ICLP'93* (1993), E. G. Omodeo and G. Rossi, Eds. Budapest.

Appendix

The following tables report some numerical values for the functions computing the minimal number of m.g.u.'s for the sample problems presented in Sect. 3. The number on the 'x' axis denotes the value for m (the first parameter in our unification problems).

(1)	1	2	3	4	5	6	7
1	1	1	1	1	1	1	1
2		2	6	14	30	62	126
3			6	36	150	540	1806
4				24	240	1560	8400
5					120	1800	16800
6						720	15120
7							5040

(2)	1	2	3	4	5	6	7
1	1	1	1	1	1	1	1
2		2	6	14	30	62	126
3			15	48	165	558	1827
4				184	680	2664	11032
5					2945	13080	59605
6						63756	320292
7							1748803

Problem (3) (numerically equal to problem (4)) would require a three dimensional matrix to represent its values. Assume $k = 3$:

(3)	1	2	3	4	5	6	7
1	7	19	61	223	877	3559	14581
2		56	195	746	3093	13808	65391
3			705	2859	12681	60231	302829
4				12226	56891	284286	1510483
5					277091	1448325	8044117
6						7888698	45590823
7							273498973

(5)	1	2	3	4	5
1	2	2	2	2	2
2	4	12	28	60	124
3	6	30	126	462	1566
4	8	56	344	1880	9368
5	10	90	730	5370	36250
6	12	132	1332	12372	106452
7	14	182	2198	24710	259574

(5)	6	7
1	2	2
2	252	508
3	5070	15966
4	43736	195224
5	228090	1359130
6	856212	6505812
7	2562182	23928758

(6)	1	2	3	4	5	6	7
1	2	2	2	2	2	2	2
2	5	12	28	60	124	252	508
3	10	42	144	486	1596	5106	16008
4	19	126	584	2584	11208	48248	205864
5	36	360	2200	11930	63000	330450	1733000
6	69	1016	8118	52740	325812	1983084	12073836
7	134	2870	29876	231518	1641444	11310530	77511140

(7)	1	2	3	4	5	6	7
1	2	4	8	16	32	64	128
2		11	30	85	248	735	2194
3			94	308	1104	4210	16538
4				1041	3920	16981	80260
5					14006	59412	303428
6						221971	1048054
7							4063382

(8)	1	2	3	4	5	6	7
1	4	9	18	35	68	133	262
2		39	131	413	1185	3459	10071
3			652	2811	11402	44983	175224
4				15937	82499	409897	1997795
5					524056	3133773	18217350
6						21998671	148144723
7							1136372140

Solutions Computed by a Naïve Algorithm (see Sect. 4)

(2)	1	2	3	4	5	6	7
1	1	1	1	1	1	1	1
2		5	13	29	61	125	253
3			73	301	1081	3613	11953
4				2069	11581	57749	268381
5					95401	673261	4306681
6						6487445	55213453
7							610093513

(8)	1	2	3	4	5	6	7
1	4	10	22	46	94	190	382
2		52	208	736	2440	7792	24328
3			1372	7516	37012	170668	754132
4				60316	418996	2653036	15780916
5					3964684	33340420	258420172
6						363503932	3587040388
7							44280657292

(7)	1	2	3	4	5	6	7
1	4	10	22	46	94	190	382
2	13	67	265	931	3073	9787	30505
3	46	424	2692	14356	69436	316324	1386172
4	193	2845	26689	201637	1343353	8259805	48109009
5	976	21046	273946	2785306	24436786	194636506	1449663106
6	5869	173215	2982457	39232711	437961529	4380170455	40526990857
7	41098	1582372	34748680	573495616	3913855304	65037766320	834652259744

Goal-Directed Query Processing in Disjunctive Logic Databases

Yuka Shimajiri, Hirohisa Seki and Hidenori Itoh

Department of Intelligence and Computer Science
Nagoya Institute of Technology
Gokiso, Showa-ku, Nagoya 466, Japan
email: {yuka@juno.ics, seki@ics, itoh@ics}.nitech.ac.jp

Abstract. We propose a goal-directed query processing method for disjunctive logic databases. The notion of "relevancy" proposed in a theorem prover called SATCHMORE in [LRW95] is useful for the satisfiability checking of a given set of clauses, but it is not sufficient for the query processing in disjunctive databases, where it is essential in some applications for the answer to a given query to be distinguished between "true", "false", or "unknown (possibly true)".

We first give a goal-directed query processing procedure based on the relevancy, but suitably modified for the above purpose, and show that our procedure is sound and complete for a class of positive disjunctive databases in the minimal model semantics. We then show that a modified version of our procedure effectively works as a correct query processing method for the possible model semantics [SI93], a recently proposed alternative semantics for disjunctive databases. We also show some experimental results of the procedure.

1 Introduction

Disjunctive logic programs have recently attracted much attention (e.g., [LMR92]). Its expressiveness is recognized as a useful basis for knowledge representation (e.g., [BG94]), and is expected to reveal new interesting applications such as theorem proving (including, among others, the new discovery of mathematical results [FSB93]) and abduction (e.g., [SI94]). While a large body of research has been recently acquired on the semantic issues on disjunctive logic programs, it seems that much work is still be needed to obtain efficient query processing procedures for answering queries in disjunctive logic programs, though there have been already several significant procedural semantics proposed in the literature.

In the case of definite programs, many efficient query processing strategies both topdown and bottom-up are well known (e.g., [BR86]). Their efficiency mainly comes from both avoiding redundancy (i.e., avoiding duplicate proofs of the repeated subgoals) and *goal-directedness* (i.e., focusing on only related subgoals necessary to solve a given goal). Unlike definite programs, disjunctions in the head of a rule, however, make the query processing much complicated. For example, a bottom-up computation like SATCHMO [MB87] and MGTP (model generation theorem prover) [FH91], though its computation mechanism is quite simple, loses the goal-directedness in query processing.

In this paper, we propose a goal-directed query processing method for disjunctive logic databases. To do that, we utilize the notion of "relevancy" which was proposed in a theorem prover called SATCHMORE [LRW95]. Though useful for the satisfiability checking of a given set of clauses, the notion of relevancy is not sufficient as it is for the query processing in disjunctive databases, where it is essential in some applications for the answer to a given query to be distinguished between "true", "false", or "unknown (possibly true)". We first give a goal-directed query processing procedure based on the relevancy, but suitably modified for the above purpose, and show that our procedure is sound and complete for a class of positive disjunctive databases in the minimal model semantics. We then show that a modified version of our procedure effectively works as a correct query processing method for the possible model semantics [SI93], a recently proposed alternative semantics for disjunctive databases.

The rest of the paper is organized as follows. In Section 2, we give a description of our framework of query processing, and some definitions on relevancy. In Section 3, we explain our query processing procedure for the minimal model semantics, and its correctness is shown. In Section 4, we show that a modified procedure also works for the possible model semantics. Section 5 shows some experimental results of our proposed procedure.

2 Query Processing for Disjunctive Databases

A *positive* disjunctive logic program is a finite set of clauses of the form:

$$A_1 \vee \ldots \vee A_l \leftarrow B_1 \wedge \ldots \wedge B_m \quad (l \geq 0, m \geq 0) \tag{1}$$

where A_i's and B_j's are atoms.

In this paper we assume that a program is function-free and *range-restricted*[1], which is usually imposed on a program in the field of deductive databases.

The meaning of a disjunctive logic program P is captured by the minimal model semantics [Min82]. That is, let $\mathcal{MM}_P$ be the set of all minimal models of P, and suppose that a ground query $\mathcal{Q}$ is given. Then, the answer to $\mathcal{Q}$ is defined to be "true", "false", or "unknown (possibly true)" as follows.

$$\begin{aligned} &\mathcal{Q} \text{ is true} &&\text{iff } \forall M_i \in \mathcal{MM}_P \ M_i \models \mathcal{Q} \\ &\mathcal{Q} \text{ is false} &&\text{iff } \forall M_i \in \mathcal{MM}_P \ M_i \not\models \mathcal{Q} \\ &\mathcal{Q} \text{ is unknown iff } \mathcal{Q} \text{ is neither true nor false.} \end{aligned}$$

In the rest of the paper, we assume that a query $\mathcal{Q}$ is ground without loss of generality.[2]

---

[1] A clause is range-restricted iff every variable in the head of the clause also occurs in the body.

[2] If $\mathcal{Q}(\mathbf{x})$ contains a tuple of free variables, then we can introduce a new atom O (with arity 0) and add a clause $O \leftarrow \mathcal{Q}(\mathbf{x})$ to P. Then, we can consider O as a new query.

2.1 Hybrid Query Processing for Disjunctive Logic Programs

There have been a lot of proposals on the proof procedures for disjunctive databases ([GM86], [YH85], [HP86], [RT88], [MR90], and [LMR92] for bibliography).

Among others, SLI resolution is, for example, one of topdown linear resolution, thus goal-directed. It is, however, not so simple as SLD-resolution, for it has somewhat complicated operations such as ancestry resolution and factoring.

On the other hand, Inoue et al. [IKH92] proposed a query processing method for disjunctive databases (wrt the stable model semantics). Their procedure is quite simple, which is based on the bottom-up model generation theorem prover MGTP. Namely, MGTP is to try to construct models of a given set P of clauses, starting with an empty set as the initial model candidate set. The set S of model candidates are successively extended, if possible, in the following way: for any $I \in S$ and any ground clauses C from P of the form: $C = A_1 \vee \cdots \vee A_m \leftarrow B_1 \wedge \cdots \wedge B_n$ $(m \geq 1)$, if $I \not\models A_1 \vee \cdots \vee A_m$ and $I \models B_1 \wedge \cdots \wedge B_n$, then remove I from S and add $I \cup \{A_j\}$ to S for every $j = 1, \ldots, m$, provided that $I \cup \{A_j\} \not\models \Gamma$ for any ground negative clause $\leftarrow \Gamma$ from P. Otherwise, (i.e., when $I \cup \{A_j\}$ does not satisfy P), such $I \cup \{A_j\}$ is removed from S. This process proceeds until none of the sets in S can be extended. Note that a model candidate will be ramified into at most m new model candidates, when a disjunctive clause C has m disjunct in its head.

Though their bottom-up proof procedure is very simple, it is not goal-directed. The potential disadvantage of losing goal-directedness will be obvious from the following example.

Example 1. Suppose that query q is given to the following program.

$$q \leftarrow r \qquad \qquad (c1)$$
$$a \vee b \leftarrow \qquad \qquad (c2)$$
$$p \vee c \vee d \leftarrow \qquad \qquad (c3)$$
$$p \vee r \leftarrow \qquad \qquad (c4)$$
$$\leftarrow p \qquad \qquad (c5)$$

Obviously, neither (c2) nor (c3) contributes to prove that the given goal q is true. However, the naive bottom-up computation would use those *irrelevant* clauses, which increases the number of model candidates generated. □

Recently, Loveland et al. [LRW95] proposed a theorem prover called SATCHMORE [LRW95] (SATCHMO with RElevancy), which tries to control bottom-up computation in the presence of disjunctions. The procedure checks the satisfiability of a given set of clauses, using a hybrid computation (i.e., combing both topdown and bottom-up). Namely, the set of clauses P is divided into its Horn clause subset P_{01} and the non-Horn subset $P_{>1}$, and it utilizes topdown SLD-resolution (Prolog) over P_{01} and bottom-up forward chaining on $P_{>1}$.

The control of the bottom-up computation on $P_{>1}$ is done using the notion of *relevancy* (defined below), which prevents a potential explosion of the search space when irrelevant non-Horn clauses are applied. Though the use of relevancy is sufficient for the satisfiability checking, it is not so for answering queries in disjunctive databases, where we would like to distinguish a given query between true, false, or unknown (possibly true).

Example 2. (Continued from Example 1) Suppose that query c is given. Though SATCHMORE can show that the set of clauses $\{(c1), \cdots, (c5)\} \cup \{\leftarrow c\}$ is satisfiable, it cannot tell whether a given goal is false or unknown (possibly true). □

In this paper, we address how the suitably modified notions of relevancy are effectively usedfor query answering in disjunctive databases.

2.2 Relevancy

In the following, we give some definitions related with *relevancy* [LRW95].

Definition 1 I-failure atom. Let P_{01} be a set of Horn clauses and I be a set of ground atoms. An atom B is an *I-failure* atom iff B is not a logical consequence of $P_{01} \cup I$, i.e., $P_{01} \cup I \not\models B$. □

Let P be a set of Horn clauses and $\leftarrow G$ be a goal. We denote by $\mathcal{FB}_P(\leftarrow G)$ the set of all the selected atoms in the failure branches, if any, of the SLD trees for $P \cup \{\leftarrow G\}$, assuming that some computation rule is given.

Definition 2 total relevancy. Let P_{01} ($P_{>1}$) be a set of Horn (non-Horn) clauses, respectively, I be a (possibly empty) set of ground atoms and P_n be a set of negative clauses. Then, an *I-failure* atom B is said to be *totally relevant* wrt $(P_n, P_{01} \cup I)$ if one of the following conditions holds:

(i) $B \in \mathcal{FB}_{P_{01} \cup I}(\leftarrow \Gamma)$, where $\leftarrow \Gamma$ is a negative clause in P_n, or

(ii) $B \in \mathcal{FB}_{P_{01} \cup I}(\leftarrow B_1 \wedge \cdots \wedge B_n)$, where there exists a non-Horn clause $C \in P_{>1}$ of the form:

$$A_1 \vee \ldots \vee A_l \leftarrow B_1 \wedge \ldots \wedge B_m \quad (l \geq 2,\ m \geq 0),$$

and *each* A_i $(l \geq i \geq 1)$ has common instances with a *totally relevant* atom wrt $(P_n, P_{01} \cup I)$. □

Definition 3 partial relevancy. Let $P_{01}, P_{>1}, I$ and P_n be the same as in Definition 2. Then, an *I-failure* atom B is said to be *partially relevant* wrt $(P_n, P_{01} \cup I)$ if one of the following conditions holds:

(i) $B \in \mathcal{FB}_{P_{01} \cup I}(\leftarrow \Gamma)$, where $\leftarrow \Gamma$ is a negative clause in P_n, or

(ii) $B \in \mathcal{FB}_{P_{01} \cup I}(\leftarrow B_1 \wedge \cdots \wedge B_n)$, where there exists a non-Horn clause $C \in P_{>1}$ of the form:

$$A_1 \vee \ldots \vee A_l \leftarrow B_1 \wedge \ldots \wedge B_m \quad (l \geq 2,\ m \geq 0),$$

and *at least one of* A_i's has common instances with a *partially relevant* atom wrt $(P_n, P_{01} \cup I)$. □

Intuitively, an atom is said to be relevant, if it will potentially contribute to prove *false* (contradiction) in $P_{01} \cup I$.

A non-Horn clause is said to be *totally relevant* (*partially relevant*) wrt $(\{\leftarrow Q\}, P_{01} \cup I)$ if each (some) atom of its head is totally relevant (partially relevant), respectively. When the parameter P_{01} is clear from context, we will simply refer to totally (partially) relevant clauses wrt I.

Moreover, a non-Horn clause $C : A_1 \vee \cdots \vee A_m \leftarrow B_1 \wedge \cdots \wedge B_n$ is said to be *violated* [MB87] wrt $P_{01} \cup I$ iff $P_{01} \cup I \vdash (B_1 \wedge \cdots \wedge B_n)\theta$ for some substitution θ and $P_{01} \cup I \not\vdash (A_1 \vee \cdots \vee A_m)\theta$.

We introduce the following notations:

$$TRV(P_n, I) = \{(\Sigma \leftarrow \Gamma)\theta \mid \Sigma \leftarrow \Gamma \text{ is a } \textit{totally} \text{ relevant clause wrt } (P_n, I) \text{ and }$$
$$P_{01} \cup I \vdash \Gamma\theta \text{ and } P_{01} \cup I \not\vdash \Sigma\theta \text{ for some ground substitution } \theta\}$$

$$PRV(P_n, I) = \{(\Sigma \leftarrow \Gamma)\theta \mid \Sigma \leftarrow \Gamma \text{ is a } \textit{partially} \text{ relevant clause wrt } (P_n, I) \text{ and }$$
$$P_{01} \cup I \vdash \Gamma\theta \text{ and } P_{01} \cup I \not\vdash \Sigma\theta \text{ for some ground substitution } \theta\}$$

Example 3. Consider the program in Example 1. Suppose that $I = \phi$. Then, $\mathcal{FB}_{P_{01} \cup I}(\leftarrow q) = \{q, r\}$, and both q and r are partially relevant atoms wrt $(P_{01}, \{\leftarrow q\})$. Therefore, $PRV(\{\leftarrow q\}, I) = \{(c4)\} = \{p \vee r \leftarrow\}$. $\square$

3 Goal-directed Query Processing with Relevancy

In this section, we give a query processing procedure, called TPR, which is shown in Figure 1. The inputs of TPR are a given ground query Q and a program $P = P_{01} \cup P_{>1}$. We assume throughout the paper that P is satisfiable. Basically, TPR consists of two phases: the first phase (from line 4 to line 15), which we call *TR-phase*, determines whether Q is true or not, while the second phase (from line 16 to line 31), which we call *PR-phase*, then determines whether it is unknown or not.

In Figure 1, S_i $(i \geq 0)$ is a set of I_j $(1 \leq j \leq |S_i|)$, where I_j is a set of ground instances of some head literals of non-Horn clauses in $P_{>1}$. I_j is a partial interpretation (model candidate) of P, and at each iteration step in the procedure, I_j is expanded by adding to it a ground instance of some head literals of non-Horn clauses in $P_{>1}$, if I_j so expanded is consistent with P_{01}.

Initially, S_0 is set to $\{\phi\}$ (1. 2).

In TR-phase, while for each I_j in S_i, either $P_{01} \cup I_j \models Q$ or $TRV(P_n \cup \{\leftarrow Q\}, I_j)$ is not empty, the following operations are repeated. If each I_j satisfies $P_{01} \cup I_j \models Q$, it follows from Lemma 12 (given in Appendix) that Q is true in all minimal models of P, thus the answer to the given query is *true*. Otherwise, the procedure generates new partial interpretations by using a totally relevant and violated clause in $P_{>1}$ wrt each I_j (1. 12).

In PR-phase, we know from Lemma 9 that there exists a model in which Q is false (i.e. $P \cup \{\leftarrow Q\}$ is satisfiable). Then, the procedure checks whether or not there exists a model M of P such that $M \models Q$. To do that, while for some I_j in S_i, either $P_{01} \cup I_j \models Q \wedge TRV(P_n, I_j) = \phi$ or $PRV(\{\leftarrow Q\}, I_j)$ is not empty, the following operations are repeated (l. 16). If there exists some $I_j \in S_i$ such that $P_{01} \cup I_j \models Q$ and $TRV(P_n, I_j) \neq \phi$ (l. 17) then the procedure checks whether or not there exists a minimal model M of P with the following function:

$$MIN(I, Q) = \{M \mid M \text{ is a minimal model of } P \text{ which is a subset of}$$
$$\text{a model } I^+ \supseteq I \text{ and includes } Q.\} \tag{2}$$

where I^+ is a model of P such that $I^+ \supseteq I$. If $MIN(I_j, Q) \neq \phi$, then the query is *unknown*. Otherwise, the procedure generates new partial interpretations by using a partially relevant and violated clause in $P_{>1}$ wrt some partial interpretation (l. 28).

If the condition in l. 16 does not hold, then we know from Lemma 11 that the query is distinguished to be *false* (l. 32).

Example 4. Consider the following set P^{ex1} of clauses.

$$P_{01}^{ex1} : \quad p_1(X) \leftarrow r_1(X) \tag{c6}$$
$$q_1(a) \leftarrow \tag{c7}$$
$$s_1(X) \leftarrow r_1(X) \wedge r_2(X) \tag{c8}$$
$$t(X) \leftarrow p_1(X) \tag{c9}$$
$$t(X) \leftarrow r_2(X) \tag{c10}$$
$$\leftarrow p_2(X) \tag{c11}$$
$$P_{>1}^{ex1} : \quad p_1(X) \vee p_2(X) \leftarrow q_1(X) \wedge q_2(X) \tag{c12}$$
$$q_2(a) \vee q_3(b) \leftarrow \tag{c13}$$
$$r_1(X) \vee r_2(X) \leftarrow q_1(X) \tag{c14}$$
$$s_2(a) \vee s_3(b) \vee s_4(c) \leftarrow \tag{c15}$$

(case 1) Suppose that a query $t(a)$ is given. First let $S_0 = \{\phi\}$. Then, $I_1^0 = \phi$, and $TRV(\{(c11), \leftarrow t(a)\}, I_1^0) = \{(c14)\}$. Since $P_{01} \cup I_1^0 \not\models t(a)$, (c14) is chosen in line 11. $P_{01} \cup \{r_1(a)\}$ and $P_{01} \cup \{r_2(a)\}$ are consistent, so $S_1 = \{\{r_1(a)\}, \{r_2(a)\}\}$ is generated. Then, $TPR(t(a), P^{ex1}) =$ true because $P_{01} \cup \{r_1(a)\} \models t(a)$ and $P_{01} \cup \{r_2(a)\} \models t(a)$.

(case 2) Next, suppose that a query $p_1(a)$ is given. No clauses are totally relevant and violated wrt $P_{01} \cup I_1^0$, where $I_1^0 = \phi$ in TR-phase. Then, we move to PR-phase. Since $PRV(\{\leftarrow a(a)\}, I_1^0) = \{(c13)\}$, clause (c13) is chosen in line 27. Then, $S_1 = \{\{q_2(a)\}, \{q_3(b)\}\}$ is generated. There exists no $I_j^1 \in S_1$ s.t. $P_{01} \cup I_j^1 \models Q$, so we then choose clause (c12) from $PRV(\{\leftarrow a(a)\}, \{q_2\}) = \{(c12)\}$, which generates $S_2 = \{\{p_1(a), q_2(a)\}, \{q_3(b)\}\}$. Now, we have that $TPR(p_1(a), P^{ex1}) =$ unknown (possibly true), because $MIN(I_1^2, p_1(a)) = \{\{p_1(a), q_2(a), q_1(a), r_1(a), s_2(a), t(a)\}, \cdots\}$ where $I_1^2 = \{p_1(a), q_2(a)\}$.

(case 3) Thirdly, suppose that a query $s_1(a)$ is given. In TR-phase, no clause is totally relevant and violated wrt $I_1^0 = \phi$. Then, we move to PR-phase, and clause (c14) is chosen from $PRV(\{\leftarrow s_1(a)\}, \{\phi\}) = \{(c14)\}$, generating $S_1 = \{\{r_1(a)\}, \{r_2(a)\}\}$. We know that $TPR(s_1(a), P^{ex1}) =$ false, because $PRV(\{\leftarrow s_1(a)\}, \{r_1(a)\}) = \phi$ and $PRV(\{\leftarrow s_1(a)\}, \{r_2(a)\}) = \phi$.

In this example, each size of the lastly generated sets S_i for the above three cases is 2. If we did not use the notion of relevancy and use only violated clauses for generating partial interpretations, the size of (the worst case of) the lastly generated sets S_i would be 12. This particular example shows that the notion of relevancy is quite effective. The other results of experiments will be given in Section 5. □

In our procedure, the notion of relevancy plays the following two important roles. First, it gives a bias in selecting non-Horn clauses in $P_{>1}$ so that it makes it possible to avoid unnecessary computation caused by choosing unpromising non-Horn clauses which will lead to the derivation of neither Q nor false (contradiction). Secondly, the absence of totally relevant clauses means that Q is no longer true. Moreover, the absence of partially relevant clauses means Q must be false. The use of total relevancy and partial relevancy *in this order* therefore enables us to distinguish the answer to a given query, as early as possible.

We can show the correctness of our TPR procedure as follows. The proof is given in Appendix A.

Theorem 4 Correctness of the procedure TPR. *Suppose that P is a given positive disjunctive program and Q is a ground atom. Then,*

$$
\begin{aligned}
&Q \text{ is true} && \text{iff } TPR(P,Q) = true \\
&Q \text{ is unknown} && \text{iff } TPR(P,Q) = unknown \\
&Q \text{ is false} && \text{iff } TPR(P,Q) = false
\end{aligned}
$$

□

Remark. We note two possible extensions of our TPR procedure.

(i) We assume that a given program is function free, that is, we allow only constants as function symbols. This restriction is only for the sake of guaranteeing the termination of the TPR procedure. If a program contains function symbols, the procedure is still sound, but sometimes not complete. To consider some syntactic condition which ensures the termination in such case is another research issue.

(ii) We assume in the above that a given query is ground. However, when a given query, say $Q(X)$, contains variables, we can compute its answer substitution also in the TPR procedure. That is, we add a new clause $Q \leftarrow Q(X)$ into a given program P, and we perform the TPR procedure with inputs $(Q, P \cup \{Q \leftarrow Q(X)\})$. If Q is either true or unknown, then each set I_j $(1 \leq j \leq k; k > 0)$ in the lastly generated set S_n in the TPR procedure contains an instance of $Q(X)$, say, $Q(t_j) \in I_j$. Then, we have $X = t_1 \vee \cdots \vee t_k$ as a computed answer substitution for query $Q(X)$.

procedure *TPR*

Input : a given query $\mathcal{Q}$, a program $P = P_{01} \cup P_{>1}$ $(P_n \subseteq P_{01})$

Output : an answer *Ans* to $\mathcal{Q}$ (*true*, *false*, or *unknown*)

```
 1   begin
 2       S0 := {φ} ; % Si is used to accumulate partial interpretations of P
 3       i := 0;
     % TR-phase
 4       while ∀Ij ∈ Si (P01 ∪ Ij |= Q or TRV(Pn ∪ {← Q}, Ij) ≠ φ )do
 5           if ∀Ij ∈ Si  P01 ∪ Ij |= Q then
 6               return Ans:=true
 7           else
 8               foreach Ij ∈ Si do
 9                   if TRV(Pn ∪ {← Q}, Ij) ≠ φ then
10                       begin
11                           choose a clause C from TRV(Pn ∪ {← Q}, Ij) ;
                             % C = A1 ∨ ··· ∨ Am ← B1 ∧ ··· ∧ Bn
12                           St := {{Ak} ∪ Ij | 1 ≤ k ≤ m, P01 ∪ Ij ∪ {Ak} : sat.};
13                           Si+1 := Si ∪ St − {Ij} ;
14                           i := i + 1
15                       end ;
     % PR-phase
16       while ∃Ij ∈ Si ( (P01 ∪ Ij |= Q ∧ TRV(Pn, Ij) = φ)
                          or PRV({← Q}, Ij) ≠ φ ) do
17           if ∃Ij ∈ Si (P01 ∪ Ij |= Q ∧ TRV(Pn, Ij) = φ) then
18               begin
19                   choose Ij s.t. P01 ∪ Ij |= Q ∧ TRV(Pn, Ij) = φ ;
20                   if MIN(Ij, Q) ≠ φ then
21                       return Ans:=unknown
22                       else Si := Si − {Ij}
23               end
24           else
25               begin
26                   choose Ij s.t. PRV({← Q}, Ij) ≠ φ ;
27                   choose a clause C from PRV({← Q}, Ij) ;
28                   St := {{Ak} ∪ Ij | 1 ≤ k ≤ m, P01 ∪ Ij ∪ {Ak} : sat.};
29                   Si+1 := Si ∪ St − {Ij} ;
30                   i := i + 1
31               end ;
32       return Ans := false
33   end
```

Fig. 1. Query Processing Procedure for the Minimal Model Semantics

4 Query Processing for the Possible Model Semantics

4.1 The Possible Model Semantics

The possible model semantics [SI93] is a recently proposed alternative semantics for disjunctive logic programs. One of the features of the possible model semantics is that it allows us to interpret disjunctions *inclusively*.

Example 5. Suppose that a query $omnivorous(jelly)$ is given to the following program P^{ex2}.

$$eat(X, plants) \vee eat(X, animals) \leftarrow animal(X) \tag{c16}$$
$$omnivorous(X) \leftarrow eat(X, plants) \wedge eat(X, animals) \tag{c17}$$
$$animal(jelly) \leftarrow \tag{c18}$$

The set $\mathcal{MM}_{P^{ex2}}$ of the minimal models of P^{ex2} is

$$\{ \{animal(jelly), eat(jelly, plants)\},$$
$$\{animal(jelly), eat(jelly, animals)\} \}$$

Then, the answer to the query is *false* according to the minimal model semantics. But if we interpret the head of the clause (c16) *inclusively*, P^{ex2} has also the following (non-minimal) model:

$$\{ animal(jelly), eat(jelly, plants), eat(jelly, animals),$$
$$omnivorous(jelly) \}$$

and $omnivorous(jelly)$ can be considered unknown(possibly true). $\square$

Given a ground disjunctive clause $C : A_1 \vee \cdots \vee A_m \leftarrow B_1 \wedge \cdots \wedge B_n$ and a non-empty subset U of $\{A_1, \ldots, A_m\}$, the split of C wrt U is defined by the set of ground Horn clauses $\{A_i \leftarrow B_1 \wedge \ldots \wedge B_n \mid A_i \in U\}$. Here, C has $2^m - 1$ splits.

Definition 5. Let P be a positive disjunctive program. Then $\mathbf{Horn}(P)$ is the set of all ground Horn programs such that each Horn program P' in $\mathbf{Horn}(P)$ is obtained by

(i) replacing each ground disjunctive clause from P with the clause in one of its splits;
(ii) keeping other (non-disjunctive) ground clauses from P.

$\square$

Definition 6. Let P be a positive program. Then, the set of *possible worlds* $\mathcal{PW}_P$ of P is defined by the set of least Herbrand models of consistent programs in $\mathbf{Horn}(P)$. $\square$

The truth value of a given query for the possible model semantics is defined similarly to the case of the minimal model semantics. Namely, suppose that a ground query Q is given. Then, the answer to Q is defined to be "true", "false", or "unknown (possibly true)" as follows.

$$Q \text{ is } true \quad \text{iff} \quad \forall M_i \in \mathcal{PM}_P \ M_i \models Q$$

$$Q \text{ is } false \quad \text{iff} \quad \forall M_i \in \mathcal{PM}_P \ M_i \not\models Q$$

$$Q \text{ is } unknown \quad \text{iff} \quad Q \text{ is neither true nor false.}$$

4.2 TPR Procedure for the Possible Model Semantics

In TPR for the minimal model semantics, only violated clauses in $P_{>1}$ are utilized to expand partial interpretations. However, since the possible model semantics enables us to interpret disjunctions inclusively, it makes insufficient to perform forward reasoning by using only violated clauses. To consider a query processing procedure for the possible model semantics, we therefore introduce the following definition:

Definition 7. Let $P = P_{01} \cup P_{>1}$ be a disjunctive program, I be a set of atoms, and Q be a ground atom. Then,

$$PRS(\{\leftarrow Q\}, I) = \{C \in P_{>1} \mid C = \Gamma \leftarrow \Delta \text{ is partially relevant wrt}$$

$$(\{\leftarrow Q\}, P_{01} \cup I) \text{ and } P_{01} \cup I \models \Delta\theta \text{ for some substitution } \theta\}$$

$\square$

In general, we have that $PRS(\{\leftarrow Q\}, I) \supseteq PRV(\{\leftarrow Q\}, I)$.

Our query processing procedure TPR_{PM} for the possible model semantics is shown in Figure 2 in Appendix B. It differs from the previous TPR for the minimal model semantics only in that (i) the set $PRS(\{\leftarrow Q\}, I_j)$ is used instead of $PRV(\{\leftarrow Q\}, I_j)$ in PR-phase and (ii) at line23 in Figure 2, a partial interpretation I_j is extended by adding some subset of (possibly instantiated version of) the head of a non-Horn clause from $PRS(\{\leftarrow Q\}, I_j)$, which apparently reflects the inclusive interpretation of disjunctions in the possible model semantics.

We can show the following correctness of our TPR_{PM} procedure. The proof is omitted for the lack of space, however, it is almost similarly done as in Theorem 4.

Theorem 8 Correctness of TPR_{PM}. *Suppose that P is a given positive disjunctive program and Q is a ground atom. Then,*

$$Q \text{ is } true \qquad \text{iff } TPR_{PM}(Q, P) = true$$

$$Q \text{ is } unknown \text{ iff } TPR_{PM}(Q, P) = unknown$$

$$Q \text{ is } false \qquad \text{iff } TPR_{PM}(Q, P) = false$$

$\square$

Example 6. Consider the same program as in Example 4.

(case 1) Suppose that a query $t(a)$ is given. The reasoning process is exactly the same as in $TPR(t(a),P^{ex1})$, so we have that $TPR_{PM}(t(a),P^{ex1})=$ true.

(case 2) Next, suppose that a query s_1 is given. No clause is totally relevant and violated wrt $I_1^0 = \phi$ in TR-phase. Then, in PR-phase, clause (c14) is chosen from $PRS(\{\leftarrow s_1(a)\},\phi) = \{(c14)\}$, which is used to generate $S_1 = \{\{r_1(a)\},\{r_1(a),r_2(a)\},\{r_2(a)\}\}$. Then, we have that $TPR_{PM}(s_1(a),P^{ex1})=$ unknown, because $P_{01} \cup \{r_1(a),r_2(a)\} \models s_1(a)$.

Note that in the case of the minimal model semantics, $TPR(s_1(a),P^{ex1})=$ false in (case3) in Example 4. The procedure $TPR(s_1(a),P^{ex1})$ does not generate partial interpretation $\{r_1(a),r_2(a)\}$, while so does $TPR_{PM}(s_1(a),P^{ex1})$. This is because we can interpret the disjunction in the clause (c14) inclusively. □

5 Some Experimental Results

We have made some experiments of our query processing procedure. Our experimental system is written in SICStus Prolog and all runtimes are actual execution times on SS-5. Table 1 shows some results for the following program.

Horn clauses : P_{01}^{ex3}

$has(a, gym) \leftarrow$
$has(b, gym) \leftarrow$
$has(c, gym) \leftarrow$
$has(d, pool) \leftarrow$
$has(e, pool) \leftarrow$
$wet(X) \leftarrow swims(X)$
$wet(X) \leftarrow rain$
$\leftarrow rain \wedge fine$

non-Horn clauses : $P_{>1}^{ex3}$

$fine \vee cloudy \vee rain \leftarrow$
$trains(X) \vee plays(X, pp) \leftarrow has(X, gym)$
$swims(X) \vee rain \leftarrow has(X, pool)$

In Table 1, we give a comparison between the TPR procedure, a naive bottom-up computation (shown as "without R") and that with only partial relevancy (shown as "PR^2"). The table shows that the TPR procedure is about 30 times as fast as a naive bottom-up computation, and that the numbers of ramifications of model candidates in TPR are much smaller than those in PR^2 and *without R*.

6 Conclusion

This paper proposed a goal-directed query processing method for disjunctive logic databases. We showed that our TPR procedure based on total relevancy and partial relevancy (in this order) is simple, yet effectively applied to query processing for the minimal model semantics. We then showed that slightly modified versions of TPR procedure also work as a query processing procedure for

Query (Ans.)	runtimes[seconds](K)			ratios
	TPR	PR^2	without R	$\frac{TPR}{without\,R}$
wet(e) (true)	0.020 (1)	0.060 (3)	0.570 (47)	0.035
swims(e)(unknown)	0.020 (1)	0.059 (2)	0.790 (54)	0.025
swims(a) (false)	0.030 (1)	0.080 (2)	0.839 (54)	0.035

Table 1. Comparison of the runtimes and the numbers of ramifications K

the possible model semantics. We note that only a few work has addressed the goal-directed query processing for disjunctive logic programs.

We assumed in this paper that a program is function free. As we noted in Remark 3, our TPR procedure is still sound for programs containing function symbols. To find some syntacticcondition which ensures the termination in such case will be interesting especially for practical applications.

In this paper, we have restricted ourselves mainly to positive disjunctive logic programs, but we believe that our TPR procedure will be extended to disjunctive logic program with negation as failure, by incorporating the notion of relevancy into the bottom-up computation procedure, say by Inoue et al. [IKH92].

Moreover, we think that more sophisticated techniques such as program analysis (e.g., [KSI93]) will be utilized in our TPR procedure, which is expected to prune unnecessary search in query processing.

One of the future work is a parallel implementation of our TPR procedure in concurrent logic programming languages like KL1.

Acknowledgements

We would like to thank C. Sakama for valuable discussions on the possible model semantics and the query processing in disjunctive logic databases. The work was partially supported by The Telecommunications Advancement Foundation (TAF).

References

[BG94] Baral, C. and Gelfond, M., Logic Programming and Knowledge Representation, in *J. of Logic Programming* vol. 19/20:73–148 (1994).

[BR86] Bancilhon, F. and Ramakrishnan. R., An Amateur's Introduction to Recursive Query Processing Strategies, *Proc. of ACM SIGMOD Intl. Conf. on Management of Data*, pp.16–52, 1986.

[FSB93] Fujita, M. Slaney, J. and Bennet, F., Automatic Generation of Some Results in Finite Algebra, *Proc. of IJCAI93*, pp.52–57, 1993.

[GM86] Grant, J. and Minker, J., Answering Queries in Indefinite Database and the Null Value Problem, in: *Advance in Computing Theory* 3, *The Theory of Databases*, P.Kanellakis(ed.), JAI Press, Greenwich:247-267, 1986.

[HP86] Hensen, L. and Prak, H., Compiling the GCWA in Indefinite Deductive Database, in: *Foundations of Deductive Database and Logic Programming*, Morgen Kaufmann, 1988.

[IKH92] Inoue, K., Koshimura, M. and Hasegawa, R. Embedding Negation as Failure into a Model Generation Theorem Prover, *Proc. of 11th Int'l Conf. on Automated Deduction*, 1992.

[KSI93] Kato, S., Seki, H. and Itoh, H. An Efficient Abductive Reasoning System Based on Program Analysis, *Proc. of the 3rd Int'l WSA*, LNCS, Vol.724: 230-241, 1993.

[Kow91] Kowalski, R., Logic Programming in Artificial Intelligence, *Proc. of IJCAI'91*: 596-603, 1991.

[LRW95] Loveland, D. W., Reed, D. D. and Willson, D. S., SATCHMORE : SATCHMO with RElevancy, *J. of Automated Reasoning*, 14: 325–351 (1995).

[LMR92] Lobo, J., Minker, J. and Rajaseker, A., *Foundations of Disjunctive Logic Programming*, MIT Press, 1992.

[MB87] Manthey, R. and Bry, F., SATCHMO: a theorem prover implemented in Prolog, *Proc. of CADE'88*, 1987.

[FH91] Fujita, H. and Hasegawa, R., A Model Generation Theorem Prover in KL1 using Ramified Stack Algorithm, *Proc. of ICLP'91*, pp. 535–548, 1991.

[Min82] Minker, J., On Indefinite Data Bases and the Closed World Assumption, *Proc. 6th Int. Conf. on Automated Deduction*, Lecture Notes in Computer Science 138, Springer-Verlag: 292-308, 1982.

[MR90] Minker, J. and Rajasekar, A. A fixpoint semantics for disjunctive logic programs. *J. of Logic Programming*, Vol. 9, No. 1: 45–74 (1990).

[RT88] Ross, K. A. and Topor, R. W., Inferring Negative Infomation from Disjunctive Databases, *J. of Automated Reasoning* 4: 397-424 (1988).

[Raj90] Rajaseker, A., Lobo, J. and Minker, J., Weak Generalized Closed World Assumption, *J. of Automated Reasoning* 5, 293-307. (1990)

[Ram91] Ramsay, A., Generated Relevant Models, *J. of Automated Reasoning* 7: 359-368 (1991).

[SI93] Sakama, C. and Inoue, K., Negation in Disjunctive Logic Programs, *Proc. of ICLP'93*, pp. 703–719 (1993)

[SI94] Sakama, C. and Inoue, K., On the Equivalence between Disjunctive and Abductive Logic Program, *Proc. of ICLP'94*, 1994.

[YH85] Yahya, A. and Henschen, L.J., Deduction in Non-Horn Databases, *J. of Automated Reasoning* 1 (2): 141-160 (1993).

Appendix

A Correctness of TPR Procedure

In this section, we show the correctness of TPR procedure wrt the minimal model semantics (**Theorem 4**).

Let P_n be the set of negative clauses in P_{01}. The following lemma shows that, if a given query Q is true (in the sense of the minimal model semantics) and it is not known to be true from $P_{01} \cup I_i$, where I_i is a currently generated set of atoms, then there always exists a totally relevant and violated clause wrt $(P_n \cup \{\leftarrow Q\}, I_i)$ in $P_{>1}$.

Lemma 9. *Suppose that $P_{01} \not\models Q$ and $P_{01} \cup P_{>1} \models Q$. Then, for any set of atoms I_i such that $P_{01} \cup I_i \not\models Q$, there exists a non-Horn clause $C \in P_{>1}$ such that C is totally relevant and violated wrt $(P_n \cup \{\leftarrow Q\}, I_i)$, i.e., $TRV(P_n \cup \{\leftarrow Q\}, I_i) \neq \phi$, where P_n is the set of negative clauses in P_{01}.*

Proof. From the assumption, $P_{01} \cup P_{>1} \cup \{\leftarrow Q\}$ is unsatisfiable. Then, the above lemma follows immediately from Lemma 13 in [LRW95], where it is shown that if $P_{01} \cup P_{>1} \cup \{\leftarrow Q\}$ is unsatisfiable and $P_{01} \cup I_i$ is satisfiable, then there exists a clause in $P_{>1}$ which is totally relevant and violated wrt $(P_n \cup \{\leftarrow Q\}, I_i)$, i.e., $TRV(P_n \cup \{\leftarrow Q\}, I_i) \neq \phi$. $\qquad\square$

Next, we show the partially relevancy counterpart of Lemma 9. To do that, we need the following technical lemma.

Lemma 10. *Let M be a minimal model of $P_{01} \cup P_{>1}$ such that $M \models Q$. Then, for all S_i $(i \geq 0)$ generated in the TPR procedure, there exists a set of atoms, $I \in S_i$, such that $M \supseteq I$.*

Proof. (i) When $i = 0$, $I = \phi$. The lemma is obvious.

(ii) We assume that the lemma holds when $i \leq k$ $(k \geq 0)$. Then, there exists a set of atoms, $I \in S_k$, such that $M \supseteq I$. We consider the following two cases.

    (a) When S_{k+1} is generated without choosing I, then the lemma holds obviously.

    (b) Otherwise, I is modified by using a non-Horn clause $C = A_1 \vee \cdots \vee A_m \leftarrow B_1 \wedge \cdots \wedge B_n$ in $TRV(P_n \cup \{\leftarrow Q\}, I)$. Since the body of C is true in M, some head atom, say A_i', is also true in M. S_{k+1} contains $I \cup \{A_i'\} \subseteq M$, thus the lemma holds.

$\qquad\qquad\qquad\qquad\qquad\qquad\qquad\qquad\qquad\qquad\qquad\qquad\qquad\square$

Lemma 11. *Suppose that Q is unknown. Suppose further that, S_i $(i \geq 0)$ generated in the PR-phase of TPR procedure satisfies the following condition:*

$$\forall I \in S_i \quad P_{01} \cup I \not\models Q,$$

Then, there exists some $I_0 \in S_i$ such that $PRV(\{\leftarrow Q\}, I_0) \neq \phi$.

Proof. Suppose that for every $I \in S_i$, $PRV(\{\leftarrow Q\}, I) = \phi$. Since Q is not false, there exists a minimal model, say M, of $P_{01} \cup P_{>1}$ such that $M \models Q$. From Lemma 10, it follows that there exists some $I_0 \in S_i$ such that $I_0 \subseteq M$.

Now, consider the following set of definite clauses:

$$P(M) = \{A \leftarrow \Gamma \mid A \vee \Sigma \leftarrow \Gamma \text{ is a ground clause from } P_{01} \cup P_{>1},$$
$$M \models A \wedge M - \{A\} \models \Gamma \wedge M - \{A\} \not\models \Sigma\}$$

Then, it is easy to see that the minimal model of $P(M)$ is equal to M. Since $PRV(\{\leftarrow Q\}, I_0) = \phi$, we can show that if a head atom of any clause in $P(M)$

is partially relevant wrt $(\{\leftarrow Q\}, I_0)$, then it has at least one partially relevant atom wrt $(\{\leftarrow Q\}, I_0)$ in its body. Therefore, we have that:

$$M \cap PRA(\{\leftarrow Q\}, I_0) = \phi$$

where $PRA(\{\leftarrow Q\}, I_0)$ is the set of partially relevant atoms wrt $(\{\leftarrow Q\}, I_0)$. Since $PRA(\{\leftarrow Q\}, I_0) \ni Q$, this means that $M \not\ni Q$, which is a contradiction. □

Lemma 12. *Let $I_j \in S_i$ be the set of atoms generated in the TR-phase of TPR procedure such that it satisfies the following condition:*

$$\forall I_j \in S_i \; P_{01} \cup I_j \models Q \tag{3}$$

Then, Q is true in P.

Proof. This follows directly from the correctness of the SATCHMORE procedure. □

Lemma 13. *Let $I_j \in S_i$ be the set of atoms generated in the PR-phase of TPR procedure such that it satisfies the following condition:*

$$\exists I_j \in S_i \; MIN(I_j, Q) \neq \phi). \tag{4}$$

Then, Q is not false in $P_{01} \cup P_{>1}$.

Proof. It is obvious from the definition of $MIN(I, Q)$. □

Theorem 4 Correctness of the procedure TPR. *Suppose that P is a given positive disjunctive program and Q is a ground atom. Then,*

$$\begin{aligned} & Q \text{ is true} & & \text{iff } TPR(P, Q) = \text{true} \\ & Q \text{ is unknown} & & \text{iff } TPR(P, Q) = \text{unknown} \\ & Q \text{ is false} & & \text{iff } TPR(P, Q) = \text{false} \end{aligned}$$

Proof.
($\Leftarrow$)

(i) Suppose that $TPR(P, Q) = $ true. Then, from Lemma 12 and the definition of TPR procedure, Q is true.
(ii) Suppose that $TPR(P, Q) = $ unknown. Then, from Lemma 9, Lemma 13 and the definition of TPR procedure, Q is unknown.
(iii) Suppose that $TPR(P, Q) = $ false. It means that we exit from both TR-phase and PR phase. Then, from Lemma 9 and Lemma 11, Q is neither true or unknown, thus Q is false.

($\Rightarrow$) Suppose that Q is true. If $TPR(P, Q) \neq$ true, it contradicts the proof of the above if-part, thus $TPR(P, Q)$ should be true. The other cases are proved similarly. □

B TPR_{PM} Procedure

Input : a given query Q, a program $P = P_{01} \cup P_{>1}(P_n \subseteq P_{01})$
Output : An answer Ans

```
1    begin
2        S_0 := {φ} ; % S_i is used to accumulate partial interpretations of P
3        i := 0;
     % TR-phase
4        while ∀I_j ∈ S_i (P_01 ∪ I_j ⊨ Q or TRV(P_n ∪ {← Q}, I_j) ≠ φ)do
5            if ∀I_j ∈ S_i P_01 ∪ I_j ⊨ Q then
6                return Ans:=true
7            else
8                foreach I_j ∈ S_i do
9                    if TRV(P_n ∪ {← Q}, I_j) ≠ φ then
10                       begin
11                           choose a clause C from TRV(P_n ∪ {← Q}, I_j) ;
                                 % C = A_1 ∨ ··· ∨ A_m ← B_1 ∧ ··· ∧ B_n
12                           S_t := {{A_k} ∪ I_j | 1 ≤ k ≤ m, P_01 ∪ I_j ∪ {A_k} : sat.};
13                           S_{i+1} := S_i ∪ S_t − {I_j} ;
14                           i := i + 1
15                       end ;
     % PR-phase
16       while ∃I_j ∈ S_i ( (P_01 ∪ I_j ⊨ Q ∧ TRV(P_n, I_j) = φ)
                              or PRS({← Q}, I_j) ≠ φ ) do
17           begin
18               if ∃I_j ∈ S_i (P_01 ∪ I_j ⊨ Q ∧ TRV(P_n, I_j) = φ) then
19                   return Ans:=unknown
20               repeat
21                   choose I_j s.t. PRS({← Q}, I_j) ≠ φ ;
22                   choose a clause C from PRS({← Q}, I_j) ;
23                   S_t := {A ∪ I_j | A ⊆ {A_1,...,A_m}, P_01 ∪ I_j ∪ A : sat.};
24                   S_{i+1} := S_i ∪ S_t − {I_j} ;
25               until S_{i+1} ≠ S_i or *no other PR clause exists than tried ones.;
26               if S_{i+1} = S_i then exit while
27               else i := i + 1
28           end ;
29       return Ans := false
30   end
```

Fig. 2. Query Processing Procedure for the Possible Model Semantics

A Dataflow Semantics for Constraint Logic Programs

Livio Colussi[1], Elena Marchiori[2], Massimo Marchiori[1]

[1] Dept. of Pure and Applied Mathematics, Via Belzoni 7, 35131 Padova, Italy
{colussi,max}@euler.math.unipd.it
[2] CWI, P.O. Box 94079, 1090 GB Amsterdam, The Netherlands
elena@cwi.nl

Abstract. This paper introduces an alternative operational model for constraint logic programs. First, a transition system is introduced, which is used to define a trace semantics $\mathcal{T}$. Next, an equivalent fixpoint semantics $\mathcal{F}$ is defined: a dataflow graph is assigned to a program, and a consequence operator on tuples of sets of constraints is given whose least fixpoint determines one set of constraints for each node of the dataflow graph. To prove that $\mathcal{F}$ and $\mathcal{T}$ are equivalent, an intermediate semantics $\mathcal{O}$ is used, which propagates a given set of constraints through the paths of the dataflow graph. Possible applications of $\mathcal{F}$ (and $\mathcal{O}$) are discussed: in particular, its incrementality is used to define a parallel execution model for clp's based on asynchronous processors assigned to the nodes of the program graph. Moreover, $\mathcal{O}$ is used to formalize the Intermittent Assertion Method of Burstall [Bur74] for constraint logic programs.

1 Introduction

In this paper a dataflow semantics for constraint logic programs (clp's for short) is introduced. The importance of dataflow semantics is well-known: they specify the 'functionality' of the program; and hence can be used to transform a program into a functional expression, preserving semantics equality. Or to reason about run-time properties of a program depending on the form of the arguments of program atoms before and after their call. From the practical point of view, dataflow semantics support efficient parallel implementations based on networks, where the nondeterminism of programs is exploited.

In this paper we consider for simplicity 'ideal' CLP systems with Prolog selection rule (cf. [JM94]). The extension of the results to more general systems is given in the last section of the paper. A clp $\mathcal{P}$ is a set of clauses together with a goal-clause. First, a transition system is introduced the configurations of which are pairs consisting of an annotated sequence of atoms and a constraint. Then an operational semantics $\mathcal{T}$ is defined, which assigns to a program $\mathcal{P}$ (with goal-clause G) and a set ϕ of constraints, the set of all partial transition traces starting in (G, α), with α in ϕ.

Next, a fixpoint semantics $\mathcal{F}$, equivalent to $\mathcal{T}$, is introduced. Its definition is based, for a program $\mathcal{P}$, on a *dataflow graph* $dg(\mathcal{P})$: this graph has program points as nodes. The arcs of $dg(\mathcal{P})$ are abstractions of the transition rules where configurations are replaced by program points. This graph is used to define the fixpoint semantics $\mathcal{F}$ of $\mathcal{P}$ w.r.t. a set of constraints: a consequence operator on tuples of sets of constraints is given, based on a predicate transformer for constraints, and the least fixpoint of this operator determines one set of constraints for each node of $dg(\mathcal{P})$. We prove that $\mathcal{F}$ and $\mathcal{T}$ are equivalent, by using a top-down semantics $\mathcal{O}$, which propagates a given set of constraints through the paths of $dg(\mathcal{P})$, by means of the above mentioned predicate transformer.

This is the first time that a fixpoint semantics for a clp viewed as set of program points is given. Related work for logic programs, includes e.g. the models of Mellish [Mel87] and Nilsson [Nil90]. However, they both give a fixpoint semantics in which the operational semantics is contained as a proper subset, while here we give an exact description of $\mathcal{T}$.

The fixpoint semantics $\mathcal{F}$ (and $\mathcal{O}$) is shown to have a number of interesting applications. In particular, the incrementality of $\mathcal{F}$ is used to define an or-parallel execution model for clp's based on asynchronous processors assigned to the nodes of the program graph. Moreover, the intermediate semantics $\mathcal{O}$ is used to formalize the Intermittent Assertion Method of Burstall [Bur74] for clp's. This latter application solves at the same time a problem addressed by the Cousots' in [CC93] on how to formalize the Intermittent Assertion Method for clp's.

The rest of the paper is organized as follows. The next section contains the terminology and the concepts used in the sequel. In Section 3 the operational semantics is given. In Section 4 the notion of dataflow graph is introduced, which is used in Section 5 to define the dataflow semantics $\mathcal{F}$. The equivalence of the two semantics is established in Section 6, where the intermediate semantics is introduced. In Section 7 properties of $\mathcal{F}$ are given. In Section 8 some possible applications are investigated. Finally, in Section 9 the results of this paper are discussed.

2 Preliminaries

Let Var be an (enumerable) set of variables, with elements denoted by $x, y, z,$ u, v, w. We shall consider the set $VAR = Var \cup Var^0 \cup \ldots \cup Var^k \cup \ldots$, where $Var^k = \{x^k \mid x \in Var\}$ contains the so-called *indexed variables* (i-variables for short) of *index* k. These special variables will be used to describe the standardization apart process, which distinguishes copies of a clause variable which are produced at different calls of that clause. Thus x^k and x^j will represent the same clause variable at two different calls. This technique is known as 'structure-sharing', because x^k and x^j share the same structure, i.e. x. For an index k and a syntactic object E, E^k denotes the object obtained from E by replacing every variable x with the i-variable x^k. We denote by $Term(VAR)$ (resp. $Term(Var)$) the set of terms built on VAR (resp. Var), with elements denoted by r, s, t.

A sequence $E_1, \ldots, E_k$ of syntactic objects is denoted by $\overline{E}$ or $\langle E_1, \ldots, E_k \rangle$, $(s_1 = t_1 \wedge \ldots \wedge s_k = t_k)$ is abbreviated by $\overline{s} = \overline{t}$, and $\tilde{x}$ represents a sequence of distinct variables.

Constraint Logic Programs

The reader is referred to [JM94] for a detailed introduction to Constraint Logic Programming. Here we present only those concepts and notation that we shall need in the sequel.

A constraint c is a (first-order) formula on $Term(VAR)$ built from primitive constraints. We shall use the symbol $\mathcal{D}$ both for the domain and the set of its elements. We write $\mathcal{D} \models c$ to denote that c is valid in all the models of $\mathcal{D}$.

A *constraint logic program* $\mathcal{P}$, simply called program or clp, is a (finite) set of clauses $H \leftarrow A_1, \ldots, A_k$ (denoted by C, D), together with one goal-clause $\leftarrow B_1, \ldots, B_m$ (denoted by G), where H and the A_i's and B_i's are atoms built on $Term(Var)$ (primitive constraints are considered to be atoms as well) and H is not a constraint. Atoms which are not constraints are also denoted by $p(\overline{s})$, and $pred(p(\overline{s}))$ denotes p; for a clause C, $pred(C)$ denotes the predicate symbol of its head. A clause whose body either is empty or contains only constraints is called *unitary*.

As in the standard operational model states are consistent constraints, i.e. $States \stackrel{\text{def}}{=} \{c \in \mathcal{D} \mid c \text{ consistent}\}$. States are denoted by c or α. We use the two following operators on states:

$$push, pop : States \rightarrow States,$$

where $push(\alpha)$ is obtained from α by increasing the index of all its i-variables by 1, and $pop(\alpha)$ is obtained from α by first replacing every i-variable of index 0 with a new fresh variable, and then by decreasing the index of all the other i-variables by 1. For instance, suppose that α is equal to $(x^1 = f(z^0) \wedge y^0 = g(x^2))$. Then $push(\alpha)$ is equal to $(x^2 = f(z^1) \wedge y^1 = g(x^3))$ and $pop(\alpha)$ to $(x^0 = f(u) \wedge v = g(x^1))$, where u and v are new fresh variables.

3 Operational Semantics

In Table 1 the operational behaviour of a clp by means of a transition system (TS) is given.

In a pair $(\overline{A}, \alpha)$, α is a state, and $\overline{A}$ is a sequence of atoms and possibly of tokens of the form pop, whose use is explained below.

The rules of TS describe the standard operational behaviour of a clp (cf. e.g. [JM94]), but for the fact that we fix a suitable standardization apart mechanism: In the standard operational semantics of (C)LP, every time a clause is called it is renamed apart, generally using indexed variables. Here if a clause is called then *push* is first applied to the state, and if it is released then *pop* is applied to the state. To mark the place at which this should happen the symbol *pop* is used. Rule **R** describes a resolution step. Note that, the way the operators *push* and *pop* are used guarantees that every time an atom is called, its variables can be indexed with index equal to 0. Then, in rule **R** the tuple of terms $push(\overline{s}^0)(= \overline{s}^1)$ is considered, because a *push* is applied to the state. Rule **S** describes the situation where an atom has concluded with success its computation, i.e. when the control

$$\mathbf{R} \quad (\langle p(\bar{s})\rangle \cdot \overline{A}, \alpha) \longrightarrow (\overline{B} \cdot \langle pop\rangle \cdot \overline{A}, push(\alpha) \wedge \bar{s}^1 = \bar{t}^0),$$
$$\text{if } C = p(\bar{t}) \leftarrow \overline{B} \text{ is in } \mathcal{P}$$
$$\text{and } push(\alpha) \wedge \bar{s}^1 = \bar{t}^0 \text{ is consistent}$$

$$\mathbf{S} \quad (\langle pop\rangle \cdot \overline{A}, \alpha) \longrightarrow (\overline{A}, pop(\alpha))$$

$$\mathbf{C} \quad (\langle d\rangle \cdot \overline{A}, \alpha) \longrightarrow (\overline{A}, \alpha \wedge d^0),$$
$$\text{if } d \text{ is a constraint}$$
$$\text{and } \alpha \wedge d^0 \text{ is consistent}$$

Table 1. Transition rules for CLP.

reaches a *pop*. In this case, the operator *pop* is applied to the state. Finally, rule C describes the execution of a constraint.

This formalization will lead to an elegant definition of the dataflow semantics. Note that we do not describe explicitly failure, because it is not relevant for our dataflow model.

To refer unambiguously to clause variables, the following non-restrictive assumption is used.

Assumption 3.1 Different clauses of a program have disjoint sets of variables.

We write $(\overline{A}, \alpha) \to (\overline{B}, \beta)$ to denote a generic transition using the rules of Table 1. We call *computation*, denoted by τ, any sequence $\langle conf_1, \ldots, conf_k, \ldots\rangle$ of configurations s.t. for $k \geq 1$ we have that $conf_k \to conf_{k+1}$. We consider an operational semantics $\mathcal{T}(\mathcal{P}, \phi)$ for a program $\mathcal{P}$ w.r.t. a set ϕ of states, called *precondition*. This semantics describes all the computations starting in (G, α) (recall that G denotes the goal-clause of $\mathcal{P}$) with α in ϕ. It is defined as follows. We use $\cdot$ for the concatenation of sequences.

Definition 3.2 (partial trace semantics) $\mathcal{T}(\mathcal{P}, \phi)$ is the least set T s.t. $\langle (G, \alpha)\rangle$ is in T, for every $\alpha \in \phi$, and if $\tau = \tau' \cdot \langle (\overline{A}, \alpha)\rangle$ is in T and $(\overline{A}, \alpha) \to (\overline{B}, \beta)$, then $\tau \cdot \langle (\overline{B}, \beta)\rangle$ is in T. □

Observe that this is a very concrete semantics: the reason is that it is not meant for the study of program equivalence, but for the study of run-time properties of clp's, and for the definition of models for parallel implementations. These applications are discussed in Section 8.

4 A Dataflow Graph for clp's

To define a dataflow semantics equivalent to $\mathcal{T}(\mathcal{P}, \phi)$, we start by introducing a dataflow graph associated with a clp, whose nodes are the program points, and whose arcs describe in an abstract way the transition rules of Table 1.

In logic programming, program points are (often implicitly) used to describe the operational observables considered. Similar e.g. to [Nil90], we view a program clause $C : H \leftarrow A_1, \ldots, A_k$ as a sequence consisting alternatingly of (labels l of) *program points* (pp's for short) and atoms,

$$H \leftarrow {}_{l_0} A_1\, {}_{l_1} \cdots {}_{l_{k-1}} A_k\, {}_{l_k}.$$

The labels l_0 and l_k indicate the *entry point* and the *exit point* of C, denoted by $entry(C)$ and $exit(C)$, respectively. For $i \in [1, k]$, l_{i-1} and l_i are called the *calling point* and *success point* of A_i, denoted by $call(A_i)$ and $success(A_i)$, respectively. Notice that $l_0 = entry(C) = call(A_1)$ and $l_k = exit(C) = success(A_k)$. In the sequel $atom(l)$ denotes the atom of the program whose calling point is equal to l. Moreover, for notational convenience the following non-restrictive assumptions are used.

Assumption 4.1 $l_0, \ldots, l_k$ are natural numbers ordered progressively; distinct clauses of a program are decorated with different pp's; the pp's form an initial segment, say $\{1, 2, \ldots, n\}$ of the natural numbers; and 1 denotes the leftmost pp of the goal-clause, called the *entry point of the program*. Finally, to refer unambiguously to program atom occurrences, all atoms occurring in a program are supposed to be distinct.

The following $CLP(\mathcal{R})$ ([JMSY92]) program *Prod* is explicitly labelled with its pp's.

```
G:  ← ₁ prod(u,v) ₂
C1: prod([x|y],z) ← ₃ z=x*w ₄ prod(y,w) ₅
C2: prod([ ],1) ← ₆
```

In the sequel, $\mathcal{P}$ denotes a program and $\{1, \ldots, n\}$ the set of its pp's. Program points are used to define the notion of dataflow graph.

Definition 4.2 (dataflow graph) The *dataflow graph* $dg(\mathcal{P})$ of $\mathcal{P}$ is the pair $(Nodes, Arcs)$ s.t. $Nodes = \{1, \ldots, n\}$ and $Arcs$ is the subset of $Nodes \times Nodes$ s.t. (i, j) is in $Arcs$ iff it satisfies one of the following conditions:

- i is $call(A)$ and j is $entry(C)$, where A is not a constraint, and $pred(A)$ and $pred(C)$ are equal;
- i is $exit(C)$ and j is $success(A)$, where $pred(A)$ and $pred(C)$ are equal;
- i is $call(A)$ and j is $success(A)$, where A is a constraint.

An element (i, j) of $Arcs$ is called *(directed) arc from i to j*. □

Arcs of $dg(\mathcal{P})$ are graphical abstractions of the transition rules of Table 1. Rule **R** is abstracted as an arc from the calling point of an atom to the entry point of a clause. Rule **S** is abstracted as an arc from the exit point of a clause to a success point of an atom. Finally, rule **C** is abstracted as an arc from the calling point of a constraint to its success point. Below the dataflow graph $dg(Prod)$ of *Prod* is pictured.

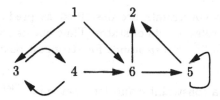

Remark 4.3 Our notion of dataflow graph differs from other graphical representations of (c)lp's, as for instance the *predicate dependency graph* [Kun87] or the *U-graph* [WS94], mainly because of the presence in $dg(\mathcal{P})$ of those arcs from exit points of clauses to success points of atoms, such as the arc from 5 to 2 in $dg(Prod)$. These arcs are crucial to obtain an exact fixpoint description of the operational semantics. For instance, in $dg(Prod)$ there is one arc from 5 to 5 and one from 5 to 2, one from 6 to 2 and one from 6 to 5.

Remark 4.4 One can refine this definition by using also semantic information, i.e. by pruning the arcs stemming from the first two conditions if $\mathcal{D} \models \neg(\overline{s} = \overline{t})$, i.e. if $p(\overline{s})$ and $p(\overline{t})$ do not 'unify', where $p(\overline{s})$ is A and $p(\overline{t})$ is (a variant of) the head of C.

A *path of* $\mathcal{P}$ is a non-empty sequence of pp's forming a (directed) path in $dg(\mathcal{P})$. Paths are denoted by π, and concatenation of paths by $\cdot$. Moreover, $path(i, j)$ denotes the set of all the paths from i to j, and $path(i)$ the set of all the paths from 1 to i.

5 Dataflow Semantics

In this section a dataflow semantics $\mathcal{F}$ for clp's is given, w.r.t. a given 'precondition' ϕ which is associated with the entry point 1 of the program. This semantics determines for every node l of $dg(\mathcal{P})$ a suitable set ϕ_l of states. In Section 6 it will be shown that $\mathcal{F}$ is equivalent to $\mathcal{T}$, i.e. that ϕ_l is the set of the final states of all partial derivations, with initial state in ϕ, ending in l. This semantics describes the run-time behaviour of a clp, i.e. the form of the body atoms of the program (goal-)clauses at the moment when they are called and after their execution. The importance of this information is well-known: it can be used for instance to determine for which class of goals a program terminates and for which class of goals the computation is sufficiently efficient. It will be shown in Section 7 that $\mathcal{F}$ enjoys two relevant properties: it is incremental and and-compositional. Incrementality allows us to compute the semantics of the union of two clp's $\mathcal{P}$ and $\mathcal{P}'$, by computing first the semantics of one of them, say $\mathcal{F}(\mathcal{P})$ of $\mathcal{P}$, and then by using $\mathcal{F}(\mathcal{P})$ to determine the semantics of their union $\mathcal{P} \cup \mathcal{P}'$. Also, from the practical point of view, the incrementality of $\mathcal{F}$ allows us to define parallel execution models of clp's based on asynchronous processors, as explained in

Section 8. And-compositionality allows us to compute the semantics of a goal $\leftarrow \overline{A}, \overline{B}$ from the semantics of $\leftarrow \overline{A}$ and of $\leftarrow \overline{B}$.

To define $\mathcal{F}$, first constraints are described as predicate transformers, by lifting the transition rules to sets of states. Thus one can view a constraint c as a map $sp.c : 2^{States} \rightarrow 2^{States}$ (sp stands for strongest postcondition) defined as follows.

Definition 5.1 For a constraint c and for a set ϕ in 2^{States},

$$sp.c.\phi = \{\alpha \wedge c \in States \mid \alpha \in \phi\}. \qquad \square$$

This definition corresponds to the rule **C** of TS. Observe that it also describes the rule **R**, by taking the constraint c to be equal to $(\overline{s}^1 = \overline{t}^0)$.

Sets of states are denoted by ϕ, ψ, where *false* stands for $\emptyset$, and $\neg \phi$ for $States \setminus \phi$. The set

$$free(x) = \{\alpha \mid \mathcal{D} \models \alpha \rightarrow \forall x.\alpha\}$$

of states will be used in the sequel, describing those states where x is a free variable. The intuition is that x is free in a state if it can be bound to any value without affecting that state. For instance, $y = z$ is in $free(x)$, because x does not occur in the formula. Also $y = z \wedge x = x$ is in $free(x)$, because $\mathcal{D} \models (y = z \wedge x = x) \rightarrow \forall x\,(y = z \wedge x = x)$. The definitions of *pop* and *push* are extended in the natural way to sets of states, where $push(\phi)$ is equal to $\{push(\alpha) \mid \alpha \in \phi\}$. Analogously for $pop(\phi)$. It is convenient to make the following assumptions on non-unitary (goal-)clauses.

Assumption 5.2 The body of every non-unitary clause does not contain two atoms with equal predicate symbol; and at least one argument of its head is a variable.

Notice that every program can be transformed into one satisfying Assumption 5.2. Although the transformation can modify the semantics of the original program (the set of pp's changes and new predicates could be introduced), it is easy to define a syntactic transformation that allows us to recover the semantics of the original program.

These assumptions are used to simplify the definition of the dataflow semantics. Because of the second one, one can fix a variable-argument of the head of a non-unitary clause C, that we call *the characteristic variable of C*, denoted by x_C. Also, a new fresh variable x_G is associated with the goal-clause G, called *the characteristic variable of G*. These variables play a crucial role in the following definitions, to be explained below.

We can introduce now, for a program $\mathcal{P}$ with set $\{1, \ldots, n\}$ of pp's, the immediate consequence operator Ψ on n-tuples of sets of states, defined w.r.t. a given set ϕ of states associated with the entry point of $\mathcal{P}$. For a node j of $dg(\mathcal{P})$, let $input(j)$ denote the set of the nodes i s.t. (i, j) is an arc of $dg(\mathcal{P})$. Because every pp is either an entry point of a clause, or a success point of an atom, it is enough to distinguish these two cases in the following definition of Ψ. In the sequel, Ψ_k denotes the k-th projection of Ψ.

Definition 5.3 For a program $\mathcal{P}$ with set $\{1, \ldots, n\}$ of pp's, and for a given set ϕ of states (the precondition), the operator $\Psi : (2^{States})^n \to (2^{States})^n$ is defined as follows. For $\overline{\psi} = (\psi_1, \ldots, \psi_n)$:

- $\Psi_1(\overline{\psi}) = \phi$;
- for $k \in [2, n]$:
  1. if k is $entry(C)$ then

  $$\Psi_k(\overline{\psi}) = \bigcup_{j \in input(k)} sp.\,(\overline{s}^1 = \overline{t}^0).\,push(\psi_j),$$

  where $p(\overline{t})$ is the head of C and $p(\overline{s})$ is $atom(j)$;
  2. if k is $success(A)$ and A is not a constraint then

  $$\Psi_k(\overline{\psi}) = \bigcup_{j \in input(k)} pop(\psi_j) \cap \neg free(x_C^0),$$

  where C is the clause containing A;
  3. if k is $success(A)$ and A is a constraint then

  $$\Psi_k(\overline{\psi}) = sp.\,A^0.\,\psi_{k-1}.$$

$\square$

Because $sp.\ c.\ \bigcup_i \psi_i = \bigcup_i (sp.\ c.\ \psi_i)$ it follows that Ψ is a continuous operator on the complete lattice $((2^{States})^n, \subseteq)$, where $\subseteq$ denotes componentwise inclusion. Hence by the Knaster-Tarski theorem it has a least fixpoint $\mu(\Psi) = \bigcup_{k=0}^{\omega} \Psi^k(\bot)$, where $\bot$ stands for the least element $(\emptyset, \ldots, \emptyset)$ of $(2^{States})^n$.

Definition 5.4 (dataflow semantics) Let ϕ be s.t. $\phi \subseteq \neg free(x_G^0)$, and $\phi \subseteq free(x_C^0)$ for every non-goal, non-unitary clause C. Then the *semantics* $\mathcal{F}(\mathcal{P}, \phi)$ of $\mathcal{P}$ with respect to ϕ is the least fixpoint $\mu(\Psi)$. $\square$

Let us comment on the above definitions. The operational intuition behind the definition of Ψ can be explained using the transition system of Table 1: let $\overline{A}$ be a generic sequence of atoms and/or *pop* tokens. Then in case 1. $entry(C)$ 'receives' those states obtained by applying rule **R** to $(\langle atom(j) \rangle \cdot \overline{A}, \alpha)$, for every α in ψ_j, and for every j s.t. the arc $(j, entry(C))$ is in the dataflow graph. In case 2. $success(A)$ 'receives' those states obtained by applying the rule **S** to $(\langle pop \rangle \cdot \overline{A}, \alpha)$, for every α in ψ_j, for every j s.t. the arc $(j, success(A))$ is in the dataflow graph. Finally, in case 3. $success(A)$ 'receives' those states obtained by applying the transition rule **C** to $(\langle A \rangle \cdot \overline{A}, \alpha)$, for every α in $\psi_{call(A)}$. In Definition 5.4 the operator Ψ is iterated ω times starting from $\bot$.

The characteristic variables of the program are used in case 2. of Definition 5.3, where the result is intersected with $\neg free(x_C^0)$, and in the two conditions in Definition 5.4. They are of crucial importance for obtaining a dataflow semantics which is equivalent to $\mathcal{T}$. In fact, they are used to rule out all those paths which are not semantic, i.e. which do not describe partial traces.

Informally, whenever a state is propagated through a semantic path the characteristic variable x_C^0 of a non-unitary clause is initially free (by assumption). Then, the index of x_C is increased and decreased by means of the applications of the *push* and *pop* operators. When C is called, then x_C^0 is bound (because by assumption it occurs in the head of C), hence x_C^0 is not free. From that moment on its index will be increased and decreased and it will become 0 *only* if the success point of an atom of the body of C is reached. Concerning the characteristic variable x_G^0 of the goal, it is initially not free (by assumption). Then, its index is increased and decreased by means of the applications of the *push* and *pop* operators and it will become 0 *only* if the success point of an atom of G is reached. In that case, for each other clause C, x_C^0 is free, because either C was never called, or x_C^0 has been replaced with a fresh variable by an application of *pop*. Observe that Assumptions 3.1 and 5.2, and those of Definition 5.4 are needed.

Example 5.5 We illustrate how $\mathcal{F}$ is determined by computing $\mathcal{F}(Prod, \phi)$, where ϕ is the set $\{(u^0 = [\,] \wedge x_G^0 = 1), (u^0 = [r] \wedge x_G^0 = 1)\}$ (with r a variable). We choose x as characteristic variable of $C1$ and the fresh variable x_G as the one of G. For every $k \geq 0$, we have that Ψ_1^k is ϕ. Then in the following steps, Ψ_1^k is not mentioned. Moreover, the other Ψ_i^k's which are omitted are assumed to be equal to $\emptyset$. Finally, the abbreviation $s_1 = s_2 = \ldots = s_m$ stands for $s_1 = s_2 \wedge \ldots \wedge s_{m-1} = s_m$, and the brackets for singleton sets are omitted.

- Ψ_3^1 is $(u^1 = [r] \wedge x_G^1 = 1 \wedge y^0 = [\,] \wedge v^1 = z^0)$;
  Ψ_6^1 is α, where α is $u^1 = [\,] \wedge x_G^1 = 1 \wedge v^1 = 1$.
- Ψ_2^2 is $pop(\alpha)$;
  Ψ_4^2 is $u^1 = [r] = [x^0] \wedge x_G^1 = 1 \wedge y^0 = [\,] \wedge v^1 = z^0 \wedge z^0 = x^0 * w^0$;
  Ψ_i^2 is Ψ_i^1, for $i = 3, 6$. Observe that while $pop(\alpha)$ is added to Ψ_2^2, it is not added to Ψ_5^2 (which remains empty), because x^0 does not occur in $pop(\alpha)$, hence $pop(\alpha)$ intersected with $\neg free(x^0)$ yields the empty set.
- Ψ_i^3 is Ψ_i^2, for $i = 2, 3, 4$;
  Ψ_6^3 is $\{\alpha, \beta\}$, where β is $u^2 = [r] = [x^1] \wedge x_G^2 = 1 \wedge y^1 = [\,] \wedge v^2 = z^1 = x^1$.
- Ψ_i^4 is equal to Ψ_i^3, for $i = 2, 3, 4, 6$;
  Ψ_5^4 is $pop(\beta)$. Observe that here $pop(\beta)$ is added to Ψ_5^4 but not to Ψ_2^4, because x_G^0 does not occur in $pop(\beta)$.
- Ψ_i^5 is Ψ_i^4, for $i = 3, \ldots, 6$;
- Ψ_2^5 is $\{pop(pop(\beta)), pop(\alpha)\}$. Observe that here $pop(pop(\beta))$ is added to Ψ_2^5, but not to Ψ_5^5, because x^0 does not occur in $pop(pop(\beta))$.
- Ψ^6 is Ψ^5.

$\square$

Remark 5.6 In order to illustrate how to compute $\mathcal{F}$, we have assumed to deal with an ideal system. However, in $CLP(\mathcal{R})$ the constraint $z = x * w$ is delayed until it becomes linear (cf. [JMSY92]). In Section 9 we shall discuss how to modify the dataflow semantics to deal with such systems, and to handle this example.

6 Equivalence of $\mathcal{T}$ and $\mathcal{F}$

To prove the equivalence of $\mathcal{T}$ and $\mathcal{F}$, an intermediate semantics $\mathcal{O}$ is introduced, which propagates sets of states through the paths of $dg(\mathcal{P})$ by means of the predicate transformer sp. This semantics is not only useful to prove the above mentioned equivalence. It also allows us to define the Burstall Intermittent Assertion Method for clp's, as will be described in Section 8.

Definition 6.1 Consider a path π in $dg(\mathcal{P})$. The *path strongest postcondition psp. π. ϕ of π w.r.t. ϕ* is inductively defined as follows:

– If π is of the form $\langle l \rangle$ then

$$psp.\ \pi.\ \phi = \phi.$$

– Otherwise, if π is of the form $\pi' \cdot \langle l_k \rangle$, where π' is $\langle l_1, \ldots, l_{k-1} \rangle$ and $k \geq 2$, then:

1. if l_k is $entry(C)$ and l_{k-1} is $call(A)$, where A is an atom, say $p(\bar{s})$, then

$$psp.\pi.\phi = sp.\ (\bar{s}^1 = \bar{t}^0).\ push(psp.\pi'.\phi),$$

where $p(\bar{t})$ is the head of C;

2. if l_k is $success(A)$ and l_{k-1} is $exit(D)$, where A is not a constraint and D is a clause, then

$$psp.\pi.\phi = pop(psp.\pi'.\phi) \cap \neg free(x_C^0),$$

where C is the clause containing A;

3. if l_k is $success(A)$, where A is a constraint, then

$$psp.\pi.\phi = sp.\ A^0.\ (psp.\pi'.\phi).$$

$\square$

Definition 6.2 Let $\mathcal{P}$ be a program with set $\{1, \ldots, n\}$ of pp's, and let ϕ be s.t. $\phi \subseteq \neg free(x_G^0)$, and $\phi \subseteq free(x_C^0)$ for every non-goal, non-unitary clause C. The *semantics $\mathcal{O}(\mathcal{P}, \phi)$ of $\mathcal{P}$* w.r.t. ϕ is the n-tuple:

$$(\phi, \cup_{\pi \in path(2)} psp.\pi.\phi, \ \ldots, \ \cup_{\pi \in path(n)} psp.\pi.\phi).$$

$\square$

Recall that $path(i)$ denotes the set of all the paths of $dg(\mathcal{P})$ from 1 to i. The operational intuition behind the definition of $psp.\pi.\phi$ can be illustrated using the transition rules of Table 1: case 1. corresponds to the application of rule **R**, case 2. to the application of rule **S** and case 3. to the application of rule **C**. Then the semantics $\mathcal{O}(\mathcal{P}, \phi)$ associates with every node of $dg(\mathcal{P})$ the union, over all the paths π from the entry point of $\mathcal{P}$ to that node, of the strongest postconditions of the π's w.r.t. ϕ. The characteristic variables have here the same function as in the definition of $\mathcal{F}$. The following example illustrates the crucial role of these variables to discriminate those paths which are not semantic.

at pp	x_G^0	x_{C1}^0	x_{C2}^0
1	not free	free	free
3	free	not free	free
4	free	not free	free
6	free	free	not free
2	free	not free	free

Table 2. Characteristic variables of index 0 through π

Example 6.3 Consider again the program *Prod*. Let π be $\langle 1, 3, 4, 6, 2 \rangle$ and let α be $x_G^0 = 0$, where 0 is a constant. The behaviour, with respect to freeness, of the characteristic variables of index 0 during the propagation of α through π is described in Table 2. Observe that, at program point 2, the i-variable x_G^0 is free. Then, Definition 6.1 is not applicable. In fact, π does not describe a computation, because it 'jumps' to the success point of the goal before finishing the execution of the called clause $C1$. To describe a computation, π has to be modified by replacing 2 with 5. In fact, x_{C1}^0 is not free at pp 5.

□

We now show that $\mathcal{T}$ and $\mathcal{F}$ are equivalent, by proving that $\mathcal{T}$ and $\mathcal{O}$ are isomorphic ($\mathcal{T} \sim \mathcal{O}$), and that $\mathcal{F}$ and $\mathcal{O}$ are equal. To define the isomorphism between $\mathcal{T}$ and $\mathcal{O}$, we use a relation *Rel* relating partial traces and paths.

We write *conf*, possibly subscripted, to denote a configuration $(\overline{A}, \alpha)$ used in the rules of TS. The relation *Rel* is defined inductively on the number of elements of a partial trace as follows.

The base case is $\langle (\langle p(\overline{s}) \rangle \cdot \overline{A}, \alpha) \rangle$ *Rel* $\langle call(p(\overline{s})) \rangle$, and the induction case is as follows. Suppose that $\tau' \cdot \langle conf_1 \rangle$ *Rel* π and that τ is $\tau' \cdot \langle conf_1, conf_2 \rangle$ (by definition this implies $conf_1 \rightarrow conf_2$). Then:

- τ *Rel* $\pi \cdot \langle entry(C) \rangle$,
  if $conf_1$ is $(\langle p(\overline{s}) \rangle \cdot \overline{A}, \alpha)$ and C is the selected clause;
- τ *Rel* $\pi \cdot \langle success(A) \rangle$,
  if $conf_1$ is $(\langle pop \rangle \cdot \overline{A}, \alpha)$, and if the atom A satisfying the following condition exists: Let π be of the form $\langle l_1, \ldots, l_k \rangle$. Then for some $i \in [1, k]$, $call(A)$ is equal to l_i, and for every B in $\mathcal{P}$, the sets $I_{call(B)}$ and $I_{success(B)}$ have the same cardinality, where $I_\star$ is the set $\{ j \mid i < j \leq k, l_j = \star \}$, for $\star$ in $\{ call(B), success(B) \}$.
- τ *Rel* $\pi \cdot \langle success(d) \rangle$,
  if $conf_1 = (\langle d \rangle \cdot \overline{A}, \alpha)$.

Informally, the isomorphism $\sim$ first extracts from an element τ of $\mathcal{T}$ of the form $\tau' \cdot \langle (\overline{A}, \beta) \rangle$ its final state β, and maps it into the l-th component ϕ_l of $\mathcal{O}$, where l is the last node of a path π s.t. τ *Rel* π holds. Vice versa, $\sim$ maps a β

in ϕ_l, with $l \in \{1, \ldots, n\}$, into the partial trace τ of T of the form $\langle (G, \alpha) \rangle \cdot \tau'$, s.t. for some π in $path(l)$, we have that τ Rel π, and $\{\beta\}$ is $psp.\pi.\{\alpha\}$.

Theorem 6.4 $(T \sim \mathcal{O})$ *Let ϕ be s.t. $\phi \subseteq \neg free(x_G^0)$, and $\phi \subseteq free(x_C^0)$ for every non-goal, non-unitary clause C. Then $T(\mathcal{P}, \phi)$ and $\mathcal{O}(\mathcal{P}, \phi)$ are isomorphic.*

Theorem 6.5 $(\mathcal{F} = \mathcal{O})$ *Let ϕ be s.t. $\phi \subseteq \neg free(x_G^0)$, and $\phi \subseteq free(x_C^0)$ for every non-goal, non-unitary clause C. Then $\mathcal{F}(\mathcal{P}, \phi) = \mathcal{O}(\mathcal{P}, \phi)$.*

This result can be proven by showing that for every $k \geq 0$, $\Psi_i^k(\bot)$ is equal to the union of the path strongest postconditions w.r.t. ϕ of all the paths π which start in 1 and have length less or equal than k.

Corollary 6.6 $(T \sim \mathcal{F})$ *Let ϕ be s.t. $\phi \subseteq \neg free(x_G^0)$, and $\phi \subseteq free(x_C^0)$ for every non-goal, non-unitary clause C. Then $\mathcal{F}(\mathcal{P}, \phi) \sim T(\mathcal{P}, \phi)$.*

7 Properties of $\mathcal{F}$

We show here that $\mathcal{F}$ enjoys some important properties, namely it is incremental, monotonic and and-compositional. Incrementality is important because, for instance, it allows us to compute the semantics of the union of two clp's $\mathcal{P}$ and $\mathcal{P}'$, by computing first the semantics of one of them, say $\mathcal{F}(\mathcal{P})$ of $\mathcal{P}$, and then by using $\mathcal{F}(\mathcal{P})$ to determine the semantics of their union $\mathcal{P} \cup \mathcal{P}'$. Also, from the practical point of view, incrementality allows us to define parallel execution models of clp's based on asynchronous processors, as explained in Section 8. And-compositionality allows us to compute the semantics of a goal $\leftarrow \overline{A}, \overline{B}$ from the semantics of $\leftarrow \overline{A}$ and of $\leftarrow \overline{B}$. The and-compositionality of $\mathcal{F}$ is used in the next section to define using $\mathcal{F}$ a goal-independent semantics.

Formally, let S be a subset of $\{1, \ldots, n\}$. We define $\Psi_S : (2^{States})^n \rightarrow (2^{States})^n$, called *the restriction of Ψ to the pp's in S*, as in Definition 5.3 except that for every pp l which is not in S, $(\Psi_S)_l(\overline{\psi})$ is set to be ψ_l.

Lemma 7.1 (Incrementality) *Let S be a subset of $\{1, \ldots, n\}$. If $\overline{\psi} \subseteq \mu\Psi$ then $\bigcup_{k=0}^{\omega} \Psi_S^k(\overline{\psi}) \subseteq \mu\Psi$.*

This lemma says that to compute $\mathcal{F}$ one can first restrict to a subset S of the pp's of the program, and iterate Ψ a number of times, using only the pp's of S; then the result $\overline{\psi}$ obtained can be incremented by iterating Ψ starting from $\overline{\psi}$ instead than $\bot$.

Lemma 7.2 (Monotonicity) *If $\phi \subseteq \phi'$ then $\mathcal{F}(\mathcal{P}, \phi) \subseteq \mathcal{F}(\mathcal{P}, \phi')$.*

A program without a goal is called *pure*.

Lemma 7.3 (And-compositionality) *Let $G = \leftarrow A_1, \ldots, A_\ell, B_1, \ldots, B_m$ and let $\mathcal{P}$ be a pure program. Suppose that:*
$$\mathcal{F}(\{\leftarrow A_1, \ldots, A_\ell\} \cup \mathcal{P}, \phi_1) = (\phi_1, \phi_2, \ldots, \phi_{\ell+1}, \phi_{\ell+2}, \ldots, \phi_{\ell+k}),$$
$$\mathcal{F}(\{\leftarrow B_1, \ldots, B_m\} \cup \mathcal{P}, \phi_{\ell+1}) = (\psi_1, \psi_2, \ldots, \psi_{m+1}, \psi_{m+2}, \ldots, \psi_{m+k}). \; Then$$
$$\mathcal{F}(\{G\} \cup \mathcal{P}, \phi_1) = (\phi_1, \ldots, \phi_{\ell+1}, \psi_2 \ldots, \psi_{m+1}, \phi_{\ell+2} \cup \psi_{m+2}, \ldots, \phi_{\ell+k} \cup \psi_{m+k}).$$

The Monotonicity lemma follows by the monotonicity of Ψ, while the proofs of the other lemmas use the intermediate semantics $\mathcal{O}$, and can be found in the full version of the paper. The Monotonicity and the And-Compositionality Lemmas are used in the next section to define a goal-independent dataflow semantics for clp's.

A Goal-Independent Semantics

$\mathcal{F}$ is defined w.r.t. a set of input states describing a set of initial bindings for the goal, hence *lifting* to sets of goals the so called goal-dependent analysis, where only one goal is considered. In logic programming other semantics, like those based on the s-semantics ([BGLM94]), perform an analysis which is goal-independent, i.e. they refer to *pure* (viz. without goal) programs. These two different kinds of analysis can be nicely reconciled, since one can (finitely) define for a pure clp $\mathcal{P}$ a goal-independent semantics $\widehat{\mathcal{F}}(\mathcal{P})$.

Let $\{G\} \cup \mathcal{P}$ be a program. Define the *restriction of* $\mathcal{F}(\{G\} \cup \mathcal{P}, \phi)$ *to* $\mathcal{P}$, written $\mathcal{F}(\{G\} \cup \mathcal{P}, \phi)_{|\mathcal{P}}$, to be the tuple obtained from $\mathcal{F}(\{G\} \cup \mathcal{P}, \phi)$ by deleting those elements which are associated with the pp's of G.

Then the *goal-independent semantics* $\widehat{\mathcal{F}}(\mathcal{P})$ of a pure clp $\mathcal{P}$ is

$$\widehat{\mathcal{F}}(\mathcal{P}) = \bigcup_{p \text{ in } pred(\mathcal{P})} \mathcal{F}(\{G_p\} \cup \mathcal{P}, \phi_{G_p})_{|\mathcal{P}},$$

where $pred(\mathcal{P})$ is the set of predicate symbols occurring in $\mathcal{P}$, G_p is $\leftarrow p(\tilde{x})$, and ϕ_{G_p} is the set $\neg free(x^0_{G_p}) \cap free(x^0_{C1}) \cap \ldots \cap free(x^0_{Ck})$, where $C1, \ldots, Ck$ are the non-unitary clauses of $\mathcal{P}$.

Then $\widehat{\mathcal{F}}$ is the best goal-independent dataflow semantics, in the following sense:

Theorem 7.4 *For every pure program* $\mathcal{P}$, $\widehat{\mathcal{F}}(\mathcal{P}) = \bigcup_{\substack{G \text{ a goal} \\ \phi \subseteq \phi_G}} \mathcal{F}(G \cup \mathcal{P}, \phi)_{|\mathcal{P}}.$

Proof. By the Monotonicity and And-compositionality Lemmas. $\square$

8 Applications

The dataflow semantics $\mathcal{F}$ allows us to view a program as a dataflow, where a node l receives states from the set $input(l)$ of all the nodes l' s.t. (l', l) is an arc of the dataflow graph. This description of the semantics of a clp is important for various reasons. $\mathcal{F}$ can be used to study run-time properties of clp's, as done e.g. in [DM88, CM91, DM93] for logic programs. For instance, we have used $\mathcal{F}(\mathcal{P}, \phi)$ in [CMM95] to develop a sound and complete method to prove termination of a clp w.r.t. a precondition ϕ. In this section we give two other possible applications of the dataflow semantics. In the first one $\mathcal{F}$ is used to define a parallel execution model based on asynchronous processors. In the second one the semantics $\mathcal{O}$ is used to define an *à la* Burstall [Bur74] intermittent assertions method for clp's.

8.1 A Parallel Execution Model

The Incrementality Lemma 7.1 for $\mathcal{F}$ suggests a possible parallel execution model $\mathcal{M}$ of clp's based on a network of processors, defined as follows:

Network Let N be the set of pp's of $\mathcal{P}$. For $l \in N$, a *processor* P_l is associated with l.

Communication among processors is realized by means of channels, as follows:

Communication Processors are connected by the following *channels*:

- $c_{env}^{entry(G)}$ from the environment *env* to $P_{entry(G)}$ and $c_{exit(G)}^{env}$ from $P_{exit(G)}$ to the environment;
- c_i^j from i to j for every i, j such that there is an arc from i to j in $dg(\mathcal{P})$.

A channel c_i^j is called an *input channel of* P_j and an *output channel of* P_i. Each channel is supposed to have a *memory* that contains a queue of states whose policy is fair (e.g. first in first out).

The execution model allows the processors to run in parallel and asynchronously:

Execution Model Processors in the network execute asynchronously the following *algorithms*:

- $P_{entry(G)}$ takes an α from $c_{env}^{entry(G)}$ and sends it to all its output channels.
- $P_{entry(C)}$ selects with *fair choice* from one of its input channels, say $c_{call(A)}^{entry(C)}$, an α, and it computes $push(\alpha) \wedge \bar{s}^1 = \bar{t}^0$, where $A = p(\bar{s})$ and $p(\bar{t})$ is the head of H; then $P_{entry(C)}$ sends $push(\alpha) \wedge \bar{s}^1 = \bar{t}^0$ to every its output channel.
- $P_{success(A)}$, where A is not a constraint and is contained in the clause C, selects with fair choice from one of its input channels, say $c_{exit(D)}^{success(A)}$, an α; then it computes $pop(\alpha)$; if $pop(\alpha)$ is in $\neg free(x_C^0)$ then $P_{success(A)}$ sends $pop(\alpha)$ to every its output channel.
- $P_{success(A)}$, where A is a constraint, takes an α from its input channel and computes $\alpha \wedge A^0$, then $P_{success(A)}$ sends $\alpha \wedge A^0$ to every its output channel.

This model describes a sound and complete implementation of $\mathcal{O}$, as stated in the following theorem.

Theorem 8.1 (Adequacy of $\mathcal{M}$) *If the input channel c_{env}^e of $\mathcal{M}$ is feed with the set of states ϕ s.t. $\phi \subseteq \neg free(x_G^0)$, and $\phi \subseteq free(x_C^0)$ for every non-goal, non-unitary clause C, then $\bigcup_{\pi \in path(l)} psp.\pi.\phi$ is the set of states that P_l in $\mathcal{M}$ sends on its output channels.*

This result can be proven using $\mathcal{O}$. For the completeness part, observe that, intuitively, since the choice of the state to be processed is fair, no state will be delayed forever.

Remark 8.2 Our execution model assigns one processor to each program point. However, because the processors work asynchronously, in case there are less processors than program points, then a single processor can be assigned to a number of pp's, which can be encoded as distinct tasks to be executed with a fair schedule discipline. This will still yield a complete and asynchronous model.

8.2 Burstall's Intermittent Assertions Method

We show how the intermittent assertions method of Burstall [Bur74] can be adapted to clp's. The advantages of the Intermittent Assertion Method, and of Temporal Logic (TL) in general, for instance to prove *liveness properties*, *termination*, *total correctness* etc. are well known (see for instance [CC93]). So far, finding a suitable presentation of the intermittent assertion method for logic programming was still an open problem ([CC93]). In this section we show how one can give a solution to this problem for clp's, by means of the intermediate semantics $\mathcal{O}$. For lack of space, the presentation is rather sketchy: We mention the main ingredients of the system, and give an example to illustrate its application. The complete specification of the corresponding formal system is the subject of another forthcoming paper.

For simplicity, assertions are denoted by ϕ, ψ, thus identifying an assertion with the set of states it denotes. Implication is interpreted as set inclusion, i.e. $\phi \Rightarrow \psi$ iff $\phi \subseteq \psi$. Also, conjunction and disjunction are interpreted set-theoretically as intersection and union, respectively. The assertion $push(\phi)$ is obtained by replacing each i-variable x^i in ϕ by the i-variable x^{i+1}; and $pop(\phi)$ is obtained by first renaming with fresh variables all the i-variables of index 0 and then replacing each remaining i-variable x^i with x^{i-1}.

Here, an 'intermittent rule' is a formula in temporal logic of the form $\Box(\phi \wedge \mathrm{at}(i) \Rightarrow \Diamond(\psi \wedge \mathrm{at}(j)))$, where $\Box$ and $\Diamond$ are the 'always' and 'sometime' operators, and $\mathrm{at}(i)$ indicates that execution is at program point i. The intended meaning of this formula is: for every state α which satisfies ϕ, there is at least one execution of the program starting in the pp i with state α, which reaches the pp j in a state which satisfies ψ. The set of proof rules we consider contains a formalization of the induction principle (Burstall's "little induction"), a suitable axiomatization of TL (cf. [Sti92, CC93]), plus the following *path rule*, which formalizes the "hand simulation" part of the method:

$$(\pi \in path(i,j) \wedge psp.\pi.\phi \neq false) \Rightarrow \Box(\phi \wedge \mathrm{at}(i) \Rightarrow \Diamond(psp.\pi.\phi \wedge \mathrm{at}(j)))$$

A *sound* and *relatively complete* proof system w.r.t. $\mathcal{F}$ can be defined using these tools.

We illustrate by means of an example how the method can be applied to prove total correctness of a clp. The following *composition rule* will be used:

$$\frac{\Box(\phi \wedge \mathrm{at}(i) \Rightarrow \Diamond(\psi \wedge \mathrm{at}(j))) \qquad \Box(\psi \wedge \mathrm{at}(j) \Rightarrow \Diamond(\chi \wedge \mathrm{at}(k)))}{\Box(\phi \wedge \mathrm{at}(i) \Rightarrow \Diamond(\chi \wedge \mathrm{at}(k)))} \tag{1}$$

It enables us to compose intermittent assertions (note that this is a particular case of the 'chain rule' which is one of the basic tools in the proof system presented in [MP83]).

Example 8.3 Consider again the program *Prod*. Let the initial assertion ϕ be $u^0 = [r_0, \ldots, r_k] \wedge \neg free(x_G^0) \wedge free(x_{C1}^0) \wedge at(1)$.

Suppose that we want to prove that *Prod* satisfies the following assertion:

$$\Box(\phi \Rightarrow \Diamond(v^0 = r_0 * \ldots * r_k \wedge at(2))) \tag{2}$$

which says that for every state α of ϕ, at least one execution of the goal $\leftarrow prod(u, v)$ starting in α terminates (i.e. reaches the pp 2) and its final state binds v to $r_0 * \ldots * r_k$. Using the path rule we obtain the following (simplified) assertions:

$$\Box(\phi \Rightarrow \Diamond(v^1 = z^0 = r_0 * w^0 \wedge y^0 = [r_1, \ldots, r_k]) \wedge at(4)))$$
  with path $\langle 1, 3, 4 \rangle$;

$$\Box(v^{k+1} = z^k = r_0 * \ldots * r_k * w^0 \wedge y^0 = [] \wedge at(4) \Rightarrow$$
$$\Diamond(v^{k+1} = z^k = r_0 * \ldots * r_k \wedge y^0 = [] \wedge at(5)))$$
  with path $\langle 4, 6, 5 \rangle$;

$$\Box(v^1 = z^0 = r_0 * \ldots * r_k \wedge at(5) \Rightarrow$$
$$\Diamond(v^0 = r_0 * \ldots * r_k \wedge at(2)))$$
  with path $\langle 5, 2 \rangle$;

The following assertions can be proven by straightforward induction:

$$\Box(v^{m+1} = z^m = r_0 * \ldots * r_m * w^0 \wedge y^0 = [r_{m+1}, \ldots, r_k] \wedge m < k \wedge at(4) \Rightarrow$$
$$\Diamond(v^{k+1} = z^k = r_0 * \ldots * r_k * w^0 \wedge y^0 = [] \wedge at(4)))$$
  using as path $\pi = \langle 4, 3, 4 \rangle$, and

$$\Box(v^{k+1} = z^k = r_0 * \ldots * r_k \wedge y^0 = [] \wedge at(5) \Rightarrow$$
$$\Diamond(v^1 = z^0 = r_0 * \ldots * r_k \wedge at(5)))$$
  using as path $\pi = \langle 5, 5 \rangle$.

Then, the repeated application of rule (1) to compose the above assertions yields (2). □

9 Discussion

In this paper an alternative operational model for clp's was proposed, where a program is viewed as a dataflow graph and a predicate transformer semantics transforms a set of states associated with a fixed node of the graph (corresponding to the entry-point of the program) into a tuple of set of states, one for each node of the graph. To the best of our knowledge, this is the first predicate transformer semantics for clp's based on dataflow graphs. The dataflow graph provides a static description of the flow of control of a program, where sets of constraints 'travel' through its arcs. The relevance of this approach was substantiated in the Applications section.

We would like to conclude this paper by giving an extension of its results to more general CLP systems. We have considered 'ideal' CLP systems. With slight modifications, the dataflow semantics $\mathcal{F}$ (and all its applications) can be adapted to deal also with 'quick-check' and 'progressive' systems (cf. [JM94]), which are those more widely implemented. This can be done as follows. States are considered to be pairs (c_1, c_2) of constraints, instead than constraints, where c_1 denotes the active part and c_2 the passive part.

$$States = \{(c_1, c_2) \mid c_1 \text{ and } c_2 \text{ are constraints s.t. } consistent(c_1)\},$$

where the test $consistent(c_1)$ checks for (an approximation of) the consistency of c_1. Then rules $\mathbf{R}$ and $\mathbf{C}$ of Table 1 have to be changed as illustrated below, where a state $\alpha = (c_1, c_2)$ is also denoted by (α_1, α_2):

$$\mathbf{R} \quad (\langle p(\bar{s}) \rangle \cdot \overline{A}, \alpha) \longrightarrow (\overline{B} \cdot \langle pop \rangle \cdot \overline{A}, infer(\alpha_1', \alpha_2' \wedge \bar{s}^1 = \bar{t}^0)),$$

with $\alpha' = push(\alpha)$, if $C = p(\bar{t}) \leftarrow \overline{B}$ is in $\mathcal{P}$.

$$\mathbf{C} \quad (\langle d \rangle \cdot \overline{A}, \alpha) \longrightarrow (\overline{A}, infer(\alpha_1, \alpha_2 \wedge d^0)),$$

if d is a constraint. Finally, the definition of sp has to be changed in:

$$sp.c.\phi = \{\alpha' \in States \mid \alpha' = infer(\alpha_1, \alpha_2 \wedge c) \text{ and } \alpha \in \phi\}.$$

The operator $infer$ computes from the current state (c_1, c_2) a new active constraint c_1' and passive constraint c_2', with the requirement that $c_1 \wedge c_2$ and $c_1' \wedge c_2'$ are equivalent constraints. The intuition is that c_1 is used to obtain from c_2 more active constraints; then c_2 is simplified to c_2'. For instance, in the example of Section 5.5, in the state of Ψ_4^2 the constraint $z^0 = x^0 * w^0$ would be passive, because the equation is not linear (cf. [JMSY92]). Then, in Ψ_6^3 this constraint is transformed by applying first $push$ to it and then $infer$. So $z^1 = x^1 * w^1$ becomes active, because w^1 is bound to 1 and hence the equation becomes linear.

Acknowledgments: We would like to thank Jan Rutten and the anonymous referees for their useful comments. The research of the second author was partially supported by the Esprit Basic Research Action 6810 (Compulog 2).

References

[BGLM94] A. Bossi, M. Gabbrielli, G. Levi, and M. Martelli. The s-semantics approach: theory and applications. *The Journal of Logic Programming*, 19,29: 149–197, 1994.

[Bur74] R.M. Burstall. Program proving as hand simulation with a little induction. *Information Processing*, 74:308–312, 1974.

[CC93] P. Cousot and R. Cousot. "A la Burstall" Intermittent Assertions Induction Principles for Proving Inevitability Properties of Programs. *Theoretical Computer Science*, 120:123–155, 1993.

[CM91] L. Colussi and E. Marchiori. Proving correctness of logic programs using axiomatic semantics. In *Proceedings of the Eight ICLP*, pages 629–644. MIT Press, 1991.

[CMM95] L. Colussi, E. Marchiori and M. Marchiori. On Termination of Constraint Logic Programs. In *Proc. First International Conference on Principles and Practice of Constraint Programming*. LNCS, Springer–Verlag, 1995. To appear.

[DM88] W. Drabent and J. Małuszyński. Inductive assertion method for logic programs. *TCS*, 59(1):133–155, 1988.

[DM93] P. Deransart and J. Małuszyński. A Grammatical View of Logic Programming. The MIT Press, 1993.

[JMSY92] J. Jaffar, S. Michaylov, P.J. Stuckey and R.H.C. Yap. The CLP($\mathcal{R}$) Language and System. *ACM TOPLAS*, 14(3):339–395, 1992.

[JM94] J. Jaffar and M.J. Maher. Constraint Logic Programming: A Survey. *JLP* 19,20: 503-581, 1994.

[Kun87] K. Kunnen. Signed Data Dependency in Logic Programs. *Computer Science Technical Report 719*, University of Wisconsin - Madison, 1987.

[Mel87] C. Mellish. Abstract interpretation of Prolog programs. In S. Abramsky and C. Hankin, editors, *Abstract Interpretation of declarative languages*, pp. 181–198. Ellis Horwood, 1987.

[MP83] Z. Manna and A. Pnueli. How to cook a proof system for your pet language. In *Proceedings 10th Annual ACM Symposium on Principles of Programming Languages (POPL)*, pp. 141–154, 1983.

[Nil90] U. Nilsson. Systematic semantics approximations of logic programs. In *Proc. PLILP*, pp. 293–306. Eds. P. Deransart and J. Małuszyński, Springer Verlag, 1990.

[Sti92] C. Stirling. Modal and Temporal Logics. In S. Abramsky, Dov M. Gabbay, and T.S.E. Maibaum, editors, *Handbook of Logic in Computer Science*, volume 2, pages 477–563, 1992.

[WS94] B. Wang and R.K. Shyamasundar. A methodology for proving termination of logic programs. *JLP* 21(1): 1–30, 1994.

Posters and Demonstrations

A Functional Framework for Deterministically Interleaved Interactive Programs

Peter Achten (presenter, in in)

Computing Science Institute, University of Nijmegen, ...
Nijmegen, The Netherlands

The I/O system of the lazy, purely functional programming language Clean [4][7], the Great I/O system [1][2], enables programmers to write Graphical User Interface (GUI) applications at a high level of abstraction. Clean has a special type system called Uniqueness Typing [6] that offers the pure functional world an interface with the imperative world by guarantee of single-threaded use of functions of interest. In this paper we start the following extensions to the I/O system: dynamic process creation, and inter-process communication by data sharing and message passing. The semantics of the system is defined in Clean itself. As extended system, the Interleaved Clean I/O system can be regarded as an operational semantics of a small Operating System.

...

Interactive processes are created dynamically with the function await (fig.). New processes run when used with other processes, and independently of their father process. The interleaving semantics of interactive processes is such that abstract event handlers are atomic operations.

Process P_i can communicate by data sharing if they have process states of State α. Such a set of interactive processes forms a process group. Shared ...

A Functional Framework for Deterministically Interleaved Interactive Programs

Peter Achten (peter88@cs.kun.nl)

Computing Science Institute, University of Nijmegen, Toernooiveld 1, 6525ED
Nijmegen, The Netherlands

The I/O system of the lazy, purely functional programming language Clean
[4][7], the Event I/O system [1][2], enables programmers to write Graphical User
Interface (GUI) applications at a high level of abstraction. Clean has a special
type system called Uniqueness Typing [3] that offers the pure functional world
an interface with the imperative world by guaranteeing single-threaded use of
destructible objects. In this paper we study the following extensions to the Event
I/O system: *dynamic process creation* and inter-process communication by *data
sharing* and *message passing*. The semantics of the system is defined in Clean.
The extended system, the Interleaved Event I/O system, can be regarded as an
operational semantics of a mini Operating System.

Interactive processes are state transition systems. The state is called the
process state, and is a structured value of type ($PState$ α β). It contains two
special, single-threaded objects of abstract type *Files* and *IOState* to do file
and GUI I/O respectively. State transitions are functions of type ($PState$ α
β)$\rightarrow$($PState$ α β). To define when a state transition is valid, each function is
associated with a GUI element. State transition functions are called *abstract
event handlers*, and their evaluation condition contexts *abstract events*. Programs
start with one process, evaluated by the pure function openIO (fig.).

```
openIO        ::(a,b)                    (InitIO a b)  World          -> World
newIO         ::(a,b)                    (InitIO a b)  (PState c d) -> PState c d
shareIO       ::a                        (InitIO a b)  (PState c b) -> PState c b
openReceiver  ::(ReceiverDef m (PState a b)) (PState a b) -> (RId m,PState a b)
asend         ::(RId m) m                              (PState a b) -> PState a b
send          ::(RId m) m                              (PState a b) -> PState a b

::  InitIO       a b :== [(PState a b) -> (PState a b)]
::  ReceiverDef m c =    Receiver (m->c->c)
```

Interactive processes are created dynamically with the function newIO (fig.).
New processes run *interleaved* with other processes, and independently of their
father process. The interleaving semantics of interactive processes is such that
abstract event handlers are *atomic operations*.

Processes P_i can communicate by *data sharing* if they have process states ps_i
= $PState$ α_i β. Such a set of interactive processes forms a *process group*. Shared

interactive processes are spawned with the function shareIO (fig.). Access to the shared data structure is atomic.

Message passing can be done between arbitrary processes. In our system messages are considered to be raisable abstract events. Message events are handled by receivers. Message passing is type-safe: correctly typed programs can not send messages of the wrong type. Creation of a receiver that handles messages of type μ yields an *identification value* of type $RId\,\mu$ (openReceiver). This value is used for asynchronous and synchronous message passing (asend and send). Messages of one source arrive in the same order as sent.

Related work: *(a) dialogue combinators* [6], *(b)* FUDGETS [5], and *(c) Kaos* [8]. Dynamic processes are not considered in *(a)*, are very limited in *(b)* and are fully supported in *(c)*. None of the systems have a concept of shared state.

Conclusions: a model of interactive processes based on state transition systems can be described well in a pure functional language. Although the definition is executable it is geared for readability. To obtain a good implementation efficiency considerations need to be incorporated. Future work is to study truly concurrent interactive processes distributed over a network of processors.

References

[1] Achten, P.M., Groningen J.H.G. van, and Plasmeijer, M.J. High Level Specification of I/O in Functional Languages. In Launchbury, J., Sansom, P. eds. *Proc. Glasgow Workshop on Functional Programming, Ayr, Scotland*, 6-8 July 1992, Workshops in Computing, Springer-Verlag, Berlin, 1993, pp. 1-17.

[2] Achten, P.M. and Plasmeijer, M.J. The ins and outs of Clean I/O. In *Journal of Functional Programming* 5(1), January 1995, Cambridge University Press, pp. 81-110.

[3] Barendsen, E. and Smetsers, J.E.W. Conventional and Uniqueness Typing in Graph Rewrite Systems. In Shyamasundar, R.K. ed. *Proc. 13th Conf. on the Foundations of Software Technology and Theoretical Computer Science*, 15-17 December 1993, Bombay, India, LNCS **761**, Springer-Verlag, Berlin, pp. 41-51.

[4] Brus, T., Eekelen, M.C.J.D. van, Leer, M.O. van, Plasmeijer, M.J., and Barendregt, H.P. Clean: A Language for Functional Graph Rewriting. In Kahn. G. ed. *Proc. 3rd International Conf. on Functional Programming Languages and Computer Architecture*, 1987, Portland, Oregon, USA, LNCS **274**, Springer-Verlag, pp. 364-384.

[5] Carlsson, M. and Hallgren, Th. FUDGETS - A Graphical User Interface in a Lazy Functional Language. In *Proc. Conf. on Functional Programming Languages and Computer Architecture*, Copenhagen, Denmark, 9-11 June 1993, ACM Press, pp. 321-330.

[6] Dwelly, A. Functions and Dynamic User Interfaces. In *Proc. 4th International Conf. on Functional Programming Languages and Computer Architectures*, Imperial College, London, September 11-13, 1989, pp. 371-381.

[7] Plasmeijer, M.J. and Eekelen, M.C.J.D. van. Functional Programming and Parallel Graph Rewriting. Addison-Wesley Publishing Company 1993.

[8] Turner, D.A. An Approach to Functional Operating Systems. In Turner, D.A. ed. *Research topics in Functional Programming*, Addison-Wesley Publishing Company 1990, pp. 199-217.

A Debugging Model for Lazy Narrowing[*]

Puri Arenas-Sánchez, Ana Gil-Luezas

Unv. Complutense de Madrid, Dpto. de Informática y Automática
Fac. de CC. Matemáticas, Av. Complutense s/n, 28040 Madrid, Spain
email: {puri,anagil}@dia.ucm.es

One of the main difficulties within all implementations of lazy narrowing consists of finding good control regimes which avoid repeated evaluations of arguments and minimize the risk of non termination. The *demand driven* control regime presented in [2] arose as an attempt of solving the mentioned problems. Nevertheless, such control regime makes quite hard to foresee the sequence of evaluations due to the delay of the evaluations of expressions until they are needed and the control of the searching of solutions.

We propose a debugging model for a programming language *BabLog* which combines lazy functional, higher order and logic programming. Assume that $DC = \bigcup_{n \in N} DC^n$ and $DF = \bigcup_{n \in N} DF^n$ are disjoint ranked alphabets of data constructor and function symbols, respectively. The set of *terms* and *expressions* over (DC, DF) is defined in the natural way. A *BabLog program* P is a finite set of function definitions, where a function symbol f is defined by a finite set of rules P_f of the following form:

$$\underbrace{f \; t_1 \ldots t_n}_{head} := condition \rightarrow \underbrace{e}_{body}$$

where $t_i \in terms$, $e \in expressions$ and *condition* is a sequence of equalities between expressions. The computation of *BabLog*'s goals (which have the same structure than *conditions*) is based on lazy narrowing under the *demand driven* control regime.

Our model, based on the standard box-oriented model for logic programs, pretends to be a methodological tool to facilitate users the understanding of the operational semantics of *BabLog*. In our opinion, such kind of tools can help users to debug programs and find easily possible programming bugs. Finally, this model can be applied to other existing lazy strategies for lazy narrowing, such that the *naïve* one, presented in [2].

The idea of applying the box model for debugging functional logic languages has been already treated in literature. Nevertheless, nothing is commented on laziness. In order to represent lazy narrowing, we propose to incorporate new kind of boxes [1], connected through their four associated ports: *redo, fail, exit* and *call*, in the natural way.

The most relevant box in our model is the hnf-box associated to the evaluation to head normal form of a *Total Function Application*[2] *tfa* $\equiv f \; e_1 \ldots e_n$ ($f \in$

---

[*] This research has been partially supported by the Spanish National Project TIC92–0793 "PDR" and the Esprit BRA Working Group Nr. 6028 "CCL".

[2] An expression may have several head normal forms. In pure functional languages that is not possible.

$\mathcal{DF}^n$). Assume that $pos(tfa)$ represents the set of positions of $f\ e_1 \ldots e_n$, defined in the natural way. A position $u \in pos(tfa)$ is *uniformly demanded* in $R \subseteq P_f$ if all heads of rules in R contain a constructor symbol at position u. In such case, we say also that the constructor symbol demands u. u is *demanded* in R if at least one head of some rule in R contains a constructor symbol at position u. Otherwise we say that u is *no demanded* in R. Now, the hnf-box for tfa can be recursively described by means of the following algorithm, in which, TFA, POS and RUL are the input parameters (note that the initial call must be done with TFA $\equiv tfa$, POS $\equiv pos(tfa)$ and RUL $\equiv P_f$).

1. If all positions in POS are *no demanded* in RUL, then the process finishes by trying in textual order all rules in RUL. This event is represented by a reduce-box which contains |RUL| different app-rule-boxes. Obviously an app-rule-box must be composed of a condition-box for solving lazily the condition of the rule at hand, and a hnf-box for evaluating its body.

2. If there is a *uniformly demanded* position in POS, then select the less one u (following the lexicographic order). Now the following processes must be performed sequentially:
   (a) the evaluation to head normal form (represented by a new hnf-box) of the expression occurring at position u in TFA (denoted by hnf_u), and
   (b) for each demanding constructor $c \in \mathcal{DC}$, the attempts of unifying hnf_u with the construction occurring at position u. Each one of these attempts is represented by a unify-box.

   All boxes generated by (a) and (b) are included in a lazy-unification-box. Finally, the *exit* port of the lazy-unification-box connects to the *call* port of the hnf-box generated by the following input parameters (where c is the constructor at hand for which (b) succeeded):
   - TFA$_c$ resulting of replacing in TFA the expression occurring at position u with the result of the unification returned by (b)
   - the set POS$_c$ given by $pos(\text{TFA}_c) - \{u\}$ and
   - the subset RUL$_c$ of RUL composed of those rules in whose heads c occurs at position u.

3. If there are no *uniformly demanded* positions but there exist *demanded* positions, then it is straightforward to reduce the problem to several subproblems in such a way that either step 1 or 2 are applicable to them.

To conclude, we have also developed an environment BabLog [1] which incorporates this debugging model and several facilities typical of debuggers. We have added additional information in most box ports referring to the position to be evaluated, the rule being tried and so on. Moreover, each box has an associated identifier and number for representing, respectively, the kind of computation and the number of boxes which are still to be proved.

References

1. P. Arenas-Sánchez, A. Gil-Luezas. *A Debugging Model for Lazy Functional Logic Languages*, Technical Report DIA 94/6, UCM, April, 1994.
2. R. Loogen, F. J. López Fraguas, M. Rodríguez Artalejo. *A Demand Driven Computation Strategy for Lazy Narrowing*, In PLILP'93, LNCS 714, pp. 184-200, 1993.

Evaluation of Expressions in a Multiparadigm Framework

Xavier Burgués and Xavier Franch

Universitat Politècnica de Catalunya, Pau Gargallo 5, 08028 Barcelona, Spain

Software development with abstract data types is usually done by first specifying the types and next implementing them using an imperative programming language. In this paper, we present a procedure for executing software which combines formal, executable specifications and imperative programs in an arbitrary manner. As a result, we may adopt a prototyping software development model oriented to testing in which a program is obtained by applying a sequence of refinement steps starting from a formal, possibly incomplete specification of the problem, being able to execute the prototypes at every stage of development.

Prototypes are composed of modules, which may be specified and/or implemented. Formal specifications define data types by means of positive conditional equations and thus it is possible to use a term-rewriting system to manipulate terms. Implementations provide a representation of types (which may include non-implemented parts), an abstraction function for each of them and code for their functions. These functions are executed with an interpreter, which calls the term-rewriting system when evaluating non-implemented operations, possibly after converting data structures to terms using the appropriate abstraction functions. So, the main task during execution is the evaluation of arbitrary expressions combining terms with data structures.

Given an expression to evaluate (which may range from a large piece of software to a simple arithmetic expression), we first substitute its variables to obtain a term t denoting a value and, then, we apply a sequence of transformations on this term to obtain another, simpler one, denoting the same value: its final form, which may be not further simplified in our scheme. The set of possible transformations applicable to any subterm $f(...)$ of t is the following:

1. (a) If the type of f (that is, the type of its result) is defined in the same module as f, substitute $f(...)$ by $abstr(F(...concr(...)...))$, being F the code for f and $abstr$ and $concr$ the abstraction and concretion functions for the type ("concretion function" is just interpretation of the term yielding a data structure).

   (b) If the type of f is not defined in the same module as f, substitute $f(...)$ by $F(...concr(...)...)$.

   In both cases, $concr$ has to be applied to those parameters whose type is defined in the same module as f, because these are the only ones which representation is available to f. For similar reasons, the abstraction function is applied to $F(...concr(...)...)$ in the first case.

2. If $f(...)$ is of the form $F(...)$, being F the code for f, substitute $f(...)$ by the result of the evaluation of $F(...)$ with the interpreter.

3. Substitute *f(...)* by another expression using a rewrite rule.
4. If *f(...)* is of the form *abstr(x)*, substitute *f(...)* by its result (a term).
5. If *f(...)* is of the form *concr(abstr(x))*, substitute *f(...)* by *x*.

Note that these transformations may be applied to a term in different ways that could lead to different final terms denoting the same value. Among different possibilities, our execution procedure is designed to enhance testing during prototyping [1]. To do so, we give priority to all functions under test, evaluating them even if it would not be necessary to obtain the final term. There exists an attribute to know when a function is under test; if it is set, it also indicates if we are testing the specification or the implementation. When it is not set, another attribute points out if the function should be executed with the term-rewriting system or with the interpreter. By default, the specification of a function is under test until an implementation for it is built, which will be under test until the attribute is unset. Then, the interpreter is chosen as the tool for executing the function. This default behaviour may be tuned.

During evaluation, subterms reaching their final form will hold a tag which will remain on the term when assigned to a variable. Thus, it will be present in the terms obtained by further substitution of the variable and it will be useful to cut branches on term traversal. Another tag will be held by subterms that do not contain non-executed operations under test, for similar reasons.

Other projects do exist that support execution of equations mixed with imperative programs. For instance, we mention the execution system of Asspegique which combines PLUSS specifications with Ada implementations. However, mixed execution in Asspegique relies on the compilation of equations into Ada and in the generation of packages for the involved modules. We think that, in the testing framework, compilation makes the easy interaction between the user and the execution system, as well as quick and interactive modifications of the rewrite rules and customization of the environment.

As future work, we want to allow mixed execution even if more than one implementation for the same type is used in different contexts of the same program. This issue has been addressed in [2] and the integration into the current work seems to be natural: no new constructions are required over the language and initial semantics still provides a valid formal framework. On the other hand, efficiency of computations should be improved; we are studying sharing of subterms with a lazy evaluation strategy and the maintenance of both terms and data structures as values of variables.

References

1. Burgués, X., Franch, X.: Evaluation of Expressions in a Multiparadigm Framework. Technical Report LSI-95-35-R, Dept. LSI (UPC), 1995.
2. Franch, X.: Combining Different Implementations of Types in a Program. In Proceedings Joint of Modular Languages Conference, Ulm (Germany), 1994.

An Expression-Or-Parallel Implementation for a Functional Logic Language

W. Hans, F. Sáenz, and St. Winkler*

RWTH Aachen, Lehrstuhl für Informatik II, D-52056 Aachen, Germany
{hans,winkler}@zeus.informatik.rwth-aachen.de

Universidad Complutense de Madrid, Dept. Informática y Automática
E-28040 Madrid, Spain, fernan@eucmvx.sim.ucm.es

Overview Functional logic languages like Babel [4] offer a high degree of (implicit) parallelism. They extend functional programming languages with principles taken from logic programming. Apart from the unification parallelism, which can be integrated into a finer layer within the unification process, expression-parallelism and or-parallelism seem to be worthwhile to be exploited on parallel architectures transparently [1, 3, 5]. It has been shown that almost linear speed-ups can be obtained on theirs machines. But their efficiency collapse when the shared memory model is simulated at low level in a distributed environment.

Our objective is to develop a computational model suited for parallel systems with loosely-coupled memories, too. As far as this kind of machines is concerned, often stack-copying and re-computation models for or-parallelism are cited in the literature, e.g. [2]. In our model we follow a process model [5] extended with retroactive parallelism and process inheritance. Each processor has to execute a certain piece of the overall computation, which has been delegated by a parent processor.

The process model naturally fits to expression-parallelism, too. The progressive approach of [3] implementing independent and-parallelism is intended for a shared memory architecture and adopts the Warren Abstract Machine in an efficient way. An extension [6] for a distributed machine lacks in any capabilities of multiprocessing which yield severe restrictions concerning the abilities for goal scheduling. Our abstract machine integrates multiprocessing that seems to be a crucial feature, when dealing with scalable architectures with many processing elements. Otherwise, several restrictions for the load distribution would impede the uniform workload. Other relevant duties include a suitable memory organization like a meshed stack, a speculative data distribution, and a distributed garbage collector.

Furthermore, the functional aspects of Babel offer more opportunities for parallelization due to some determinative characteristics and the compositional style of programming with its simpler data-flow.

Results The parallel system is mapped onto a loosely-coupled Transputer network with up to 64 processor elements, each one having 1MB main memory. In order to

*This work was supported by the Spanish PRONTIC project TIC92-0793-C02-01 and by the German DFG-grant In 20/6-1.

test the implementation, we measured the efficiency of the implementation on one processor. A comparison with [6] which provides a parallel emulator implementation for Prolog in a similar environment, shows that we get double speed.

	program	2	4	8	16	32	48	60
Expr.-Par	f-queen 11	1.97	3.93	7.85	15.49	30.95	45.54	–
	nfib 30	1.97	3.95	7.82	14.67	26.56	35.80	–
	Hanoi 22	1.99	3.98	7.94	14.62	26.66	33.94	–
Or.-Par.	queens12	1.98	3.93	7.84	15.57	28.62	–	50.83
	naive-sort9	1.98	3.95	7.85	15.68	31.36	–	46.95
	puzzle	1.97	3.93	7.79	14.41	25.71	–	38.08
	knapsack	1.95	3.90	5.00	6.75	9.78	-	14.77

Table 1: Speed-ups on the Transputer system

Table 1 shows the runtime results for typical functional programs and for nondeterministic search problems. In the upper part of the table we considered the following deterministic programs: a functional version of the classical n-queens problem with 11 queens, the slightly modified Fibonacci, and the towers of Hanoi examples. The or-parallel programs are the n-queens problem in the logical version, the naive-sort, the knapsack and the puzzle problem. Note that the examples show almost linear speed-ups concerning topologies with up to 16 processors and often even with 32 processors. Break-ins can be observed when the system is scaled beyond the frontier of processor elements for nowadays shared memory architectures.

Currently, we are porting the Transputer implementation to the Fujitsu AP1000 and to a shared memory machine on a Sequent Symmetry which may give new insights for this kind of parallelism.

References

[1] K. A. M. Ali and R. Karlsson. The Muse Or-Parallel Prolog Model and its Performance. In *North American Conference on Logic Programming*, pages 757–776. MIT Press, 1990.

[2] L. Araujo and J.J. Ruz. PDP: Prolog Distributed Processor for Independent AND/OR Parallel Execution of Prolog. In *International Conference on Logic Programming (ICLP'94)*, pages 142–156. The MIT Press, 1994.

[3] M.V. Hermenegildo. *An Abstract Machine based Execution Model for Computer Architecture Design and Efficient Implementation of Logic Programs in Parallel*. PhD thesis, The University of Texas at Austin, 1986.

[4] J.J. Moreno-Navarro and M. Rodríguez-Artalejo. BABEL: A Functional and Logic Programming Language Based on a Constructor Discipline and Narrowing. In *Conference on Algebraic and Logic Programming, LNCS 343*, pages 223–232. Springer-Verlag, 1988.

[5] B. Ramkumar and L. Kale. Compiled Execution of the Reduce-Or Process Model on Multiprocessors. In *North American Conference on Logic Programming*, pages 313–331. MIT Press, 1989.

[6] A. Verden and H. Glaser. An AND–Parallel Distributed Prolog Executor. In P. Kacsuk and M.J. Wise, editors, *Implementations of Distributed Prolog*, Series in Parallel Computing, chapter 7, pages 143–158. Wiley, 1992.

Attributed Models of Executable Specifications*

Merik MERISTE[1] and Jaan PENJAM[2]

[1] University of Tartu, Liivi 2, EE2400 Tartu, Estonia
[2] Tallinn Technical University/Institute of Cybernetics
Akadeemia tee 21, EE0026 Tallinn, Estonia
e-mail: {merik|jaan}.cs.ioc.ee

Motivation and Definition. Regular and context-free structures are classical and efficiently implementable models of data structures. Automated computing is often as successful as adequately regular and/or context-free (surface or deep) substructures are extracted from the rest structure of data. The question of formal models integrating these structures with others and supporting restructuring of data, analysis and implementation are of great methodological importance. The modifications of models known in the theory of formal languages and automata (finite and pushdown automata, state transition systems etc.), and, corresponding declarative formalisms (attribute grammars, graph grammars etc.) are widely used to achieve satisfactory solutions in the field.

In this paper, attributed automata (AA) are considered. AA were introduced as a formalism for executable specifications using regular syntax with attributes representing contextual relations as well as semantic properties of underlying concepts. AA can be treated as a generalization of a state transition network with attributes and computation rules attached to states and transitions respectively.

Definition 1. *An attributed automaton* is a transition network $M = (S, T)$, where:

- S is a set of *states* with two distinguished subsets: $S_0 \subseteq S$ the initial states, and $S_f \subseteq S$ the final states;
  Every state $s \in S$ is associated with an attribute a_s that is a variable over the domain A_s;
- $T \subseteq S \times S$ is a set of *transitions*;
  Every transition $t = (s, s') \in T$ is associated with
  - *enabling predicate* $P_t : A_s \longrightarrow \text{bool}$, and
  - *transformation function* $f_t : A_s \longrightarrow A_{s'}$.

Starting in an initial state with initial attribute value $x \in A_{s_0}$, the functioning of the automaton M is considered as successive change of the current state. Transition from one state to another is possible only if the corresponding enabling predicate is true. Every transition is accompanied by evaluation of the attribute of the next state using associated transformation function.

AA can be used for implementation of transformational as well as reactive systems.

---

* Full version of the paper is available as Research Report CS80/95, Department of Computer Science, Institute of Cybernetics, Tallinn 1995.
URL: ftp://greta.cs.ioc/pub/CS_Reports/CS80_95.ps.Z

Expressive Power of Attributed Automata. In general, the formal power of the attributed model is that of the Turing machine. In practice, the special cases of AA are interesting to compare with respect to the expressive power. The class of so the called Attributed Language Recognizers (ALR) permits a key for indirect estimation of expressive power of the formalism. In this case we need a specialization of the AA where attributes of states have one component I over domain Σ^* (strings in the input alphabet Σ) while other attributes run over natural numbers. ALR can accept some context-sensitive languages as it is shown in Fig. 1.

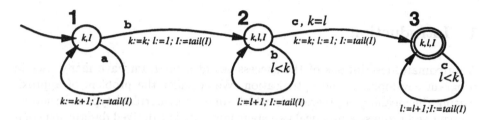

Fig. 1. Recognizer of the language $\mathcal{L} = \{a^n b^n c^n \mid n > 0\}$.

Composition of Attributed Automata. The AA model provides additional tools for restructuring large state transition systems allowing to reduce the number of states in a transition system as well as the conceptual complexity of their specification. Due to their properties, AA have interesting applications in pattern recognition [1], formal language processing [2] and in specification of distributed systems [3].

The serial composition and superposition of AA above form together with "minimization" studied in [2] a fundamental set of compositions for transformational AA. For reactive AA some another types of compositions are useful like parallel compositions with different kinds of syncronization, disjoint union, external and internal choice, action hiding relabeling, and their timed equivalents (for real-time applications). Some preliminary version of composition of reactive AA is given in [3].

References

1. M.Juhola, et al. *Syntactic Recognition of ECG Signals by Attributed Finite Automata*, 21 p., To appear in *Pattern Recognition*.
2. M.Meriste, et al. *Attributed Automata and Language Recognizers*. In Programming Languages and Software Tools. Visegrád, Hungary, June 9-10, 1995, 114 – 121.
3. J.Penjam. *Attributed Automata: a Formal Model for Protocol Specification*. Research Report TRITA-IT-94:30, IOC-CS-71/94. KTH and IoC, 1994, - 49 pp.

Approximating Constraint Logic Programs Using Polymorphic Types and Regular Descriptions *

Hüseyin Sağlam and John P. Gallagher

Department of Computer Science, University of Bristol,
Queen's Building, University Walk, Bristol BS8 1TR, U.K.

1 Introduction

Approximate descriptions of the success set of a program have many uses in program development and optimisation. We consider the problem of approximating the meaning of programs in which some (polymorphic) type information is given and propose a practical two-stage method. The derived description combines polymorphic type terms, including union types, with regular descriptions of untyped terms. The method allows both goal-dependent and goal-independent analyses.

Our work combines aspects of two different notions of types, classical semantic types and success types, in logic programming. We consider the interaction between type information and regular term descriptions inferred by abstract interpretation. Our aim is to infer an approximation of the success set of partly-typed programs by means of a mixture of polymorphic type terms and regular descriptions.

2 Approximation of Partly-Typed Programs

Two kinds of application are considered.

1. Analysis of Prolog programs in which types are imposed on selected symbols.
2. Analysis of CLP programs in which the system types are taken into account and combined with untyped or user-typed symbols.

The approximation method consists of two stages:

1. A program is reduced to an abstract one, by replacing all typed symbols by their types, using the type declarations as rewrite rules (*type reduction*).
2. An established algorithm for regular approximation [3] is applied to the transformed program to approximate the (well-typed) computations of the original program. The algorithm is guaranteed to terminate without using artificial techniques such as depth-k bounds on (type) terms.

---

* Work partly supported by ESPRIT Project PRINCE (5246)

Example Consider the following example adapted from a program discussed in [1]. The predicate `map_weight/2` takes a list of words (here represented as lists of the atoms `a`, `b` and `c`) and computes a list in which each word is immediately followed by its weight.

```
char_weight(a,1).    char_weight(b,2).    char_weight(c,3).
word_weight([], 0).
word_weight([X|Y],W):-char_weight(X,S),word_weight(Y,V),W is S+V.
map_weight([],[]).
map_weight([X|Z],[X|[Y|W]]):-word_weight(X,Y),map_weight(Z,W).
```

We use the following type declarations.

```
[] --> list( bottom )    [ X | list(Y) ] --> list(union(X,Y))
0 --> num   1 --> num    2 --> num   . . .    num + num --> num
```

Note that, unlike the example in [1], we assume no types for the symbols `a`, `b` and `c`. We assume that the builtin predicate `X is Y` is approximated by the fact `num is num`. The following approximation for `map_weight/2` is obtained by the regular approximation procedure.

```
map_weight(X1,X2):-t53(X1),t54(X2).
t53(list(X1)):-t47(X1).  t33(c):-true.       t54(list(X1)):-t51(X1).
t47(list(X1)):-t33(X1).  t33(a):-true.       t51(list(X1)):-t33(X1).
t47(bottom):-true.       t33(b):-true.       t51(num):-true.
                         t33(bottom):-true.  t51(bottom):-true.
```

This is a regular unary logic program. It can be interpreted, using a classical assumption that union types are tuple-distributive, and removing `bottom`, as `map_weight`$(list(list(\{a, b, c\})), list(num \cup list(\{a, b, c\})))$.

3 Discussion

In the paper we show some experimental results and show how well it scales up when applied to larger programs. The most closely related work to ours are [1] and [2]. One important difference is that they both ensure termination using the depth-k cutting of terms. We use a widening operator [3] to ensure termination and apply our method to partly-typed programs.

References

1. R. Barbuti and R. Giacobazzi. A bottom-up polymorphic type inference in logic programming. *Science of Computer Programming*, 19:281–313, 1992.
2. M. Codish and B. Demoen. Deriving polymorphic type dependencies for logic programs using multiple incarnations of prop. In *Proceedings of the First Symposium on Static Analysis*, pages 281–297, Springer-Verlag, September 1994.
3. J. Gallagher and D.A. de Waal. Fast and precise regular approximations of logic programs. In *Proceedings of the Eleventh International Conference on Logic Programming*, pages 599–613, MIT Press, 1994.

Performance of a Data-Parallel Concurrent Constraint Programming System*

Bo-Ming Tong** and Ho-Fung Leung

Department of Computer Science and Engineering
The Chinese University of Hong Kong
Shatin, New Territories, Hong Kong
email: bmtong@cs.cuhk.hk, lhf@cs.cuhk.hk

Finite domain constraints [3] are very effective in solving a class of integer problems which find its applications in various areas like scheduling and operations research. The advantage over conventional approaches is that the problem is specified declaratively. The actual computation is carried out by *logic inference* and *constraint satisfaction*. Solving a set of finite domain constraints is an intractable problem and we propose to use massively parallel computers to obtain satisfactory performance. In our previous papers [1, 2], we have shown that finite domain concurrent constraint languages can be implemented on massively parallel SIMD machines. The resulting system, Firebird, runs on a DECmpp 12000 Sx-100 massively parallel computer with 8,192 processor elements.

In Firebird, execution consists of two alternating derivation steps, *indeterministic derivation* and *nondeterministic derivation*. In an *indeterministic derivation step*, execution consists of guard tests, commitment, output unification and spawning in the same manner as committed-choice logic programming languages. In a *nondeterministic derivation step*, a choice point based on one of the domain variables in the system is set up and all possible values in its domain are attempted in an or-parallel manner. The domain variable used in a nondeterministic derivation step is said to be *labeled* [3] and each or-parallel branch is called a *partition*. Data-parallelism falls out naturally from nondeterministic derivation as follows. The labeled domain variable becomes a vector of all possible values of its domain. Subsequently, a SIMD computer may perform computation on the elements of these vectors in parallel.

In our full paper, we present some preliminary performance results of our massively parallel implementation of Firebird. We conclude that

1. Our figures indicate that a speedup of 121 is possible when we compare the performance using 8,192 processor elements and the performance using a single processor element of the same machine. It seems at first glance that a speedup of 121 is disappointing given the 8,192 processor elements used to obtain it, but we argue that the shared-memory architecture is simply incapable of scaling to that level. In fact, the design of our Data-Parallel Abstract

---

* This research is partially supported by RGC Earmarked Grant CUHK 70/93E. The full version of this paper can be obtained from:
  ftp://ftp.cs.cuhk.hk/pub/techreports/94/tr-94-19.ps.gz
** Present address: Department of Computer Science, University of Arizona, Tucson, AZ 85721, U.S.A. email: bmtong@cs.arizona.edu

degree of parallelism

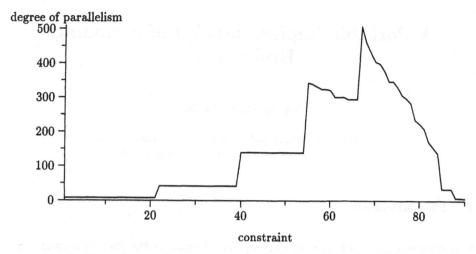

Fig. 1. Execution trace of 8-queens, number of processor elements=8,192

Machine [2] is optimized for a maximum degree of parallelism, and is most suitable in situations where resources are repleted and top performance is sought after.

2. Data-parallel implementations need specific optimizations like the *on-demand creation of bit vectors, eager nondeterministic derivation* and *priority scheduling*. We measured their effects on performance. Very little research has been done on optimizations specific to data-parallel implementations and it is an important direction of future work.

3. Our results indicate that most of the execution time is spent on constraint solving. The overheads of nondeterministic derivation and parallel backtracking is negligible. Improving the compiler technology and implementation techniques for constraint solving is very important.

4. In Figure 1 we observe the characteristics of or-parallelism when the number of processor elements is large. More investigation is needed and a thorough understanding of large-scale or-parallelism is crucial to massively parallel implementations.

References

1. B. M. Tong and H. F. Leung. Concurrent constraint logic programming on massively parallel SIMD computers. In D. Miller, editor, *Logic Programming: Proceedings of the 1993 International Symposium*, pages 388–402, Vancouver, Canada, October 1993. The MIT Press.
2. B. M. Tong and H. F. Leung. Implementation of a data-parallel concurrent constraint programming system. In *Proceedings of the First International Symposium on Parallel Symbolic Computation*, pages 382–393, Linz, Austria, September 1994. World Scientific.
3. P. Van Hentenryck. *Constraint Satisfaction in Logic Programming*. MIT Press, 1989.

A Portable Implementation of Memoing Evaluation

Jens E. Wunderwald

Technische Universität München, Germany
E-mail: wunderwa@informatik.tu-muenchen.de

1 Introduction

Memoing evaluation of logic programs [War92] enlarges the class of terminating programs and sometimes improves performance. The main idea of memoing evaluation is to store all calls and answers for so-called tabled predicates. Traditional implementation techniques for memoing evaluation include meta-interpretation and WAM modification. We propose a program transformation approach, compromising portability and performance.

This program transformation proceeds in two stages: magic sets rewriting (MSR) and bottom-up to top-down rewriting (BUTDR). BUTDR is a new technique, application of MSR to Prolog is also new.

The transformation accepts essentially the same kind of programs as the XSB system [SSW94]. Tabled predicates must have precise modes: an argument indicated as "b" in the mode declaration must always be called with a fully instantiated term, a "f"-argument with a free variable. Cut usage within tabled predicates is restricted, tabled predicates can use only stratified negation.

2 Applying Magic Sets Rewriting to Prolog

MSR [BR91] mimics Prolog's data flow from the query to the base relations in logic programming languages which evaluate bottom-up. As Prolog clauses are Horn clauses, just like bottom-up rules, MSR can be applied to them. The resulting clauses are dedicated for bottom-up evaluation. We use the left-to-right total sideways information passing strategy [BR91] inherent in Prolog.

MSR produces two kinds of relations: magic relations define the calls to tabled predicates, memo relations the answers. At the end of an evaluation, magic and memo relations hold the same information as XSB's tables [War92].

MSR does not change untabled predicates. A call to a untabled predicate in the body of a tabled predicate results in a built-in predicate of the resulting bottom-up rules. To connect Prolog rules (defining untabled predicates) and bottom-up rules (defining tabled predicates) a call to a tabled predicate gives an update of a seed base tuple of the corresponding magic relation, followed by a read access to the corresponding memo relation.

3 Adding Bottom-up rules to Prolog

For evaluating a program resulting from the first transformation step, we must equip Prolog with the ability to evaluate bottom-up rules. Unlike most bottom-up systems [BR86], BUTDR evaluates bottom-up rules tuple-by-tuple, thereby avoiding the impedance mismatch of combing bottom-up rules and Prolog rules. BUTDR can be seen as a variant of semi-naive evaluation [BR86] but tuple orientation leads to a simplified evaluation scheme.

Bottom-up relations are materialized. On addition of a new bottom-up tuple, all new consequences are computed and materialized. This process is called updating and proceeds as follows: first the new tuple is materialized. Then the bodies of bottom-up rules are searched for atoms unifying with the tuple. The tuple is (anti-)joined with the relations of the other body literals. This may produce new tuples for the rule's head which are updated recursively. For programs with finite fixpoint this process terminates because new tuples undergo a duplicate test before being updated.

This semi-naive, tuple-oriented bottom-up evaluation is implemented by a program transformation which produces a propagation rule for every body atom of a bottom-up rule and a materialization procedure for each bottom-up relation. One can use the most appropriate materialization mechanism provided by the underlying Prolog system, the asserted database being the default. The choice of a materialization technique can be made at the granularity of relations. BUTDR does not change the Prolog rules of the input program.

4 Implementation

As a final result we get a Prolog program runnable by any standard Prolog. The propagation and materialization rules should not be regarded as logic clauses, they make use of Prolog as an procedural language. The transformations are implemented as a part of the Sisyphos system [Wun94], which runs on top of several Prologs.

References

[BR86] F. Bancilhon and R. Ramakrishnan. An amateur's introduction to recursive query processing strategies, invited paper. In *Proceedings of SIGMOD*. 1986.

[BR91] C. Beeri and R. Ramakrishnan. On the power of magic. *The Journal of Logic Programming*, 10:255–300, January 1991.

[SSW94] K. Sagonas, T. Swift, and D. S. Warren. XSB as an efficient deductive database engine. In *Proceedings of SIGMOD*. 1994.

[War92] D. S. Warren. Memoing for logic programs with applications to abstract interpretation and partial deduction. In *Communications of ACM*. 1992.

[Wun94] J. E. Wunderwald. Logikprogrammieren mit frei wählbarer Auswertungsrichtung. In *Proceedings of the Tenth Logic Programming Workshop*, Zürich, 94.

Author Index

Springer-Verlag
and the Environment

We at Springer-Verlag firmly believe that an international science publisher has a special obligation to the environment, and our corporate policies consistently reflect this conviction.

We also expect our business partners – paper mills, printers, packaging manufacturers, etc. – to commit themselves to using environmentally friendly materials and production processes.

The paper in this book is made from low- or no-chlorine pulp and is acid free, in conformance with international standards for paper permanency.

Lecture Notes in Computer Science

For information about Vols. 1–912

please contact your bookseller or Springer-Verlag